Introduction to Elementary Particle Physics

The third edition of this successful textbook has been redesigned to reflect the progress of the field in the last decade, including the latest studies of the Higgs boson, quark–gluon plasma, progress in flavour and neutrino physics and the discovery of gravitational waves. It provides undergraduate students with complete coverage of the basic elements of the Standard Model of particle physics, assuming the reader has done only introductory courses in nuclear physics, special relativity and quantum mechanics. Examples of fundamental experiments are highlighted before discussions of the theory, giving students an appreciation of how experiment and theory interplay in the development of physics. The author examines leptons, hadrons and quarks, before presenting the dynamics and the surprising properties of the charges of the different forces, concluding with a discussion on neutrino properties beyond the Standard Model. This title is also available as Open Access on Cambridge Core.

Alessandro Bettini is Emeritus Professor of Physics at the University of Padua, Italy, and a research associate of the Italian National Institute for Nuclear Physics (INFN). He has served the INFN as Director of the Padova section, as Vice-president and as Director of the Gran Sasso Laboratory (LNGS). He has also served the International Union of Pure and Applied Physics (IUPAP) as Chair of the Particle and Nuclear Astrophysics and Gravitation International Committee and the Spanish Government as Director of the Canfranc Underground Laboratory (LSC). He is the author of more than 200 scientific publications, 8 university textbooks and 2 books for the general public. He is a fellow of the European Physical Society, 'Socio Benemerito' of the Italian Physical Society (SIF) and a member of the Accademia Galileiana di Scienze, Lattere e Arti.

Introduction to
Elementary Particle Physics

ALESSANDRO BETTINI

Università degli Studi di Padova

Third Edition

CAMBRIDGE
UNIVERSITY PRESS

Shaftesbury Road, Cambridge CB2 8EA, United Kingdom

One Liberty Plaza, 20th Floor, New York, NY 10006, USA

477 Williamstown Road, Port Melbourne, VIC 3207, Australia

314–321, 3rd Floor, Plot 3, Splendor Forum, Jasola District Centre, New Delhi – 110025, India

103 Penang Road, #05–06/07, Visioncrest Commercial, Singapore 238467

Cambridge University Press is part of Cambridge University Press & Assessment,
a department of the University of Cambridge.

We share the University's mission to contribute to society through the pursuit of
education, learning and research at the highest international levels of excellence.

www.cambridge.org
Information on this title: www.cambridge.org/highereducation/9781009440738

DOI: 10.1017/9781009440745

First published 2008
Second edition 2014
Third edition 2024

A catalogue record for this publication is available from the British Library

Library of Congress Cataloging-in-Publication Data
Names: Bettini, Alessandro, 1939– author.
Title: Introduction to elementary particle physics / Alessandro Bettini,
Università degli Studi di Padova, Italy.
Description: 3rd edition. | Cambridge, United Kingdom; New York, NY, USA :
Cambridge University Press, [2024] | Includes bibliographical references and index.
Identifiers: LCCN 2024004811 | ISBN 9781009440738 (hardback) |
ISBN 9781009440745 (ebook)
Subjects: LCSH: Standard model (Nuclear physics) | Standard model (Nuclear
physics) – Problems, exercises, etc.
Classification: LCC QC794.6.S75 B48 2024 | DDC 539.7/2–dc23/eng/20240314
LC record available at https://lccn.loc.gov/2024004811

ISBN 978-1-009-44073-8 Hardback

Additional resources for this publication at www.cambridge.org/bettini-3

Contents

Preface

This textbook is a presentation of subnuclear physics, at an introductory level, for undergraduate physics students, including those not specializing in the field. The first edition was published in 2008 and the second in 2014. This third edition, while keeping the design principles, is fully updated with the latest experimental and theoretical results, including on Higgs boson physics and on CP violation in the quark and the neutrino sectors. The reader will find presentations of the relativity principle, of the Dirac and Majorana equations and of quantum chromodynamics calculations on the lattice. The electron and muon magnetic moments as extreme precision tests of the Standard Model are also discussed, as is the quark–gluon plasma as a laboratory tool to study the Universe soon after the Big Bang; there is also a new chapter on gravitational waves outlining the information they provide on the graviton.

The Standard Model is the theory of the fundamental constituents of matter, describing all the fundamental interactions (excluding gravitation) as gauge quantum field theories. The reader is assumed to have already taken, at an introductory level, nuclear physics, special relativity and quantum mechanics. Knowledge of angular momentum, its composition rules and the underlying group theoretical concepts is also assumed at a working level. No prior knowledge of elementary particles or of quantum field theories is assumed. While the book will not lead to an in-depth knowledge of these theories, it will convey the basic physics elements and their beauty at an elementary level. 'Elementary' means that only knowledge of elementary concepts (in relativistic quantum mechanics) is assumed. However, it does not mean a superficial discussion. In particular, I have tried not to cut corners and I have avoided hiding difficulties, whenever they arise, following the Einstein dictum: 'make everything as simple as possible, but not simpler'. As in the previous editions, I have included only well-established elements with the exception of the final outlook, in which I survey the main challenges of the present frontier.

The text now contains more material than can be accommodated in a single undergraduate course. However, several chapters are quite independent from one another, leaving the instructor a range of choices.

The majority of the texts on elementary particles place special emphasis on theoretical aspects. However, physics is an experimental science and only experiment can decide which of the possible theoretical schemes has been chosen by Nature. Moreover, the progress of our understanding is often due to the discovery of unexpected phenomena. I have tried to select examples of basic experiments first, and then to go on to the theoretical picture.

A direct approach to the subject would start from leptons and quarks and their interactions and explain the properties of hadrons as consequences. A historical approach would also discuss the development of ideas. The former is shorter, but is lacking in depth. I tried to arrive at a balance between the two views.

The necessary experimental and theoretical tools are presented in the first chapter. Even if the students have already met special relativity, the theory is presented here from a more advanced point of view, independent of electromagnetism. Indeed, the laws of all the interactions, not only those of electromagnetism, must respect the relativity principle. In addition, students are guided to practical exercises in the use of relativistic invariants and Lorentz transformations. Chapter 1 also contains a summary of the artificial and natural sources of high-energy particles and of detectors. In Chapter 2, I present the Dirac Lagrangian and derivations of the Dirac equation, which had been assumed to be known in the previous editions. The Majorana equation for completely neutral fermions is also introduced here. As in the previous editions, the symmetries of elementary particles are discussed in Chapter 3.

The elementary fermions fall into two categories: the leptons, which can be found free, and the quarks, which always live inside the hadrons. Hadrons are non-elementary, compound structures, rather like nuclei. The 'oldest' elementary particles are presented in Chapter 2, with a partly historical approach, while the hadron 'zoo', and how hadrons are described by the quark model, is discussed in Chapter 4.

There is a fundamental difference between hadrons on the one hand and atoms and nuclei on the other. While the electrons in atoms and nucleons in nuclei move at non-relativistic speeds, the quarks in the nucleons move almost at the speed of light. Actually, their rest energies are much smaller than their total energies. Subnuclear physics is fundamentally relativistic quantum mechanics.

The second part of the book is dedicated to the fundamental interactions and to the Standard Model. The approach is substantially more direct, but the main historical steps are recalled. The most important experiments that prove the crucial aspects of the theory are discussed in some detail.

Chapter 5 deals with quantum electrodynamics (QED), the prototype of all the relativistic gauge theories. I show how the concept of gauge invariance evolves from classical electromagnetism, where it plays a secondary role, to the quantum theory in which it becomes dominant. I introduce the Feynman diagram at a non-quantitative level as a tool to analyse scattering and decay probabilities; I discuss the experiments that led to QED and test its main aspects. New to this edition, the magnetic moments of the electron and the muon are discussed from both the experimental and the theoretical points of view, as examples of the extreme precision reached by both.

Chapter 6 presents quantum chromodynamics (QCD), its analogies and its radical differences from QED, entering into the magic 'coloured' world. New to this edition is a discussion of QCD on the lattice, a theoretical tool enabling calculations at low energies where the perturbative approach fails, and of how the most powerful super-computers that recently became available allow precise predictions to be obtained. Amongst these, I discuss the hadron mass spectrum and the deconfinement phase transition to the quark–gluon plasma, the phase of matter 1 μs after the Big Bang.

In Chapter 7 the weak interaction is described. I start with the Fermi theory, before discussing the experiments that showed that parity and charge conjugation are violated and that led to the V–A theory of the 'charged currents'. We continue with lepton universality and quark mixing. The chapter ends with the discovery of the weak neutral currents. Chapter 8 discusses in detail the neutral meson oscillations and the CP violation in quark interactions. The recent discoveries in the beauty and charm sectors are included for the first time.

In Chapter 9, electroweak unification is discussed in detail, both in its theoretical principles and in the experimental proofs: from the measurements of the weak mixing angle, to the discovery of the vector bosons, to the precision tests at the electron–positron (Large Electron–Positron (LEP)) and proton–antiproton (Tevatron) colliders. Finally, after introducing the Brout–Englert–Higgs (BEH) theory, the long search for the Higgs boson and how it was discovered are discussed. The measurements of its principal properties, determined after the previous edition of this book, are now presented.

In Chapter 10, we turn to physics beyond the Standard Model (SM). Neutrinos are massive and can change their lepton flavour, which is different from what is assumed in the SM. Actually neutrino mixing, masses, oscillations and adiabatic flavour conversion in matter make a beautiful set of phenomena that can be properly described at an elementary level, using only the basic concepts of quantum mechanics. Possible CP violation in the neutrino sector is also presented.

Chapter 11 is new. Even though we do not have a quantum theory of gravitation, after the discovery in 2016 of gravitational waves (GW) and the subsequent great progress of the field, we now know elements close to particle physics. I discuss the measurement of the velocity of GW and of a limit on the graviton mass.

Chapter 12 contains a short discussion on the limits of the Standard Model and on facts beyond it. I briefly discuss gravity, dark matter, dark energy, supersymmetry, strong CP violation, absence of antimatter in the universe and structural theoretical problems.

Problems

Numbers in physics are important; the ability to calculate a theoretical prediction on an observable or an experimental resolution is a fundamental characteristic of any physicist. More than 260 numerical examples and problems are presented. The simplest ones are included in the main text under the form of questions. Other problems covering a range of difficulty are given at the end of the chapters. In every case the student can arrive at the solution without studying further theoretical material. Physics rather than mathematics is emphasized.

The physical constants and the principal characteristics of the particles are not given explicitly in the text of some problems. The student is expected to look for them in the tables given in the Appendices. Solutions to about half of the problems are given at the end of the book.

Appendices

One appendix contains the dates of the main discoveries in particle physics, both experimental and theoretical. It is intended to give a bird's-eye view of the history of the field. However, keep in mind that the choice of the discoveries is partially arbitrary and that history is always a complex non-linear phenomenon. Discoveries are seldom due to a single person and never happen instantaneously.

Tables of the Clebsch–Gordan coefficients, of the spherical harmonics and of the rotation functions in the simplest cases are included in the appendices. They are needed for some of the problems. Other tables give the main properties of gauge bosons, of leptons, of quarks and of the ground levels of the hadronic spectrum.

The principal source of the data in the tables is the most recent edition of 'Review of Particle Physics' (Workman *et al.* 2022), PDG for short. Its website, http://pdg .lbl.gov/, is a very useful resource for the reader. It includes not only the complete data on elementary particles, but also short reviews of topics such as tests of the Standard Model, searches for hypothetical particles, particle detectors and probability and statistical methods.

Reference Material on the Internet

The URLs cited in this work were correct at the time of going to press, but the publisher and the author make no undertaking that the citations remain live or accurate or appropriate.

Acknowledgements

It is a pleasure to thank the following for their help in preparing this third edition: Roberto Onofrio, for his very careful reading of the second edition, suggesting important changes (including corrections of errors) and proposing new elements; Harry Varvoglis, who, when translating to Greek, found a number of typos; Luisa Cifarelli and Massimo Cerdonio for comments and discussions; and Aldo Ianni, Salvatore Mele and Alessandro Tarabini for relevant figures.

I am indebted to the following authors, institutions and laboratories for the permission to reproduce or adapt the following photographs and diagrams:

CERN for Figs. 7.24, 9.6 and 9.25. CERN and ATLAS and CMS Collaborations for
    Figs. 9. 45 (a) and (b), 9.46, 9.47, 9.49, 9.50, 9.55, 9.56, 9.59, 9.60, 9.62 and 9.65

INFN for Figs. 1.16 and 10.10

W. Hanlon for Fig. 1.13

Kamioka Observatory, Institute for Cosmic Ray Research, University of Tokyo and
    Y. Suzuki for Fig. 1.19

Kamioka Observatory, Institute for Cosmic Ray Research, University of Tokyo for
    Fig. 1.20

Lawrence Berkeley National Laboratory for Fig. 1.21

The NEXT Collaboration for Fig. 1.27

Brookhaven National Laboratory for Fig. 4.21

The Physical Society of Japan and Professor K. Niu for Fig. 4.29 from K. Niu
    et al. Prog. Theor. Phys. **46** (1971) 1644, Fig. 3(a)

J. Jackson and the Fermi National Laboratory for Fig. 4.33

Springer, D. Plane and the OPAL Collaboration for Fig. 5.32 from G. Abbiendi
    et al. Euro. Phys. J. **C33** (2004) 173, Fig. 17

Stanford Linear Accelerator Center for Fig. 6.8(b)

Springer, E. Gallo and the ZEUS Collaboration for Fig. 6.13 from S. Chekanov
    et al. Eur. Phys. J. **C21** (2001) 443, Fig. 13

S. Mele, CERN for Fig. 6.23(a)

D. Leinweber, CSSM, University of Adelaide for Fig. 6.32

L. Grodznis for the photo in Fig. 7.6

Figs. 9.15 and 9.20 are reprinted with permission of John Wiley & Sons, Inc. and the
    author J. W. Rohlf of *Modern Physics from α to Z^0*, 1994, Figs. 18.17 and 18.21

A. Tarabini and CMS for Fig. 9.62

'Hepdata' for Fig. 9.64

(1) A. Ianni for Figs. 10.14 and 10.25

Nobel Foundation for: Fig. 2.2 from C. Powell, Nobel Lecture 1950, Fig. 5; Fig. 2.7 from F. Reines, Nobel Lecture 1995, Fig. 5; Fig. 4.16(a) from L. Alvarez, Nobel Lecture 1968, 19; Figs. 4.26, 4.27 and 4. 28 from B. Richter, Nobel Lecture 1976, Figs. 5, 6 and 18; Fig. 4.30 from L. Lederman, Nobel Lecture 1988, Fig. 12; Fig. 6.8(a) from R. E. Taylor, Nobel Lecture 1990, Fig. 14; Fig. 6.10 from J. Friedman, Nobel Lecture 1990, Fig. 1; Figs. 8.4 and 8.5(b) from M. Fitch, Nobel Lecture 1980, Fig. 3; Figs. 9.12, 9.14 and 9.19 (a) and (b) from C. Rubbia, Nobel lecture 1984, Figs. 8a, 16(a) and (b), 25 and 26

Particle Data Group and the Institute of Physics for Figs. 1.9, 1.11, 1.12, 4.4, 5.27, 6.3, 6.12, 6.23(b), 9.30, 9.33, 10.13

Elsevier for: Fig. 5.35 (a) and (b) from P. Achard *et al. Phys. Lett.* **B623** 26–36 (2005), Figs. 3 and 5; Figs. 6.2 and 6.6 from B. Naroska, *Phys. Rep.* **148** (1987) 67–215, Fig. 5.1 (a) and (b); Fig. 6.7 from S. L. Wu *Phys. Rep.* **107** (1984) 59–324, Fig. 3.23; Fig. 7.7 from F. Koks and J. van Klinken, *Nucl. Phys.* **A272** (1976) 61, Fig. 1; Fig. 8.2 from S. Gjesdal *et al. Phys. Lett.* **B52** (1974) 113, Fig. 1; Fig. 9.9 from D. Geiregat *et al. Phys. Lett.* **B259** (1991) 499, Fig. 1; Figs. 9.17, 9.18(b) and 9.21 from C. Albajar *et al. Z. Phys.* **C44** (1989) 15, Figs. 16(a) and 48; Fig. 9.22 from C. Albajar *et al. Z. Phys.* **C36** (1987), Fig. 3a

The American Physical Society for: Fig. 4.6 from L. Alvarez *et al. Phys. Rev. Lett.* **10** (1963) 184, Fig. 1; Fig. 4.14 from J. Orear *et al. Phys. Rev.* **102** (1956) 1676, Fig. 2; Fig. 4.15(a) from A. Pevsner *et al. Phys. Rev. Lett.* **7** (1961) 421, Fig. 2; Fig. 4.15(b) from C. Alff *et al. Phys. Rev. Lett.* **9** (1962) 325 (Author OK R. J. Plano); Fig. 7.3 from C. S. Wu *et al. Phys. Rev.* **105** (1957) 1413, Fig. 2; Fig. 8.11 from I. Adachi *et al. Phys. Rev. Lett.* **108** (2012) 171802, Fig. 2 (Author OK M. Nakao) and B. Aubert *et al. Phys. Rev.* **D79** (2009) 072009, Fig. 2 (Author OK M. Giorgi).

1 Preliminary Notions

Elementary particles are at the deepest level of the structure of matter. Students have already met the upper levels, namely the molecules, the atoms and the nuclei. These structures are small and their physics is properly described by non-relativistic quantum mechanics, by the Schrödinger equation, which is not relativistic. Indeed, the speeds of electrons in a molecule or in an atom and of the protons and neutrons in a nucleus are much smaller than the speed of light.

Protons and neutrons contain quarks, which have very small masses, corresponding to rest energies much smaller than their total energy, and their speed is close to that of light. The structure of the nucleons, and more generally of the hadrons that we shall discuss, is described by relativistic quantum mechanics, in which the relevant equation is the Dirac equation. Even if most readers have already met it, we shall provide an introduction.

The study of elementary particles requires experiments with beams accelerated at very high energies. There are two reasons for this: (a) the creation of new particles requires an initial energy large enough to be converted in the mass–energy of the new particles; (b) to study the internal structure of an object we must probe it with adequate resolving power, which increases with the momentum, and hence also with the energy of the probe, as we shall discuss.

In this chapter the student will learn the basic notions and instruments that will be necessary for further study.

We shall start by recalling the relativity principle and showing how the Lorentz transformations, certainly known to the reader, can be demonstrated as consequences of basic properties of space-time, independent of electromagnetism, as in the usual presentations. This is relevant, considering that Lorentz invariance is a common property of all the fundamental interactions. We then clearly discuss the fundamental concepts of energy, momentum and mass, the relations amongst them and their transformations between reference systems, in particular the laboratory and centre of mass frames.

Students are urged to work on several numerical problems, which may be found at the end of the chapter, together with an introduction to the methods to solve them. This is the only way to master, in particular, relativistic kinematics.

Experiments on elementary particles study their collisions and decays. This chapter continues by introducing the basic concepts of collisions and decays. We shall then introduce the different types of particles (hadrons, quarks, leptons and bosons) and their fundamental interactions. Here and throughout, we proceed, when appropriate, by successive approximations. Indeed, this is the way in which science itself makes its progress.

The basic components of a collision experiment are a beam of high-energy particles, protons, antiprotons, electrons, photons, neutrinos, etc., and a target on which they collide. The student will find in this chapter a basic description of the sources of such particles, which are naturally occurring cosmic rays, used in the first years of the research, and different types of accelerators. The products of a collision or of a decay, which are also elementary particles, are detected and their properties (energy, momentum, charge) measured with suitable 'detectors'. The progress of our knowledge is fully linked to the experimental 'art' of detector design and development. Detectors are made of matter, solid, liquid or gaseous. Consequently, a fair degree of knowledge of the interactions of charged and neutral high-energy particles with matter, with its atoms and molecules, is necessary to understand how detectors work; this is introduced in this chapter. This chapter also introduces the principal types of detector and the principles of their operation. In later chapters, the detectors' systems as implemented in important experiments will be described. We shall see here, in particular, how to measure the energy, momentum and mass of a particle, in the different energy ranges and situations in which they are met.

1.1 The Principle of Relativity

The principle of relativity is a fundamental law of nature, closely related to the law of inertia. It states that a continuous infinite number of reference frames exists in space-time that are equivalent, in the sense that the physical laws have the same expression when referred to any of them; in other words there is no physical effect that can distinguish between them. These are the frames in which the law of inertia holds, and are called inertial frames. Given two inertial frames $S(t,x,y,z)$ and $S'(t',x',y',z')$, the equations that connect the spatial and temporal coordinates in the two frames are called *transformations*. These are the *Lorentz transformations*.

To be inertial, the two reference frames should be in rectilinear uniform relative motion. We take the corresponding axes of the two frames in the same directions, and consider, as usual, the case in which the relative motion is in the x direction. We choose the axes as represented in Fig. 1.1. At a certain moment, which we take as $t' = t = 0$, the origins and the axes coincide. The frame S' moves relative to S with speed **V**.

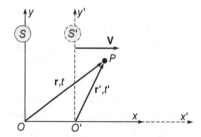

Fig. 1.1. Two reference frames in rectilinear relative motion.

We introduce the following two dimensionless quantities relative to the motion in S of the origin of S'

$$\beta \equiv \frac{V}{c} \qquad (1.1)$$

and

$$\gamma = \frac{1}{\sqrt{1-\beta^2}}, \qquad (1.2)$$

called the 'Lorentz factor'. An event is defined by the four-vector of the coordinates $(ct, \mathbf{r})$. Its components in the two frames (t, x, y, z) and (t', x', y', z') are linked by the Lorentz transformations (Lorentz 1904; Poincaré 1905), in the considered particular case of relative motion in the x direction, as

$$\begin{aligned} x' &= \gamma(x - \beta ct) \\ y' &= y \\ z' &= z \\ ct' &= \gamma(ct - \beta x). \end{aligned} \qquad (1.3)$$

The set of all the Lorentz transformations, when joined to the rotations of the axes, form a group that H. Poincaré, who first recognized this property in 1905, called the proper Lorentz group. The group contains the parameter c, a constant with the dimensions of the velocity. A physical entity moving at speed c in a reference frame moves with the same speed in any other one. In other words, c is invariant under Lorentz transformations. It is the propagation speed of light and gravitational waves (Poincaré 1905).

Elementary particles usually move at high velocities, close to the limiting speed, the speed of light. Their physics is the realm of special relativity. It is consequently worth revisiting their derivation from a more general point of view than that usually presented, even if the Lorentz transformations are well known to the reader.

Historically, the Lorentz transformations were constructed as a consequence of the experimentally established fact that the speed of light (in a vacuum) is independent of the speed of the inertial reference frame in which it is measured. Indeed, the existence of an invariant speed was one of the two Einstein postulates. This is *not*, however, a necessary assumption and, indeed, Poincaré had already shown that the invariance of the speed of light is a consequence of the relativity principle and of Maxwell equations.

A weak point of the purely historical approach, important as it is, is that it tends to hide the logical structure of special relativity and to overemphasize the role of the electromagnetism in the foundations of the theory. More than one century after its creation, we know that *all* the fundamental interactions, not only the electromagnetic one, but also the gravitational, the strong and the weak interactions, obey the relativity principle. All the physics laws are covariant under Lorentz transformations. The fields of the fundamental interactions in quantum mechanics are mediated by 'quanta'. The quantum of light is the photon. Its velocity is the velocity of light, which implies that the mass of the photon is zero. However, it would be logically possible that the photon would be massive. In this case, the Lorentz transformations would not change, but the parameter c appearing in the equations would not be the speed of light and the latter would not be invariant. Indeed, this is the case for the weak interaction, the quanta of which, called Z^0 and $W^\pm$ bosons, have mass and do not move at the speed of light. If photons were massive, the demonstration based on the invariance of the speed of light would not hold. But the final result would still be valid.

From the logic point of view, we must ask ourselves the following questions. Can we establish the theory of relativity independently of electromagnetism? What assumptions are needed for that? In 1911, von Ignatowsky (von Ignatowsky 1911) showed that the Lorentz transformations in which the c is an invariant velocity that must be determined experimentally (and the Galilei transformation in the case that $c \to \infty$) are the only admissible ones under the following quite general assumptions on the structure of space-time.

(1) Space-time is isotropic and homogeneous.
(2) At least one inertial reference frame exists.
(3) The relativity principle holds, namely there is no privileged reference frame.
(4) The transformations form a group.
(5) Causality. If a causality relationship exists between two events A and B in space-time, A being the cause and B the effect, then the time of B must be larger (future) than the time of A in all the inertial reference frames. We must assume that the transformations are such that at least a class of events exists in which the sign of the time interval, and hence the nature of a possible causal relationship, is not changed by the transformation.

The very assumption of space-time as a unique manifold implies that its four coordinates must have the same physical dimensions. To reduce time to the space dimension, we must multiply t by a constant having the dimension of a velocity, c, and this constant must be universal, independent of the reference frame. Electromagnetism has no privileged role; it enters in the game when we establish that c is the speed of light.

Simplified versions of the von Ignatowsky derivation have been published (Lévy-Leblond 1976; Pelissetto & Testa 2015). We shall follow here Lévy-Leblond, with further simplifications, and find the transformations (1.3) between the two inertial frames in Fig. 1.1. The non-interested reader can go directly to the next section.

Homogeneity of Space-Time. The transformation properties of the space-time intervals $(\Delta x, \Delta t)$ depend only on the intervals and not on the locations of the extremes.

Inertia. First, the transformations must be linear, because the law of inertia requires that any rectilinear motion in S should be rectilinear in S' too. In addition, any uniform motion in one frame must be uniform in the other, implying that x' should depend on $x - Vt$ rather than from x and t separately. The transformations must then have the form

$$x' = \gamma(V)(x - Vt) \tag{1.4}$$

$$t' = \gamma(V)\left[\lambda(V)t - \mu(V)x\right],$$

where γ, λ and μ are functions of V to be determined.

Isotropy of Space. We assume that all directions in space are equivalent. Hence, in particular, the transformation laws of the coordinates (x,t) of an event in S to its coordinates (x',t') in S' must also have the form of Eq. (1.4) if we invert the x direction, namely from $(-x,t)$ to $(-x',t')$. Clearly the relative velocity is now $-V$. We then must have

$$-x' = \gamma(-V)(-x+Vt)$$
$$t' = \gamma(-V)\left[\lambda(-V)t + \mu(-V)x\right].$$

Comparing with Eq. (1.4), this implies that

$$\gamma(-V) = \gamma(V) \tag{1.5a}$$

$$\lambda(-V) = \lambda(V) \tag{1.5b}$$

$$\mu(-V) = -\mu(V). \tag{1.5c}$$

The Group Conditions.

(1) An **identity transformation** must exist, namely a value of V must exist for which $x' = x$ and $t' = t$. This value is clearly $V = 0$, and we have

$$\gamma(0) = 1 \tag{1.6a}$$

$$\lambda(0) = 1 \tag{1.6b}$$

$$\mu(0) = 0. \tag{1.6c}$$

(2) **Inverse transformation.** Equation (1.4) gives the transformations from (x,t) to (x',t'), with velocity V. The inverse transformation from (x',t') to (x,t) must exist, with a certain velocity W and the same structure. Clearly $W = -V$, but we do not use that for the moment and write the inverse transformations as

$$x = \gamma(W)(x' - Wt') \tag{1.7}$$

$$t = \gamma(W)\left[\lambda(W)t' - \mu(W)x'\right].$$

Inverting (1.4) we obtain

$$x = \frac{1}{\gamma(V)}\left(1 - \frac{V\mu(V)}{\lambda(V)}\right)^{-1}\left(x' + \frac{V}{\lambda(V)}t'\right) \tag{1.8}$$

$$t = \frac{1}{\gamma(V)}\left(1 - \frac{V\mu(V)}{\lambda(V)}\right)^{-1}\left(\frac{1}{\lambda(V)}t' + \frac{\mu(V)}{\lambda(V)}x'\right).$$

Equating (1.8) and (1.7) we obtain

$$W = -V / \lambda(V) \tag{1.9a}$$

$$\lambda(W) = 1 / \lambda(V) \tag{1.9b}$$

$$\mu(W) = -\mu(V)/\lambda(V) \tag{1.9c}$$

$$\gamma(W) = \left[1/\gamma(V)\right]\left[1 - V\mu(V)/\lambda(V)\right]^{-1}. \tag{1.9d}$$

Considering that $W = -V$, (1.9b) with (1.9a) gives

$$\lambda(V) = 1. \tag{1.10}$$

Equation (1.9c) then simply restates Eq. (1.5c). Finally, Eq. (1.9d) gives a relation between the two remaining functions:

$$\left[\gamma(V)\right]^2\left[1 - V\mu(V)\right] = 1. \tag{1.11}$$

3. **The product of two transformations is a transformation.** Take then a first transformation

$$x_1 = \gamma(V_1)(x - V_1 t) \tag{1.12}$$

$$t_1 = \gamma(V_1)\left[t - \mu(V_1)x\right]$$

and a second one

$$x_2 = \gamma(V_2)(x_1 - V_2 t_1) \tag{1.13}$$

$$t_2 = \gamma(V_2)\left[t_1 - \mu(V_2)x_1\right].$$

The product transformation is

$$x_2 = \gamma(V_1)\gamma(V_2)\left[1 + \mu(V_1)V_2\right]\left(x - \frac{V_1 + V_2}{1 + \mu(V_1)V_2}t\right) \tag{1.14a}$$

$$t_2 = \gamma(V_1)\gamma(V_2)\left[1 + \mu(V_2)V_1\right]\left(t - \frac{V_1 + V_2}{1 + \mu(V_2)V_1}x\right). \tag{1.14b}$$

These expressions must have the form of Eq. (1.4) with a value of the parameter V to be found, namely it must have the form

$$x_2 = \gamma(V)(x - Vt) \tag{1.15}$$

$$t_2 = \gamma(V)\left[\lambda(V)t - \mu(V)x\right].$$

Identifying the factors in square brackets in Eq. (1.14a) and (1.14b), we have

$$\mu(V_1)V_2 = \mu(V_2)V_1, \tag{1.16}$$

which means that μ is proportional to V, namely that

$$\mu(V) = \alpha V \tag{1.17}$$

where α is a constant. Finally, Eq. (1.11) now gives us the last unknown function as

$$\gamma(V) = \left(1 - \alpha V^2\right)^{-1/2}, \tag{1.18}$$

where the sign of the square root has been chosen to respect Eq. (1.6a). From Eq. (1.14), the 'law of addition of velocities' also follows:

$$U = \frac{V_1 + V_2}{1 + \alpha V_1 V_2}. \tag{1.19}$$

Three cases are possible depending on α being positive, negative or null. Let us look at each of them.

Case $\alpha < 0$. Considering that α has the dimensions of an inverse square velocity we can define the constant κ, having the dimension of velocity, as $\alpha = -\kappa^{-2}$ The transformations are

$$x' = \frac{x - Vt}{\left(1 + V^2/\kappa^2\right)^{1/2}} \quad t' = \frac{t + Vx/\kappa^2}{\left(1 + V^2/\kappa^2\right)^{1/2}}. \tag{1.20}$$

The law of addition of velocities is

$$U = \frac{V_1 + V_2}{1 - V_1 V_2/\kappa^2}. \tag{1.21}$$

Case $\alpha = 0$. The transformations are

$$x' = x - Vt \quad t' = t. \tag{1.22}$$

And the law of addition of velocities is

$$U = V_1 + V_2. \tag{1.23}$$

Clearly, these are the Galilei transformations.

Case $\alpha > 0$. We define the constant c with the dimension of velocity as $\alpha = c^{-2}$ and obtain the transformations

$$x' = \frac{x - Vt}{\left(1 - V^2/c^2\right)^{1/2}} \quad t' = \frac{t - Vx/c^2}{\left(1 - V^2/c^2\right)^{1/2}}. \tag{1.24}$$

The law of addition of velocities is

$$U = \frac{V_1 + V_2}{1 + V_1 V_2/c^2}, \tag{1.25}$$

which are the Lorentz transformations. The constant c appears to be the upper bound of any possible velocity; in the case of $\alpha < 0$, on the contrary, velocities can have any value.

Causality. At least a class of events should exist in which the sign of the time interval is not changed by the transformation, namely Δt and $\Delta t'$ should have the same sign. This is clearly the case for the Galilei transformations (1.22), for any time interval. It is also true for the Lorentz transformation, due to the limitation of the velocities, for those intervals such that $|\Delta x/\Delta t| \leq c$ (the 'time-like' ones). It is not the case for $\alpha < 0$, due to the unlimited range of velocities. Indeed in this case

$$\Delta t' = \frac{\Delta t + V \Delta x / \kappa^2}{\left(1 + V^2 / \kappa^2\right)^{\frac{1}{2}}}. \tag{1.26}$$

For every given $(\Delta x, \Delta t)$ it is always possible to find a V, such that $\Delta t'$ has the opposite sign of Δt. No casual relation is possible.

In conclusion, the above general assumptions are sufficient to establish the Lorentz transformations as the only possible ones. The universal constant c turns out to be the speed either of zero-mass particles, or, which is equivalent, of the waves of the fields of force of infinite range (i.e. electromagnetic and gravitational fields) (see Chapter 11). The degenerate case, for $c \to \infty$, the Galilei transformation, separates time from space and is not compatible with the assumption of space-time as a single manifold. As already stated, the latter directly requires the existence of an invariant velocity.

1.2 Mass, Energy, Linear Momentum

In this section, and in the next three, we recall a few simple properties of relativistic kinematics and dynamics.

The same transformations in Eq. (1.3) of the space-time vector are valid for any four-vector. Of special importance is the energy–momentum vector $(E/c, \mathbf{p})$ of a free particle

$$\begin{aligned} p'_x &= \gamma \left(p_x - \beta \frac{E}{c} \right) \\ p'_y &= p_y \\ p'_z &= p_z \\ \frac{E'}{c} &= \gamma \left(\frac{E}{c} - \beta p_x \right). \end{aligned} \tag{1.27}$$

Notice that the same 'Lorentz factor' γ appears both in the geometric transformations (1.3) and in those of dynamic quantities (1.27).

The transformations that give the components in S as functions of those in S', the inverse of (1.3) and (1.27), can be most simply obtained by changing the sign of the speed $\mathbf{V}$.

The norm of the energy–momentum is, as for all the four-vectors, an invariant; it is the square of the mass of the system multiplied by the invariant factor c^4

$$m^2 c^4 = E^2 - p^2 c^2. \tag{1.28}$$

This is a fundamental expression: it is the definition of the mass. It is, we repeat, valid only for a free body, but is, on the other hand, completely general for point-like bodies, such as elementary particles, and for composite systems, such as nuclei or atoms, even in the presence of internal forces.

The most general relationship between the linear momentum (we shall call it simply momentum) **p**, the energy E and the speed **v** is

$$\mathbf{p} = \frac{E}{c^2}\mathbf{v},$$ (1.29)

which is valid both for bodies with zero and non-zero mass.

For massless particles, (1.28) can be written as

$$pc = E.$$ (1.30)

The photon mass is exactly zero. Neutrinos have non-zero but extremely small masses in comparison with the other particles. In the kinematic expressions involving neutrinos, their mass can usually be neglected.

If $m \neq 0$, the energy can be written as

$$E = m\gamma c^2,$$ (1.31)

and (1.29) takes the equivalent form

$$\mathbf{p} = m\gamma\mathbf{v}.$$ (1.32)

We call the reader's attention to the fact that one can find in the literature, and not only in that addressed to the general public, concepts that arose when the theory was not yet well understood and that are useless and misleading. One of these is the 'relativistic mass', which is the product $m\gamma$, and the dependence of mass on velocity. The mass is a Lorentz invariant, independent of the speed; the 'relativistic mass' is simply the energy divided by c^2 and as such the fourth component of a four-vector; this, of course, is if $m \neq 0$, while for $m = 0$ relativistic mass has no meaning at all. Another related term to be avoided is the 'rest mass', namely the 'relativistic mass' at rest, which is simply the mass (see Okun (1989) in 'Further Reading').

The concept of mass applies, to be precise, only to the eigenstates of the free Hamiltonian, just as only monochromatic waves have a well-defined frequency. Even the barely more complicated wave, the dichromatic wave, does not have a well-defined frequency. We shall see in Chapter 8 that there are two-state quantum systems, such as K^0, B^0, B_s^0 and D^0, which are naturally produced in states different from stationary states. For the former states it is not proper to speak of mass and of lifetime.

As we shall see, the nucleons (as protons and neutrons are collectively called) and more generally the hadrons, which we shall define later in this chapter, are made up of quarks. The quarks are never free; consequently, the definition of quark mass presents difficulties, which we shall discuss later.

Example 1.1   Consider a source emitting a photon with energy E_0 in the frame of the source. Take the x-axis along the direction of the photon. What is the energy E of the photon in a frame in which the source moves in the x direction at speed $v = \beta c$? Compare this with the Doppler effect.

Call S' the frame of the source. Remembering that photon energy and momentum are proportional, we have $p'_x = p' = E_0/c$. The inverse of the last equation in (1.27) gives

$$\frac{E}{c} = \gamma \left(\frac{E_0}{c} + \beta p'_x \right) = \gamma \frac{E_0}{c} (1 + \beta)$$

and we have

$$\frac{E}{E_0} = \gamma(1 + \beta) = \sqrt{\frac{1+\beta}{1-\beta}}.$$

The Doppler effect theory tells us that, if a source emits a light wave of frequency ν_0, an observer who sees the source approaching at speed $\upsilon = \beta c$ measures the frequency ν, such that $\dfrac{\nu}{\nu_0} = \sqrt{\dfrac{1+\beta}{1-\beta}}$. This is no wonder; in fact, quantum mechanics tells us that $E = h\nu$. $\square$

1.3 The Law of Motion of a Particle

The 'relativistic' law of motion of a particle was found in 1905 by Poincaré (but published the following year) (Poincaré 1906) and independently by Planck in 1906 (Planck 1906). As in Newtonian mechanics, a force $\mathbf{F}$ acting on a particle of mass $m \neq 0$ results in a variation in time of its momentum. Newton's law in the form $\mathbf{F} = d\mathbf{p}/dt$ (the form used by Newton himself) is also valid at high speed, provided the momentum is expressed by Eq. (1.32). The expression $\mathbf{F} = m\mathbf{a}$, used by Einstein in 1905, on the contrary, is wrong. It is convenient to write explicitly

$$\mathbf{F} = \frac{d\mathbf{p}}{dt} = m\gamma\mathbf{a} + m\frac{d\gamma}{dt}\mathbf{v}. \tag{1.33}$$

Taking the derivative, we obtain

$$m\frac{d\gamma}{dt}\mathbf{v} = m\frac{d\left(1 - \dfrac{\upsilon^2}{c^2}\right)^{-1/2}}{dt}\mathbf{v} = -m\frac{1}{2}\left(1 - \frac{\upsilon^2}{c^2}\right)^{-3/2}\left(-2\frac{\upsilon}{c^2}a_t\right)\mathbf{v} = m\gamma^3(\mathbf{a}\cdot\boldsymbol{\beta})\boldsymbol{\beta}.$$

Hence

$$\mathbf{F} = m\gamma\mathbf{a} + m\gamma^3(\mathbf{a}\cdot\boldsymbol{\beta})\boldsymbol{\beta}. \tag{1.34}$$

We see that the force is the sum of two terms, one parallel to the acceleration and one parallel to the velocity. Therefore, we cannot define any 'mass' as the ratio between acceleration and force. At high speeds, the mass is not the inertia to motion.

To solve for the acceleration, we take the scalar product of the two members of Eq. (1.34) with $\boldsymbol{\beta}$. We obtain

$$\mathbf{F} \cdot \boldsymbol{\beta} = m\gamma\, \mathbf{a} \cdot \boldsymbol{\beta} + m\gamma^3 \beta^2\, \mathbf{a} \cdot \boldsymbol{\beta} = m\gamma\,(1 + \gamma^2\beta^2)\, \mathbf{a} \cdot \boldsymbol{\beta} = m\gamma^3\, \mathbf{a} \cdot \boldsymbol{\beta}.$$

Hence

$$\mathbf{a} \cdot \boldsymbol{\beta} = \frac{\mathbf{F} \cdot \boldsymbol{\beta}}{m\gamma^3}$$

and, by substitution into (1.34),

$$\mathbf{F} - (\mathbf{F} \cdot \boldsymbol{\beta})\boldsymbol{\beta} = m\gamma\,\mathbf{a}.$$

The acceleration is the sum of two terms, one parallel to the force, and one parallel to the speed.

Force and acceleration have the same direction in two cases only: (1) force and velocity are parallel: $\mathbf{F} = m\gamma^3\mathbf{a}$; (2) force and velocity are perpendicular: $\mathbf{F} = m\gamma\mathbf{a}$. Notice that the proportionality factors are different.

In order to have simpler expressions in subnuclear physics, the so-called 'natural units' are used. We shall discuss them in Section 1.5, but we anticipate here one definition: without changing the SI unit of time, we define the unit of length in such a way that $c = 1$. In other words, the unit length is the distance the light travels in a second in a vacuum, namely 299 792 458 m, a very long distance. With this choice, in particular, mass, energy and momentum have the same physical dimensions. We shall often use as their unit the electronvolt (eV) and its multiples.

1.4 The Mass of a System of Particles: Kinematic Invariants

The mass of a system of particles is often called 'invariant mass', but the adjective is useless; the mass is always invariant. The expression is simple only if the particles of the system do not interact amongst themselves. In this case, for n particles of energies E_i and momenta $\mathbf{p}_i$, the mass is

$$m = \sqrt{E^2 - P^2} = \sqrt{\left(\sum_{i=1}^{n} E_i\right)^2 - \left(\sum_{i=1}^{n} \mathbf{p}_i\right)^2}. \tag{1.35}$$

Consider the square of the mass, which we shall indicate by s, obviously an invariant quantity

$$s = E^2 - P^2 = \left(\sum_{i=1}^{n} E_i\right)^2 - \left(\sum_{i=1}^{n} \mathbf{p}_i\right)^2. \tag{1.36}$$

Notice that s cannot be negative

$$s \geq 0. \tag{1.37}$$

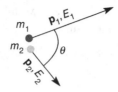

Fig. 1.2. System of two non-interacting particles.

Let us see its expression in the 'centre of mass' (CM) frame, which is defined as the reference in which the total momentum is zero. We see immediately that

$$s = \left(\sum_{i=1}^{n} E_i^* \right)^2, \qquad (1.38)$$

where E_i^* are the energies in the CM. In words, the mass of a system of non-interacting particles is also its energy in the CM frame.

Consider now a system made up of two non-interacting particles. It is the simplest system, and also a very important one. Fig. 1.2 defines the kinematic variables.

The expression of s is

$$s = (E_1 + E_2)^2 - (\mathbf{p}_1 + \mathbf{p}_2)^2 = m_1^2 + m_2^2 + 2E_1 E_2 - 2\mathbf{p}_1 \cdot \mathbf{p}_2 \qquad (1.39)$$

and, in terms of the velocity, $\beta = \mathbf{p} / E$

$$s = m_1^2 + m_2^2 + 2E_1 E_2 (1 - \boldsymbol{\beta}_1 \cdot \boldsymbol{\beta}_2). \qquad (1.40)$$

Clearly in this case, and as is also true in general, the mass of a system is not the sum of the masses of its constituents, even if these do not interact. It is also clear from Eq. (1.35) that energy and momentum conservation implies that the mass is a conserved quantity: in a reaction such as a collision or decay, the mass of the initial system is always equal to that of the final system. For the same reason, the sum of the masses of the bodies present in the initial state is generally different from the sum of the masses of the final bodies.

Example 1.2   Find the expressions for the mass of the system of two photons of the same energy E, if they move in equal or in different directions.

The energy and the momentum of the photon are equal, because its mass is zero, $p = E$. The total energy $E_{\text{tot}} = 2E$.

If the photons have the same direction, then the total momentum is $p_{\text{tot}} = 2E$ and therefore the mass is $m = 0$.

If the velocities of the photons are opposite, $E_{\text{tot}} = 2E$, $\mathbf{p}_{\text{tot}} = 0$, and hence $m = 2E$.

In general, if θ is the angle between the velocities, $p_{\text{tot}}^2 = 2p^2 + 2p^2 \cos\theta = 2E^2(1 + \cos\theta)$ and hence $m^2 = 2E^2(1 - \cos\theta)$.

Notice that the same arguments apply to massive particles in the ultra-relativistic limit, $E \gg m$. $\square$

Notice that the system does not contain any matter, but only energy. Contrary to intuition, mass is not a measure of the quantity of matter in a body.

Now consider one of the basic processes of subnuclear physics: collisions. In the initial state, two particles, a and b, are present; in the final state we may have two particles (not necessarily a and b) or more. Call these $c, d, e, \ldots$. The process is

$$a + b \rightarrow c + d + e + \cdots. \tag{1.41}$$

If the final state contains only the initial particles, then the collision is said to be elastic:

$$a + b \rightarrow a + b. \tag{1.42}$$

We specify that the excited state of a particle must be considered as a different particle.

The time spent by the particles in interaction, the collision time, is extremely short and we shall think of it as instantaneous. Therefore, the particles in both the initial and final states can be considered as free.

We shall consider two reference frames, the CM frame already defined above and the laboratory frame (L). The latter is the frame in which, before the collision, one of the particles (called the target) is at rest, while the other (called the beam) moves against it. Let a be the beam particle, m_a its mass, $\mathbf{p}_a$ its momentum and E_a its energy; let b be the target mass and m_b its mass. Fig. 1.3 shows the system in the initial state.

In L, s is given by

$$s = (E_a + m_b)^2 - p_a^2 = m_a^2 + m_b^2 + 2m_b E_a. \tag{1.43}$$

In practice, the energy of the projectile is often, but not always, much larger than both the projectile and the target masses. If this is the case, we can approximate Eq. (1.43) by

$$s \simeq 2m_b E_a \qquad (E_a \gg m_a, m_b). \tag{1.44}$$

We are often interested in producing new types of particles in the collision, and therefore we are also interested in the energy available for such a process. This is obviously the total energy in the CM, which, as seen in (1.44), grows proportionally to the square root of the beam energy.

Let us now consider the CM frame, in which the two momenta are equal and opposite, as in Fig. 1.4. If the energies are much larger than the masses, $E_a^* \gg m_a$ and $E_b^* \gg m_b$, the energies are approximately equal to the momenta: $E_a^* \approx p_a^*$ and $E_b^* \approx p_b^*$. Hence, they are equal to each other, and we call them simply E^*. The total energy squared is

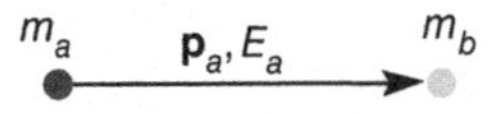

Fig. 1.3.   The laboratory frame (L).

Fig. 1.4.   The centre of mass reference frame (CM).

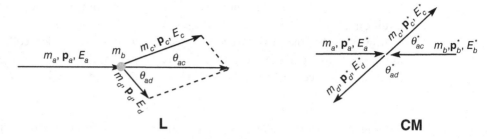

Two-body scattering in the L and CM frames.

$$s = \left(E_a^* + E_b^* \right)^2 \approx \left(2E^* \right)^2, \tag{1.45}$$

where the approximation at the last member is valid for $E^* \gg m_a, m_b$. We see that the total centre of mass energy is proportional to the energy of the colliding particles. In the CM frame, all the energy is available for the production of new particles; in the L frame only part of it is available, because momentum must be conserved.

Now let us consider a collision with two particles in the final state: this is two-body scattering

$$a + b \rightarrow c + d. \tag{1.46}$$

Fig. 1.5 shows the initial and final kinematics in the L and CM frames. Notice in particular that, in the CM frame, the final momentum is, in general, different from the initial momentum; they are equal in absolute value only if the scattering is elastic.

Because s is an invariant, it is equal in the two frames; because it is conserved, it is equal in the initial and final states. We have generically in any reference frame

$$\begin{aligned} s &= \left(E_a + E_b \right)^2 - \left(\mathbf{p}_a + \mathbf{p}_b \right)^2 \\ &= \left(E_c + E_d \right)^2 - \left(\mathbf{p}_c + \mathbf{p}_d \right)^2 . \end{aligned} \tag{1.47}$$

These properties are useful to solve a number of kinematic problems, as we shall see in the 'Problems' section later in this chapter.

In two-body scattering, there are two other important kinematic variables that have the dimensions of the square of an energy: the $a - c$ four-momentum transfer t, and the $a - d$ four-momentum transfer u. The first is defined as

$$t = \left(E_c - E_a \right)^2 - \left(\mathbf{p}_c - \mathbf{p}_a \right)^2 . \tag{1.48}$$

It is easy to see that the energy and momentum conservation implies that

$$\begin{aligned} t &= \left(E_c - E_a \right)^2 - \left(\mathbf{p}_c - \mathbf{p}_a \right)^2 \\ &= \left(E_d - E_b \right)^2 - \left(\mathbf{p}_d - \mathbf{p}_b \right)^2 . \end{aligned} \tag{1.49}$$

In a similar way

$$
\begin{aligned}
u &= \left(E_d - E_a\right)^2 - \left(\mathbf{p}_d - \mathbf{p}_a\right)^2 \\
&= \left(E_c - E_b\right)^2 - \left(\mathbf{p}_c - \mathbf{p}_b\right)^2 .
\end{aligned}
\tag{1.50}
$$

The three variables are not independent. It is easy to show (see Problems) that

$$
s + t + u = m_a^2 + m_b^2 + m_c^2 + m_d^2 .
\tag{1.51}
$$

Notice, finally, that

$$
t \le 0, \quad u \le 0.
\tag{1.52}
$$

1.5 Systems of Interacting Particles

Let us now consider a system of interacting particles. We immediately stress that its total energy is not, in general, the sum of the energies of the single particles, $E \ne \sum_{i=1}^{n} E_i$, because the field responsible for the interaction itself contains energy. Similarly, the total momentum is not the sum of the momenta of the particles, $\mathbf{P} \ne \sum_{i=1}^{n} \mathbf{p}_i$, because the field contains momentum. In conclusion, Eq. (1.35) does not in general give the mass of the system. We shall restrict ourselves to a few important examples in which the calculation is simple.

Let us first consider a particle moving in an external, given field. This means that we can consider the field independent of the motion of the particle.

Let us start with an atomic electron of charge q_e at a distance r from a nucleus of charge Zq_e. The mass of the nucleus is much larger than that of the electron, hence it is not disturbed by the electron motion. The electron then moves in a constant potential $\phi = -\dfrac{1}{4\pi\varepsilon_0}\dfrac{Zq_e}{r}$. The electron energy (in SI units) is

$$
E = \sqrt{m_e^2 c^4 + p^2 c^2} - \frac{1}{4\pi\varepsilon_0}\frac{Zq_e^2}{r} \approx \frac{p^2}{2m_e} - \frac{1}{4\pi\varepsilon_0}\frac{Zq_e^2}{r},
$$

where, in the last expression, we have taken into account that the atomic electron speeds are much smaller than c. The final expression is valid in non-relativistic situations, as is the case in an atom, and it is the Newtonian expression of the energy, apart from the irrelevant constant $m_e c^2$.

Let us now consider a system composed of an electron and a positron. The positron, as we shall see, is the antiparticle of the electron. It has the same mass and opposite charge. In the hydrogen atom, the mass of the proton is much larger than that of the electron, so we can approximate the motion of the latter as taking place in a stationary field. This is not the case for positronium, for which we must consider not only the two particles, but also the electromagnetic field in which they move, which, in turn, depends on their motion. If the energies are high enough, quantum processes happen at an appreciable frequency: the electron and the positron can annihilate each other by producing photons; inversely, a photon of the field can 'materialize' in a positron–electron pair. In these circumstances, we can no longer speak of potential.

In conclusion, the concept of potential is non-relativistic: we can use it if the speeds are small in comparison to c or, in other words, if kinetic energies are much smaller than the masses. It is correct for the electrons in the atoms, to first approximation, but not for the quarks in the nucleons.

Example 1.3 Consider the fundamental level of the hydrogen atom. The energy needed to separate the electron from the proton is $\Delta E = -13.6$ eV. The mass of the atom is smaller than the sum of the masses of its constituents by this quantity: $m_H + \Delta E = m_p + m_e$. The relative mass difference is

$$-\frac{m_H - m_p - m_e}{m_H} = \frac{13.6}{9.388 \times 10^8} = 1.4 \times 10^{-8}.$$

This quantity is extremely small, justifying the non-relativistic approximation. □

Example 1.4 The processes we have mentioned above of electron–positron annihilation and pair production can take place only in the presence of another body. If not, energy and momentum cannot be conserved simultaneously. Let us now consider the following processes.

- $\gamma \rightarrow e^+ + e^-$. Let E_+ be the energy and let $\mathbf{p}_+$ be the momentum of e^+, and E_- and $\mathbf{p}_-$ those of e^-. In the initial state, $s = 0$; in the final state, $s = (E_+ + E_-)^2 - (\mathbf{p}_+ + \mathbf{p}_-)^2 = 2m_e^2 + 2(E_+ E_- - p_+ p_- \cos\theta) > 2m_e^2 > 0$. This reaction cannot occur.
- $e^+ + e^- \rightarrow \gamma$. This is just the inverse reaction, and it cannot occur either.
- $\gamma + e^- \rightarrow e^-$. Let the initial electron be at rest. Let E_γ be the energy of the photon, and let E_f and $\mathbf{p}_f$ be the energy and the momentum of the final electron. Initially, $s = (E_\gamma + m_e)^2 - p_\gamma^2 = 2m_e E_\gamma + m_e^2$; in the final state, $s = E_f^2 - p_f^2 = m_e^2$. Setting the two expressions equal, we obtain $2m_e E_\gamma = 0$, which is false. The same is true for the inverse process $e^- \rightarrow e^- + \gamma$. This process happens in the Coulomb field of the nucleus, in which the electron accelerates and radiates a photon. The process is known by the German word bremsstrahlung. □

Example 1.5 Macroscopically inelastic collision. Consider two bodies of the same mass m moving initially one against the other with the same speed υ (for example, two wax spheres). The two collide and remain attached in a single body of mass M.

The total energy does not vary, but the initial kinetic energy has disappeared. Actually, the rest energy has increased by the same amount. The energy conservation is expressed as $2\gamma mc^2 = Mc^2$. The mass of the composite body is $M > 2m$, but changes by just a little.

Let us see by how much, as a percentage, for a speed of $v = 300$ m / s. This is rather high by macroscopic standards, but small compared to c, $\beta = v / c = 10^{-6}$. By series expansion, we have $M = 2\gamma m = \dfrac{2m}{\sqrt{1-\beta^2}} \approx 2m\left(1+\dfrac{1}{2}\beta^2\right)$. The relative mass difference is $\dfrac{M-2m}{2m} \approx \dfrac{1}{2}\beta^2 \approx 10^{-12}$. It is so small that we cannot measure it directly; we do it indirectly by measuring the increase in temperature with a thermometer. □

Example 1.6 Nuclear masses. Let us consider a ^{4}He nucleus, which has a mass of $m_{\mathrm{He}} = 3727.41$ MeV. Recalling that $m_p = 938.27$ MeV and $m_n = 939.57$ MeV, the mass defect is $\Delta E = \left(2m_p + 2m_n\right) - m_{\mathrm{He}} = 28.3$ MeV or, in relative terms,

$$\frac{\Delta E}{m_{\mathrm{He}}} = \frac{28.3}{3727.41} = 0.8\%.$$

In general, the mass defects in the nuclei are much larger than in the atoms; indeed, they are bound by a much stronger interaction. □

1.6 Measurement Units

Two systems of units are used in this textbook: the 'International System of Units' (SI) and the 'natural units' (NU).

A major reform of the SI was approved by the Conférence Générale des Poids et Mesures (CGPM in November 2018 with effect from 20 May 2019. The new SI is based on seven 'fundamental constants', the values of which are fixed to a numerical value and a symbol by definition as follows:

- the unperturbed ground state hyperfine transition frequency of the ^{133}Cs atom $\Delta\nu_{\mathrm{Cs}}$, is 9 192 631 770 Hz
- the speed of light in a vacuum, c, is 299 792 458 m/s
- the Planck constant, h, is 6.626 070 15×10^{-34} J s
- the elementary charge, e, is 1.626 176 634×10^{-19} C
- the Boltzmann constant, k, is 1.380 649×10^{-23} J / K
- the Avogadro constant, N_A, is 6.022 140 76×10^{23} mol^{-1}
- the luminous efficacy of monochromatic radiation of frequency $540×10^{12}$ Hz, K_{cd}, is 633 lm / W

where the hertz, joule, coulomb, lumen and watt, with unit symbols Hz, J, C, lm and W, respectively, are related to the base units second, metre, kilogram, ampere, kelvin, mole and candela, with unit symbols s, m, kg, A, K, mol and cd, respectively, according to Hz = s^{-1}, J = kg m^{2}s^{-2}, C = A s, lm = cd m^{2}m^{-2} = cd sr and W = kg m^{2}s^{-3}.

Details on the SI can be found on the brochure of the Bureau International des Poids et Mesures (BIPM) at www.bipm.org/en/publications/si-brochure.

As we already did, we shall often use the electronvolt (eV) in place of the Joule as the unit of energy.

It is useful to memorize approximate values of the reduced Planck constant and the speed of light in $\hbar$ and c as follows:

$$\hbar = 6.58 \times 10^{-16} \text{ eVs}, \tag{1.53}$$

$$c = 3 \times 10^{23} \text{ fm s}^{-1}, \tag{1.54}$$

$$\hbar c = 197 \text{ MeV fm (or GeV am)}. \tag{1.55}$$

As we have already done, we keep the second as the unit of time and define the unit of length such that $c = 1$. Therefore, in dimensional equations we shall have $[L] = [T]$. We now define the unit of mass in such a way as to have $\hbar = 1$.

Mass, energy and momentum have the same dimensions: $[M] = [E] = [P] = [L^{-1}]$. For unit conversions, the following relationships are useful:

$$1 \text{ MeV} = 1.52 \times 10^{21} \text{ s}^{-1}; 1 \text{ MeV}^{-1} = 197 \text{ fm}.$$
$$1 \text{ s} = 3 \times 10^{23} \text{ fm}; 1 \text{ s}^{-1} = 6.5 \times 10^{-16} \text{ eV}; 1 \text{ ps}^{-1} = 0.65 \text{ meV}.$$
$$1 \text{ m} = 5.07 \times 10^{6} \text{ eV}^{-1}; 1 \text{ m}^{-1} = 1.97 \times 10^{-7} \text{ eV}.$$

The square of the electron charge is related to the fine structure constant α by the relation

$$\frac{q_e^2}{4\pi\varepsilon_0} = \alpha hc \approx 2.3 \times 10^{-28} \text{ Jm} \tag{1.56}$$

Being dimensionless, α has the same value in all unit systems (notice that, unfortunately, one can still find the Heaviside–Lorentz units in the literature, in which $\varepsilon_0 = \mu_0 = 1$):

$$\alpha = \frac{q_e^2}{4\pi\varepsilon_0} \approx \frac{1}{137} \tag{1.57}$$

Notice that the symbol m can mean both the mass and the rest energy mc^2, but remember that the first is Lorentz-invariant, the second is the fourth component of a four-vector. The same symbol may also mean the reciprocal of the Compton wavelength λ_C times 2π, $\lambda_C = \frac{2\pi\hbar}{mc} = \frac{h}{mc}$.

Example 1.7 Measuring the lifetime of the π^0 meson, one obtains $\tau_{\pi^0} = 8.4 \times 10^{-17}$ s; what is its width? Measuring the width of the η meson, one obtains $\Gamma_\eta = 1.3$ keV; what is its lifetime? We simply use the uncertainty principle.

$$\Gamma_{\pi^0} = \hbar/\tau_{\pi^0} = (6.6 \times 10^{-16} \text{ eV s}) / (8.4 \times 10^{-17} \text{ s}) = 8 \text{ eV};$$
$$\tau_\eta = \hbar/\Gamma_\eta = (6.6 \times 10^{-16} \text{ eV s}) / (1300 \text{ eV}) = 5 \times 10^{-19} \text{ s}.$$

In conclusion, lifetime and width are completely correlated. It is sufficient to measure one of the two. The width of the π^0 particle is too small to be measured, and so we measure its lifetime, and vice versa in the case of the η particle. $\square$

Example 1.8   Evaluate the Compton wavelength of the proton.

$$\lambda_p = 2\pi / m = (6.28 / 938)\,\text{MeV}^{-1} = 6.7 \times 10^{-3}\,\text{MeV}^{-1}$$
$$= 6.7 \times 10^{-3} \times 197\,\text{fm} = 1.32\,\text{fm}. \qquad \square$$

1.7 Collisions and Decays

As we have already stated, subnuclear physics deals with two types of processes: collisions and decays. In both cases the transition amplitude is given by the matrix element of the interaction Hamiltonian H_{int} between final $\langle f |$ and initial $| i \rangle$ states

$$M_{fi} = \langle f \mid H_{\text{int}} \mid i \rangle \qquad (1.58)$$

We shall now recall the basic concepts and relations.

Collisions

Consider the collision $a + b \to c + d$. Depending on what we measure, we can define the final state with more or fewer details: we can specify or not specify the directions of c and d, we can specify or not specify their polarizations, we can say that particle c moves in a given solid angle around a certain direction without specifying the rest, etc. In each case, when computing the **cross-section** of the observed process, we must integrate on the non-observed variables, for example the angles if we want the total cross-section.

Given the two initial particles a and b, we can have different particles in the final state. Each of these processes is called a 'channel' and its cross-section is called the 'partial cross-section' of that channel. The sum of all the partial cross-sections is the total cross-section.

Decays

Consider, for example, the three-body decay $a \to b + c + d$: again, the final state can be defined with more or fewer details, depending on what is measured. Here the quantity to compute is the decay rate in the measured final state. Integrating over all the possible kinematic configurations, one obtains the partial decay rate Γ_{bcd}, or partial width of a into the $b\,c\,d$ channel. The sum of all the partial decay rates is the **total width** of a. The latter, as we have anticipated in Example 1.7, is the reciprocal of the lifetime: $\Gamma = 1/\tau$.

The **branching ratio** of a into $b\,c\,d$ is the ratio $R_{bcd} = \Gamma_{bcd} / \Gamma$

For both collisions and decays, one calculates the number of interactions per unit time, normalizing in the first case to one target particle and one beam particle, and in the second case to one decaying particle.

Let us start with the collisions, more specifically with 'fixed target' collisions. There are two elements.

(1) The beam, which contains particles of a definite type moving, approximately, in the same direction and with a certain energy spectrum. The beam intensity I_b is the number of incident particles per unit time, the beam flux Φ_b is the intensity per unit normal section.

(2) The target, which is made of matter. It contains the scattering centres of interest to us, which may be the nuclei, the nucleons, the quarks or the electrons, depending on the case. Let n_t be the number of scattering centres per unit volume and let N_t be their total number (if the beam section is smaller than that of the target, N_t is the number of centres in the beam section).

The interaction rate R_i is the number of interactions per unit time (the quantity that we measure). By definition of the cross-section σ of the process, we have

$$R_i = \sigma N_t \Phi_b = W N_t, \tag{1.59}$$

where W is the rate per particle in the target. To be rigorous, one should consider that the incident flux diminishes with increasing penetration depth in the target, due to the interactions of the beam particles. We shall consider this issue soon. We find N_t by recalling that the number of nucleons in a gram of matter is, in any case, with sufficient accuracy, the Avogadro number N_A. Consequently, if M is the target mass in kg, we must multiply by 10^3, obtaining

$$N_{\text{nucleons}} \approx M\left(\text{kg}\right)\left(10^3 \text{ kg g}^{-1}\right)N_A. \tag{1.60}$$

If the targets are nuclei of mass number A

$$N_{\text{nuclei}} \approx \frac{M\left(\text{kg}\right)\left(10^3 \text{ kg g}^{-1}\right)N_A}{A\left(\text{mol g}^{-1}\right)}. \tag{1.61}$$

The cross-section has the dimensions of a surface. In nuclear physics one uses as a unit the barn $\left(= 10^{-28} \text{ m}^2\right)$. Its order of magnitude is the geometrical section of a nucleus with $A \approx 100$. In subnuclear physics the cross-sections are smaller and submultiples are used: mb, μb, pb, etc.

In NU, the following relationships are useful:

$$1 \text{ mb} = 2.5 \text{ GeV}^{-2}, 1 \text{ GeV}^{-2} = 389 \text{ μb}. \tag{1.62}$$

Consider a beam of initial intensity I_0 entering a long target of density $\rho\left(\text{kg m}^{-3}\right)$. Let z be the distance travelled by the beam in the target, measured from its entrance point. We want to find the beam intensity $I(z)$ as a function of this distance. Consider a generic infinitesimal layer between z and $z + dz$. If dR_i is the total number of interactions per unit time in the layer, the variation of the intensity in crossing the layer is $dI(z) = -dR_i$. If Σ is the normal section of the target, $\Phi_b(z) = I(z)/\Sigma$ is the flux and σ_{tot} is the total cross-section, we have

$$dI(z) = -dR_i = -\sigma_{\text{tot}}\Phi_b(z)dN_t = -\sigma_{\text{tot}}\frac{I(z)}{\Sigma}n_t\Sigma dz$$

or

$$\frac{dI(z)}{I(z)} = -\sigma_{\text{tot}}n_t dz.$$

In conclusion, we have

$$I(z) = I_0 e^{-n_t\sigma_{\text{tot}}z}. \tag{1.63}$$

The 'absorption length', defined as the distance at which the beam intensity is reduced by the factor $1/e$, is

$$\mathcal{L}_{\text{abs}} = \frac{1}{(n_t\sigma_{\text{tot}})}. \tag{1.64}$$

Another related quantity is the 'luminosity' $\mathcal{L}[\text{m}^{-2}\text{s}^{-1}]$, often given in $[\text{cm}^{-2}\text{s}^{-1}]$, defined as the number of collisions per unit time and unit cross-section:

$$\mathcal{L} = \frac{R_i}{\sigma}. \tag{1.65}$$

Let I be the number of incident particles per unit time and Σ the beam section; then $I = \Phi_b\Sigma$. Equation (1.59) gives

$$\mathcal{L} = \frac{R_i}{\sigma} = \Phi_b N_t = \frac{IN_t}{\Sigma} \tag{1.66}$$

We see that the luminosity is given by the product of the number of incident particles in a second multiplied by the number of target particles divided by the beam section. This expression is somewhat misleading because the number of particles in the target seen by the beam depends on its section. We then express the luminosity in terms of the number of target particles per unit volume n_t and in terms of the length l of the target $(N_t = n_t\Sigma l)$. Equation (1.66) becomes

$$\mathcal{L} = In_t l = I\rho N_A 10^3 l, \tag{1.67}$$

where ρ is the target density.

Example 1.9 An accelerator produces a beam of intensity $I = 10^{13}\,\text{s}^{-1}$. The target is made up of liquid hydrogen ($\rho = 60\ \text{kg m}^{-3}$) and is $l = 10$ cm long. Evaluate its luminosity.

$$\begin{aligned}\mathcal{L} &= In_t l = \mathcal{L} = I\rho N_A 10^3 l \\ &= 10^{13}\times 60\times 10^3\times 0.1\times 6\times 10^{23} \\ &= 3.6\times 10^{40}\ \text{m}^{-2}\ \text{s}^{-1} \qquad \square\end{aligned}$$

We shall now recall a few concepts that should already be known to the reader. We start with the Fermi 'golden rule', which gives the interaction rate W per target particle

$$W = 2\pi \, | \, M_{fi} \, |^2 \, \rho(E) \qquad (1.68)$$

where E is the total energy and $\rho(E)$ is the phase-space volume (or simply the phase-space) available in the final state.

There are two possible expressions of phase-space: the 'non-relativistic' expression used in atomic and nuclear physics, and the 'relativistic' one used in subnuclear physics. Obviously the rates W must be identical, implying that the matrix element M is different in the two cases. In the non-relativistic formalism, neither the phase-space nor the matrix element are Lorentz-invariant. Both factors are invariant in the relativistic formalism, a fact that makes things simpler.

We recall that in the non-relativistic formalism the probability that a particle i has the position $\mathbf{r}_i$ is given by the square modulus of its wave function, $| \psi(\mathbf{r}_i) |^2$. This is normalized by putting its integral over the entire volume equal to one.

The volume element dV is a scalar in three dimensions, but not in space-time. Under a Lorentz transformation $\mathbf{r} \to \mathbf{r}'$, the volume element changes as $dV \to dV' = \gamma \, dV$. Therefore, the probability density $| \psi(\mathbf{r}_i) |^2$ transforms as $| \psi(\mathbf{r}_i) |^2 \to | \psi'(\mathbf{r}_i) |^2 = | \psi(\mathbf{r}_i) |^2 / \gamma$. To have a Lorentz-invariant probability density, we profit from the energy transformation $E \to E' = \gamma E$ and define the probability density as $| (2E)^{-1/2} \psi(\mathbf{r}_i) |^2$.

The number of phase-space states per unit volume is $d^3 p_i / h^3$ for each particle i in the final state. With n particles in the final state, the volume of the phase-space is therefore

$$\rho_n(E) = (2\pi)^4 \int \prod_{i=1}^{n} \frac{d^3 p_i}{(h)^3 \, 2E_i} \delta\left(\sum_{i=1}^{n} E_i - E \right) \delta^3\left(\sum_{i=1}^{n} \mathbf{p}_i - \mathbf{P} \right) \qquad (1.69)$$

or, in NU (be careful! $\hbar = 1$ implies $h = 2\pi$)

$$\rho_n(E) = (2\pi)^4 \int \prod_{i=1}^{n} \frac{d^3 p_i}{2E_i \, (2\pi)^3} \delta\left(\sum_{i=1}^{n} E_i - E \right) \delta^3\left(\sum_{i=1}^{n} \mathbf{p}_i - \mathbf{P} \right), \qquad (1.70)$$

where δ is the Dirac function. Now we consider the collision of two particles, say a and b, resulting in a final state with n particles. We shall give the expression for the cross-section.

The cross-section is normalized to one incident particle; therefore, we must divide by the incident flux. In the laboratory frame, the target particles b are at rest, the beam particles a move with a speed of, say, β_a. The flux is the number of particles inside a cylinder of unitary base and height β_a.

Let us consider, more generally, a frame in which particles b also move, with velocity β_b, which we shall assume is parallel to β_a. The flux of particles b is their number inside a cylinder of unitary base of height β_b. The total flux is the number of particles in a cylinder of height $\beta_a - \beta_b$ (i.e. the difference between the speeds, which is not, as is often written, the relative speed).

If E_a and E_b are the initial energies, the normalization factors of the initial particles are $1/(2E_a)$ and $1/(2E_b)$. It is easy to show, but we shall give only the result, that the cross-section is

$$\sigma = \frac{1}{2E_a 2E_b |\beta_a - \beta_b|} \int |M_{fi}|^2 (2\pi)^4 \prod_{i=1}^n \frac{d^3 p_i}{(2\pi)^3 2E_i} \delta\left(\sum_{i=1}^n E_i - E\right) \delta^3\left(\sum_{i=1}^n \mathbf{p}_i - \mathbf{P}\right). \quad (1.71)$$

The case of a decay is simpler, because in the initial state there is only one particle of energy E. The probability of transition per unit time to the final state f of n particles is

$$\Gamma_{if} = \frac{1}{2E} \int |M_{fi}|^2 (2\pi)^4 \prod_{i=1}^n \frac{d^3 p_i}{(2\pi)^3 2E_i} \delta\left(\sum_{i=1}^n E_i - E\right) \delta^3\left(\sum_{i=1}^n \mathbf{p}_i - \mathbf{P}\right). \quad (1.72)$$

With these expressions, we can calculate the measurable quantities, cross-sections and decay rates, once the matrix elements are known. The Standard Model gives the rules to evaluate all the matrix elements in terms of a set of constants. Even if we do not have the theoretical instruments for such calculations, we shall be able to understand the physical essence of the principal predictions of the model and to study their experimental verification.

Now let us consider an important case, the **two-body phase-space**.

Let c and d be the two final-state particles of a collision or decay. We choose the CM frame, in which calculations are easiest. The initial momentum is zero and the initial energy, E, is also known. Then the momenta of the final particles are equal and opposite, $\mathbf{p}_c = -\mathbf{p}_d$. The masses being fixed, the final energies E_c and E_d, with $E = E_c + E_d$, and the absolute values of the momenta are also fixed. Consequently, the matrix element depends only on the angles. We must evaluate the integral

$$\int |M_{fi}|^2 \frac{d^3 p_c}{(2\pi)^3 2E_c} \frac{d^3 p_d}{(2\pi)^3 2E_d} (2\pi)^4 \delta(E_c + E_d - E) \delta^3(\mathbf{p}_c - \mathbf{p}_d).$$

Consider the phase-space integral

$$\rho_2 = \int \frac{d^3 p_c}{(2\pi)^3 2E_c} \frac{d^3 p_d}{(2\pi)^3 2E_d} (2\pi)^4 \delta(E_c + E_d - E) \delta^3(\mathbf{p}_c - \mathbf{p}_d).$$

Integrating over $d^3 p_d$ we obtain

$$\rho_2 = \frac{1}{(4\pi)^2} \int \frac{d^3 p_c}{E_c E_d(p_c)} \delta(E_c + E_d(p_c) - E) = \frac{1}{(4\pi)^2} \int \frac{p_f^2 dp_f d\Omega_f}{E_c E_d(p_f)} \delta(E_c + E_d(p_f) - E).$$

Using the remaining δ-function, we obtain straightforwardly

$$\frac{1}{(4\pi)^2} \frac{p_f^2}{E_c E_d(p_f)} \frac{dp_f}{d(E_c + E_d(p_f))} d\Omega_f = \frac{1}{(4\pi)^2} \frac{p_f^2}{E_c E_d(p_f)} \frac{1}{\frac{d}{dp_f}(E_c + E_d(p_f))} d\Omega_f.$$

But $\dfrac{dE_c}{dp_f} = \dfrac{p_f}{E_c}$ and $\dfrac{dE_d}{dp_f} = \dfrac{p_f}{E_d}$,

hence $\dfrac{1}{(4\pi)^2} \dfrac{p_f^2}{E_c E_d} \dfrac{1}{\dfrac{p_f}{E_c} + \dfrac{p_f}{E_d}} d\Omega_f = \dfrac{p_f}{E} \dfrac{d\Omega_f}{(4\pi)^2}.$

Finally, (1.72) gives

$$\Gamma_{a,cd} = \frac{1}{2m} \frac{p_f}{E} \int \left| M_{a,cd} \right|^2 \frac{d\Omega_f}{(4\pi)^2}. \tag{1.73}$$

By integrating the above equation on the angles and recalling that $E = m$, we obtain

$$\Gamma_{a,cd} = \frac{p_f}{8\pi m^2} n \overline{\left| M_{a,cd} \right|^2}, \tag{1.74}$$

where the angular average of the absolute square of the matrix element appears.

Now let us consider the cross-section of the process $a + b \to c + d$, in the centre of mass frame. Again let E_a and E_b be the initial energies, and let E_c and E_d be the final ones. The total energy is $E = E_a + E_b = E_c + E_d$. Let $\mathbf{p}_i = \mathbf{p}_a = -\mathbf{p}_b$ be the initial momenta and let $\mathbf{p}_f = \mathbf{p}_c = -\mathbf{p}_d$ be the final ones.

Let us restrict ourselves to the case in which neither the beam nor the target are polarized and in which the final polarizations are not measured. Therefore, in the evaluation of the cross-section we must sum over the final spin states and average over the initial ones. Using (1.71) we have

$$\frac{d\sigma}{d\Omega_f} = \frac{1}{2E_a 2E_b |\beta_a - \beta_b|} \overline{\sum_{\text{initial}} \sum_{\text{final}} \left| M_{fi} \right|^2} \frac{1}{(4\pi)^2} \frac{p_f}{E}. \tag{1.75}$$

We evaluate the difference between the speeds

$$|\beta_a - \beta_b| = \beta_a + \beta_b = \frac{p_i}{E_a} + \frac{p_i}{E_b} = \frac{p_i E}{E_a E_b}.$$

Hence

$$\frac{d\sigma}{d\Omega_f} = \frac{1}{(8\pi)^2} \frac{1}{E^2} \frac{p_f}{p_i} \overline{\sum_{\text{initial}} \sum_{\text{final}} \left| M_{fi} \right|^2}. \tag{1.76}$$

The average over the initial spin states is the sum over them divided by their number. If s_a and s_b are the spins of the colliding particles, then the spin multiplicities are $2s_a + 1$ and $2s_b + 1$. Hence

$$\frac{d\sigma}{d\Omega_f} = \frac{1}{(8\pi)^2} \frac{1}{E^2} \frac{p_f}{p_i} \frac{1}{(2s_a + 1)(2s_b + 1)} \sum_{\text{initial}} \sum_{\text{final}} \left| M_{fi} \right|^2. \tag{1.77}$$

1.8 Scattering Experiments

Scattering experiments are fundamental in physics because they are the tools for studying the internal structure of the objects, crystals and liquids, molecules and atoms, nuclei and nucleons. As well known from optics, a beam of a given wavelength has a resolving power inversely proportional to that wavelength. The momentum $\mathbf{p}$, the wave-vector $\mathbf{k}$ and the wavelength λ are related, in natural units, by the relations

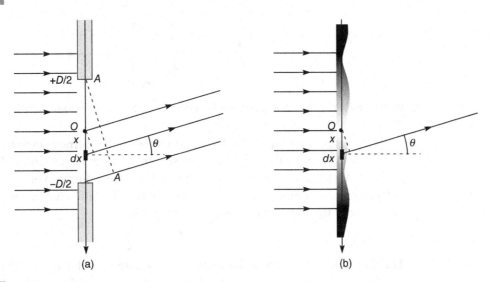

Fig. 1.6. Schematics of a diffraction experiment.

$$\mathbf{p} = \mathbf{k} \qquad p = \frac{2\pi}{\lambda}. \qquad (1.78)$$

We start by considering an experiment in optics, the Fraunhofer diffraction from a slit. Fig. 1.6 shows a slit of width D along the x direction opened from $x = -D/2$ to $x = +D/2$. A monochromatic plane light wave illuminates perpendicularly the plane of the slit, as shown in Fig. 1.6(a).

To obtain the diffracted amplitude at a certain angle θ, we must sum coherently the contributions dA of all the elements dx of the slit. The phase delay at θ for the element at x is $\phi = k\, x \sin\theta$. We make the important observation that $k \sin\theta = k_x$, namely the component of the wave vector perpendicular to the beam, and write $\phi = k_x\, x$. Putting the incident amplitude equal to 1, the diffracted amplitude is

$$A = \int_{-D/2}^{+D/2} e^{-ik_x x} dx.$$

To be more general, we consider the plane with the slit to be a screen with amplitude transparency (i.e. the ratio between the transmitted and incident amplitude) equal to $T(x)$. Clearly this is a rectangular function, namely $T(x) = 1$ for $-D/2 \le x \le +D/2$ and $T(x) = 0$ for $x < D/2$ and $> D/2$. We write

$$A = \int_{-\infty}^{+\infty} T(x) e^{-ik_x x} dx. \qquad (1.79)$$

This result is general for the diffraction by a plane having an amplitude transparency varying as a function of x. In Fig. 1.6(b) we show a schematic example in which the incident wave encounters a plate made, say, of glass or plastic, having a refraction index $n \neq 1$. Both its transparency and its thickness vary as functions of x. Consequently, the plate modifies both the amplitude and the phase of the transmitted wave. The amplitude transparency $T(x)$ is then a complex function with an absolute value and a phase.

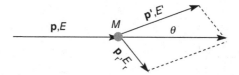

Kinematic variables for the elastic scattering of a particle of mass m by a particle of mass M in the laboratory frame.

We recognize from Eq. (1.79) that the amplitude of the diffracted wave is the spatial Fourier transform of the amplitude transparency of the target screen. The Fourier conjugate variable is the wave-vector component in the x direction or, more generally, of its transverse component $\mathbf{k}_T$. Finally, we obtain the diffracted intensity by taking the square of the square modulus of the amplitude

$$I = \left| \int_{-\infty}^{+\infty} T(\mathbf{r}) e^{-i\mathbf{k}_T \cdot \mathbf{r}} dx \right|^2 . \tag{1.80}$$

The diffracted, or scattered, intensity is the square modulus of the spatial Fourier transform of the amplitude transparency of the target. The Fourier transform variable is the transverse component of the wave-vector (or of the momentum).

Notice that if the diameter of the target is $D < \lambda / 2$, the maximum phase variation $D\phi$ becomes so small that it is unobservable. We can conclude that a target appears point-like if its diameter $D \leq \lambda / 2 = \pi / \mathbf{k}_T$.

Question 1.1 Sketch the absolute square of the Fourier transform of a rectangular function and compare it with the Fraunhofer slit diffraction pattern. □

In a similar manner, we use probes of adequate resolving power to study the structure of microscopic objects, such as a nucleus or a nucleon.

The probes are particle beams. The situation is simple if these are point-like particles such as electrons and neutrinos. The two are complementary: the electrons 'see' the electric charges inside the nucleon, while neutrinos 'see' the weak charges.

We must now study the kinematics of the collisions. We start with the elastic scattering of small-mass spinless particles, say electrons, neglecting their spin, of mass m_e, against, as a target, a large-mass particle, say a nucleus, of mass M. Fig. 1.7 defines the kinematic variables in the laboratory frame.

For elastic scattering, the knowledge of the momentum and energy of the incident particle and the measurement of the energy and direction of the scattered particle completely determine the event. Let us see.

The four-momenta and their norms are

$$p_\mu = (E, \mathbf{p}); \quad p'_\mu = (E', \mathbf{p}'); \quad p_\mu p^\mu = p'_\mu p'^\mu = m_e^2$$
$$P_\mu = (M, \mathbf{0}); \quad P'_\mu = (E_r, \mathbf{p}_r); \quad P_\mu P^\mu = P'_\mu P'^\mu = M^2. \tag{1.81}$$

Energy and momentum conservation give

$$p_\mu + P_\mu = p'_\mu + P'_\mu \Rightarrow p_\mu p^\mu + P_\mu P^\mu + 2p_\mu P^\mu = p'_\mu p'^\mu + P'_\mu P'^\mu + 2p'_\mu P'^\mu. \tag{1.82}$$

Taking into account that $p_\mu p^\mu = p'_\mu p'^\mu = m_e^2$ and that $P_\mu P^\mu = P'_\mu P'^\mu = M^2$, this gives $p_\mu P^\mu = p'_\mu P'^\mu$, which is $EM - 0 = E'E_r - \mathbf{p}' \cdot \mathbf{p}_r$. Considering that $E_r = E + M - E'$ and that $\mathbf{p}_r = \mathbf{p} - \mathbf{p}'$, we have

$$EM = E'(E + M - E') - \mathbf{p}' \cdot (\mathbf{p} - \mathbf{p}') = E'E + E'M - pp' \cos\theta - m_e^2 .$$

That is the required relationship. It becomes very simple if the electron energy is high enough. We then neglect the term m_e^2 and take the momenta to be equal to the energies, obtaining

$$E' = \frac{E}{1 + \dfrac{E}{M}(1 - \cos\theta)} = \frac{E}{1 + \dfrac{2E}{M}\sin^2\dfrac{\theta}{2}} . \tag{1.83}$$

This important relationship shows, in particular, that the energy transferred to the target $E - E'$ becomes negligible for a large target mass, namely if $E/M \ll 1$. However, the momentum transfer is not negligible.

Now consider the scattering of an electron of charge q_e by the electrostatic potential of a nucleus $\phi(r)$. Let us calculate the transition matrix element. The incident and scattered electrons in Fig. 1.7 are free, hence their wave functions are plane waves. Neglecting uninteresting constants, we have

$$\langle \psi_f | q_e\phi(r) | \psi_i \rangle \propto q_e \int \exp(i\mathbf{p}' \cdot \mathbf{r})\phi(r)\exp(-i\mathbf{p} \cdot \mathbf{r})dV \tag{1.84}$$
$$= q_e \int \exp[i\mathbf{q} \cdot \mathbf{r}]\phi(r)dV,$$

where

$$\mathbf{q} = \mathbf{p}' - \mathbf{p}$$

is the three-momentum transfer between the incident and the scattered particles.

We see that the scattering amplitude is proportional to the Fourier transform of the potential. The momentum transfer is the variable conjugate to the distance from the centre:

$$\langle \psi_f | q_e\phi(r) | \psi_i \rangle \propto q_e \int \exp[i\mathbf{q} \cdot \mathbf{r}]\phi(r)dV . \tag{1.85}$$

We want to find the dependence of the matrix element on the charge density (the electric charge in this case), which we call $\rho(r)$. This is the source of the potential and, according to the electrostatic equation, we write

$$\Delta^2\phi = -\frac{\rho}{\varepsilon_0} . \tag{1.86}$$

We now use the relationship $\nabla^2 \exp(i\mathbf{q} \cdot \mathbf{r}) = -q^2 \exp(i\mathbf{q} \cdot \mathbf{r})$ and the identity $\int \phi\nabla^2 [\exp(i\mathbf{q} \cdot \mathbf{r})]dV = \int [\exp(i\mathbf{q} \cdot \mathbf{r})]\nabla^2\phi dV$, obtaining

$$\langle \psi_f | q_e\phi(r) | \psi_i \rangle \propto \frac{q_e}{\varepsilon_0}\frac{1}{q^2} \int \rho(r)\exp[i\mathbf{q} \cdot \mathbf{r}]dV . \tag{1.87}$$

It is now easy to calculate the cross-section, but since we are not interested in the proof, we simply give the result:

$$\frac{d\sigma}{d\Omega} = \frac{q_e^2}{(2\pi)^2 \varepsilon_0^2}\frac{E'^2}{|\mathbf{q}|^2} \int \rho(r)\exp(i\mathbf{q} \cdot \mathbf{r})dV . \tag{1.88}$$

Let us now call $f(\mathbf{r})$ the target charge density normalized to one, namely

$$f(\mathbf{r}) = \frac{1}{Zq_e}\rho(\mathbf{r}) \tag{1.89}$$

and $F(\mathbf{q})$ its Fourier transform

$$F(\mathbf{q}) = \int f(\mathbf{r})\exp(i\mathbf{q}\cdot\mathbf{r})dV. \tag{1.90}$$

Then we can write (1.88) as

$$\frac{d\sigma}{d\Omega} = 4Z^2\alpha^2\frac{E'^2}{|\mathbf{q}|^4}|F(\mathbf{q})|^2. \tag{1.91}$$

We can say that the intensity scattered by an immobile target is proportional to the square of the Fourier transform of its charge distribution.

Rutherford Cross-section

Let us consider a point-like target with charge Zq_e at the origin. The charge density function is $Zq_e\delta(0)$. Its transform is a constant. If zq_e is the beam charge, Eq. (1.91) becomes

$$\left(\frac{d\sigma}{d\Omega}\right)_{\text{Rutherford}} = 4z^2Z^2\alpha^2\frac{E'^2}{|\mathbf{q}|^4} \tag{1.92}$$

This is the well-known Rutherford cross-section.

Rutherford found this expression to interpret the Geiger and Marsden experiment. The probe was a beam of α particles, with kinetic energy E_k of a few MeV. Therefore, we can write $E_k = p^2/2m$. Let us now find the cross-section as a function of E_k and of the scattering angle.

Looking at Fig. 1.8, we see that, from $p' = p$ and $E' = E$, it follows that $q = 2p\sin\theta/2$. We can also set $E = m$, obtaining the well-known expression

$$\frac{d\sigma}{d\Omega} = \frac{z^2Z^2\alpha^2}{16E_k^2}\frac{1}{\sin^4\theta/2}. \tag{1.93}$$

The cross-section is independent of the azimuth ϕ. We integrate over ϕ recalling that $d\Omega = d\phi\, d\cos\theta$, obtaining

$$\frac{d\sigma}{d\cos\theta} = \frac{\pi}{8}\frac{z^2Z^2\alpha^2}{E_k^2}\frac{1}{\sin^4\dfrac{\theta}{2}} = \frac{\pi}{2}\frac{z^2Z^2\alpha^2}{E_k^2}\frac{1}{(1-\cos\theta)^2}. \tag{1.94}$$

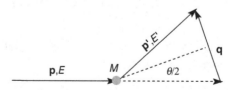

Geometric relation between three-momentum transfer and scattering angle in Rutherford scattering.

Notice the divergence for $\theta \to 0$. This is a consequence of the divergence of the assumed potential for $r \to 0$, a situation that is never found in practice.

Mott Cross-section

Neglecting the electron spin effects is a good approximation at low energies. However, if the speed of the electron is high, the spin effects become important. The expression of the cross-section for electron scattering in the Coulomb potential of an infinite mass target, which implies that $E' = E$, is due to Mott. It is given by Eq. (1.92) multiplied by $\cos^2 \left(\theta / 2 \right)$

$$\left(\frac{d\sigma}{d\Omega} \right)_{\text{Mott}} = \left(\frac{d\sigma}{d\Omega} \right)_{\text{Rutherford}} \cos^2 \frac{\theta}{2} = 4z^2 Z^2 \alpha^2 \frac{E'^2}{|\mathbf{q}|^4} \cos^2 \frac{\theta}{2}. \tag{1.95}$$

The Mott cross-section decreases with increasing angle faster than the Rutherford cross-section and becomes zero at 180°. We shall give the reason for this in Section 5.6.

We finally consider the ultra-relativistic case, in which the projectile mass is negligible compared with its energy. The recoil energy cannot be neglected any more, and $E' < E$. The expression valid for a point target is

$$\left(\frac{d\sigma}{d\Omega} \right)_{\text{Point}} = \frac{E'}{E} \left(\frac{d\sigma}{d\Omega} \right)_{\text{Mott}}. \tag{1.96}$$

1.9 Hadrons, Leptons, Quarks and Bosons

The particles can be classified, depending on their characteristics, in different groups. Here we shall give the names of these groups and summarize their properties.

The particles of a given type, the electrons for example, are indistinguishable. Take, for example, a fast proton hitting a stationary one. After the collision, which we assume to be elastic, there are two protons moving, in general, in different directions with different energies. It is pointless to try to identify one of these as, say, the incident proton.

First of all, we can distinguish the particles of integer spin, in units $\hbar, (0, \hbar, 2\hbar, \ldots)$, that follow Bose statistics and are called **bosons**, and the half-integer spin particles $\left(\frac{1}{2}\hbar, \frac{3}{2}\hbar, \frac{5}{2}\hbar, \ldots \right)$ that follow Fermi–Dirac statistics and are called **fermions**. Bosons are, in particular, the mediators of the fundamental interactions, like the photon, and the 'Higgs boson'. We recall that the wave function of a system of identical bosons is symmetric under the exchange of any pair of them, while the wave function of a system of identical fermions is antisymmetric.

Matter is made up of atoms. Atoms are made of **electrons** and **nuclei** bound by the electromagnetic force, whose quantum is the **photon** (from the Greek word $\varphi \tilde{\omega} \varsigma$ = light).

Photons are massless. Their charge is zero and therefore they do not interact between themselves. Their spin is equal to one; they are bosons. The electrons have negative electric charge and spin 1/2; they are fermions. Their mass is small, $m_e = 0.511$ MeV, in comparison with that of the nuclei. As far as we know they do not have any structure; they are elementary.

Nuclei contain most of the mass of the atoms, and hence of the matter. They are positively charged and made of protons and neutrons. **Protons** (from πρῶτον, *the first* in Greek) and **neutrons** have similar masses, slightly less than a GeV. The charge of the proton is positive, opposite and exactly equal to the electron charge; neutrons are globally neutral, but contain charges, as shown, for example, by their non-zero magnetic moment. As anticipated, protons and neutrons are collectively called **nucleons**. Nucleons have spin 1/2; they are fermions. Protons are stable, within the limits of present measurements; the reason is that they have another conserved 'charge' beyond the electric charge, the 'baryonic number', that we shall discuss in Chapter 3.

In 1935, Yukawa formulated a theory of the strong interactions between nucleons. Nucleons are bound in nuclei by the exchange of a zero-spin particle, the quantum of the nuclear force. Given the finite range of this force, its mediator must be massive, and given the value of the range, about 10^{-15} m, its mass should be intermediate between the electron and the proton masses; therefore, it was called **meson** (= *that which is in the middle*). More specifically, it is the π meson, also called **pion**. We shall describe its properties in the next chapter. Pions come in three charge states: π^+, π^- and π^0. Unexpectedly, from 1946 onwards, other mesons were discovered in cosmic radiation, the K-mesons, which come in two different charge doublets K^+ and K^0 and their antiparticles, K^- and $\bar{K}^0$.

In the same period other particles were discovered that, like the nucleons, have half-integer spin and baryonic number. They are somewhat more massive than nucleons and are called **baryons** (= *that which is heavy or massive*). Notice that nucleons are included in this category.

Baryons have half-integer spins (1/2, 3/2, ...); mesons have integer spins (0, 1, ...). Both classes are not point-like, but they have a structure and are composite objects. The components of both of them are the **quarks**. In a first approximation, the baryons are made up of three quarks, the mesons of a quark and an antiquark. Quarks interact via one of the fundamental forces, the strong force, that is mediated by the **gluons** (from *glue*). As we shall see, there are eight different gluons; all are massless and have spin one. Baryons and mesons have a similar structure and are collectively called **hadrons** (from the Greek ἁδρός, meaning *thick*, *strong*). All hadrons are unstable, with the exception of the lightest one, the proton.

By shooting a beam of electrons or photons at an atom we can extract the electrons it contains, provided the beam energy is large enough. Analogously we can break a nucleus into its constituents by bombarding it, for example, with sufficiently energetic protons. The two situations are similar, with quantitative, not qualitative, differences: in the first case a few eV are sufficient, in the second several MeV are needed. On the contrary, nobody has ever succeeded in breaking a hadron and extracting the quarks, whatever the energy and the type of the bombarding particles. We have been forced to conclude that quarks do not exist in a free state; they exist only inside the hadrons. We shall see how the Standard Model explains this property, which is called 'quark confinement'.

The spin of the quarks is 1/2. There are three quarks with electric charge +2/3 (in units of the elementary charge), called up-type, and three with charge −1/3 called down-type. In order of increasing mass, the up-type are: 'up' u, 'charm' c and 'top' t; the down-type are: 'down' d, 'strange' s and 'beauty' b. Nucleons, hence nuclei, are made up of up quarks and down quarks: uud the proton, udd the neutron.

The electrons are also members of a class of particles of similar properties, the **leptons** (*light* in Greek, but there are also leptons heavier than baryons). Their spin is 1/2. There are three charged leptons, the electron e, the muon μ and the tau τ, and three neutral leptons, the **neutrinos**, one for each of the charged ones. The electron is stable, the μ and the τ decay in lighter particles, and neutrinos do not decay.

For every particle there is an antiparticle with the same mass and the same lifetime and, if it has any charge, these have opposite values: the positron for the electron, the antiproton, the antiquarks, etc.

One last consideration: astrophysical and cosmological observations have shown that the 'ordinary' matter, baryons and leptons, makes up only a small fraction of the total mass of the Universe, no more than 20%. We do not know what the rest is made of. There is still a lot to understand beyond the Standard Model.

1.10 The Fundamental Interactions

Each of the interactions is characterized by one, or more, 'charge' that, like the electric charge, is the source and the receptor of the interaction. The Standard Model, SM for brief, is the theory that describes all the fundamental interactions, except gravitation. For the latter, we do not yet have a microscopic theory, but only a macroscopic approximation, general relativity.

The source and the receptor of the gravitational interaction is the energy–momentum tensor; consequently, this interaction is felt by all particles. However, gravity is extremely weak at all the energy scales experimentally accessible in particle physics and we shall neglect its effects.

Let us find the orders of magnitude by the following dimensional argument. The fundamental constants, the Newton constant G_N for gravity, the speed of light c, the Lorentz transformations, the Planck constant h and quantum mechanics can be combined in an expression with the dimensions of a mass that is called the Planck mass:

$$M_P = \sqrt{\frac{\hbar c}{G_N}} = \frac{1.06 \times 10^{-34}\ \text{Js} \times 3 \times 10^8\ \text{ms}^{-1}}{6.67 \times 10^{-11}\ \text{m}^3\text{kg}^{-1}\text{s}^{-2}} \tag{1.97}$$
$$= 2.18 \times 10^{-8}\ \text{kg} = 1.22 \times 10^{19}\ \text{GeV}.$$

It is enormous, not only in comparison with the energy scale of the nature around us on Earth (eV) but also of nuclear (MeV) and subnuclear (GeV) physics. We shall never be able to build an accelerator to reach such an energy scale. We must search for quantum features of gravity in the violent phenomena occurring naturally in the Universe.

The gravitational interaction is accurately described at the macroscopic level by general relativity, which, however is not the final theory, which should be a quantum theory. We do not have a quantum theory of gravity. In a hypothetical one, the mediator, the graviton, would be a boson of zero mass, since we know that the range of the gravitational force is infinite, and spin 2, due to the tensor character of the gravitational waves, as we shall see in Chapter 11.

All the known particles have weak interactions, with the exception of photons and gluons. Weak interactions are responsible for the decay of particles and nuclei over relatively long lifetimes, in particular beta decay. The presence of a neutral particle not detected in nuclear beta decay was assumed in Pauli in 1930 as a 'desperate remedy' to explain the observed continuum electron energy spectrum. Three years later Fermi (1933) laid down the first theory of a new force, the weak interaction. In a beta decay, two particles are emitted, an electron and a neutrino, the latter having a very small or zero mass. Fermi considered this process as being similar to the emission of a photon by an excited atom. As with the photon, the electron + neutrino pair is created at the moment of the decay, rather than being present in the decaying system. We know now that the weak interaction is mediated by three spin-one bosons, W^+, W^- and Z^0; their masses are rather large, in comparison to, say, the proton mass (in round numbers $M_W \approx 80$ GeV, $M_Z \approx 90$ GeV). Their existence becomes evident at energies comparable to those masses.

All charged particles have electromagnetic interactions. This interaction is transmitted by the photon, which is massless.

Quarks and gluons have strong interactions; the leptons do not. The corresponding charges are called 'colours'. The interaction amongst quarks in a hadron is mediated by gluons and confined inside the hadron. If two hadrons – two nucleons, for example – come close enough (typically 1 fm), they interact via the 'tails' of the colour field that, so to speak, leaks out of the hadron. The phenomenon is analogous to the Van der Waals force that is due to the electromagnetic field leaking out from the molecule. Therefore, the nuclear (Yukawa) forces are not fundamental.

As we have said, charged leptons more massive than the electron are unstable; the lifetime of the muon is about 2 μs, that of the τ is 0.3 ps. These are large values on the scale of elementary particles, and are characteristic of weak interactions.

All mesons are unstable: the lifetimes of $\pi^\pm$ and of $K^\pm$ are 26 ns and 12 ns, respectively; they are weak decays. In the 1960s, other larger-mass mesons were discovered; they have strong decays and extremely short lifetimes, of the order of $10^{-23} - 10^{-24}$ s.

All baryons, except for the proton, are unstable. The neutron has a beta decay into a proton with a lifetime of 878 s. This is exceptionally long even for the weak interaction standard because of the very small mass difference between neutrons and protons. Some of the other baryons, the less-massive ones, decay weakly with lifetimes of the order of a tenth of a nanosecond; others, the more massive ones, have strong decays with lifetimes of $10^{-23} - 10^{-24}$ s.

Example 1.10   Consider an electron and a proton separated by a distance r. Evaluate the ratio between the electrostatic and gravitational forces. Does it depend on r?

$$F_{\text{electrost.}}(ep) = \frac{1}{4\pi\varepsilon_0}\frac{q_e^2}{r^2} \qquad F_{\text{gravit.}}(ep) = G_N \frac{m_e m_p}{r^2};$$

$$\frac{F_{\text{electrost.}}(ep)}{F_{\text{gravit.}}(ep)} = \frac{q_e^2}{4\pi\varepsilon_0 G_N m_e m_p} = \frac{(1.6\times10^{-19})^2}{4\pi\times8.8\times10^{-12}\times6.67\times10^{-11}\times9.1\times10^{-31}\times1.7\times10^{-27}} \approx 10^{39},$$

independent of r. □

1.11 The Passage of Radiation through Matter

The Standard Model has been developed and tested by a number of experiments, some of which we shall describe. This discussion is not possible without some knowledge of the physics of the passage of radiation through matter, of the main particle detectors and of the sources of high-energy particles.

When a high-energy charged particle or a photon passes through matter, it loses energy that excites and ionizes the molecules of the material. It is through experimental observation of these alterations of the medium that elementary particles are detected. Experimental physicists have developed a wealth of detectors aimed at measuring different characteristics of the particles (energy, charge, speed, position, etc.). This broad and very interesting field is treated in specialized courses and books. Here we shall summarize only the main conclusions relevant for the experiments discussed in the text.

Ionization Loss

The energy loss of a relativistic charged particle more massive than the electron passing through matter is due to its interaction with the atomic electrons. The process results in a trail of ion–electron pairs along the path of the particle. These free charges can be detected. Electrons also lose energy through bremsstrahlung in the Coulomb fields of the nuclei.

The expression of the average energy loss to ionization per unit length of charged particles other than electrons is known as the Bethe–Bloch equation (Bethe 1930). Here we give an approximate expression, which is enough for our purposes. If z is the charge of the particle, ρ is the density of the medium, Z is its atomic number and A its atomic mass, the equation is

$$-\frac{dE}{dx} = K\frac{\rho Z}{A}\frac{z^2}{\beta^2}\left[\ln\left(\frac{2mc^2\gamma^2\beta^2}{I} - \beta^2\right)\right], \qquad (1.98)$$

where m is the electron mass (the hit particle), the constant K is given by

$$K = \frac{4\pi\alpha^2(\hbar c)^2 N_A\left(10^3\,\text{kg}\right)}{mc^2} = 30.7\ \text{keV m}^2\ \text{kg}^{-1}, \qquad (1.99)$$

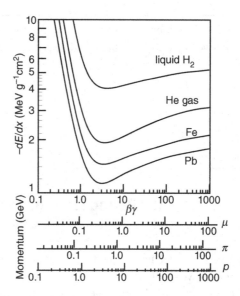

Fig. 1.9. Specific average ionization loss per unit density (in $g\,cm^{-3}$) for relativistic particles of unit charge. Adapted by permission of Particle Data Group, Lawrence Berkeley National Lab.

and I is an average ionization potential. For $Z > 20$, $I \approx 12Z$ eV. The energy loss per unit density of the medium and unit track length is a universal function of $\beta\gamma$, in a very rough approximation, but there are important differences in the different media, as shown in Fig. 1.9. The curves are drawn for particles of charge $z = 1$; for larger charges, multiply by z^2.

All the curves decrease rapidly at small momenta (roughly as $1/\beta^2$), reach a shallow minimum for $\beta\gamma = 3 - 4$ and then increase very slowly. The energy loss of a minimum ionizing particle (mip) is $\left(0.1 - 0.2 \text{ MeV m}^2\text{kg}^{-1}\right)\rho$.

The Bethe–Bloch formula is valid only in the energy interval corresponding to approximately $0.05 < \beta\gamma < 500$. At lower momenta, the particle speed is comparable to the speed of the atomic electrons. In these conditions a (possibly large) fraction of the energy loss is due to the excitation of atomic and molecular levels, rather than to ionization. This fraction must be detected as light, coming from the de-excitation of those levels or, in a crystal, as phonons.

At energies larger than a few hundred GeV for pions or muons, much larger for protons, another type of energy loss becomes more important than ionization: the bremsstrahlung losses in the nuclear fields. Consequently, the specific energy loss dE/dx grows at energies larger than or around 1 TeV for muons and pions, above several TeV for kaons and then for protons, and so on for increasing particle mass. Fig. 1.10 shows a set of measurements of the ionization losses as functions of the momentum for different particles as measured by the ALICE experiment (see Section 6.11) at the CERN Large Hadron Collider (LHC).

Notice that the Bethe–Bloch formula gives the average energy loss, while the measured quantity is the energy loss for a given length. The latter is a random variable

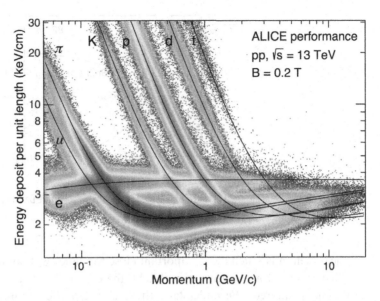

Fig. 1.10. Specific energy loss (dE / dx) versus momentum in the Time Projection Chamber (TPC, see Section 1.12) of the ALICE experiment at the CERN LHC in pp collisions at 13 TeV CM energy. ALICE-PUBLIC-2015-004. CC BY 4.0 Fig. 1. https://cds.cern.ch/record/2047855/files/13tev.pdf.

with a frequency function centred on the expectation value given by the Bethe–Bloch equation. The variance, called the straggling, is quite large. Notice in Fig. 1.10 the dispersion around the average values. The mere observation of the track produced by a charged particle does not allow us to establish its nature, namely if it is a proton, a pion, an electron, etc. This can be done by measuring the specific energy loss dE / dx along the track and the momentum.

Energy Loss of the Electrons

Fig. 1.10 shows that electrons behave differently from other particles. As anticipated, electrons and positrons, owing to their small mass, lose energy not only by ionization but also by bremsstrahlung in the nuclear Coulomb field. This already happens at several MeV.

As we have seen in Example 1.4, the process $e^- \rightarrow e^- + \gamma$ cannot take place in a vacuum, but can happen near a nucleus. The reaction is

$$e^- + N \rightarrow e^- + N + \gamma, \tag{1.100}$$

where N is the nucleus. The case for positrons is similar:

$$e^+ + N \rightarrow e^+ + N + \gamma. \tag{1.101}$$

Classically, the power radiated by an accelerating charge is proportional to the square of its acceleration. In quantum mechanics, the situation is similar: the probability of radiating a photon is proportional to the acceleration squared. Therefore, this

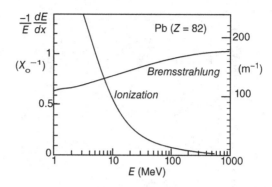

Fig. 1.11. Relative energy loss of electrons in lead (Pb). Adapted by permission of Particle Data Group, Lawrence Berkeley National Lab.

phenomenon is much more important close to a nucleus than to an atomic electron. Furthermore, for a given external field, the probability is inversely proportional to the mass squared. We understand that, for the particle immediately more massive than the electron, the muon that is about 200 times heavier, the bremsstrahlung loss becomes important at energies larger by four orders of magnitude.

Comparing different materials, the radiation loss is more important if Z is larger. More specifically, the materials are characterized by their radiation length X_0. The radiation length is defined as the distance over which the electron energy decreases to $1/e$ of its initial value due to radiation, namely

$$-dE / E = dx / X_0. \tag{1.102}$$

The radiation length is roughly inversely proportional to Z, and hence to the density. A few typical values are as follows: air at normal temperature and pressure (n.t.p.) $X_0 \approx 300$ m, water $X_0 \approx 0.36$ m, C $X_0 \approx 0.2$ m, Fe $X_0 \approx 2$ cm and Pb $X_0 \approx 5.6$ mm.

In Fig. 1.11 we show the electron energy loss in Pb; in other materials the behaviour is similar. At low energies the ionization loss dominates; at high energies the radiation loss becomes more important. The crossover, when the two losses are equal, is called the critical energy. With a good approximation, it is given by

$$E_c = 600 \text{ MeV} / Z. \tag{1.103}$$

For example, the critical energy of Pb, which has $Z = 82$, is $E_c = 7$ MeV.

Energy Loss of the Photons

At energies of the order of dozens of electronvolts (eV), the photons lose energy mainly by the photoelectric effect on atomic electrons. Above a few keV, the Compton effect becomes important. When the production threshold of the electron–positron pairs is crossed, at 1.022 MeV, this channel rapidly becomes dominant. The situation is shown in Fig. 1.12 for the case of Pb.

In the pair production process

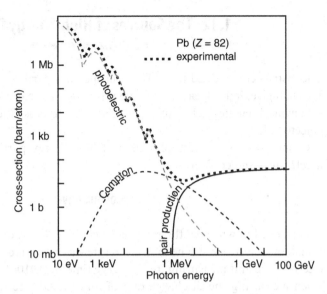

Fig. 1.12. Photon cross-sections in lead versus energy; total and calculated contributions of the three principal processes. Adapted by permission of Particle Data Group, Lawrence Berkeley National Lab.

$$\gamma + N \rightarrow N + e^- + e^+, \tag{1.104}$$

a photon disappears: it is absorbed. The attenuation length of the material is defined as the length that attenuates the intensity of a photon beam to $1/e$ of its initial value. The attenuation length is closely related to the absorption length, being equal to $(9/7)X_0$. Therefore, X_0 determines the general characteristics of the propagations of electrons, positrons and photons.

Energy Loss of the Hadrons

High-energy hadrons passing through matter do not lose energy by ionization only. Eventually they interact with a nucleus by strong interaction. This leads to the disappearance of the incoming particle, the production of secondary hadrons and the destruction of the nucleus. At energies larger than several GeV, the total cross-sections of different hadrons become equal within a factor 2 or 3. For example, at 100 GeV, the cross-sections $\pi^+ p, \pi^- p, \pi^+ n, \pi^- n$ are all about 25 mb, whereas those for pp and pn are about 40 mb. The collision length λ_0 of a material is defined as the distance over which a neutron beam (particles that do not have electromagnetic interactions) attenuates by $1/e$ in that material.

Typical values of pion collision length are as follows: air at n.t.p. $\lambda_0 \approx 750$ m; water $\lambda_0 \approx 0.86$ m; C $\lambda_0 \approx 0.39$ m; Fe $\lambda_0 \approx 0.14$ m; and Pb $\lambda_0 \approx 0.12$ m. Comparing with the radiation length, we see that collision lengths are larger and do not depend heavily on the material, provided this is solid or liquid. These observations are important in the design of calorimeters (see Section 1.12).

1.12 The Sources of High-Energy Particles

The instruments needed to study the elementary particles are sources and detectors. We shall give, in both cases, only the pieces of information that are necessary for the following discussions. In this section, we shall discuss the sources; in the next, the detectors.

Cosmic rays are a natural source of high-energy particles; artificial sources are accelerators and colliders.

Cosmic Rays

It has been known since the nineteenth century that radioactivity produces ionization in the atmosphere. The ionization rate was measured by charging a well-isolated electroscope and measuring its discharge time. In 1910–11, D. Pacini (Pacini 1912), when measuring the discharge rate of an isolated electrometer, on the surface of the sea, far enough from land (300 m), and even under water (3 m), discovered the existence of an ionization source different from the radioactivity of the ground. He could not establish, however, whether this source was in the atmosphere or above it. In 1912, V. F. Hess, flying at high altitudes with aerostatic balloons (Hess 1912), up to 5.2 km, found that the flux of ionizing radiation decreased up to about 1 km and then steadily increased to reach a value double that on the ground at the maximum height of his flights. This established the extraterrestrial origin of the radiation, which was later called 'cosmic' rays. Fermi formulated the first theory of the acceleration mechanism in 1949 (Fermi 1949), but the issue is still under development. Until the early 1950s, when the first high-energy accelerators were built, cosmic rays were the only source of particles with energy larger than 1 GeV. The study of the cosmic radiation remains, even today, fundamental for both subnuclear physics and astrophysics.

We know rather well the energy spectrum of cosmic rays, which is shown in Fig. 1.13, in the compilation of W. Hanlon (Ph.D. Dissertation, Utah University 2008). It extends up to 100 EeV $\left(10^{20}\,\text{eV}\right)$, 12 orders of magnitude on the energy scale and 32 orders of magnitudes on the flux scale. To make a comparison, notice that the highest-energy accelerator, the LHC at CERN, has a centre of mass energy of 14 TeV, corresponding to 'only' 0.1 EeV in the laboratory frame. At these extreme energies, the flux is very low, typically one particle per square kilometre per century. The Pierre Auger observatory in Argentina has an active surface of 3000 km^2 and is exploring the energy range >EeV.

The initial discoveries in particle physics, which we shall discuss in the next chapter, used the spectrum around a few GeV, where the flux is largest, tens of particles per square metre per second. In this region the primary composition, namely at the top of the atmosphere, consists of 85% protons, 12% alpha particles, 1% heavier nuclei and 2% electrons.

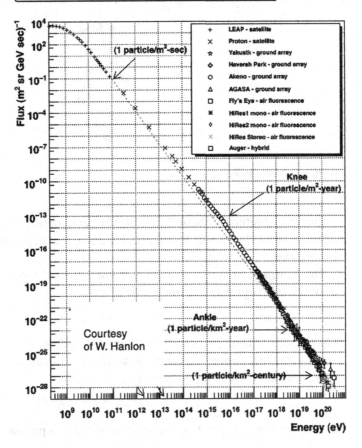

Cosmic Ray Spectra of Various Experiments

Fig. 1.13. The cosmic rays flux. Courtesy of W. Hanlon. https://web.physics.utah.edu/~whanlon/spectrum.html.

A proton or a nucleus penetrating the atmosphere eventually collides with a nucleus of the air. This strong interaction produces πs, less frequently K-mesons and, even more rarely, other hadrons. The hadrons produced in the first collision generally have enough energy to produce other hadrons in a further collision, and so on. The average distance between collisions is the collision length $(\lambda_0 = 750 \text{ m at n.t.p.})$. The primary particle gives the origin to a 'hadronic shower': the number of particles in the shower initially grows, then, when the average energy becomes too small to produce new particles, decreases. This is because the particles of the shower are unstable. The charged pions, which have a lifetime of only 26 ns, decay through the reactions

$$\pi^+ \rightarrow \mu^+ + \nu_\mu \qquad \pi^- \rightarrow \mu^- + \bar{\nu}_\mu. \tag{1.105}$$

The muons, in turn, decay as

$$\mu^+ \rightarrow e^+ + \bar{\nu}_\mu + \nu_e \qquad \mu^- \rightarrow e^- + \nu_\mu + \bar{\nu}_e. \tag{1.106}$$

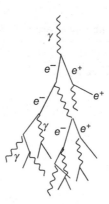

Fig. 1.14.   Sketch of an electromagnetic shower.

The muon lifetime is about $2\ \mu s$, much larger than that of the pions. Therefore, the composition of the shower becomes richer and richer in muons while travelling in the atmosphere.

The hadronic collisions produce not only charged pions but also π^0. The latter decay quickly with the electromagnetic reaction

$$\pi^0 \rightarrow \gamma + \gamma. \tag{1.107}$$

The photons, in turn, give rise to an 'electromagnetic shower', which overlaps geometrically with the hadronic shower but has different characteristics. Actually the photons interact with the nuclei producing a pair:

$$\gamma + N \rightarrow e^+ + e^- + N. \tag{1.108}$$

The electron and the positron, in turn, can produce a photon by bremsstrahlung:

$$e^\pm + N \rightarrow e^\pm + N + \gamma. \tag{1.109}$$

In addition, the new photon can produce a pair, and so on. The average distance between such events is the radiation length, which for air at n.t.p. is $X_0 = 300$ m. Fig. 1.14 shows the situation schematically.

In the first part of the shower, the number of electrons, positrons and photons increases, while their average energy diminishes. When the average energy of the electrons decreases below the critical energy, the number of particles in the shower has reached its maximum and gradually decreases.

In 1933, B. Rossi discovered that the cosmic radiation has two components: a 'soft' component that is absorbed by a material of modest thickness, for example a few centimetres of lead, and a 'hard' component that penetrates through a material of large thickness (Rossi 1933). From the above discussion we understand that the soft component is the electromagnetic one, and the hard component is made up mostly of muons.

There is actually a third component, which is extremely difficult to detect: the neutrinos and antineutrinos produced mainly in the reaction (1.105) and (1.106). Neutrinos have only weak interactions and can cross the whole Earth without being absorbed.

Consequently, observing them requires detectors with sensitive masses of a thousand tons or more. These observations have led, in the past few years, to the discovery that neutrinos have non-zero masses.

Accelerators

Several types of accelerator have been developed. We shall discuss here only the synchrotron, the acceleration scheme that has made the most important contributions to subnuclear physics. Synchrotrons can be built to accelerate protons or electrons. Schematically, in a synchrotron, the particles travel in a pipe, in which a high vacuum is established. The 'beam pipe' runs inside the gaps of dipole magnets forming a ring. The orbit of a particle of momentum p in a uniform magnetic field $\mathbf{B}$ has a circumference with radius R. These three quantities are related by an equation that we shall often use (see Problem 1.27):

$$p(\text{GeV}) = 0.3B(\text{T})R(\text{m}). \tag{1.110}$$

Other fundamental components are the accelerating cavities, in which a radio-frequency electromagnetic field (RF) is tuned to give a push to the bunches of particles every time they go through. Actually, the beam does not continuously fill the circumference of the pipe, but is divided in bunches, in order to allow the synchronization of their arrival with the phase of the RF.

In the structure we have briefly described, the particle orbit is unstable; such an accelerator cannot work. The stability can be guaranteed by the 'principle of phase stability', independently discovered by V. Veksler in 1944 (Veksler 1944) in the former USSR and by E. McMillan in 1945 in the USA (McMillan 1945). In practice, stability is reached by alternating magnetic elements that focus and defocus in the orbit plane (Courant & Snyder 1958). The following analogy can help. If you place a rigid stick upwards vertically on a horizontal support, it will fall; the equilibrium is unstable. However, if you place it on your hand and move the hand quickly from left to right and back again the stick will not fall.

The first proton synchrotron was the Cosmotron, operational at the Brookhaven National Laboratory in the USA in 1952, with 3 GeV energy. Two years later, the Bevatron was commissioned at Berkeley, also in the USA. The proton energy was 7 GeV, designed to be enough to produce antiprotons. In 1960, two 30 GeV proton synchrotrons became operational, the Proton Synchrotron at CERN, the European Laboratory at Geneva, and the Alternate Gradient Synchrotron (AGS) at Brookhaven.

The search for new physics has demanded that the energy frontier be moved towards higher and higher values. To build a higher-energy synchrotron one needs to increase the length of the ring or increase the magnetic field, or both. The next generation of proton synchrotrons was ready at the end of the 1960s: the Super Proton Synchrotron (SPS) at CERN (450 GeV) and the Main Ring at the Fermi National Accelerator Laboratory (FNAL) near Chicago (500 GeV). The radius of each is about 1 km.

The synchrotrons of the next generation reach higher energies by using field intensities of several tesla with superconducting magnets. These are the Tevatron at FNAL,

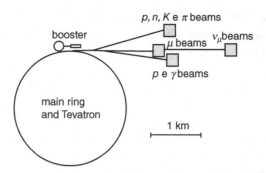

Fig. 1.15. The Tevatron beams. The squares represent schematically the locations of the experimental halls.

built in the same tunnel of the Main Ring with maximum energy of 1 TeV, and the proton ring of the HERA complex at DESY (Hamburg in Germany) with 0.8 TeV.

The high-energy experiments generally use so-called secondary beams. The primary proton beam, once accelerated at maximum energy, is extracted from the ring and driven onto a target. The strong interactions of the protons with the nuclei of the target produce all types of hadrons. Beyond the target, a number of devices are used to select one type of particle, possibly within a certain energy range. In such a way, one can build beams of pions, K-mesons, neutrons, antiprotons, muons and neutrinos. A typical experiment steers the secondary beam of interest into a secondary target where the interactions to be studied are produced. The target is followed by a set of detectors to measure the characteristics of these interactions. These experiments are said to be on a 'fixed target', as opposed to those at the storage rings that we shall soon discuss. Fig. 1.15 shows, as an example, the secondary beam configuration at FNAL in the 1980s.

Colliders

The ultimate technique to reach higher energy scales are the colliders, or storage rings as they are also called. Consider a fixed target experiment with a target particle of mass m_t and a beam of energy E_b, and an experiment using two beams colliding from opposite directions in the CM frame, each of energy E^*. Equations (1.45) and (1.47) give the condition needed to have the same total centre of mass energy in the two cases

$$E^* = \sqrt{m_t E_b / 2}. \qquad (1.111)$$

We see that to increase the centre of mass energy at a fixed target by an order of magnitude, we must increase the beam energy by two orders, with colliding beams by only one.

In a collider, two accelerated beams are stored in circular vacuum pipes in which they rotate in opposite directions. The beam orbits are defined by magnets providing the requested magnetic fields and radio-frequency cavities with powerful electric fields to accelerate the beam particles and, once the operational energy is reached, to compensate for radiation losses. The two rings intercept each other at a few positions

ADA at Frascati. Courtesy of Archivio Audio-Video, INFN-LNF.

along the circumference. The phases of the bunches circulating in the two rings are adjusted to make them meet at the intersection points, where the detectors are located. Then, if the number of particles in the bunches is sufficient, many collisions happen at every crossing. Notice that the same particles cross repeatedly a very large number of times. The basic operation of a collider starts with a preliminary acceleration of the beams, using ancillary accelerators. The beams are then transferred in bunches in the storage rings. The filling process continues until the beams have reached the foreseen intensity and are further accelerated to the final energy. The machine regime is then stable and the experiments can collect data for several hours.

Historically, the first collider consisted of two electron storage rings, tangent in one intersection point in which e^-e^- collisions $(500\ \text{MeV} + 500\ \text{MeV})$ took place. Proposed by G. K. O'Neill in 1957, it was built at Stanford in the following few years. Later, B. Touschek at Frascati in Italy advanced the proposal of a particle–antiparticle collider, electrons against positrons. When matter and antimatter collide, they often annihilate; matter disappears in a state of pure energy. Moreover, this state has well-defined quantum numbers: those of the photon. In addition, in a particle–antiparticle collider (e^+e^- and later $\bar{p}p$), the structure of the accumulator can be simplified. As particles and antiparticles have opposite charges and exactly the same mass, a single magnetic structure is sufficient to keep the two beams circulating in opposite directions. As a first test, in 1960, Touschek proposed building (Touschek 1960) a small storage ring $(250\ \text{MeV} + 250\ \text{MeV})$, which was called ADA (*Anello Di Accumulazione*, 'storage ring' in Italian). The next year, ADA was working (Fig. 1.16).

The development of a facility suitable for experimentation was an international effort, mainly by the groups led by F. Amman in Frascati, G. I. Budker in Novosibirsk and B. Richter in Stanford. Then, around the world, many e^+e^- rings of increasing energy and luminosity were built. Their contribution to particle physics was, and

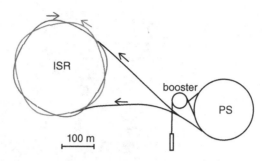

Fig. 1.17. The CERN machines in the 1970s.

still is, enormous. The maximum energy for an electron–positron collider, more then 200 GeV, was reached with the Large Electron–Positron Collider (LEP) at CERN. Its length was 27 km.

The first pp storage ring became operational at CERN in 1971: it was called ISR (Intersecting Storage Rings) and is shown in Fig. 1.17. The protons are first accelerated up to 3.5 GeV in the small synchrotron called the 'booster', transferred to the proton synchrotron (PS) and accelerated up to 31 GeV. Finally they are transferred in bunches, alternately, to the two storage rings.

The centre of mass energy is very important but it is useless if the interaction rate is too small. Another important parameter is the luminosity. We can think of the collision as taking place between two gas clouds, the bunches, that have densities much lower than that of condensed matter. To overcome this problem it is necessary:

(1) to focus both beams in the intersection point to reduce their transverse dimensions as much as possible, in practice to a few μm or less
(2) to reduce the random motion of the particles in the bunch. The fundamental technique, called 'stochastic cooling' was developed at CERN by S. van der Meer in 1970.

The luminosity is proportional to the product of the numbers of particles, n_1 and n_2, in the two beams. Notice that in a proton–antiproton collider the number of antiprotons is smaller than that of protons, due to the energetic cost to produce them. The luminosity is also proportional to the number of crossings in a second f and inversely proportional to the section at the intersection point Σ

$$\mathcal{L} = f \cdot \frac{n_1 n_2}{\Sigma}. \tag{1.112}$$

In 1976, C. Rubbia, C. P. McIntire and D. Cline (Rubbia *et al.* 1976) proposed transforming the CERN SPS from a simple synchrotron to a proton–antiproton collider. The enterprise had limited costs, because the magnetic structure was left substantially as it was, although it was necessary to improve the vacuum considerably. It was also necessary to develop further the stochastic cooling techniques, which had been already employed in the ISR. Finally the CM energy ($\sqrt{s} = 540$ GeV) and the luminosity $\left(L = 10^{28}\,\mathrm{cm}^{-2}\mathrm{s}^{-1}\right)$ necessary for the discovery of the bosons W and Z, the mediators of the weak interactions, were reached.

In 1987, a proton–antiproton ring, the Tevatron, based on the same principles, became operational at FNAL. Its energy was larger, $\sqrt{s} = 2\,\mathrm{TeV}$, as was its luminosity $\mathcal{L} = 10^{31} - 10^{32}\,\mathrm{cm}^{-2}\mathrm{s}^{-1}$.

In 2011, the LHC started operations at CERN. It was built in the 27 km long tunnel that previously hosted the LEP. It can accelerate and store in two pipes counter-rotating beams of the same particles, protons or heavy ions such as Pb or Xe, or of different ones, such as protons against heavy ions. The ring is made of superconducting magnets built with the most advanced technology to obtain an 8 T field. The maximum CM energy is 13 TeV; the luminosity is $\mathcal{L} = 10^{34}\,\mathrm{cm}^{-2}\mathrm{s}^{-1}$.

Example 1.11 We saw in Example 1.9 that a secondary beam from an accelerator of typical intensity $I = 10^{13}\,\mathrm{s}^{-1}$ impinging on a liquid hydrogen target $l = 10$ cm long gives a luminosity $\mathcal{L} = 3.6 \times 10^{40}\,\mathrm{cm}^{-2}\mathrm{s}^{-1}$. We now see that this is much higher than that of the highest luminosity colliders. Calculate the luminosity for such a beam on a gas target, for example air in normal conditions $\left(\rho = 1\,\mathrm{kg\,m}^{-3}\right)$. We obtain

$$\mathcal{L} = I\rho N_A l / 10^3 = 10^{13} \times 10^3 \times 0.1 \times 6 \times 10^{23} = 6 \times 10^{38}\,\mathrm{m}^{-2}\mathrm{s}^{-1}\;\square$$

HERA, which was operational at the DESY laboratory at Hamburg from 1991 to 2008, is a third type of collider. It is made up of two rings, one for electrons, or positrons, that are accelerated up to 30 GeV, and one for protons that reach 920 GeV energy. The scattering of the point-like electrons on the protons informs us on the deep internal structure of the latter. The high centre of mass energy available in the head-on collision makes HERA the 'microscope' with the highest existing resolving power.

1.13 Particle Detectors

The progress in our understanding of the fundamental laws of nature is directly linked to our ability to develop instruments to detect particles and measure their characteristics, with ever-increasing precision and sensitivity. Here we shall give only a summary of the principal classes of detectors.

The quantities that we can measure directly are the electric charge, the magnetic moment, the lifetime, the velocity, the momentum and the energy. The kinematic quantities are linked by the fundamental equations

$$\mathbf{p} = m\gamma\beta \tag{1.113}$$

$$E = m\gamma \tag{1.114}$$

$$m^2 = E^2 - p^2. \tag{1.115}$$

We cannot measure the mass directly; to do so we measure two quantities: energy and momentum, momentum and velocity, etc.

Let us review the principal detectors.

Scintillation Detectors

The scintillator counter was invented by S. Curran in 1944, when he was working on the Manhattan Project (Curran & Baker 1944). It was made of ZnS coupled to a photomultiplier. The work was declared secret.

There are several types of scintillator counters or, simply, 'scintillators'. We shall restrict ourselves to the plastic and organic liquid ones.

Scintillator counters are made with transparent plastic plates with a thickness of a centimetre or so and of the required area (up to square metres in size). The material is doped with molecules that emit light at the passage of an ionizing particle. The light is guided by a light guide glued, on a side of the plate, to the photocathode of a photomultiplier. One typically obtains 10 000 photons per MeV of energy deposit. Therefore the efficiency can be high. The time resolution is very good and can reach one tenth of a nanosecond or even less. Two counters at a certain distance on the path of a particle are used to measure its time of flight between them and, knowing the distance, its velocity. Plastic counters are also used as the sensitive elements in the 'calorimeters', as we shall see. A drawback of plastic (and crystal) scintillators is that their light attenuation length is not large. Consequently, when assembled in large volumes, the light collection efficiency is poor.

In 1947, Broser and Kallmann discovered (Broser & Kallmann 1947) that naphthalene emits fluorescent light under ionizing radiation. In the next few years, different groups (Ageno *et al.* 1950; Kallmann 1950; Reynolds *et al.* 1950) discovered that binary and ternary mixtures of organic liquids and aromatic molecules had high scintillation yield, that is, high numbers of photons per unit of energy loss (of the order of 10 000 photons per MeV), and long (up to tens of metres) attenuation lengths. These discoveries opened the possibility of building large scintillator detectors at affordable costs. The liquid scintillator technique has been, and is, of enormous importance, in particular for the study of neutrinos, including their discovery (Section 2.4).

Nuclear Emulsions

Photographic emulsions are made of an emulsion sheet deposited on a transparent, plastic or glass support. The emulsions contain grains of silver halides, the sensitive element. Once exposed to light, the emulsions are developed with a chemical process that reduces to metallic Ag only those grains that have absorbed photons. It was known as early as 1910 that ionizing radiation produces similar effects. Therefore, a photographic plate, once developed, shows as trails of silver grains the tracks of the charged particles that have gone through it.

In practice, normal photographic emulsions are not suitable for scientific experiments because of both their small thickness and low efficiency. The development of emulsions as a scientific instrument, the 'nuclear emulsion', is mainly due to C. F. Powell and G. Occhialini at Bristol in co-operation with Ilford Laboratories, immediately after World War II. In 1948, Kodak developed the first emulsion sensitive to minimum ionizing particles; it was with these that Lattes, Muirhead, Occhialini and Powell discovered the pion (Chapter 2).

Nuclear emulsions have a practically infinite 'memory'; they integrate all the events during the time they are exposed. This is often a drawback. On the positive side, they have an extremely fine granularity, of the order of several micrometres (μm). The coordinates of points along the track are measured with sub-micrometric precision.

Emulsions are a 'complete' instrument: the measurement of the 'grain density' (their number per unit length) gives the specific ionization dE/dx, hence $\beta\gamma$, the 'range', that is, the total track length to the stop point (if present), gives the initial energy. Charged particles in matter undergo 'multiple scattering', a series of scatters in the electric fields of the nuclei, usually through small angles. Multiple scattering is a function of the momentum, thereby giving us a mean to measure this quantity.

On the other hand, the extraction of the information from the emulsion is a slow and time-consuming process. With the advent of accelerators, bubble chambers and, later, time projection chambers replaced the emulsions as visualizing devices. But emulsions remain, even today, unsurpassed in spatial resolution and are still used when this is mandatory.

Cherenkov Detectors

In 1934, P. A. Cherenkov (Cherenkov 1934) and S. I. Vavilov (Vavilov 1934) discovered that gamma rays from radium induce luminous emission in solutions. The light was due to the Compton electrons produced by the gamma rays, as discovered by Cherenkov, who experimentally elucidated all the characteristics of the phenomenon. I. M. Frank and I. E. Tamm gave the theoretical explanation in 1937 (Frank & Tamm 1937).

If a charged particle moves in a transparent material with a speed υ larger than the phase velocity of light, namely if $\upsilon > c/n$, where n is the refractive index, it generates a wave similar to the shock wave made by a supersonic jet in the atmosphere. Another, visible, analogy is the wave produced by a duck moving on the surface of a pond. The wave front is a triangle with the vertex at the duck, moving forward rigidly with it. The rays of Cherenkov light are directed normally to the V-shaped wave, as shown in Fig. 1.18(a).

The wave is the envelope of the elementary spherical waves emitted by the moving source at subsequent moments. In Fig. 1.18(b) we show the elementary wave emitted t seconds before. Its radius is then $OB = ct/n$; in the meantime the particle has moved by $OA = \upsilon t$. Hence

$$\theta = \cos^{-1}\left(\frac{1}{\beta n}\right), \tag{1.116}$$

where $\beta = \upsilon/c$.

Question 1.2 Calculate the Cherenkov vertex angle for an electron and a muon of 1, 10 and 100 GeV energy in water. □

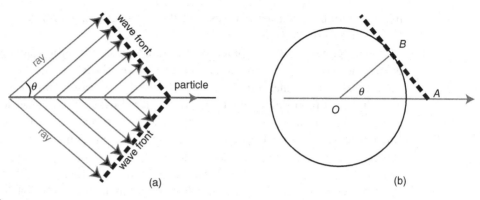

Fig. 1.18. The Cherenkov wave geometry.

The spectrum of the Cherenkov light is continuous with important fractions in the visible and in the ultraviolet.

Consider the surface limiting the material in which the particle travels. Its intersection with the light cone is a circle or, more generally, an ellipse, called the 'Cherenkov ring'. We can detect the ring by covering the surface with photomultipliers (PMs). If the particle travels, say, towards that surface, the photomultipliers see a ring gradually shrinking in time. From this information, we determine the trajectory of the particle. The space resolution is given by the integration time of the PMs, 30 cm for a typical value of 1 ns.

From the radius of the ring, we measure the angle at the vertex of the cone, and hence the particle speed. The thickness of the ring, if greater than the experimental resolution, gives information on the nature of the particle. Multiple scattering is small for muons, which travel substantially straight, but is quite sizeable for electrons that scatter much more. As a consequence, the electrons' rings are thicker, giving a handle to identify them.

Example 1.12 Super-Kamiokande is a large Cherenkov detector based on the technique described. It contains 50 000 t of pure water. Fig. 1.19 shows a photo taken while it was being filled. The PMs, being inspected by the people on the boat in the picture, cover the entire surface. Their diameter is half a metre. The detector, in a laboratory under the Japanese Alps, is dedicated to the search for astrophysical neutrinos and proton decay.

Fig. 1.20 shows examples of events consisting of a single charged track, a muon, sharper, in Fig. 1.20(a), and an electron, more diffuse, in Fig. 1.20(b). The dots correspond to the PMs that gave a signal; the colour, in the original, codes the arrival time. □

The Cherenkov counters are much simpler devices of much smaller dimensions. The light is collected by one PM, or by a few, possibly using mirrors. In its simplest version, the counter gives a 'yes' if the speed of the particle is $\beta > 1/n$, a 'no' in the opposite case. In more sophisticated versions, one measures the angle of the cone, and hence the speed.

Fig. 1.19.   Inside Super-Kamiokande, being filled with water. People on the boat are checking the photomultipliers. Courtesy of Kamioka Observatory – Institute of Cosmic Ray Research, University of Tokyo.

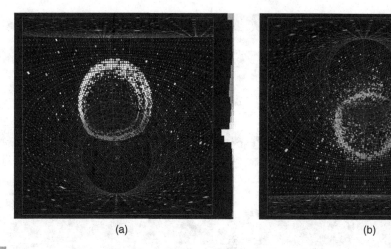

(a)   (b)

Fig. 1.20. Cherenkov rings in Super-Kamiokande, (a) of a muon; (b) of an electron. Kamioka Observatory, ICRR (Institute for Cosmic Ray Research), The University of Tokyo.

Example 1.13   Determine for a water-Cherenkov ($n = 1.33$): (1) the threshold energy for electrons and muons; (2) the radiation angle for an electron of 300 MeV; (3) whether a K^+ meson with a momentum of 550 MeV gives light.

(1) Threshold energy for an electron:

$$E = \gamma m = \frac{m}{\sqrt{1 - \left(1/n\right)^2}} = \frac{0.511\,\text{MeV}}{\sqrt{1 - (1/1.33)^2}} = 0.775\,\text{MeV}.$$

Threshold energy for a muon (μ):

$$E = \frac{106\,\text{MeV}}{\sqrt{1 - (1/1.33)^2}} = 213\,\text{MeV}$$

(2) The electron is above threshold. The angle is

$$\theta = \cos^{-1}\left(\frac{1}{\beta n}\right) = \cos^{-1}\left(\frac{1}{1.33}\right) = 41.2°.$$

(3) Threshold energy for a K^+:

$$E = \frac{497.6\,\text{MeV}}{\sqrt{1 - (1/1.33)^2}} = 755\,\text{MeV}.$$

The corresponding momentum is:

$$p = \sqrt{E^2 - m_K^2} = \sqrt{755^2 - 497.6^2} = 567\,\text{MeV}.$$

Therefore at 550 MeV a K^+ does not make light. $\square$

Cloud Chambers

In 1895, C. T. R. Wilson, fascinated by atmospheric optical phenomena, the glories and the coronae he had admired from the observatory that existed on top of Ben Nevis in Scotland, started laboratory research on cloud formation. He built a container, with a glass window, filled with air and saturated water vapour. The volume could be suddenly expanded, bringing the vapour to a supersaturated state. Very soon, Wilson understood that condensation nuclei other than dust particles were present in the air. He thought that maybe they were electrically charged atoms (that is ions). The hypothesis was confirmed by irradiating the volume with X-rays, which had recently been discovered. By the end of 1911, Wilson had developed his device to the point of observing the first tracks of α and β particles (Wilson 1912). Actually, an ionizing particle crossing the chamber leaves a trail of ions, which seed, when the chamber is expanded, as many droplets. By flashing light and taking a picture, one can record the track. By 1923 the Wilson chamber had been fully developed (Wilson 1933).

If the chamber is immersed in a magnetic field **B**, the tracks are curved. Measuring the curvature radius R, one determines the momentum p by Eq. (1.110).

The expansion of the Wilson chamber can be triggered. If we want, for example, to observe charged particles coming from above and crossing the chamber, we put one Geiger counter (see later) above and another below the chamber. We send the two electronic signals to a coincidence circuit, which commands the expansion. Blackett and Occhialini discovered positron–electron pairs in cosmic radiation with this method in 1933. The coincidence circuit had been invented by B. Rossi in 1930 (Rossi 1930).

Bubble Chambers

The bubble chamber was invented by D. Glaser in 1952 (Glaser 1952), but it became a scientific instrument only with L. Alvarez (Nobel lecture) (see Example 1.14). The working principle is similar to that of the cloud chamber, with the difference that the fluid is a liquid which becomes super-heated during expansion. Along the tracks, a trail of gas bubbles is generated.

Unlike the cloud chamber, the bubble chamber must be expanded before the arrival of the particle to be detected. Therefore, the bubble chambers cannot be used to detect random events such as cosmic rays, but are a perfect instrument at an accelerator facility, where the arrival time of the beam is known exactly in advance.

The bubble chamber acts at the same time both as target and as detector. From this point of view, the advantage over the cloud chamber is the higher density of the liquids, compared with gases, which makes the interaction probability larger. Different liquids can be used, depending on the type of experiment: hydrogen to have a target nucleus as simple as a proton, deuterium to study interactions on neutrons, liquids with high atomic numbers to study the small cross-section interactions of neutrinos.

Historically, bubble chambers have been exposed to all available beams (protons, antiprotons, pions, K-mesons, muons, photons and neutrinos). In a bubble chamber, all the charged tracks are visible. Gamma rays can also be detected if they 'materialize' into e^+e^- pairs. The 'heavy liquid' bubble chambers are filled with a high-Z liquid (for example a freon) to increase the probability of the process. All bubble chambers are in a magnetic field to provide the measurement of the momenta.

Bubble chambers made enormous contributions to particle physics: from the discovery of unstable hadrons, to the development of the quark model, to neutrino physics and the discovery of 'neutral' currents, to the study of the structure of nucleons.

Example 1.14   The Alvarez bubble chambers.

The development of bubble chamber technology and of the related analysis tools took place at Berkeley in the 1950s in the group led by L. Alvarez. The principal device was a large hydrogen bubble chamber 72′ long, 20′ wide and 15′ deep. The chamber could be filled with liquid hydrogen if the targets of the interaction were to be protons or with deuterium if they were to be neutrons. The uniform magnetic field had the intensity of 1.5 T.

In the example shown in Fig. 1.21 (Alvarez 1972) one sees, in a 10″ bubble chamber, seven beam tracks, which are approximately parallel and enter from the left (three more are due to an interaction before the chamber). The beam particles are π^- produced at the Bevatron.

The small curls one sees coming out of the tracks are due to atomic electrons that received an energy high enough to produce a visible track during the ionization process. Moving in the liquid they gradually lose energy and the radius of their orbit decreases accordingly. They are called 'δ-rays'.

The second beam track, counting from below, disappears soon after entering. A pion has interacted with a proton with all neutrals in the final state. The point where this happens is called the primary vertex. A careful study shows that the primary interaction is

$$\pi^- + p \rightarrow K^0 + \Lambda^0 \tag{1.117}$$

followed by the two decays

$$K^0 \rightarrow \pi^+ + \pi^- \tag{1.118}$$

$$\Lambda^0 \rightarrow \pi^- + p. \tag{1.119}$$

We see in the picture two V-shaped events, called V^0s, the decays (secondary vertices) of two neutral particles into two charged particles. Both are clearly coming from the primary vertex. One of the tracks is a proton, as can be understood by the fact that it is positive and has a large bubble density, corresponding to a large dE/dx, and hence to a low speed.

For every expansion, three pictures are taken with three cameras in different positions, obtaining a stereoscopic view of the events. The quantitative analysis implies the following steps:

• The measurement of the coordinates of the three vertices and of a number of points along each of the tracks in the three pictures.

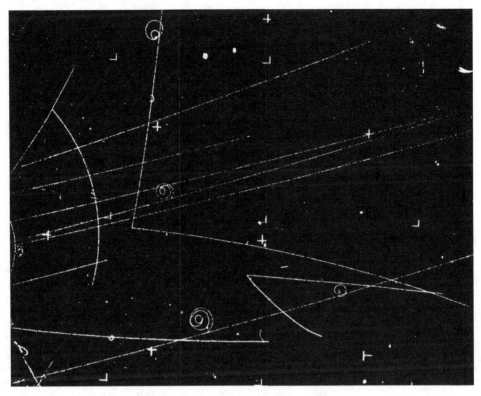

A picture of the 10″ bubble chamber. Courtesy Lawrence Berkeley National Laboratory.

- The spatial reconstruction of the tracks, obtaining their directions and curvatures, namely their momenta.
- The kinematic 'fit'. For each track, one calculates the energy, assuming in turn the different possible masses (proton or pion, for example). The procedure then constrains the measured quantities imposing energy and momentum conservation at each vertex. The problem is overdetermined. In this example, one finds that reactions (1.117, 1.118, 1.119) 'fit' the data.

Notice that the known quantities are sufficient to allow the reconstruction of the event even in the presence of one (but not of more) neutral unseen particles. If the reaction had been $\pi^- + p \to K^0 + \Lambda^0 + \pi^0$, we could have reconstructed the momentum and energy of the undetected π^0. $\square$

The resolution in the measurement of the coordinates is typically one tenth of the bubble radius. The latter ranges from about 1 mm in the heavy liquid chambers, to 0.1 mm in the hydrogen chambers, to about 10 μm in the rapid cycling hydrogen chamber LEBC (Allison *et al.* 1974b) that was used to detect picosecond lifetime particles such as the charmed mesons.

Example 1.15 In general, the curvature radius R of a track in a magnetic field in a cloud chamber is computed by finding the circle that best fits a set of points measured along the track. Knowing the field B, Eq. (1.110) gives the momentum p. How can we proceed if we measure only three points as in Fig. 1.22?

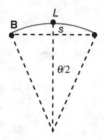

Fig. 1.22. Geometry of the track of a charged particle in magnetic field.

The measurements give directly the sagitta s. This can be expressed, with reference to Fig. 1.22, as $s = R(1 - \cos(\theta/2)) \approx R\theta^2/8$. Furthermore $\theta \approx L/R$, and we obtain

$$s \approx \frac{L^2}{8R} = 0.3\frac{BL^2}{8p},\qquad(1.120)$$

which gives us p. □

Ionization Detectors

An ionization detector contains two, or more, electrodes and a fluid, liquid or gas, in between. The ion pairs produced by the passage of a charged particle drift towards the electrodes in the electric field generated by the voltage applied to the electrodes. Electrons drift faster than ions and the intensity of their current is consequently larger.

For low electric field intensity, the electron current intensity is proportional to the primary ionization. Its measurement at one of the electrodes determines the value of dE/dx that gives a measurement of the factor $\beta\gamma$, hence of the velocity, of the particle. If we know the mass of the particle, we can calculate its momentum; if we do not, we can measure the momentum independently and determine the mass.

At higher field intensities, the process of secondary ionization sets in, giving the possibility of amplifying the initial charge. At very high fields (say MV m^{-1}), the amplification process becomes catastrophic, producing a discharge in the detector.

The Geiger Counter

The simplest ionization counter is shown schematically in Fig. 1.23. It was invented by H. Geiger in 1908 at Manchester and later modified by W. Mueller (Geiger & Mueller 1928). The counter consists of a metal tube, usually grounded, bearing a central, insulated, metallic wire, with a diameter of the order of 100 μm. A high potential, of the order of 1000 V, is applied to the wire. The tube is filled with a gas mixture, typically Ar and alcohol (to quench the discharge).

The electrons produced by the passage of a charged particle drift towards the wire where they enter a very intense field, owing to its $1/r$ dependence. They accelerate and produce secondary ionization. An avalanche process propagates along the anode and triggers the discharge of the capacitance. The process is independent of the charge deposited by the particle; consequently, the response is of the yes/no type. The time resolution is limited to about a microsecond by the variation from discharge to discharge of the temporal evolution of the avalanche. After the discharge, the tube is not sensitive, becoming so only after the power supply has charged back the capacitance.

Proportional Counters

The gas-filled proportional counter was invented by S. Curran in 1948 in Glasgow (Curran *et al.* 1948). The simplest geometry is a cylinder similar to that in Fig. 1.23, but the device is operated at a lower voltage and its anode wire is much thinner. Like in a Geiger counter, the primary electrons drift towards the anode wire and enter the intense electric field in its surroundings, where they are accelerated and produce an avalanche of ions and secondary electrons. The process takes place in a small-radius region around the anode. Unlike the Geiger regime, the avalanche does not propagate along the wire. The mechanism is called proportional charge amplification, because the total separated charge is proportional to the charge of the primary electrons.

The voltage pulse on the amplifier input capacitance C is the result of the motion of the charges of both signs. The electrons move towards the anode, inducing a voltage drop, which is fast because they have high drift velocity. The positive ions move away from the anode, further decreasing the voltage, but at a slower rate, because they have lower velocities. In a cylindrical counter, in practice the contribution of the electrons is small (Curran & Craggs 1949; Rossi & Staub 1949). Let us consider a typical counter with anode and cathode radiuses of, respectively, $r_a = 20$ μm and $r_c = 20$ mm. The largest fraction of the avalanche develops within a distance from the anode surface of a few times the electron free path, which is of a few micrometres at n.t.p. We make the approximation that all the ionization charge is produced in a point at 10 μm from the surface, or at $r_0 = 30$ μm from the axis. It can be shown that the contributions of

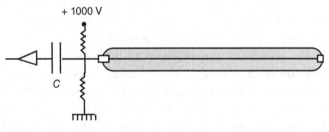

Fig. 1.23. The Geiger counter.

the positive and the negative charges are proportional to the potential differences between r_0 and the cathode and anode, respectively (Rossi & Staub 1949). Consequently, the electrons that are very close to the anode do not contribute much. The ratio between the contributions of the ions and the electrons is

$$R = \frac{\Delta V^+}{\Delta V^-} = \frac{\ln\left(\dfrac{r_c}{r_0}\right)}{\ln\left(\dfrac{r_0}{r_a}\right)} \cong 16$$

Proportional chambers with several parallel anode wires have been used since the 1950s in nuclear physics experiments, mainly for energy measurements. In these geometries, pulses are induced also on the wires near to the one of the avalanches. However, the pulses are caused by ions approaching rather than going away from them and are positive rather than negative. Consequently, they can be easily discarded by the read-out electronics, without need of any electrical shielding, even if the distance between anode wires is small.

Multi-wire Proportional Chambers

Multi-wire proportional chambers (MWPCs) were developed for tracking purposes by G. Charpak starting in 1967 (Charpak *et al.* 1968; Charpak 1992), based on the abovementioned concepts. At that time, integrated circuits had become commercially available, making the electronic read-out of thousands of wires affordable.

The MWPC scheme is shown in Fig. 1.24. The anode is a plane of metal wires (thickness from 10 μm to 30 μm), drawn parallel and equally spaced with a pitch of typically 2 mm. The anode plane is enclosed between two cathode planes, which are parallel and at the same distance of several millimetres, as shown in Fig. 1.24.

The MWPCs are used mainly in experiments on secondary beams at an accelerator, in which the particles to be detected leave the target within a limited solid angle around the forward direction. The chambers are positioned perpendicularly to the average direction. This technique allows large areas (several square metres) to be covered with detectors whose data can be transferred directly to a computer, unlike bubble chambers. Fig. 1.24 shows the inclined trajectory of a particle. The electric field shape divides the volume of the chamber into cells, one for each sensitive wire. The ionization electrons produced in the track segment belonging to a given cell will drift towards the anode wire of that cell, following the field lines. In the neighbourhood of the anode wire, the charge amplification process described above takes place. Typical amplification factors are of the order of 10^5.

The coordinate perpendicular to the wires, x in Fig. 1.24, is determined by the position of the wire that gives a signal above threshold. The coordinate z normal to the plane is known by construction. To measure the third coordinate y, a second chamber (at least) is needed with wires in the x direction. The spatial resolution is the variance of a uniform distribution with the width of the spacing. For example, for 2 mm pitch, $\sigma = 2/\sqrt{12} = 0.6$ mm.

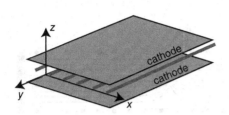
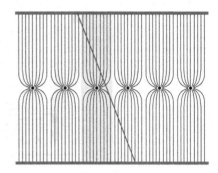

Geometry of the MWPC.

Read-out electronics is of fundamental importance for the MWPCs, as well as for the drift and time projection chambers and the silicon detectors that we shall soon discuss. Each wire is serviced by an electronic channel, which includes an analogue stage and a digital stage. The analogue section performs charge amplification for negative polarity pulses and pulse shape discrimination. It is followed by analogue to digital conversion, delaying, logical processing and storing in digital memories. Typically, thousands of such electronic channels are necessary. Integrated circuits containing hundreds of transistors per chip became commercially available at low cost starting in 1968, as a result of the contributions of scientists operating in companies and of the economic stimulus to them from the aerospace programmes of the US Government. The integration scale, that is, the number of transistors per chip, has increased since then at a constant rate.

The success of the MWPCs was largely due to the effort, principally made at CERN over several years, dedicated to the development of custom integrated circuits. The first MWPCs, outside R&D prototypes, with associated electronics, were built by G. Amato and G. Petrucci in 1968 for a beam profile analysing system (Amato & Petrucci 1968).

Drift Chambers

Drift chambers (DCs) are similar to MWPCs, but provide two coordinates. One coordinate, as in the MWPC, is given by the position of the wire giving the signal; the second, perpendicular to the wire in the plane of the chamber, is obtained by measuring the time taken by the electron to reach it (drift time). The chambers are positioned perpendicularly to the average direction of the tracks. The distance between one of the cathodes and the anode is typically several centimetres. Fig. 1.25 shows the field geometry originally developed at Heidelberg by J. Heintze and A. H. Walenta in 1971 (Walenta *et al.* 1971). The chamber consists of a number of cells along the x-axis. The 'field wires' on the two sides of the cell are polarized at a gradually diminishing potential to obtain a uniform electric field.

In a uniform field, and with the correct choice of gas mixture, one obtains a constant drift velocity. Given a typical value of drift velocity of 50 mm μs^{-1}, measuring

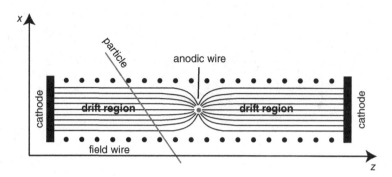

Fig. 1.25.   Drift chamber geometry.

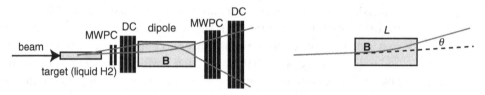

Fig. 1.26.   A simple spectrometer.

the drift time with a 4 ns precision one obtains a spatial resolution in z of 200 μm, that is, about three times better than in an MWPC.

One can also measure the induced charge integrating the current from the wire, obtaining a quantity proportional to the primary ionization charge, and so determining dE/dx.

Fig. 1.26 shows an example of the use of MWPCs and DCs in a fixed target spectrometer, used to measure the momenta and the sign of the charges of the particles. A dipole magnet deflects each particle by an angle inversely proportional to its momentum, towards one or the other side depending on the sign of its charge. The poles of the magnet are located above and below the plane of the drawing, at the position of the rectangle. Fig. 1.26 shows two tracks of opposite sign. One measures the track directions before and after the magnet as accurately as possible using MWPCs and drift chambers. The angle between the directions and the known value of the field gives the momenta.

The geometry is shown on the right of Fig. 1.26. To simplify, we assume **B** to be uniform in the magnet, of length L, and zero outside it. We also consider only small deflection angles. With these approximations the angle is $\theta \approx L/R$ and, recalling (1.110),

$$\theta \approx 0.3 \frac{BL}{p} \qquad (1.121)$$

The quantity BL, more generally $\int B dl$, is called the 'bending power' with reference to the magnet, or 'rigidity' with reference to the particle. Consider for example a magnet of bending power $\int Bdl = 1\,\text{Tm}$. A particle of momentum $p = 30\,\text{GeV}$ is bent by 100 mrad, corresponding to a lateral shift, for example at 5 m after the magnet, of 50 mm. This shift can be measured with good precision with a resolution of 100 μm.

The dependence on momentum of the deflection angle makes a dipole magnet a dispersive element similar to a prism in the case of light.

Time Projection Chambers

Time projection chambers (TPCs) have sensitive volumes of cubic metres in size, and give three-dimensional images of the tracks. Their development was due to W. W. Allison *et al.* at Oxford (Allison *et al.* 1972, 1974a) in the UK and to D. Nygren in the USA (Clark *et al.* 1976; Marx & Nygren 1978; Nygren 1981). In a typical geometry, the anode and the cathode are parallel planes at a few metres distance. The anode is made of parallel sense wires at, say, 2 mm spacing. We call y their direction. An ionizing particle crossing the chamber produces a trail of ion–electron pairs. The electrons drift to the anode and each of the sense wires 'sees', amplified, the ionization charge of a segment of the track in the direction, x, perpendicular to the wires, as in a drift chamber. The electric pulse is amplified and processed. The drift time of the electrons to the anode is also measured, giving the coordinate z in the direction of the electric field. The coordinate along the wires is determined by measuring the charge at both ends. The ratio of the two gives y with a resolution that is typically 10% of the wire length. The measurement of the specific ionization along the track provides particle identification information.

Cylindrical TPCs of different design are practically always used in collider experiments, in which the tracks leave the interaction point in all the directions. These 'central detectors' are immersed in a magnetic field to allow the momenta to be measured. This is not possible, however, at LHC due to its high luminosity.

Both DCs and TPCs can be also be operated in the electro-luminescence (EL) mode, which is useful for events with low energy and a single or few tracks. The method was first introduced for DCs by Conde and Policarpo (1967) and for TPCs by Nygren (2009) and Gómez-Cadenas *et al.* (2012). In this mode, the electric field in the neighbourhood of the sense wires is lower. Charge amplification does not happen, but the ionization electrons gain enough energy to excite the gas molecule that emits light, in proper media. In Xe gas, about 1000 photons per drifting electron can be obtained. The absence of fluctuations in the avalanche gain, as in the charge amplification in conventional chambers, leads to excellent energy resolution.

For tracking, electro-luminescence photons are detected by a bi-dimensional array of photo-sensors in the x–y plane, together with the drift time, which gives the z coordinate. Fig. 1.27 shows the projection on the three coordinate planes of the reconstructed image of an electron as seen in a prototype EL TPC for the NEXT experiment. The chamber contained 5 kg of gas Xe at a pressure of 1 MPa. The photo-sensors were Si photomultipliers of $1\,\text{mm}^2$ area in a square pattern with 10 mm spacing. The electron, whose initial energy is 2.4 MeV, comes to rest in the chamber. The grey level along the track is proportional to the specific ionization.

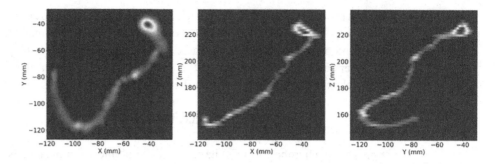

Fig. 1.27. The $x - y$, $x - z$ and $y - z$ projections of the track of a 2.4 MeV energy electron, reconstructed in a TPC with EL read-out. (courtesy NEXT Collaboration).

Notice the multiple scattering suffered by the low-energy electron. The specific ionization is close to the minimum along most of the track. Most of the energy is lost in the final part of the range. The phenomenon is known as the Bragg peak of the specific ionization.

Silicon Micro-strip Detectors

Micro-strip detectors were developed in the 1980s. They are based on a Si wafer, a hundred micrometres or so thick and with surfaces of several square centimetres. A ladder of many n–p diodes is built on the surface of the wafer in the shape of parallel strips with a pitch of tens of micrometres. The strips are the equivalent of the anode wires in an MWPC and are read out by charge amplifiers. The device is reverse biased and is fully depleted. A charged particle produces electron–hole pairs that drift and are collected at the strips. The spatial resolution is very good, of the order of 10 μm.

The Si detectors play an essential role in the detection and study of charmed and beauty particles. These have lifetimes of the order of a few tenths of a picosecond and are produced with typical energies of a few GeV and decay within millimetres from the production point. To separate the production and decay vertices, devices are built made up of a number, typically four or five, of micro-strip layers. The detectors are located just after the target in a fixed-target experiment, and around the interaction point in a collider experiment. We shall see how important this 'vertex detector' is in the discussion of the discovery of the top quark in Section 4.10 and of the physics of the B mesons in Section 8.6.

Calorimeters

In subnuclear physics, the devices used to measure the energy of a particle or a group of particles are called calorimeters. The measurement is destructive, as all the energy must be released in the detector. One can distinguish two types of calorimeters: electromagnetic and hadronic.

Electromagnetic Calorimeters

Electrons, positrons and high-energy photons travelling in a material produce an electromagnetic shower, as discussed in Section 1.10. We simply recall the two basic processes: bremsstrahlung

$$e^{\pm} + N \rightarrow e^{\pm} + N + \gamma \tag{1.122}$$

and pair production

$$\gamma + N \rightarrow e^{+} + e^{-} + N. \tag{1.123}$$

The average distance between such events is about the radiation length of the material.

In a calorimeter, one uses the fact that the total length of the charged tracks is proportional to their initial energy. This length is, in turn, proportional to the ionization charge. This charge, or a quantity proportional to it, is measured.

In Fig. 1.28, an electromagnetic shower in a cloud chamber is shown. The longitudinal dimensions of the shower are limited by a series of lead plates 12.7 mm thick. The initial particle is a photon, as recognized from the absence of tracks in the first sector. The shower initiates in the first plate and completely develops in the chamber. The absorption is due to practically only the lead, for which $X_0 = 5.6$ mm, which is much shorter than that of the gas in the chamber. The total Pb thickness is $8 \times 12.7 = 101.6$ mm, corresponding to 18 radiation lengths. In general, a calorimeter must be deep enough to completely absorb the shower: 15–25 radiation lengths, depending on the energy.

The calorimeter that we have described is of the 'sampling' type, because only a fraction of the deposited energy is detected. The larger part, which is deposited in the lead, is not measured. Calorimeters of this type are built by assembling sandwiches of Pb plates (typically 1 mm thick) alternated with plastic scintillator plates (several mm thick). The scintillation light (proportional to the ionization charge deposited in the detector) is collected and measured. The energy resolution is ultimately determined by the number, N, of the shower particles that are detected. The fluctuation is $\sqrt{N}$. Therefore, the resolution $\sigma(E)$ is proportional to $\sqrt{E}$. The relative resolution improves as the energy increases. Typical values are

$$\frac{\sigma(E)}{E} = \frac{15\% - 18\%}{\sqrt{E(\text{GeV})}}. \tag{1.124}$$

Fully active homogeneous calorimeters exist too and are widely used. They consist of arrays of similar elements. Each element is a prism, in general of square cross-section, made of a transparent medium, such as lead-glass, namely a glass with a high content of Pb. The prisms are long enough to contain the entire shower. The Cherenkov light produced by the charged particles of the shower is proportional to the total energy of the electron, positron or photon that initiated the shower. It is read out by a photomultiplier at the exit face of the prism. Statistical fluctuations due to sampling are eliminated.

Notice that the transverse dimensions of the shower are important too in several experimental situations. It is usual to define the Molière radius r_M as the radius of the

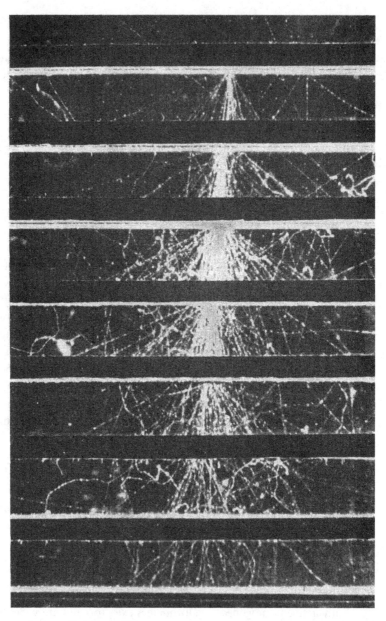

Fig. 1.28.   An electromagnetic shower (from Rossi 1952).

cylinder containing 90% of the energy of the shower. As $r_M = X_0 \left(21.2\ \text{MeV} / E_c \right)$, where X_0 is the radiation length and E_c the critical energy of the medium, to have narrow showers one uses high-density mediums. For example the CMS experiment at LHC (Section 9.14) uses crystals of lead tungstate, having diameters of the order of the Molière radius, that is $r_M = 21$ mm.

These basic elements are assembled in arrays covering the entire solid angle requested by the experiment and pointing to the interaction region.

Hadronic Calorimeters

Hadronic calorimeters are used to measure the energy of a hadron or a group of hadrons. As we shall see in Chapter 6, quarks appear in a high-energy collision as a hadronic 'jet', namely as a group of hadrons travelling within a narrow solid angle. Hadronic calorimeters are the main instrument for measuring the jet energy, which is essentially the quark energy.

Hadronic calorimeters are, in principle, similar to electromagnetic ones. The main difference is that the average distance between interactions is the interaction λ_0.

A common type of hadronic calorimeter is made like a sandwich of metal plates (Fe for example) and plastic scintillators. To absorb the shower completely, 10–15 interaction lengths $(\lambda_0 = 17 \text{ cm for Fe})$ are needed. Typical values of the resolution are

$$\frac{\sigma(E)}{E} = \frac{40\% - 60\%}{\sqrt{E(\text{GeV})}}. \tag{1.125}$$

Problems

Introduction

A common problem is the transformation of a kinematic quantity between the CM and the L frames. There are two basic ways to proceed: either by explicitly performing the Lorentz transformations or by using invariant quantities, namely s, t or u. Depending on the case, one or the other, or a combination of the two, may be more convenient.

Let us find some useful expressions for a generic two-body scattering

$$a + b \rightarrow c + d.$$

We start with s expressed in the initial state and in the L frame

$$s = (E_a + m_b)^2 - p_a^2 = m_a^2 + m_b^2 + 2E_a m_b.$$

If s and the masses are known, the beam energy is

$$E_a = \frac{s - m_a^2 - m_b^2}{2m_b}. \tag{P1.1}$$

Now consider the quantities in the CM. From energy conservation we have

$$E_a^* = \sqrt{p_a^{*2} + m_a^2} = \sqrt{s} - E_b^*$$
$$p_a^{*2} + m_a^2 = s - 2E_b^*\sqrt{s} + E_b^{*2}$$
$$2E_b^*\sqrt{s} = s + \left(E_b^{*2} - p_a^{*2}\right) - m_a^2 = s + m_b^2 - m_a^2.$$

And we obtain

$$E_b^* = \frac{s + m_b^2 + m_a^2}{2\sqrt{s}}.$$ (P1.2)

By analogy, for the other particle we write

$$E_a^* = \frac{s + m_a^2 - m_b^2}{2\sqrt{s}}.$$ (P1.3)

From the energies, we immediately have the CM initial momentum

$$p_a^* = p_b^* = \sqrt{E_{a/b}^{*2} - m_{a/b}^2}.$$ (P1.4)

The same arguments in the final state give

$$E_c^* = \frac{s + m_c^2 - m_d^2}{2\sqrt{s}}$$ (P1.5)

$$E_d^* = \frac{s + m_d^2 - m_c^2}{2\sqrt{s}}$$ (P1.6)

$$p_c^* = p_d^* = \sqrt{E_{c/d}^{*2} - m_{c/d}^2}.$$ (P1.7)

Now consider t, and write explicitly Eq. (1.49)

$$\begin{aligned} t &= m_c^2 + m_a^2 + 2 p_a p_c \cos\theta_{ac} - 2 E_a E_c \\ &= m_d^2 + m_b^2 + 2 p_b p_d \cos\theta_{bd} - 2 E_b E_d. \end{aligned}$$ (P1.8)

In the CM frame we extract the expressions of the angles

$$\cos\theta_{ac}^* = \frac{t - m_a^2 - m_c^2 + 2 m_a m_c}{2 p_a^* p_c^*}$$ (P1.9)

$$\cos\theta_{bd}^* = \frac{t - m_b^2 - m_d^2 + 2 E_b^* E_d^*}{2 p_b^* p_d^*}.$$ (P1.10)

In the L frame, where $p_b = 0, t$ has a very simple expression

$$t = m_b^2 + m_d^2 - 2 m_b E_d$$ (P1.11)

that gives E_d, if t is known

$$E_d = \frac{m_b^2 + m_d^2 - t}{2 m_b}.$$ (P1.12)

We can find E_c by using energy conservation

$$E_c = m_b + E_a - E_d = \frac{s + t - m_a^2 - m_d^2}{2 m_b}.$$ (P1.13)

Finally, let us also write u explicitly as

$$\begin{aligned} u &= m_d^2 + m_a^2 + 2 p_a p_d \cos\theta_{ad} - 2 E_a E_d \\ &= m_c^2 + m_b^2 + 2 p_b p_c \cos\theta_{bc} - 2 E_b E_c. \end{aligned}$$ (P1.14)

In the L frame, the expression of u is also simple

$$u = m_b^2 + m_c^2 - 2m_b E_c,　　　　　　(P1.15)$$

which gives E_c if u is known.

From (P1.13) and (P1.15), Eq. (1.28) follows immediately.

1.1 Estimate the energy of a Boeing 747 (mass $M = 400$ t) at cruising speed (850 km h^{-1}) and compare it with the energy released in a mosquito–antimosquito annihilation.

1.2 Three protons have momenta equal in absolute value of 3 GeV and directions at $120°$ from one another. What is the mass of the system?

1.3 Consider the weak interaction lifetimes of $\pi^{\pm}: \tau_{\pi} = 26$ ns, of $K^{\pm}: \tau_K = 12$ ns and of the $\Lambda: \tau_{\Lambda} = 0.26$ ns and compute their widths.

1.4 Consider the strong interaction total widths of the following mesons: $\rho, \Gamma_{\rho} = 149$ MeV; $\omega, \Gamma_{\omega} = 8.5$ MeV; $\phi, \Gamma_{\phi} = 4.3$ MeV; $K^*, \Gamma_{K^*} = 51$ MeV; $J/\psi, \Gamma_{J/\psi} = 93$ keV; and of the baryon $\Delta, \Gamma_{\Delta} = 118$ MeV, and compute their lifetimes.

1.5 An accelerator produces an electron beam with energy $E = 20$ GeV. The electrons scattered at $\theta = 6°$ are detected. Neglecting their recoil motion, what is the minimum structure in the proton that can be resolved?

1.6 In the collision of two protons, the final state contains a particle of mass m besides the protons.

　(a) Give an expression for the minimum (threshold) energy E_p for the process to happen and for the corresponding momentum p_p if the target proton is at rest.

　(b) Give the expression of the minimum energy E_p^* for the process to happen and of the corresponding momentum p_p^* if the two protons collide with equal and opposite velocities.

　(c) How large are the threshold energies in cases (a) and (b) if the produced particle is a pion? How large is the kinetic energy in the first case?

1.7 Consider the process $\gamma + p \rightarrow p + \pi^0$ (π^0 photoproduction) with the proton at rest.

　(a) Find the minimum energy of the photon E_{γ}.

　　The Universe is filled by 'background electromagnetic radiation' at the temperature of $T = 3$ K, and photons with energy $E_{\gamma, 3K} \approx 1$ meV.

　(b) Find the minimum energy E_p of the cosmic-ray protons needed to induce π^0 photoproduction.

　(c) If the cross-section, just above threshold, is $\sigma = 0.6$ mb and the background photon density is $\rho \approx 10^8$ m^{-3}, find the attenuation length. Is it small or large on the cosmological scale?

1.8 The Universe contains two types of electromagnetic radiation:

　(a) the 'micro-wave background' at $T = 3$ K, corresponding to photon energies $E_{\gamma, 3K} \approx 1$ meV,

　(b) the extragalactic background light (EBL) due to the stars, with a spectrum that is mainly in the infrared.

　　The Universe is opaque to photons whose energy is such that the cross-section for pair production $\gamma + \gamma \rightarrow e^+ + e^-$ is large. This already happens just above threshold (see Fig. 1.13). Compute the two threshold energies, assuming in the second case the photon wavelength $\lambda = 1$ μm.

1.9 The Bevatron was designed to have sufficient energy to produce antiprotons. What is the minimum energy of the proton beam for such a process? Take into account that, because of baryonic number conservation (see Section 2.7), the reaction is $p + p \rightarrow p + p + \bar{p} + p$.

1.10 In the LHC collider at CERN, two proton beams collide head on with energies $E_p = 7$ TeV. What energy would be needed to obtain the same centre of mass energy with a proton beam on a fixed hydrogen target? How does it compare with cosmic ray energies?

1.11 Consider a particle of mass M decaying into two bodies of masses m_1 and m_2. Give the expressions of the energies and of the momenta of the decay products in the CM frame.

1.12 Evaluate the energies and momenta in the CM frame of the two final particles of the decays $\Lambda \rightarrow p\pi^-$, $\Xi^- \rightarrow \Lambda\pi^-$.

1.13 Find the expressions of the energies and momenta of the final particles of the decay $M \rightarrow m_1 + m_2$ in the CM; the mass of m_2 is zero.

1.14 In a monochromatic π beam with momentum p_π, a fraction of the pions decays in flight as $\pi \rightarrow \mu + v_\mu$. We observe that, in some cases, the muons move backwards. Find the maximum value of p_π for this to happen.

1.15 A Λ hyperon decays as $\Lambda \rightarrow p + \pi^-$; its momentum in the L frame is $p_\Lambda = 2$ GeV. Take the direction of the Λ in the L frame as the x-axis. In the CM frame, the angle of the proton direction with x is $\theta_p^* = 30°$. Find
    (a) the energy and momentum of the Λ and the π in the CM frame;
    (b) the Lorentz parameters for the L–CM transformation;
    (c) the energy and momentum of the π, and the angle and momentum of the p in the L frame.

1.16 Consider the collision of a ball on an equal ball at rest. Compute the angle between the two final directions at non-relativistic speeds.

1.17 A proton with momentum $p_1 = 3$ GeV elastically scatters on a proton at rest. The scattering angle of one of the protons in the CM is $\theta_{ac}^* = 10°$. Find
    (a) the kinematic quantities in the L frame;
    (b) the kinematic quantities in the CM frame;
    (c) the angle between the directions of the final protons in the L frame; is it 90°?

1.18 A 'charmed' meson D^0 decays $D^0 \rightarrow K^-\pi^+$ at a distance from the production point $d = 3$ mm long. Measuring the total energy of the decay products, one finds $E = 30$ GeV. How long did the D live in proper time? How large is the π^+ momentum in the D rest-frame?

1.19 The primary beam of a synchrotron is extracted and used to produce a secondary monochromatic π^- beam. One observes that, at a distance $l = 20$ m from the production target, 10% of the pions have decayed. Find the momentum and energy of the pions.

1.20 A π^- beam is brought to rest in a liquid hydrogen target. Here π^0 are produced by the 'charge exchange' reaction $\pi^- + p \rightarrow \pi^0 + n$. Find the energy of the π^0, the kinetic energy of the n, the velocity of the π^0 and the distance travelled by the π^0 in a lifetime.

1.21 Consider an electron beam of energy $E = 2$ GeV hitting an iron target (assume it is made of pure ^{56}Fe). How large is the maximum four-momentum transfer?

1.22 Geiger and Marsden observed that the alpha particles, after having hit a thin metal foil, bounced back not too infrequently. Calculate the ratio between the scattering probabilities for $\theta > 90°$ and for $\theta > 10°$.

1.23 An α particle beam of kinetic energy $E = 6$ MeV and intensity $R_i = 10^3\,\text{s}^{-1}$ goes through a gold foil $\left(Z = 79, A = 197, \rho = 1.93 \times 10^4\,\text{kg m}^{-3}\right)$ of thickness $t = 1\,\mu$m. Calculate the number of particles per unit time scattered at angles larger than 0.1 rad.

1.24 Electrons with 10 GeV energy are scattered by protons initially at rest at 30°. Find the maximum energy of the scattered electron.

1.25 If $E = 20$ GeV electrons scatter elastically emerging with energy $E' = 8$ GeV, find the scattering angle.

1.26 Find the ratio between the Mott and Rutherford cross-sections for the scattering of the same particles at the same energy at 90°.

1.27 A particle of mass m, charge $q = 1.6 \times 10^{-19}$ C and momentum p moves in a circular orbit at a constant speed (in absolute value) in the magnetic field **B** normal to the orbit. Find the relationship between m, p and B.

1.28 We wish to measure the total $\pi^+ p$ cross-section at 20 GeV incident momentum. We build a liquid hydrogen target $(\rho = 60\text{ kg m}^{-3})$ that is $l = 1$ m long. We measure the flux before and after the target with two scintillation counters. Measurements are made with the target empty and with the target full. By normalizing the fluxes after the target to the same incident flux, we obtain in the two cases $N_0 = 7.5 \times 10^5$ and $N_H = 6.9 \times 10^5$ respectively. Find the cross-section and its statistical error (ignoring the uncertainty of the normalization).

1.29 In the experiment of O. Chamberlain *et al.* in which the antiproton was discovered, the antiproton momentum was approximately $p = 1.2$ GeV. What is the minimum refraction index needed in order to have the antiprotons above the threshold in a Cherenkov counter? How wide is the Cherenkov angle if $n = 1.5$?

1.30 Consider two particles with masses m_1 and m_2 and the same momentum p. Evaluate the difference Δt between the times taken to cross the distance L. Let us define the base with two scintillator counters and measure Δt with 300 ps resolution. How much must L be if we want to distinguish π from K at two standard deviations, if their momentum is 4 GeV?

1.31 A Cherenkov counter containing nitrogen gas at pressure Π is located on a charged particle beam with momentum $p = 20$ GeV. The dependence of the refraction index on the pressure Π is given by the law $n - 1 = 3 \times 10^{-9} \Pi\,(\text{Pa})$. The Cherenkov detector must see the π and not the K. In which range must the pressure be?

1.32 Superman is travelling along an avenue on Metropolis at high speed. At a crossroads, seeing that the lights are green, he continues. However, he is stopped by the police, claiming he had crossed on red. Assuming both to be right, what was Superman's speed?

1.33 Considering the Cherenkov effect in water $(n = 1.33)$, determine: (1) the minimum velocity of a charged particle for emitting radiation, (2) the minimum kinetic energy for a proton and a pion to do so and (3) the Cherenkov angle for a pion with energy $E_\pi = 400$ MeV.

1.34 Consider a Cherenkov apparatus to be operated as a threshold counter. The pressure of the N_2 gas it contains can be varied. The index n depends on the pressure π, measured in pascals, as $n = 1 + 3 \times 10^{-9} \pi$. A beam composed of π^+s, K^+s and protons all with momentum p crosses the counter. Knowing that π^+s are above threshold for $\pi \geq 5.2 \times 10^3$ Pa: (a) find the momentum p; (b) find the minimum pressure at which the K^+s are above threshold; (c) find the same for the protons. Hint: p is much larger than the mass of each species.

1.35 (1) What is the maximum energy of a cosmic-ray proton to remain confined in the Solar System? Assume $R = 10^{13}$ m as the radius of the system and an average magnetic field $B = 1$ nT.

(2) What is the maximum energy to remain confined in the Milky Way $\left(R = 10^{21} \text{m}, B = 0.05 \text{ nT} \right)$?

1.36 Portable neutron generators are commercially available based on the $d - t$ fusion. These devices contain a source of deuterium ions, an accelerator that accelerates the ions up to about $T_d = 130$ keV kinetic energy and a target in which tritium nuclei are chemically bound in a metallic compound (hydride). The fusion reaction $d + t \rightarrow n + {}^4$He has a maximum cross-section (5 barn) at that energy.

(1) Calculate the neutron kinetic energy $[m_d = 1875.6, m_t = 2808.9, m_\alpha = 3727.4]$.

(2) Neutrons are emitted isotropically. If the neutron production rate is $I = 3 \times 10^{10}$ s^{-1}, what is the neutron flux at the distance $R = 1$ m from the source?

(3) By measuring the direction and time of the α-particle, the direction of the neutron and the time of its production can be 'tagged'. The neutrons can be used to study some materials. A neutron collides with a nucleus of that material producing a characteristic, prompt γ-ray. By measuring the time interval between the α-particle and γ-ray signals, one can determine the position of the nucleus. Calculate the resolution on the time of flight measurement necessary to locate the nucleus within $\Delta z = 5$ cm.

1.37 Neutrons originated from radioactive elements in the ambient have kinetic energies up to a few MeV. We want to detect such neutrons with a TPC containing ^{40}Ar. If the neutron energy is low enough, the internal structure of the nucleus is not resolved and it appears as a single object to the neutron (coherent scattering). Assuming a nuclear radius $R_A = 4$ fm, what is the minimum neutron kinetic energy needed to resolve the nucleus structure? What is the maximum recoil energy of the nucleus in a collision with the limit initial kinetic energy $[m_{Ar} = 37.2$ GeV$]$?

1.38 Consider the head-on collision of two photons, γ_1 and γ_2, of energies $E_1 > E_2$ respectively. If γ_1 is produced by a laser of wavelength $\lambda = 694$ nm, what is the minimum value of E_1 to produce a positron–electron pair? Compute the velocity of the CM system at threshold as $1 - \beta$. Calculate the mass of the two photons if they move in the same direction.

Summary

In this chapter, students have learnt the basic tools that will be necessary to understand the material in the following chapters. In particular:

- the Lorentz transformations from general characteristics of space-time
- the meaning of mass, energy and momentum; their Lorentz transformation properties; and the kinematic relativistic invariants
- the laboratory (L) and centre of mass (CM) frames
- the SI and the natural units (NU)
- the concepts of cross-section, luminosity, decay rates (total and partial), branching ratios and phase-space volume
- the basic aspects of a scattering experiment
- the names of the elementary particles types and of their fundamental interactions
- the ways particles, both charged and photons, lose their energy, and are detected, travelling through matter
- the sources of high-energy particles: cosmic rays, accelerators and colliders
- the basic types of particle detectors, tracking and calorimeters.

Further Reading

Alvarez, L. (1968) Nobel Lecture; *Recent Developments in Particle Physics*

Blackett, P. M. S. (1948) Nobel Lecture; *Cloud Chamber Researches in Nuclear Physics and Cosmic Radiation*

Bonolis, L. (2005) Bruno Touschek vs. machine builders: AdA, the first matter-antimatter collider. *Riv. del Nuov. Cim.* 28 (11) 1–60

Glaser, D. A. (1960) Nobel Lecture; *Elementary Particles and Bubble Chamber*

Grupen, C. & Shwartz, B. (2008) *Particle Detectors.* Cambridge University Press

Hess, V. F. (1936) Nobel Lecture; *Unsolved Problems in Physics: Tasks for the Immediate Future in Cosmic Ray Studies*

Kleinknecht, K. (1998) *Detectors for Particle Radiation.* Cambridge University Press

Lederman, L. M. (1991) The Tevatron. *Sci. Am.* 264 (3) 48

Meyers, S. & Picasso, E. (1990) The LEP collider. *Sci. Am.* 263 (1) 54

Okun, L. B. (1989) The concept of mass. *Phys. Today* June, 31

Rees, J. R. (1989) The Stanford linear collider. *Sci. Am.* 261 (4) 58

Rohlf, J. W. (1994) *Modern Physics from α to Z^0*. John Wiley & Sons, chapter 16

van der Meer, S. (1984) Nobel Lecture; *Stochastic Cooling and the Accumulation of Antiprotons*

Wilson, C. R. T. (1925) Nobel Lecture; *On the Cloud Method of Making Visible Ions and the Tracks of Ionising Particles*

2 Nucleons, Leptons and Mesons

Only a few elementary particles are stable: the electron, the proton and the photon. Neutrinos do not decay, but can change from one type to another. Many more are unstable. The particles that decay by weak interactions live long enough to travel macroscopic distances between their production and decay points. Therefore, we can detect them by observing their tracks or measuring their time of flight, which is the time taken by a particle to cross a known distance. Distances range from a fraction of a millimetre to several metres, depending on their lifetime and energy. In this chapter, we shall study the simplest properties of these particles and discuss the corresponding experimental discoveries. In the next chapter we shall present to the reader the symmetry properties of the interactions and the corresponding selection rules, and in Chapter 4 we shall discuss the hadron resonances, that is, the particles that decay via strong interactions with lifetimes too short to allow them to travel over observable distances.

The development of experimental sciences is never linear; rather, it follows complicated paths reaching partial truths, making errors that are later corrected by new experiments, often with completely unexpected results, and gradually reaching the correct conclusions. The study of at least a few aspects of such a development requires some effort but it is worth it to gain a deeper understanding of the resulting physical laws. This is why in this chapter we shall initially follow a historical approach.

We shall start by recalling the Yukawa assumption of a meson, the pion, as the mediator of the nuclear forces. We know now that the pion is not a fundamental particle and that the fundamental strong interaction is mediated, at a deeper level, by the gluons. However, Yukawa's idea was at the root of particle physics in the 1930s and this developed with a series of surprises, as we shall see in this chapter.

We shall see how the first particle discovered in the cosmic rays and that looked like the pion was found instead to be a lepton, the muon, similar, except for its larger mass, to the electron. Nobody had expected that. Experiments at high altitudes on cosmic rays led finally to the discovery of the pion soon after World War II. However, more experiments showed other surprises: Nature was much richer, with more mesons and more baryons having such funny behaviour that they were deemed 'strange' particles.

Once a particle has been discovered, its quantum numbers, charge, magnetic moment, spin and parity must be measured. We shall discuss these, as an important example, for the charged pion. We shall then discuss the discoveries of the charged leptons and of the neutrinos.

While most readers already know the fundamental relativistic wave equation, which P. A. M. Dirac found for the electron in 1928, and which is valid for the spin 1/2 particles in general, we shall include, for the benefit of those who do not, a brief introduction to the Dirac Lagrangian, and discuss how the equation is obtained from it. We shall then summarize the basic consequences. One of Dirac's fundamental predictions was the existence for each fermion of an antiparticle with the same mass but opposite 'charge'. The discussion on how the positron (the anti-electron) and the antiproton were discovered concludes the chapter.

Students are strongly encouraged to solve several of the problems at the end of the chapter as a necessary tool for gaining a deep and quantitative understanding of the subjects studied in the chapter. In physics, numbers are important.

2.1 The Muon and the Pion

As already mentioned, in 1935 H. Yukawa (Yukawa 1935) formulated a theory of the strong interactions between nucleons inside nuclei. The mediator of the interaction is the π meson, or pion. It must have three charge states, positive, negative and neutral, because the nuclear force exists between protons, between neutrons and between protons and neutrons. As the nuclear force has a finite range $\lambda \approx 1$ fm, Yukawa assumed a potential between nucleons of the form

$$\phi(r) \propto \frac{e^{-r/\lambda}}{r}. \tag{2.1}$$

From the uncertainty principle, the mass m of the mediator is inversely proportional to the range of the force. In NU, $m = 1/\lambda$. With $\lambda = 1$ fm, we obtain $m \approx 200$ MeV.

Two years later, Anderson and Neddermeyer (Anderson & Neddermeyer 1937) and Street and Stevenson (Street & Stevenson 1937), discovered that the particles of the penetrating component of cosmic rays have masses of just this order of magnitude. Apparently, the Yukawa particle had been discovered, but the conclusion was wrong.

In 1942, Rossi and Nereson (Rossi & Nereson) measured the lifetime of penetrating particles to be $\tau = 2.3 \pm 0.2$ µs.

The crucial experiment showing that the penetrating particle was not the π meson was carried out in 1946 in Rome by M. Conversi, E. Pancini and O. Piccioni (Conversi et al. 1947). The experiment was aimed at investigating whether the absorption of positive and negative particles in a material was the same or different. Actually, a negative particle can be captured by a nucleus and, if it is the quantum of nuclear forces, quickly interacts with it rather than decaying. On the contrary, a positive particle is repelled by the nuclei and will in any case decay, as in a vacuum.

The two iron blocks F_1 and F_2 in the upper part of Fig. 2.1 are magnetized in opposite directions normal to the drawing and are used to focus the particles of one sign or, inverting their positions, the other. The 'trigger logic' of the experiment is as follows. The Geiger counters A and B, above and below the magnetized blocks, must discharge at the same instant ('fast' coincidence); one of the C counters under the absorber must fire not immediately but later, after a delay Δt in the range of 1 µs $< \Delta t <$ 4.5 µs ('delayed' coincidence). This logic guarantees the following: first that the energy of the

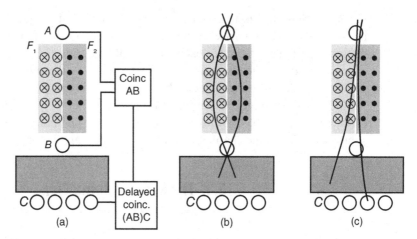

A sketch of the experiment of Conversi, Pancini and Piccioni.

particle is large enough to cross the blocks and small enough to stop in the absorber; second that, in this energy range and with the chosen geometry, only particles of one sign can hit both A and B; and finally that the particle decays in a time compatible with the lifetime value of Rossi and Nereson.

Fig. 2.1(b) shows the trajectory of two particles of the 'right' sign in the right energy range, which discharge A and B but not C; Fig. 2.1(c) shows two particles of the wrong sign. Neither of them gives a trigger signal because one discharges A and not B, the other discharges both but also C.

In a first experiment in 1945, the authors used an iron absorber. The result was that the positive particles decayed as in a vacuum; the negative particles did not decay, exactly as expected for the Yukawa particle.

The authors repeated the experiment in 1946 with a carbon absorber, finding, to their surprise, that the particles of both signs decayed (Conversi *et al.* 1947). A systematic search showed that, in materials with low atomic numbers, the penetrating particles are not absorbed by nuclei. However, calculation soon showed that the pions should interact so strongly as to be absorbed by any nucleus, even by small ones. In conclusion, the penetrating particles of the cosmic rays are not the Yukawa mesons.

In the same years, G. Occhialini and C. F. Powell, working at Bristol, exposed emulsion stacks at high altitudes in the mountains (up to 5500 m in the Andes). In 1947 they published, with Lattes and Muirhead, the observation of events in which a more massive particle decays into a less massive one (Lattes *et al.* 1947). The interpretation of this is that two particles are present in cosmic rays: the first is the π, the second, which was called μ or muon, is the penetrating particle. They observed that the muon range was equal in all the events (about 600 μm), showing that the pion decays at rest into two bodies: the muon and a neutral undetected particle.

The final proof came in 1949, when the Bristol group, using the new Kodak emulsions sensitive to minimum ionizing particles, detected events in which the complete chain of decays $\pi\mu e$ was visible. An example is shown in Fig. 2.2.

Fig. 2.2.   A $\pi\mu e$ decay chain observed in emulsions. (C. Powell. Nobel lecture 1950 Fig. 5: C. Powell, © The Nobel Foundation 1950 Figure 5 in this link www.nobelprize.org/uploads/2018/06/powell-lecture.pdf).

We now know that the charged pion decays are

$$\pi^+ \rightarrow \mu^+ + \nu_\mu \qquad \pi^- \rightarrow \mu^- + \bar{\nu}_\mu \tag{2.2}$$

and those of the muons are

$$\mu^+ \rightarrow e^+ + \nu_e + \bar{\nu}_\mu \qquad \mu^- \rightarrow e^- + \nu_\mu + \bar{\nu}_e. \tag{2.3}$$

In these expressions we have specified the types of neutrinos, something that was completely unknown at the time. We shall discuss neutrinos in Section 2.4.

Other experiments showed directly that pions interact strongly with nuclei, transforming a proton into a neutron and vice versa:

$$\pi^+ + {}_Z^A N \rightarrow {}_Z^{A-1} N + p \qquad \pi^- + {}_Z^A N \rightarrow {}_{Z-1}^{A-1} N + n. \tag{2.4}$$

In conclusion, the pions are the Yukawa particles. It took a quarter of a century to understand that the Yukawa force is not the fundamental strong nuclear interaction and that the pion is a composite particle. The fundamental interaction occurs between the quarks, mediated by the gluons, as we shall see in Chapter 6.

We shall dedicate Section 2.3 to the measurement of the pion quantum numbers. We summarize here that pions exist in three charge states: π^+, π^0 and π^-. The π^+ and the π^- are each the antiparticle of the other, while the π^0 is its own antiparticle. The π^0 practically always (99%) decays in the channel $\pi^0 \rightarrow \gamma\gamma$.

One mystery remained, however: the μ or muon. It was identical to the electron, except for its mass, 106 MeV, about 200 times bigger. What is the reason for a heavier brother of the electron? 'Who ordered that?' asked Rabi. Even today, we have no answer.

2.2 Strange Mesons and Hyperons

Nature had other surprises in store.

In 1944, Leprince-Ringuet and l'Héritier (Leprince-Ringuet & l'Héritier 1944), working in a laboratory on the Alps with a 'triggered' cloud chamber in a magnetic field $B = 0.25$ T, discovered a particle with a mass of 506 ± 61 MeV.

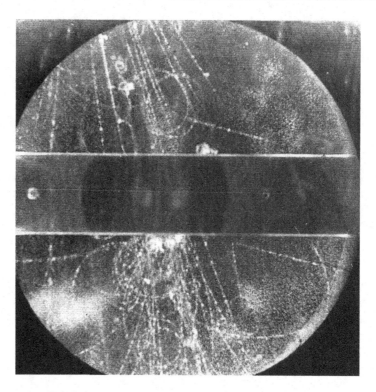

Others surprises were to follow. Soon after the discovery of the pion, in several laboratories in the UK, France and the USA, cosmic ray events were found in which particles with masses similar to that of Leprince-Ringuet decayed, apparently, into pions. Some were neutral and decayed into two charged particles (plus possibly some neutral ones) and were called V^0 because of the shape of their tracks (see Fig. 2.3), others were charged, decaying into a charged daughter particle (plus neutrals) and were named θ, and still others decayed into three charged particles, called τ.

It took a decade to establish that θ and τ are exactly the same particle, while the V^0s are its neutral counterparts. These particles are the K-mesons, also called 'kaons'.

In 1947, Rochester and Butler published the observation of the associated production of a pair of such unstable particles (Rochester & Butler 1947). It was soon proved experimentally that those particles are always produced in pairs; the masses of the two partners turned out to be different, one about 500 MeV (a K-meson), the other greater than that of the nucleon. The more massive ones were observed to decay into a nucleon and a pion. These particles belong to the class of the hyperons. The lightest are the Λ^0 and the Σ s that have three charge states, Σ^+, Σ^0 and Σ^-. We discussed in Section 1.13(a) clear example of associated production seen many years later in a bubble chamber. Fig. 1.21 shows that of $\pi^- + p \to K^0 + \Lambda^0$, followed by the decays $K^0 \to \pi^+ + \pi^-$ and $\Lambda^0 \to p + \pi^-$.

	Q	S	$m\,(\mathrm{MeV})$	$\tau\,(\mathrm{ps})$	Principal decays (BR in %)
Table 2.1. The K-mesons (n.a. means 'not applicable').					
$K+$	$+1$	$+1$	493.7	12.4	$\mu^{+}\nu_{\mu}(63), \pi^{+}\pi^{+}\pi^{-}(21), \pi^{+}\pi^{0}(5.6)$
K^{0}	0	$+1$	(497.6)	n.a.	
K^{-}	-1	-1	493.7	12.4	$\mu^{-}\bar{\nu}_{\mu}, \pi^{-}\pi^{-}\pi^{+}, \pi^{-}\pi^{0}$
$\bar{K}^{0}$	0	-1	(497.6)	n.a.	

The new particles showed very strange behaviour. There were two puzzles (plus a third to be discussed later). Why were they always produced in pairs? Why were they produced by 'fast' strong interaction processes, as demonstrated by the large cross-section, while they decayed only 'slowly' with lifetimes typical of weak interactions? In other words, why do fully hadronic decays such as $\Lambda^{0} \rightarrow p + \pi^{-}$ not proceed strongly? The new particles were called 'strange particles'.

The solution was given by Nishijima (Nakato & Nishijima 1953) and independently by Gell-Mann (Gell-Mann 1953). They introduced a new quantum number that Gell-Mann called 'strangeness' S, which is additive, like electric charge. Strangeness is conserved by strong and electromagnetic interactions, but not by weak interactions. The 'old' hadrons, the nucleons and the pions, have $S = 0$, the hyperons have $S = -1$, and the K-mesons have $S = +1$; their antiparticles have opposite values.

The production by strong interactions from an initial state with $S = 0$ can happen only if two particles of opposite strangeness are produced. The lowest-mass strange particles, the K-mesons and the hyperons, can decay for energetic reasons only into non-strange final states; therefore, they cannot decay strongly.

If the mass of a strange meson or of a hyperon is large enough, final states of the same strangeness are energetically accessible. This happens if the sum of the masses of the daughters is smaller than that of the mother particle. These particles exist and decay by strong interactions with extremely short lifetimes, of the order of $10^{-24}\,$s. In practice, they decay at the same point where they are produced and do not leave an observable track. We shall see in Chapter 4 how to detect them.

We shall not describe the experimental work done with cosmic rays and later with beams from accelerators; rather we shall summarize the main conclusions on the metastable strange particles, which we define as those that are stable against strong interactions and decay weakly or electromagnetically.

The K-mesons are the only metastable strange mesons. There are four of them. Table 2.1 gives their characteristics; in the last column the principal decay channels of the charged states are given with their approximate branching ratios (BR). The K-mesons have spin zero.

There are two charged kaons, the K^{+} with $S = +1$, and its antiparticle, the K^{-}, which has the same mass, the same lifetime and opposite charge and strangeness. The decay channels of one contain the antiparticles of the corresponding channels of the other.

We now anticipate a fundamental law of physics, CPT invariance. CPT is the product of three operations, time reversal (T) parity (P) (i.e. the inversion of the

	Q	S	m (MeV)	τ (ps)	$c\tau$ (mm)	Principal decays (BR in %)
Table 2.2. The metastable strange hyperons.						
Λ	0	-1	1115.7	263	79	$p\pi^-(64), n\pi^0(36)$
Σ^+	$+1$	-1	1189.4	80	24	$p\pi^0(51.6), n\pi^+(48.3)$
Σ^0	0	-1	1192.6	7.4×10^{-8}	2.2×10^{-8}	$\Lambda\gamma(100)$
Σ^-	-1	-1	1197.4	148	44.4	$n\pi^-(99.8)$
Ξ^0	0	-2	1314.9	290	87	$\Lambda\pi^0(99.5)$
Ξ^-	-1	-2	1321.7	164	49	$\Lambda\pi^-(99.9)$

coordinate axes) and particle–antiparticle conjugation (C). CPT invariance implies that particle and antiparticle have the same mass, lifetime and spin, and all 'charges' of opposite values.

While the neutral pion is its own antiparticle, the neutral kaon is not; K^0 and $\bar{K}^0$ are distinguished because of their opposite strangeness. We anticipate that K^0 and $\bar{K}^0$ form an extremely interesting quantum two-state system, which we shall study in Chapter 8. We mention here only that they are not the eigenstates of the mass and of the lifetime.

Now let us consider the metastable hyperons. Three types of hyperon were discovered in cosmic rays, some with more than one charge status (six states in total). These are (see Table 2.2) the Λ^0, three Σs all with strangeness $S = -1$ and two Ξ s with strangeness $S = -2$. All have spin $J = 1/2$. In the last column, the principal decays are shown. All but one are weak.

The neutral Σ^0 hyperon has a mass larger than the Λ, and the same strangeness. Therefore, the Gell-Mann and Nishijima scheme foresaw the decay $\Sigma^0 \to \Lambda^0 + \gamma$. This prediction was experimentally confirmed.

Notice that all the weak lifetimes of the hyperons are of the order of 100 ps; the electromagnetic lifetime of the Σ^0 is nine orders of magnitude smaller.

As we have already said, hadrons are not elementary objects; they contain quarks. We shall discuss this issue in Chapter 4. We already anticipated that the 'old' hadrons contain two types of quark, u and d. Their strangeness is zero. The strange hadrons contain one or more quarks s or antiquark $\bar{s}$. The quark s has strangeness $S = -1$ (pay attention to the sign!), its antiquark $\bar{s}$, has strangeness $S = +1$. The $S = +1$ hadrons, such as K^+, K^0, $\bar{\Lambda}$ and the $\bar{\Sigma}$s, contain one $\bar{s}$; those with $S = -1$, such as as K^-, $\bar{K}^0$, Λ and the Σ s, contain one s quark; the Ξs with $S = -2$ contain two s quarks, etc.

2.3 The Quantum Numbers of the Charged Pion

For every particle, we must measure all the relevant characteristics: mass, lifetime, spin, charge, strangeness, branching ratios for its decays in different channels and, as we shall discuss in the next chapter, intrinsic parity and, if completely neutral,

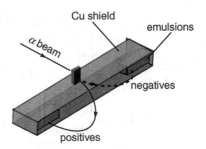

Fig. 2.4. A sketch of Burfening *et al.*'s equipment for the pion mass measurement.

charge conjugation. This enormous amount of work took several decades of the past century. We shall discuss here only some measurements of the quantum numbers of the charged pion.

The Mass

The first accelerator with sufficient energy to produce pions was the 184-inch Berkeley cyclotron, which could accelerate α particles up to a kinetic energy of $E_k = 380$ MeV.

To determine the mass, two kinematic quantities must be measured, for example the energy E and the momentum p. The mass is then given by

$$m^2 = E^2 - p^2.$$

We show in Fig. 2.4 a sketch of the set-up of the pion mass measurement by Burfening and collaborators in 1951 (Burfening *et al.* 1951). Two emulsion stacks, duly screened from background radiation, are located in the cyclotron vacuum chamber, below the plane of the orbit of the accelerated α particles. When the α particles reach their final orbit, they hit a small target and produce pions of both signs. The pions are deflected by the magnetic field of the cyclotron on either one side or the other, depending on their sign, and penetrate the corresponding emulsion stack. After exposure, the emulsions are developed, and the entrance point and direction of each pion track are measured. These, together with the known position of the target, give the pion momentum. The measurement of its range gives its energy.

The result of the measurement was

$$m_{\pi^+} = 141.5 \pm 0.6 \text{ MeV} \qquad m_{\pi^-} = 140.8 \pm 0.7 \text{ MeV}. \tag{2.5}$$

The two values are equal within the errors. The present value is

$$m_{\pi^\pm} = 139.57039 \pm 0.00018 \text{ MeV}. \tag{2.6}$$

The Lifetime

To measure decay times of the order of several nanoseconds with good resolution, we need electronic techniques and fast detectors. The first measurement with such

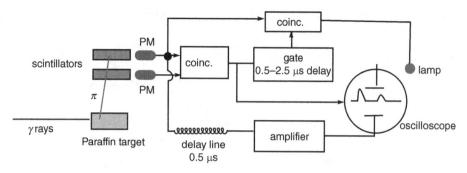

Fig. 2.5. A sketch of the detection scheme in the pion lifetime experiment of Chamberlain *et al.*

techniques was due to O. Chamberlain and collaborators, as shown in Fig. 2.5 (Chamberlain *et al.* 1950).

The 340 MeV γ beam from the Berkeley electron synchrotron hit a paraffin (a proton-rich material) target and produced pions by the reaction

$$\gamma + p \to \pi^+ + n. \tag{2.7}$$

Two scintillator counters were located one after the other on one side of the target. The logic of the experiment required a meson to cross the first scintillator and stop in the second. The positive particles were not absorbed by the nuclei and decayed at rest. The dominant decay channel is

$$\pi^+ \to \mu^+ + \nu_\mu. \tag{2.8}$$

The μ loses all its energy in ionization, stops and, after an average time of 2.2 μs, decays as

$$\mu^+ \to e^+ + \nu_e + \bar{\nu}_\mu. \tag{2.9}$$

To implement this logic, the electric pulses from the two photomultipliers that read the scintillators were sent to a coincidence circuit; this established that a particle had crossed the first counter and reached the second. A 'gate' circuit established the presence of a second pulse, from the second counter, with a delay of between 0.5 μs and 2.5 μs meaning that a μ decayed. This confirmed that the primary particle was a π^+.

The signals from the second scintillator were sent, delayed by 0.5 μs (through the 'delay line' in the Fig. 2.5) to an oscilloscope, whose sweep was triggered by the output of the fast coincidence. The gate signal, if present, lit a lamp located near the scope screen. The screen and lamp were photographed. The pictures show two pulses, one due to the arrival of the π and one due to its decay. They were well separated if their distance was >22 ns.

In total, 554 events were collected. As expected, the distribution of the times was exponential. The lifetime measurement gave $\tau = 26.5 \pm 1.2$ ns. The present value is $\tau = 26.033 \pm 0.0005$ ns.

The Spin

A particle of spin s has $2s+1$ degrees of freedom. As the probability of a reaction depends on the number of degrees of freedom, we can determine the spin by measuring such reaction probabilities. More specifically, we shall consider the ratio of the cross-sections of two processes, one the inverse of the other, at the same CM energy

$$\pi^+ + d \rightarrow p + p \qquad (2.10)$$

$$p + p \rightarrow \pi^+ + d. \qquad (2.11)$$

We call them π^+ absorption and production respectively. Writing both reactions generically as $a+b \rightarrow c+d$, Eq. (1.77) gives the cross-sections in the CM system. As we are interested in the ratio of the cross-sections at the same energy, we can neglect the common factors, included the energy E. We obtain

$$\frac{d\sigma}{d\Omega}(a+b \rightarrow c+d) \propto \frac{p_f}{p_i} \frac{1}{(2s_a+1)(2s_b+1)} \sum_{f,i} |M_{fi}|^2, \qquad (2.12)$$

where the sum is over all the spin states, initial and final. The initial and final momenta are different in the two processes, but since the energy is the same, the initial momentum in one case is equal to the final momentum in the other. We can then write for the absorption $p_i = p_\pi$ and $p_f = p_p$, for the production $p_f = p_\pi$ and $p_i = p_p$, with the same values of p_π and p_p. For the absorption process, we now write

$$\frac{d\sigma}{d\Omega}(\pi^+ d \rightarrow pp) \propto \frac{p_p}{p_\pi} \frac{1}{(2s_\pi+1)(2s_d+1)} \frac{1}{2} \sum_{f,i} |M_{fi}|^2. \qquad (2.13)$$

Pay attention to the factor 1/2, which must be introduced to cancel the double counting implicit in the integration over the solid angle with two identical particles in the final state.

For the production process, we now write

$$\frac{d\sigma}{d\Omega}(pp \rightarrow \pi^+ d) \propto \frac{p_\pi}{p_p} \frac{1}{(2s_p+1)^2} \sum_{f,i} |M_{fi}|^2. \qquad (2.14)$$

We give here, without proof, the 'detailed balance principle', which is a consequence of the time reversal invariance, which is satisfied by the strong interactions (see next chapter). The principle implies the equality

$$\sum_{f,i} |M_{fi}|^2 = \sum_{f,i} |M_{if}|^2.$$

Using this equation and knowing the spin of the proton, $s_p = \frac{1}{2}$, and of the deuteron, $s_d = 1$, we obtain

$$\frac{\sigma(\pi^+ d \rightarrow pp)}{\sigma(pp \rightarrow \pi^+ d)} = \frac{(2s_p+1)^2}{2(2s_\pi+1)(2s_d+1)} \frac{p_p^2}{p_\pi^2} = \frac{2}{3(2s_\pi+1)} \frac{p_p^2}{p_\pi^2}. \qquad (2.15)$$

The absorption cross-section was measured by Durbin and co-workers (Durbin *et al.* 1951) and by Clark and co-workers (Clark *et al.* 1951) at the laboratory kinetic energy $T_\pi = 24$ MeV. The production cross-section was measured by Cartwright and colleagues (Cartwright *et al.* 1953) at the laboratory kinetic energy $T_p = 341$ MeV. The CM energies are almost equal in both cases. From the measured values, one obtains $2s_\pi + 1 = 0.97 \pm 0.31$, hence $s_\pi = 0$.

The Neutral Pion

For the π^0, we shall give only the present values of the mass and of the lifetime.

The mass of the neutral pion is smaller than that of the charged one by about 4.5 MeV

$$m_{\pi^0} = 134.9768 \pm 0.0005 \text{ MeV} \tag{2.16}$$

The π^0 decays by electromagnetic interaction predominantly (99.8%) in the channel

$$\pi^0 \to \gamma\gamma. \tag{2.17}$$

Therefore, its lifetime is much shorter than that of the charged pions

$$\tau_{\pi^0} = (8.43 \pm 0.13) \times 10^{-17} \text{ s} \tag{2.18}$$

2.4 Charged Leptons and Neutrinos

We know about three charged leptons with identical characteristics. They differ in their masses and lifetimes, as shown in Table 2.3.

We give a few historical hints.

The electron was the first elementary particle to be discovered, by J. J. Thomson in 1897, in the Cavendish Laboratory at Cambridge. At that time, the cathode rays that had been discovered by Plücker in 1857 were thought to be waves, propagating in the ether. Thomson and his collaborators succeeded in deflecting the rays not only, as already known, by a magnetic field, but also by an electric field. By letting the rays through crossed electric and magnetic fields and adjusting the field intensities for null deflection, they measured the mass to charge ratio m/q_e and found it to have a universal value (Thomson 1897).

The muon, as we have seen, was discovered in cosmic rays by Anderson and Neddermeyer (Anderson & Neddermeyer 1937), and independently by Street and Stevenson (Street & Stevenson 1937); it was identified as a lepton by Conversi, Pancini and Piccioni in 1946 (Conversi *et al.* 1947).

The possibility of a third family of leptons, made up of the heavy lepton H_l and its neutrino v_{Hl}, with a structure similar to the two known ones, was advanced by A. Zichichi, who, in 1967, developed the search method that we shall now describe, built the experiment and searched for the H_l at ADONE (Bernardini *et al.* 1967). The

Table 2.3. The charged leptons.		
	m (MeV)	τ
e	0.511	$> 6.6 \times 10^{28}$ yr
μ	105.6	2.2 μs
τ	1777	290 fs

H_l did indeed exist, but with a mass too large for the maximum energy reachable with ADONE. It was discovered, with the same method, at the SPEAR electron–positron collider in 1975 by M. Perl *et al.* (Perl *et al.* 1975). It was called τ, from the Greek word *triton*, meaning the third. The method was as follows.

As we shall see in the next chapter, the conservation of the lepton flavours forbids the processes $e^+ e^- \to e^+ \mu^-$ and $e^+ e^- \to e^- \mu^+$. If a heavy lepton exists, the following reaction occurs

$$e^+ + e^- \to \tau^+ + \tau^-, \tag{2.19a}$$

followed by the decays

$$\tau^+ \to e^+ + \nu_e + \bar{\nu}_\tau \qquad \tau^- \to \mu^- + \bar{\nu}_\mu + \nu_\tau, \tag{2.19b}$$

and is charge conjugated, resulting in the observation of $e^- \mu^+$ or $e^+ \mu^-$ pairs and apparent violation of the lepton flavours. The principal background is due to the pions that are produced much more frequently than the $e\mu$ pairs. Consequently, the experiment must provide the necessary discrimination power. Moreover, an important signature of the sought events is the presence of undetected neutrinos. Therefore, the two tracks and the direction of the beams do not belong to the same plane, owing to the momenta of the unseen neutrinos. Such non-coplanar $e\mu$ pairs were finally found at SPEAR, when energy above threshold became available.

The neutrino was introduced as a 'desperate remedy', by W. Pauli in 1930 to explain the apparent violation of energy, momentum and angular momentum conservations in beta decays.

The neutrino was discovered by F. Reines and C. Cowan (Cowan *et al.* 1956) in 1956 at the Savannah River reactor, when only one type of 'ghost particle' was assumed to exist. To be precise, they discovered the electron antineutrino, the one produced in the fission reactions. On 15 June, Pauli, having received a telegram announcing the discovery, responded to Reines and Cowan, 'Thanks for the message. Everything comes to him who knows how to wait.'

The muon neutrino (ν_μ) was discovered, that is, identified as a particle different from ν_e, by L. Lederman, M. Schwartz and J. Steinberger in 1962 and collaborators (Danby *et al.* 1962) at the proton accelerator AGS at Brookhaven. We shall briefly describe this experiment too.

The tau neutrino (ν_τ) was discovered by K. Niwa and collaborators (Kodama *et al.* 2001) with the emulsion technique at the Tevatron proton accelerator at FNAL in 2000.

We shall now describe the discovery of the electron neutrino. The theory of beta decay was formulated by Fermi in 1933 (see Chapter 7), including the proposal to exploit the inverse beta decay reaction on a nucleon, $v + {}^{Z}_{A}N \rightarrow e^{\mp} + {}^{Z \pm 1}_{A}N$ for neutrino detection. In this expression the neutrino flavour is not specified because only one type was assumed to exist. The following year, Bethe and Peierls (1934) evaluated that the cross-section of the process was $\sigma < 10^{-44} \, \text{cm}^2$ at a few MeV of energy *corresponding to a penetrating power of 10^{16} km in solid matter* (1000 light years). The authors concluded that *it is therefore absolutely impossible to observe processes of this kind with the neutrinos created in nuclear transformations.* While the calculation was correct, the conclusion was not. The small cross-section can be compensated by a large enough neutrino flux. For example, with $\Phi = 10^{17} \, \text{m}^{-2}\text{s}^{-1}$, the flux at 10 m from the core of a 1 GW nuclear reactor (which did not exist in 1934), one has one neutrino interaction every 100 s in a 1 m^3 detector.

When nuclear reactors became available, in 1946, B. Pontecorvo elaborated on the idea to detect neutrinos with radiochemical method using the reaction

$$v + {}^{37}\text{Cl} \rightarrow {}^{37}\text{Ar} + e^-$$

in which the produced ${}^{37}\text{Ar}$ is metastable $\left(T_{1/2} = 34.1 \, \text{d}\right)$. After exposing a quantity of ${}^{37}\text{Cl}$ to neutrinos from a nuclear reactor for a time comparable with its half-life, one should extract the few argon atoms and count their decays. However, Pontecorvo overestimated by two orders of magnitude the cross-section and in 1949 L. Alvarez (Alvarez 1949) came back to the issue, correctly evaluated the cross-section and discussed the crucial problem of the experiment, namely the control of the backgrounds from environmental radioactivity. Indeed, the produced Ar atoms, which are the signal and have to be extracted, are only one in 10^{30} of Cl atoms. In addition, the background sources must be controlled in such a way that the number of Ar atoms produced by them must be much smaller than the number of those produced by the signal. He did not proceed with the experiment, which was finally tried in 1958 by Raymond Davis at the Savannah River nuclear reactor (0.7 GW). He used a target of 18.5 t of C_2Cl_4. The calculated sensitivity was twenty times larger than the minimum needed for detection. The experiment was sensitive enough, but no event was detected. The conclusion is that the neutral particle produced in β^- decay, which we call $\bar{v}_e$ as different from v_e, does not induce the Cl-Ar reaction. As we shall see in Chapter 10, however, Davis later detected the neutrinos coming from the Sun with this method and found a surprise.

F. Reines and C. Cowan discovered the $\bar{v}_e$ at the same reactor, chosen by Reines also because a massive building, located underground, a dozen metres under the core, was available for the experiment. The $\bar{v}_e$ flux was about $\Phi = 10^{17} \, \text{m}^{-2}\text{s}^{-1}$. Neutrinos were searched via inverse beta decay on protons, for which the expected cross-section is

$$\sigma\left(\bar{v}_e + p \rightarrow e^+ + n\right) \approx 10^{-47} \left(E_v \, / \, \text{MeV}\right)^2 \text{m}^2. \tag{2.20}$$

Notice that, at low energy, the cross-section grows with the square of the energy.

An easily available material containing many protons is water. Let us evaluate, in order of magnitude, the quantity of water needed to have, for example, a rate of

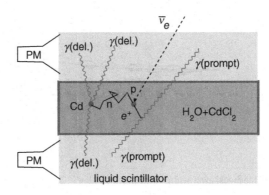

Fig. 2.6. A sketch of the detection scheme of the Savannah River experiment.

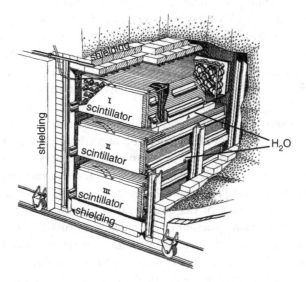

Fig. 2.7. Sketch of the equipment of the Savannah River experiment. (F. Reines, Nobel Lecture 1995. Fig. 5 (part): F. Reines, © The Nobel Foundation 1995 Figure 5 this link www.nobelprize.org/uploads/2018/06/reines-lecture.pdf).

10^{-3} Hz for reaction (2.20), on free protons. Taking a typical energy $E_\nu = 1$ MeV, the rate per target proton is $W_1 = \Phi\sigma = 10^{-30}\,\text{s}^{-1}$. Consequently, we need 10^{27} protons. Since a mole of H_2O contains $2N_A \approx 10^{24}$ free protons, we need 1000 moles, hence 18 kg. In practice, much more is needed, taking all inefficiencies into account. Reines worked with 200 kg of water.

The main difficulty of the experiment is not the rate but the discrimination of the signal from the possibly much more frequent background sources that can simulate that signal. There are three principal causes: the neutrons that are to be found everywhere near a reactor, cosmic rays and the natural radioactivity of the material surrounding the detector and in the water itself.

Fig. 2.6 is a sketch of the detector scheme used in 1955 and Fig. 2.7 is a sketch of the equipment. It shows one of the two 100 l water containers sandwiched between two

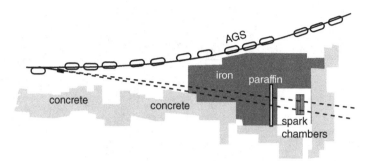

Sketch of the Brookhaven neutrino experiment.

liquid scintillator chambers, a technique that had been recently developed as we saw in Section 1.13. An antineutrino from the reactor interacts with a proton, producing a neutron and a positron. The positron annihilates immediately with an electron, producing two gamma rays in opposite directions, both with 511 keV energy. The Compton electrons produced by these gamma rays are detected in the liquid scintillators, giving two simultaneous signals. This signature of the positron is not easily emulated by background effects.

A second powerful discrimination is given by the detection of the neutron. Water is a good moderator and the neutron slows down in several microseconds. Forty kilos of Cd, a nucleus with a very high cross-section for thermal neutron capture, are dissolved in the water. A Cd nucleus captures the neutron, resulting in an excited state that soon emits gamma rays that are detected by the scintillators as a delayed coincidence.

The reduction of the cosmic ray background, due to the underground location, and the accurate design of the shielding structures were essential for the success of the experiment. Accurate control measurements showed that the observed event rate of $W = 3.0 \pm 0.2$ events per hour could not be due to background events. This was the experimental discovery of the neutrino, one quarter of a century after the Pauli hypothesis.

The muon neutrino was discovered, as already mentioned, at the AGS proton accelerator in 1962. The idea to create intense neutrino beams at high-energy proton accelerators was developed first by M. A. Markov and his students starting in 1958 in unpublished works (Markov 1985). Pontecorvo (1959) discussed the possibility of establishing whether neutrinos produced in the pion decay were different from those in beta decay by searching to see whether the former induce inverse beta decay, but the idea turned out not to be practical. Finally, and independently, a muon–neutrino beam of sufficient intensity to overcome the problem of the extremely small neutrino cross-section was proposed by M. Schwartz (1960).

Fig. 2.8 is a sketch of the experiment, which was led by L. Lederman, M. Schwartz and J. Steinberger. The intense proton beam circulating in the accelerator pipe is aimed at an internal beryllium target. Here a wealth of pions, of both signs, are produced. The pions decay as

$$\pi^+ \to \mu^+ + \nu \qquad \pi^- \to \mu^- + \bar{\nu}. \tag{2.21}$$

In these reactions, the neutrino and the antineutrino are produced in association with a muon. In the beta decays, neutrinos are produced in association with electrons. The aim of the experiment was to clarify if these neutrinos are different or not. For this reason, we have not specified the type of the neutrinos in the above expressions.

To select only the neutrinos, a 'filter' made of iron, 13.5 m long, is located after the target. It absorbs all particles, charged and neutral, apart from the neutrinos. The concrete blocks seen in Fig. 2.8 are needed to protect people from the intense radiation present near the target. To detect the neutrino interactions, one needs a device working both as target and as tracking detector. Calculations show that its mass must be about 10 t, too much, at that time, for a bubble chamber. The spark chamber technique, invented by M. Conversi and A. Gozzini in 1955 (Conversi & Gozzini 1955) and developed by S. Fukui and S. Myamoto (Fukui & Myamoto 1959) was chosen. A spark chamber element consists of a pair of parallel metal plates separated by a small gap (a few mm) filled with a suitable gas mixture. The chamber is made sensitive by suddenly applying a voltage to the plates after the passage of the particle(s), generating a high electric field $\left(\approx 1 \text{ MV m}^{-1}\right)$. The resulting discharge is located at the position of the ionization trail and appears as a luminous spark that is photographed.

The neutrino detector consisted of a series of 10 modules of 9 spark chambers each. The aluminium plates had an area of $1.1 \times 1.1 \text{ m}^2$ and a thickness of 2.5 cm, amounting to a total mass of 10 t.

After exposing the chambers to the neutrinos, photographs were scanned searching for muons that form the reactions

$$v + n \rightarrow \mu^- + p; \qquad \bar{v} + p \rightarrow \mu^+ + n \qquad (2.22)$$

and electrons that from

$$v + n \rightarrow e^- + p; \qquad \bar{v} + p \rightarrow e^+ + n. \qquad (2.23)$$

The two particles are easily distinguished because, for Eq. (2.22), the photograph shows a long penetrating track, and, for Eq. (2.23), the photograph shows an electromagnetic shower. Many muon events were observed, but no electron event was seen. The conclusion was that neutrinos produced in association with a muon produce, when they interact, muons, not electrons. It appears that two types of neutrinos exist, one associated with the electron, the other with the muon. This property is called 'lepton flavour'. The electron and the electron neutrinos have positive electron flavour $\mathcal{L}_e = +1$, the positron and the electron antineutrino have negative electron flavour $\mathcal{L}_e = -1$; all of them have zero muonic flavour. The μ^- and the v_μ have positive muonic flavour $\mathcal{L}_\mu = +1$, the μ^+ and the $\bar{v}_\mu$ have negative muonic flavour $\mathcal{L}_\mu = -1$; all have zero electronic flavour. Electronic, muonic (and tauonic) flavours are also called electronic, muonic (and tauonic) lepton numbers.

Experiments aimed at determining the masses of the neutrinos by measuring kinematic quantities such as energies and momenta of the particles present in reactions with neutrinos have produced only upper limits for those masses. These limits are much smaller than the masses of the charged leptons. Neutrinos are assumed to be rigorously massless in the Standard Model. As we shall see in Chapter 10, however, neutrinos do have a non-zero mass, even if very small.

2.5 The Dirac Equation

In 1928, P. A. M. Dirac wrote the fundamental relativistic wave equation of the electron and, more generally, of all the spin 1/2 particles. We summarize here the Dirac theory, without theoretical rigor.

We start by recalling that, in non-relativistic conditions, the wave function ψ of a free particle of mass m and momentum $\mathbf{p}$ is the Schrödinger equation. The corresponding Hamiltonian is the non-relativistic expression of the energy–momentum relation:

$$E = \frac{\mathbf{p}^2}{2m},$$

(2.24)

The relativistic expression is quadratic:

$$E^2 = \mathbf{p}^2 + m^2.$$

(2.25)

From this, through the usual substitutions

$$E \to i\frac{\partial}{\partial t}, \qquad \mathbf{p} \to -i\nabla$$

(2.26)

we obtain the Klein–Gordon equation

$$\nabla^2\psi - \frac{\partial^2\psi}{\partial t^2} = m^2\psi$$

(2.27)

However, as Dirac noticed, the presence of a second-order time derivative, corresponding to an allowed choice of the initial conditions for both ψ and $\partial\psi / \partial t$, leads, possibly, to negative probabilities. To avoid that, we need an equation containing only the first-order time derivative, namely of the type $i\dfrac{\partial}{\partial t}\psi = H\psi$. For the Hamiltonian H, Dirac made the ansatz (expressions with repeated Latin indices are implicitly summed on that index from 1 to 3):

$$H = \left(\alpha_k p_k + \beta m\right),$$

(2.28)

where four coefficients $\alpha_1, \alpha_2, \alpha_3$ and β satisfying Eq. (2.25) must be found. This imposes that

$$H^2 = p^2 + m^2 = \left(\alpha_k\alpha_k + \beta m\right)\left(\alpha_j\alpha_j + \beta m\right) = p_i p_i + m^2,$$

where we used different repeated indices to avoid ambiguities. Developing, we have:

$$\alpha_k p_k \, \alpha_j p_j + \alpha_k p_k \, \beta m + \beta m \alpha_j p_j + \beta m \beta m = p_i p_i + m^2.$$

The terms proportional to m^2 give $\beta^2 m^2 = m^2$, hence

$$\beta^2 = 1. \tag{2.29}$$

The terms proportional to m give $\alpha_k p_k \beta m + \beta m \alpha_j p_j = 0$. Being $\alpha_k p_k$ and $\alpha_j p_j$ just the same, we can write $\alpha_k p_k \beta m + \beta m \alpha_k p_k = 0$, and, simplifying, $(\alpha_k \beta + \beta \alpha_k) p_k = 0$. This must be true for any p_k, and we finally obtain: $\alpha_k \beta + \beta \alpha_k = 0$. Namely β and each of the a_k anticommute. Using for the 'anticommutator' the notation

$$\{A, B\} = AB + BA, \tag{2.30}$$

the conditions are

$$\{\alpha_k, \beta\} = 0. \tag{2.31}$$

Now think about this condition. If β and α_k were numbers, it would be $\alpha_k \beta = \beta \alpha_k$, meaning that at least one of the factors must be zero, but this contradicts our ansatz. The idea of Dirac was that the condition can be satisfied if the coefficients are matrices.

Consider now the terms without the m factor, namely $\alpha_k p_k \alpha_j p_j = p_i p_i$, which we can write, with a useful trick of renaming some indices, as

$$\frac{1}{2}\left(\alpha_k p_k \alpha_j p_j + \alpha_j p_k \alpha_k p_j\right) = p_i p_i$$

or

$$\frac{1}{2}\left(\alpha_k \alpha_j + \alpha_j \alpha_k\right) p_k p_j = p_i p_i$$

which finally gives

$$\{\alpha_k, \alpha_j\} = 2\delta_{kj}. \tag{2.32}$$

The α_k and β matrices can be constructed using the Pauli matrices

$$\sigma_1 = \begin{pmatrix} 0 & 1 \\ 1 & 0 \end{pmatrix}, \sigma_2 = \begin{pmatrix} 0 & -i \\ i & 0 \end{pmatrix}, \sigma_3 = \begin{pmatrix} 1 & 0 \\ 0 & -1 \end{pmatrix}, \tag{2.33}$$

which anti-commute as required. However, it is not possible to build four independent rank 2 matrices with three rank 2 ones. This is possible if the rank is 4. And Dirac chose:

$$\alpha_i = \begin{pmatrix} 0 & \sigma_i \\ \sigma_i & 0 \end{pmatrix}, \beta = \begin{pmatrix} 1 & 0 \\ 0 & -1 \end{pmatrix}. \tag{2.34}$$

With the substitution (2.26) in our ansatz (2.28), we obtain

$$i\frac{\partial}{\partial t}\psi = \left(-i\alpha_k \frac{\partial}{\partial x_k} + \beta m\right)\psi. \tag{2.35}$$

We define the 'Dirac gamma matrices' as

$$\gamma^0 = \beta = \begin{pmatrix} 1 & 0 \\ 0 & -1 \end{pmatrix}, \quad \gamma^i = \beta \alpha_i = \begin{pmatrix} 0 & \sigma_i \\ -\sigma_i & 0 \end{pmatrix}. \tag{2.36}$$

Multiplying (2.35) by β from the left, we finally obtain the Dirac equation

$$\left(i\gamma^{\mu}\partial_{\mu} - m\right)\psi = 0, \tag{2.37}$$

where now the understood sum on Greek repeated indices is on the space-time coordinates $\left(x_0 = ct, x_1, x_2, x_3\right)$. It can be proved, as Dirac did, that the equation is Lorentz covariant, namely it respects the relativity principle.

We observe that the choice (2.36) of the gamma matrices is not unique, being called the 'Dirac representation' of them. Any choice satisfying the anticommutation condition

$$\left\{\gamma^{\mu}, \gamma^{\nu}\right\} = 2g^{\mu\nu}, \tag{2.38}$$

where $g^{\mu\nu}$ is the metric tensor, can be used.

Notice that the Pauli matrices are the mathematical representation of the spin; consequently, their appearance in the relativistic equation of the electron means that the spin is an intrinsic property of the particles, fundamentally required by Lorentz invariance.

The Dirac equation predicts the electron gyromagnetic ratio to be

$$g = 2. \tag{2.39}$$

We recall that this dimensionless quantity is defined by the relation between the spin s and the intrinsic magnetic moment μ_e from

$$\mu_e = g\mu_B s, \tag{2.40}$$

where μ_B is the Bohr magneton

$$\mu_B = \frac{q_e \hbar}{2m_e} = 5.788 \times 10^{-11} \text{ MeV T}^{-1}. \tag{2.41}$$

We shall see in Chapter 5 how small corrections of the Dirac gyromagnetic ratio have been observed, and are correctly explained by quantum electrodynamics.

Being the Hamiltonian a 4×4 matrix, the wave function ψ must be a 'matrix' with 4 rows and 1 column, and, to respect Lorentz covariance, it must transform as a 'bispinor'. It is a complex, four-component entity that we can write as

$$\psi(x) = \begin{pmatrix} \psi_1 \\ \psi_2 \\ \psi_3 \\ \psi_4 \end{pmatrix} = \begin{pmatrix} \varphi \\ \chi \end{pmatrix}; \quad \varphi = \begin{pmatrix} \varphi_1 \\ \varphi_2 \end{pmatrix} \quad \chi = \begin{pmatrix} \chi_1 \\ \chi_2 \end{pmatrix}. \tag{2.42}$$

The two spinors φ and χ represent the particle and the antiparticle; the two components of each of them represent the two states of the third component of the spin $s_z = +1/2$ and $s_z = -1/2$. In 1933, Anderson (1933) discovered the positron, as we shall see in the next section.

The 'adjoint bispinor' is a 'matrix' with one row and four columns, defined as

$$\bar{\psi} = \psi^+ \gamma^0 = \left(\psi_1^*, \psi_2^*, -\psi_3^*, -\psi_4^*\right), \tag{2.43}$$

which obeys the 'adjoint Dirac equation'

$$\left(i\gamma^{\mu}\partial_{\mu}+m\right)\bar{\psi}=0. \tag{2.44}$$

Now let us consider the solutions corresponding to free particles with mass m and definite four-momentum p_{μ}, namely the plane wave $\psi\left(x\right)=ue^{-ip^{\mu}x_{\mu}}$, where u is

$$u=\begin{pmatrix} u_1 \\ u_2 \\ u_3 \\ u_4 \end{pmatrix}. \tag{2.45}$$

Equation (2.37) gives

$$\left(\gamma_{\mu}p^{\mu}-m\right)u=0. \tag{2.46a}$$

And for the adjoint bispinor

$$\bar{u}=u^{+}\gamma^{0}=\begin{pmatrix} u_1^* & u_2^* & -u_3^* & -u_4^* \end{pmatrix}, \tag{2.47}$$

the equation is

$$\bar{u}\left(\gamma_{\mu}p^{\mu}+m\right)=0. \tag{2.46b}$$

A fifth important matrix is $\gamma^5=i\gamma^0\gamma^1\gamma^2\gamma^3$, which in the Dirac representation is

$$\gamma_5=\begin{pmatrix} 0 & 1 \\ 1 & 0 \end{pmatrix}. \tag{2.48}$$

This matrix has the following properties, which are immediately verified:

$$\left(\gamma_5\right)^2=1 \qquad \gamma_5^{\dagger}=\gamma_5 \tag{2.49}$$

In total, five combinations of a spinor, its adjoint and gamma matrices producing covariant quantities are possible. They are called 'bilinear covariants' and have the following transformation properties:

$$\begin{aligned} &\bar{\psi}\psi && \text{scalar}(S) \\ &\bar{\psi}\gamma_5\psi && \text{pseudoscalar}(PS) \\ &\bar{\psi}\gamma_{\mu}\psi && \text{vector}(V) \\ &\bar{\psi}\gamma_5\gamma_{\mu}\psi && \text{axial vector}(A) \\ &\bar{\psi}\left(\gamma_{\alpha}\gamma_{\beta}-\gamma_{\beta}\gamma_{\alpha}\right)\psi && \text{tensor}(T). \end{aligned} \tag{2.50}$$

These quantities are important because, in principle, each of them may appear in an interaction Lagrangian. Nature has chosen, however, to use only two of them, the vector and the axial vector, as we shall see.

The Dirac equation (2.37) its adjoint (2.44) can be derived from the 'Dirac Lagrangian density', which is a Lorentz-invariant quantity, given by

$$\mathcal{L}=i\bar{\psi}\gamma^{\mu}\partial_{\mu}\psi-m\bar{\psi}\psi \tag{2.51}$$

The Dirac equation, like the equation of motion in mechanics, can be derived from the Lagrangian with the relativistic Euler–Lagrange equation, having as generalized coordinates and their four-gradients both $\left(\psi,\partial_{\mu}\psi\right)$ and $\left(\bar{\psi},\partial_{\mu}\bar{\psi}\right)$, namely:

$$\frac{\partial \mathcal{L}}{\partial \psi} - \partial_\mu \frac{\partial \mathcal{L}}{\partial\left(\partial_\mu \psi\right)} = 0 \quad \text{and} \quad \frac{\partial \mathcal{L}}{\partial \bar{\psi}} - \partial_\mu \frac{\partial \mathcal{L}}{\partial\left(\partial_\mu \bar{\psi}\right)} = 0 \qquad (2.52)$$

Starting with the latter, we have $\dfrac{\partial \mathcal{L}}{\partial \bar{\psi}} = i\gamma^\mu \partial_\mu \psi - m\psi$ and $\dfrac{\partial \mathcal{L}}{\partial\left(\partial_\mu \bar{\psi}\right)} = 0$, hence the Dirac

equation $\left(i\gamma^\mu \partial_\mu - m\right)\psi = 0$. For the former we have $\dfrac{\partial \mathcal{L}}{\partial \psi} = -m\bar{\psi}$ and $\dfrac{\partial \mathcal{L}}{\partial\left(\partial_\mu \psi\right)} = i\bar{\psi}\gamma^\mu$.

Deriving the latter, we have $\partial_\mu \dfrac{\partial \mathcal{L}}{\partial\left(\partial_\mu \psi\right)} = i\gamma^\mu \partial_\mu \bar{\psi}$, and finally $\left(i\gamma^\mu \partial_\mu + m\right)\bar{\psi} = 0$, that is

the adjoint Dirac equation.

2.6 The Positron

In 1930, C. D. Anderson built a large cloud chamber, $17 \times 17 \times 3$ cm^3, immersed in a uniform field, generated, up to about 2 T, by an electromagnet that was purposely designed. He exposed the chamber to cosmic rays. The chamber did not have a trigger (namely a logic to command the expansion when a particle had crossed the chamber) and, consequently, only a small fraction of the pictures contained interesting events. Nevertheless, he observed tracks both negative and positive that turned out to be at the ionization minimum from the number of droplets per unit length. Clearly, the negative tracks were electrons, but could the positive tracks be protons, namely the only known positive particles?

By measuring the curvatures of the tracks, in the known magnetic field, Anderson determined their momenta and, assuming they were protons, their energy. With this assumption, several tracks had a rather low kinetic energy, sometimes less than 500 MeV. If this were the case, the ionization had to be much larger than the minimum. Those tracks could not be due to protons.

Cosmic rays come from above, but the particles that appeared to be positive, if moving downwards, could have been negative going upwards, perhaps originating from an interaction in the material under the chamber. This was a rather extreme hypothesis because of the relatively small number of such tracks. Still the issue had to be settled by determining the direction of motion without ambiguity. To accomplish this purpose, a plate of lead, 6 mm thick, was inserted horizontally across the chamber. The direction of motion of the particles could then be ascertained due to the lower energy, and consequently larger curvature, after they had traversed the plate and suffered energy loss.

Fig. 2.9 is a historical photo taken 2 August 1932. It shows a single minimum ionizing track with a direction that is clearly upward (!). Knowing the direction of the field, Anderson concluded that the track was positive. Measuring the curvatures at the two sides of the plate, he obtained the momenta $p_1 = 63$ MeV and $p_2 = 23$ MeV (with $B = 1.5$ T). The expected energy loss could easily be calculated from the corresponding energy before the plate. Assuming the proton mass, the kinetic energy after

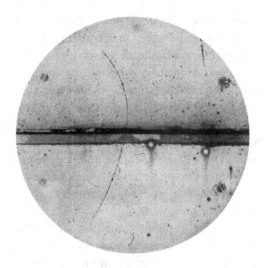

Fig. 2.9. The first positron track. Image in the public domain https://commons.wikimedia.org/wiki/File:PositronDiscovery .jpg?uselang=it#file.

the plate would be $E_{k2} = 280$ keV. This corresponds to a range in the gas of the chamber of 5 mm, to be compared to the observed one of 50 mm. The difference is too large to be due to a fluctuation. On the contrary, assuming the electron mass, the expected range was compatible with 50 mm.

From the measurement of several events of the same type, Anderson (1933) concluded that the mass of the positive particles was equal to the electron mass to within 20% and published the discovery of the positron in March 1933.

At the same time, Blackett and Occhialini were also working with a Wilson chamber in a magnetic field. The device had the added advantage of being triggered by the coincidence of Geiger counters at the passage of a cosmic ray (the coincidence circuit is due to Rossi (1930)) and of being equipped with two cameras to allow the spatial reconstruction of the tracks. They observed several pairs of tracks of opposite signs at the ionization minimum originating from the same point. Measuring the curvature and the droplet density, they measured the masses, which were equal to that of the electron. In conclusion, Blackett and Occhialini not only confirmed, in the spring of 1933, the discovery of the positron, but also discovered the production of e^+e^- pairs (Blackett & Occhialini 1933).

2.7 The Antiproton

A quarter of a century after the discovery of the positron, a fundamental question was still unanswered: does the antiparticle of the proton exist? From a theoretical point of view, the Dirac equation did not give a unique answer, because, in retrospect, the proton, unlike the electron, is not a simple particle; its magnetic moment, in particular, is not as foreseen by the Dirac equation. In addition, the partner of the proton, the neutron, has a magnetic moment even if neutral.

Searches for antiprotons in cosmic rays had been performed but had failed to provide conclusive evidence. We now know that they exist, but are very rare. It became clear that the really necessary instrument was an accelerator with sufficient energy to produce antiprotons. Such a proton synchrotron was designed and built at Berkeley under the leadership of E. Lawrence and E. McMillan, with a maximum proton energy of 7 GeV. In American terminology, the GeV was then called the BeV (from billion, meaning one thousand million) and the accelerator was called Bevatron. After it became operational in 1954, experiments at the Bevatron took the lead in subnuclear physics for several years.

As we shall see in the next chapter, the baryon number, defined as the difference between the number of nucleons and the number of antinucleons, is conserved in all interactions. Therefore, a reaction must produce a proton–antiproton pair and cannot produce an antiproton alone. The simplest one is

$$p + p \rightarrow p + p + \overline{p} + p.$$

The threshold energy (see Problem 1.9) is

$$E_p(\text{thr}) = 7m_p = 6.6\,\text{GeV}.$$

The next instrument was the detector that was built in 1955 by O. Chamberlain, E. Segrè, C. Wiegand and T. Ypsilantis (Chamberlain *et al.* 1955). The 7.2 GeV proton beam extracted from the Bevatron collided with an external target, producing a number of secondary particles. The main difficulty of the experiment was to detect the very few antiprotons that may be present amongst these secondaries. From calculations, only one antiproton every 100 000 pions was expected.

To distinguish protons from pions, one can take advantage of the large difference between their masses. As usual, this requires that two quantities be measured or defined. The choice was to build a spectrometer to define the momentum p accurately and to measure the speed υ. Then the mass is given by

$$m = \frac{p}{\upsilon}\sqrt{1 - \frac{v^2}{c^2}}. \tag{2.53}$$

We shall exploit the analogy between a spectrometer for particles and a spectrometer for light.

The spectrometer had two stages. Fig. 2.10 is a sketch of the first stage. The particles produced in the target, both positive and negative, have a broad momentum spectrum. The first stage is designed to select negative particles with a momentum defined within a narrow band. The trajectory of one of these particles is drawn in Fig. 2.10. The magnet is a dipole, which deflects the particles of an angle that, for the given magnetic field, is inversely proportional to the particle momentum (see Eq. (1.121)). Just as a prism disperses white light into its colours, the dipole disperses a non-monoenergetic beam into its components. An open slit in a thick absorber transmits only the particles with a certain momentum, within a narrow

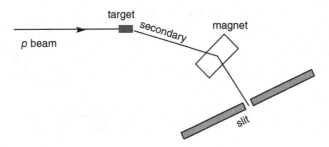

Fig. 2.10.   Sketch of the first stage of a spectrometer, without focussing.

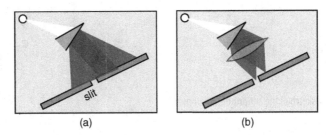

Fig. 2.11.   Principle of a focussing spectrometer.

range. Fig. 2.11(a) shows the analogy with light. The sign of the accepted particle is decided by the polarity of the magnet.

However, as pointed out by O. Piccioni, this scheme does not work; every spectrometer, for particles as for light, must contain focussing elements. The reason becomes clear if we compare Figs. 2.11(a) and 2.11(b). If we use only a prism we do select a colour, but we transmit an extremely low intensity. As is well known in optics, to have appreciable intensity we must use a lens to produce an image of the source in the slit.

Fig. 2.12 is a sketch of the final configuration, including the second stage, which we shall now discuss. Summarizing, the first stage produces a secondary source of negative particles of well-defined momentum. The chosen central value of the momentum is $p = 1.19$ GeV. The corresponding speeds of the pions and antiprotons are

$$\beta_\pi = \frac{p}{E_\pi} = \frac{1.19}{\sqrt{1.19^2 + 0.14^2}} = 0.99$$

$$\beta_p = \frac{p}{E_p} = \frac{1.19}{\sqrt{1.19^2 + 0.938^2}} = 0.78.$$

The time taken by a particle to travel 12 m between two scintillator counters S_1 and S_2, called the time of flight, is measured by the coincidence of the signals from the two counters. The times of flight expected from the above evaluated speeds are $t_{o\pi} = 40$ ns and $t_{op} = 51$ ns, for pions and antiprotons, respectively. The difference $\Delta t = 11$ ns is easily measurable. The resolution was ±1 ns.

A possible source of error comes from random coincidences. Sometimes two pulses separated by 11 ns might result from the passage of a pion in S_1, and a different one

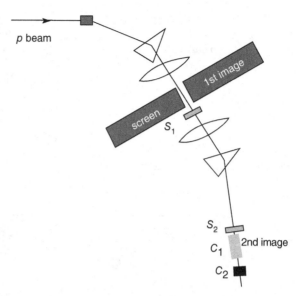

Fig. 2.12.   A sketch of the antiproton experiment.

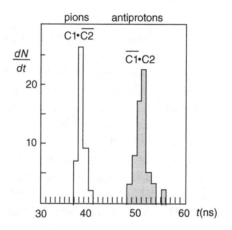

Fig. 2.13.   Time of flight distribution between S_1 and S_2; data from the Chamberlain *et al.* experiment.

in S_2. Two Cherenkov counters were used to overcome the problem. C_1 is used in the threshold mode, with the threshold set at $\beta = 0.99$, to see the pions but not the anti-protons. C_2 has a lower threshold and both particles produce light, but at different angles. A spherical mirror focusses the antiproton light onto the photomultiplier and not that of the pions; in such a way, C_2 sees only the antiprotons. In conclusion, the pions are identified by the coincidence $C_1 \bar{C}_2$, the antiprotons by $C_2 \bar{C}_1$.

Fig. 2.13 shows the time of flight distribution for the two categories. The presence of antiprotons (about 50) is clearly proved.

We now know that an antiparticle exists for every particle, both for fermions and for bosons, even if they are not always different from the particle itself.

2.8 Helicity and Chirality

We discuss here two important properties of the Dirac bispinor. It describes a fermion and its antiparticle and can be written in terms of two corresponding two-component spinors ϕ and χ as in Eq. (2.42)

There are three possibilities in order to give a physical meaning to the two components of φ and χ, depending on the quantity we wish to define.

Polarization

If we deal with polarization, the two states have a defined third component of the spin on a certain axis: they are eigenstates of the Pauli matrix σ_z. The axis must be physically defined, typically by the direction of a magnetic or electric field. Taking, for example, φ, the two eigenstates are $\varphi^+ = \begin{pmatrix} 1 \\ 0 \end{pmatrix}$ and $\varphi^- = \begin{pmatrix} 0 \\ 1 \end{pmatrix}$:

$$\frac{1}{2}\sigma_z \begin{pmatrix} 1 \\ 0 \end{pmatrix} = \frac{1}{2}\begin{pmatrix} 1 & 0 \\ 0 & -1 \end{pmatrix}\begin{pmatrix} 1 \\ 0 \end{pmatrix} = +\frac{1}{2}\begin{pmatrix} 1 \\ 0 \end{pmatrix}$$
$$\frac{1}{2}\sigma_z \begin{pmatrix} 0 \\ 1 \end{pmatrix} = \frac{1}{2}\begin{pmatrix} 1 & 0 \\ 0 & -1 \end{pmatrix}\begin{pmatrix} 0 \\ 1 \end{pmatrix} = -\frac{1}{2}\begin{pmatrix} 0 \\ 1 \end{pmatrix}$$
(2.54)

Helicity

Even in the absence of an external field, the velocity of the particle defines a direction in any reference frame different from the rest frame. The states of definite helicity are the eigenstates of the third component of the spin in that direction. If $\mathbf{p}$ is the momentum of the particle, the helicity operator is

$$h = \frac{1}{2}\frac{\mathbf{p}\cdot\boldsymbol{\sigma}}{p}$$
(2.55)

The two helicity eigenvalues are $+1/2$ if the spin is in the direction of the motion, and $-1/2$ if in the opposite direction. The helicity eigenstates are two-component spinors, describing a particle, a fermion or an antifermion.

We call the attention of the reader to the importance of the following properties of the helicity operator.

Let us first consider its behaviour under the Lorentz group. Being proportional to a scalar product, helicity is invariant under rotations. However, being the product between a vector $(\mathbf{p})$ and an axial vector $(\boldsymbol{\sigma})$, it changes sign $(h \rightarrow -h)$ under inversion of the space axes, namely under the parity operation. Helicity is not invariant also under Lorentz transformations between references with different velocities (called boosts). Consider, for example, a particle with negative helicity in a certain reference frame. If the particle is massive, namely if its speed is less than c, we can always find another frame in which the particle travels in the opposite direction. In this frame, the helicity is positive. Only if a fermion were rigorously massless would its velocity be frame-independent and its helicity Lorentz-invariant.

Let us now see when helicity is conserved and when it is not. It can be shown that helicity commutes with the Dirac Hamiltonian, namely for non-interacting fermions. Consequently, the helicity of non-interacting fermions does not change with time.

Chirality

Two different projections of the solutions of the Dirac equation are important in particle physics. They are called 'chiral' states (from χειρ, meaning 'hand' in Greek) and are the eigenstates of γ_5, with two possible eigenvalues, +1 and −1; these are called right (R) and left (L) respectively. If ψ is a solution of the Dirac equation, the projectors of the positive (R) and negative (L) chirality are $R = \frac{1}{2}(1+\gamma_5)$ and $L = \frac{1}{2}(1-\gamma_5)$ and the corresponding states

$$\psi_L = L\psi = \frac{1}{2}(1-\gamma_5)\psi; \quad \psi_R = R\psi = \frac{1}{2}(1+\gamma_5)\psi \tag{2.56}$$

The conjugated states are

$$\bar{\psi}_L = \bar{\psi}\frac{1}{2}(1+\gamma_5) \qquad \bar{\psi}_R = \bar{\psi}\frac{1}{2}(1-\gamma_5). \tag{2.57}$$

Question 2.1 Verify that these operators are projectors, namely that applying one of them twice gives the same result as applying it once and that applying both of them results in the null state. □

Let us verify the above statement that these are chirality eigenstates, namely that

$$\gamma_5\psi_L = -\psi_L; \quad \gamma_5\psi_R = +\psi_R \tag{2.58}$$

For the left one, $\gamma_5\psi_L = \frac{1}{2}(1-\gamma_5)\psi_L = \frac{1}{2}(\gamma_5 - \gamma_5\gamma_5)\psi_L = \frac{1}{2}(\gamma_5 - 1)\psi_L = -\psi_L$, where we took account of Eq. (2.49), and similarly for the right one.

The properties of chirality are somewhat opposite to those of helicity. As we have seen, helicity is conserved for a free particle, but is not Lorentz-invariant, whereas chirality is Lorentz-invariant (as can be shown), but not conserved, even for a free spinor. If a particle is created with, for example, negative chirality, it may not be so at later times. The Dirac Lagrangian can be written as the sum of two similar terms, one with negative and one with positive chirality projections, ψ_L and ψ_R, respectively, namely as $\mathcal{L} = \mathcal{L}(\psi_L) + \mathcal{L}(\psi_R)$. The mass terms explicitly break the chiral symmetry $m_f\bar{\psi}_f\psi_f = m_f\bar{\psi}_{fL}\psi_{fR} + m_f\bar{\psi}_{fR}\psi_{fL}$. In ultra-relativistic conditions, namely at energies much larger than the mass, $E \gg m$, or, even better, for massless particles, the numbers of left and right particles are separately conserved. This remains valid for all the interaction Lagrangians of the SM we shall consider in the next chapters.

We shall now study the helicity content of the chiral projections. We start by observing that the two two-component spinors ϕ and χ of a solution ψ of the Dirac equation are completely correlated. Actually

$$\left(\gamma_\mu p^\mu - m\right)\psi = \left(E\gamma_0 - \mathbf{p}\cdot\gamma - m\right)\psi = 0, \tag{2.59a}$$

or explicitly

$$\begin{pmatrix} E-m & -\mathbf{p}\cdot\sigma \\ \mathbf{p}\cdot\sigma & -(E+m) \end{pmatrix}\begin{pmatrix} \varphi \\ \chi \end{pmatrix} = \begin{pmatrix} 0 \\ 0 \end{pmatrix}, \tag{2.59b}$$

which gives the two relationships

$$\varphi = \frac{\mathbf{p}\cdot\sigma}{E-m}\chi \qquad \chi = \frac{\mathbf{p}\cdot\sigma}{E+m}\varphi. \tag{2.60}$$

Consider now the chiral states, for example the left one

$$\psi_L = \frac{1}{2}(1-\gamma_5)\ \psi = \frac{1}{2}\begin{pmatrix} 1 & -1 \\ -1 & 1 \end{pmatrix}\begin{pmatrix} \varphi \\ \chi \end{pmatrix} = \frac{1}{2}\begin{pmatrix} \varphi-\chi \\ \chi-\varphi \end{pmatrix}. \tag{2.61}$$

We see that it contains the combination $\varphi-\chi$ and not $\varphi+\chi$. Consider the upper component (i. e. the fermion, as opposed to the antifermion). Let us take the z-axis in the direction of motion and write (2.61) in terms of the helicity eigenstates φ^+ and φ^-. We have

$$\frac{1}{2}(\varphi-\chi) = \frac{1}{2}\left(1-\frac{\mathbf{p}\cdot\sigma}{E+m}\right)\varphi = \frac{1}{2}\left(1-\frac{p_z}{E+m}\right)\varphi^+ + \frac{1}{2}\left(1+\frac{p_z}{E+m}\right)\varphi^-. \tag{2.62}$$

We see that the upper component of the left bispinor is not an eigenstate of helicity. Consider two extreme conditions. If the momentum is very small, the two helicity components have equal probability. On the contrary, for ultra-relativistic energies, namely for $E \gg m$, as is often the case, we can write $p_z = E$ and neglect m in the sum $E+m$, obtaining

$$\frac{1}{2}(\varphi-\chi) \approx \frac{1}{2}\frac{m}{E}\varphi^+ + \varphi^- \tag{2.63}$$

This expression shows that, for a massless particle, the upper component of the left bispinor is an eigenstate of the helicity too, with the negative eigenvalue. If the particle is massive, the 'wrong' helicity component, the positive one φ^+, is small when $E \gg m$.

Let us now consider the antifermion in ψ_L. We have

$$\frac{1}{2}(\chi-\varphi) = \frac{1}{2}\left(1-\frac{\mathbf{p}\cdot\sigma}{E-m}\right)\chi = \frac{1}{2}\left(1-\frac{p_z}{E-m}\right)\chi^+ + \frac{1}{2}\left(1+\frac{p_z}{E-m}\right)\chi^-. \tag{2.64}$$

If the momentum is very small, the two helicity components of the negative chirality antifermion are equal, as in the case of the fermion. At high energies however, the situation is different. The antiparticle can be formally considered as a negative energy solution of the Dirac equation. We can write $E = -p_z$, and we have

$$\frac{1}{2}(\chi - \varphi) \approx \chi^+ + \frac{1}{2}\frac{m}{E}\chi^- \tag{2.65}$$

We see that the massless antiparticle of a negative chirality is the helicity eigenstate with positive eigenvalue. Again, if the particle is massive, the 'wrong' helicity component is very small when $E \gg m$.

Unfortunately, one often encounters in the literature the use of the 'left' and 'right' terms also for negative and positive helicity. We shall avoid this confusing language: with left and right we shall always mean negative and positive chirality, respectively.

In summary, the helicity content of a particle of definite chirality depends on the energy of the particle. At ultra-relativistic energies, spin 1/2 particles of negative chirality have a dominant negative helicity component, and fermion antiparticles of negative chirality have a dominant positive helicity component. The opposite is true for positive chirality particles and, respectively, antiparticles. The following observation is in order. The reader might have had the impression of an asymmetry between matter and antimatter from the above discussion. However, this is not true; the definition of which is the particle and which is the antiparticle component of the Dirac bispinor is completely arbitrary.

Chirality cannot be directly experimentally determined; rather, the helicity of the particle or of the antiparticle is measured. To be precise, we measure the expectation value of the helicity. Considering, for example, the particle spinor of the left bispinor, let Π_+ and Π_- be the probabilities of finding it in each of the two helicity states (spin in the direction of motion or opposite to it). These probabilities are the squares of the two amplitudes in (2.62). The helicity expectation value is then

$$h = \frac{\Pi_+ - \Pi_-}{\Pi_+ + \Pi_-} = \frac{(E + m - p)^2 - (E + m + p)^2}{(E + m - p)^2 + (E + m + p)^2} = \frac{-p}{E} = -\beta. \tag{2.66}$$

The expectation value of the helicity, or simply the helicity, of a fermion of negative chirality is the opposite of the ratio between its speed and the speed of light. A similar calculation shows that the helicity of a positive chirality fermion is $+\beta$.

Question 2.2 Demonstrate the last statement, after having found the equation analogous to (2.62) for a right fermion. □

2.9 The Majorana Fermion

In the Standard Model, the wave functions, and the corresponding fields, of all the spin 1/2 elementary particles are described by the four degrees of freedom solution of the Dirac equation we have discussed. However, the extension of the Dirac theory to neutrinos is not supported by experimental proofs. An alternative solution, which neutrinos might obey, was advanced in 1937 by E. Majorana (Majorana 1937).

Majorana enquired whether the existence of a spin 1/2 particle that is its own anti-particle would be possible. This is indeed the case of some integer spin particles, such as the photon, which is the same as the 'antiphoton', the neutral pion, etc. To be so, the particle must be 'completely neutral', namely all its possible 'charges', to be discussed in the next chapter, (electromagnetic, colour, weak, lepton and baryon number,…) should be zero.

To be its own antiparticle, the wave function, and the corresponding field, should have two components, rather than four, corresponding to the two helicity states, namely it should be real, rather than complex. This condition can be obtained if four gamma matrices exist, which satisfy Eq. (2.38) and are purely imaginary (purely real would work as well). They do, and Majorana found them. Expressed as tensor products of the Pauli matrices, they are

$$\tilde{\gamma}^0 = \sigma_2 \otimes \sigma_1, \quad \tilde{\gamma}^1 = i\sigma_1 \otimes 1, \quad \tilde{\gamma}^2 = i\sigma_3 \otimes 1, \quad \tilde{\gamma}^3 = i\sigma_2 \otimes \sigma_2 \qquad (2.67)$$

or explicitly

$$\tilde{\gamma}^0 = \begin{pmatrix} 0 & 0 & 0 & -i \\ 0 & 0 & -i & 0 \\ 0 & i & 0 & 0 \\ i & 0 & 0 & 0 \end{pmatrix}$$

$$\tilde{\gamma}^1 = \begin{pmatrix} 0 & 0 & i & 0 \\ 0 & 0 & 0 & i \\ i & 0 & 0 & 0 \\ o & i & 0 & 0 \end{pmatrix}$$

$$\tilde{\gamma}^2 = \begin{pmatrix} i & 0 & 0 & 0 \\ 0 & i & 0 & 0 \\ 0 & 0 & -i & 0 \\ 0 & 0 & 0 & -i \end{pmatrix} \qquad (2.68)$$

$$\tilde{\gamma}^3 = \begin{pmatrix} 0 & 0 & 0 & -i \\ 0 & 0 & i & 0 \\ 0 & i & 0 & 0 \\ -i & 0 & 0 & 0 \end{pmatrix}$$

The 'Majorana equation' is then the Dirac equation with these gamma matrices

$$\left(i\tilde{\gamma}^\mu \partial_\mu - m\right)\tilde{\psi} = 0 \qquad (2.69)$$

An important characteristic of the Majorana fermions is that they cannot be massless. Indeed a massless fermion must be an eigenstate of γ_5, positive or negative, namely it must satisfy one or the other of Eq. (2.58), which is impossible because $\gamma^5 = i\tilde{\gamma}^0 \tilde{\gamma}^1 \tilde{\gamma}^2 \tilde{\gamma}^3$ is purely imaginary.

Problems

2.1   Compute the energies and momenta in the CM system of the decay products of $\pi \to \mu + \nu$.

2.2   Consider the decay $K \to \mu + \nu$. Find
(a) the energy and momenta of the μ and the ν in the reference of the K at rest;
(b) the maximum μ momentum in a frame in which the K momentum is 5 GeV.

2.3   A π^0 decays emitting one gamma of energy $E_1 = 150$ MeV in the forward direction. What is the direction of the second gamma? What is its energy E_2? What is the speed of the π^0?

2.4   Two μ are produced by a cosmic ray collision at an altitude of 30 km. The two energies are $E_1 = 5$ GeV and $E_2 = 5$ TeV. What are the distances at which each of the muons sees the surface of the Earth in its rest reference frame? What are the distances travelled in the Earth reference frame in a lifetime?

2.5   A π^+ is produced at an altitude of 30 km by a cosmic ray collision with energy $E_\pi = 5$ GeV. What is the distance at which the pion sees the surface of the Earth in its rest reference frame? What is the distance travelled in the Earth reference frame in a lifetime?

2.6   A photon converts into an $e^+ e^-$ pair in a cloud chamber with magnetic field $B = 0.2$ T. In this case, two tracks are observed with the same radius $\rho = 20$ cm. The initial angle between the tracks is zero. Find the energy of the photon.

2.7   Consider the following particles and their lifetimes.
$\rho^0 : 5 \times 10^{-24}$ s, $K^+ : 1.2 \times 10^{-8}$ s, $\eta^0 : 5 \times 10^{-19}$ s, $\mu^- : 2 \times 10^{-6}$ s, $\pi^0 : 8 \times 10^{-17}$ s.
Guess which interaction leads to the following decays:
$\rho^0 \to \pi^+ + \pi^-; K^+ \to \pi^0 + \pi^+; \eta^0 \to \pi^+ + \pi^- + \pi^0; \mu^- \to e^- + \bar{\nu}_e + \nu_\mu$ and $\pi^0 \to \gamma + \gamma$.

2.8   Consider the decay $\pi^0 \to \gamma\gamma$ in the CM. Assume a Cartesian coordinate system x^*, y^*, z^*, and the polar coordinates ρ^*, θ^*, ϕ^*. In this reference frame, the decay is isotropic. Give the expression of the probability per unit solid angle, $P(\cos\theta^*, \phi^*) = dN / d\Omega^*$ of observing a photon in the direction θ^*, ϕ^*. Then consider the L reference frame, in which the π^0 travels in the direction $z = z^*$ with momentum p and write the probability per unit solid angle, $P(\cos\theta, \phi)$ of observing a photon in the direction θ, ϕ.

2.9   Chamberlain and co-workers used scintillators to measure the pion lifetime. Why did they not use Geiger counters?

2.10   Calculate the ratio between the magnetic moments of the electron and the μ and between the electron and the τ.

2.11   We calculated the energy threshold for the reaction $p + p \to p + p + \bar{p} + p$ on free protons as targets in Problem 1.9. Repeat the calculation for protons that are bound in a nucleus and have a Fermi momentum of $p_f = 150$ MeV. For the incident proton use the approximation $p_p \approx E_p$.

2.12   We wish to produce a monochromatic beam with momentum $p = 20$ GeV and a momentum spread $\Delta p / p = 1\%$. The beam is 2 mm wide and we have a magnet with a bending power of $BL = 4$ Tm and a slit $d = 2$ mm wide. Calculate the distance l between magnet and slit.

2.13 A hydrogen bubble chamber was exposed to a π^- beam of 3 GeV momentum. We observe an interaction with secondaries that are all neutral and two V^0s pointing to the primary vertex. Measuring the two tracks of one of them, we find for the positive: $p^- = 121$ MeV, $\theta^- = -18.2°$ and $\phi^- = 15°$, and for the negative: $p^+ = 1900$ MeV, $\theta^+ = 20.2°$ and $\phi^+ = -15°$. We know that θ and ϕ are the polar angles in a reference frame with polar axis z in the beam direction. What is the nature of the particle? Assume that the measurement errors give a $\pm 4\%$ resolution on the reconstructed mass of the V^0.

2.14 Calculate the energy thresholds (E_ν) for the processes
$(1)\, \nu_e + n \rightarrow e^- + p, \; (2)\, \nu_\mu + n \rightarrow \mu^- + p$ and $(3)\, \nu_\tau + n \rightarrow \tau^- + p$.

2.15 A photon of energy $E_\gamma = 511$ keV is scattered backwards by an electron at rest. What is the value of the energy of the scattered photon? What is the value if the target electron were moving against the photon with energy $T_e = 511$ keV?

2.16 In the SLAC linear accelerator, electrons were accelerated up to the energy $E_{el} = 20$ GeV. To produce a high-energy photon beam, a laser beam was backscattered at 180° by the electron beam. Assuming the laser wavelength $\lambda = 694$ nm, what was the scattered photon energy?

2.17 In the event discovered by Anderson, a positive track emerged from the lead plate with measured momentum $p = 23$ MeV. Calculate its kinetic energy assuming it to be: (a) a proton; and (b) a positron.

2.18 In 1933, Blackett and Occhialini discovered several e^+e^- in the electromagnetic showers from cosmic rays in a Wilson chamber. The magnetic field was $B = 0.3$ T. In one event, both the negative and the positive tracks described arcs of radius $R = 14$ cm. Calculate their energies.

2.19 Consider a π^+ beam of momentum $p = 200$ GeV at a proton accelerator facility. We can build a muon beam by letting the pion decay in a vacuum pipe. Calculate the energy range of the muons.

Summary

In this chapter we have introduced, by following the history, the elementary particles that are stable or decay by weak interactions, with lifetimes long enough to produce observable tracks. In particular we have seen

- the pions and the strange particles, mesons and hyperons, discovered in cosmic rays
- the measurements of the quantum numbers of the charged pion
- the charged leptons, e, μ and τ, and the three corresponding neutrinos, ν_e, ν_μ and ν_τ
- the Dirac equation and the bilinear covariants
- the discovery of the positron, the first antiparticle of a fundamental fermion
- the beginning of high-energy accelerator experiments
- the discovery of the antiproton, the first antiparticle of a composite fermion
- the Majorana equation describing a completely neutral spinor.

Further Reading

Anderson, C. D. (1936) Nobel Lecture; *The Production and Properties of Positrons*

Chamberlain, O. (1959) Nobel Lecture; *The Early Antiproton Work*

Lederman, L. (1963) The two-neutrino experiment. *Sci. Am.* 208 (3) 60

Lederman, L. (1988) Nobel Lecture; *Observations in Particle Physics from Two Neutrinos to the Standard Model*

Lemmerich, J. (1998) *The history of the discovery of the electron.* Proceedings of the XVIII international symposium on 'Lepton photon interactions 1997'. World Scientific.

Perkins, D. H. (1998) *The discovery of the pion at Bristol in 1947.* Proceedings of 'Physics in collision 17', 1997. World Scientific

Powell, C. B. (1950) Nobel Lecture; *The Cosmic Radiation*

Reines, F. (1995) Nobel Lecture; *The Neutrino: From Poltergeist to Particle*

Rossi, B. (1952) *High-Energy Particles.* Prentice-Hall

Schwartz, M. (1988) Nobel Lecture; *The First High Energy Neutrino Experiment*

3 Symmetries

'Symmetry' comes from the Greek word συμμετρος, meaning well ordered. Symmetry is present in many fields. Several objects in nature are geometrically symmetrical, for example a butterfly, a flower, a fullerene molecule.

Symmetry is used in physics in several different, abstract ways, in its mathematical description. In modern physics, symmetries are a powerful tool to constrain the form of equations, in particular the Lagrangian that describes the system. These are assumed to be invariant under the transformation of a given group, which may be discrete or a continuous Lie group. This approach is of fundamental importance, in particular, for particle physics.

This chapter begins with a simple classification of the various types of symmetry, introducing the concepts that will be used in later sections. The reader should already be familiar with Lorentz invariance, meaning the invariance, or, better, the covariance, of the equation of motion under the Poincaré group, as well as Noether's theorem, at least in classical physics. The situation is not different in quantum physics.

The concept of spontaneous symmetry breaking is also introduced. It will evolve into the Higgs mechanism, which gives origin to the masses of the vector bosons that mediate the weak interactions of quarks and of charged leptons.

This chapter continues by discussing discrete symmetries, in particular the parity and the particle–antiparticle conjugation operations and corresponding quantum numbers. We discuss the important example of the pions, both charged and neutral. The method to determine the spin-parity of the latter anticipates that used for the Higgs boson.

We shall introduce the baryon number and the lepton number, which are similar to, but are not, charges. We shall also show how the decay of the charged pion gives information on the symmetry of the weak interaction.

We then discuss an important example of dynamical symmetry. The reader should have already met the charge independence of nuclear forces and the corresponding isospin invariance. The symmetry corresponds to the invariance of the Lagrangian under rigid rotations in an 'internal' space, formally the analogue of the rotations of the three space axes. In hadron spectroscopy it is useful to consider, instead of the group of rotations, called $SO(3)$, the unitary group $SU(2)$, the two being equivalent. This is because, as we shall see in Chapter 4, the symmetry will be later enlarged to $SU(3)$ to include the strange particles. The students should already be familiar with the concepts of unitary group and of their simplest representations, from the study of angular momentum. Once more they will be encouraged to solve a number of numerical problems to gain practice in working with selection rules and the composition of $SU(2)$ representations.

3.1 Symmetries

The fundamental space-time symmetry requires the equations of the evolution of a physical system to be invariant, under the Poincaré group (the Lorentz group plus space-time translations). Electromagnetic, weak and strong interactions are described by quantum field theories, each ruled by a Lagrangian density, which we shall call simply Lagrangian for short, when this cannot induce ambiguity. In any case the Lagrangian is invariant under the transformations of the Poincaré group.

This means that the equations have the same form in terms of the space-time coordinates in two reference frames connected by a Lorentz transformation, a rigid rotation or a translation. The Noether theorem establishes that the covariance of the equation of motion under a continuous transformation with n parameters implies the existence of n conserved quantities. We assume the reader knows that total energy–momentum and angular momentum are the conserved quantities, the 'integrals of motion', corresponding to the invariance under the Poincaré group.

Rules that limit the possibilities of an initial state transforming into another state in a quantum process (collision or decay) are called **conservation laws** and are expressed in terms of the **quantum numbers** of those states.

Here, we summarize the types of symmetry that we shall encounter in this book. The concepts will be further developed when needed.

Gauge Symmetries and Conserved Additive 'Charges'

Quantum numbers are called additive if the total value for a system is the sum of the values of its components. The 'charges' of all fundamental interactions fall into this category: the electric charge, the colour charges and the weak charges. The charges are absolutely conserved, as far as we know. The conservation of each of them corresponds to the invariance of the Lagrangian of that interaction under the transformations of a unitary group. The group is called 'gauge group' and the invariance of the Lagrangian is called 'gauge invariance'. The gauge group of the electromagnetic interaction is $U(1)$, that of the strong interaction is $SU(3)$ and that of the electroweak interaction $SU(2) \otimes U(1)$. Other quantum numbers in this category are the quark flavours, the baryon number, the lepton flavours and the lepton numbers. They do not correspond to a gauge symmetry and are not necessarily conserved (actually, quark and lepton flavours are not).

Space-time and gauge symmetries are of fundamental importance in building the basic laws of nature, the Standard Model.

Dynamical Symmetries

In this category, the transformations are continuous and belong to a unitary group. These symmetries allow us to classify a number of particles in 'multiplets', the members of which have similar behaviour. An example of this is the charge independence of nuclear forces. The corresponding symmetry is the invariance under the transformations of the group $SU(2)$ and isotopic spin conservation. In general, dynamical symmetries determine the structure of the energy (mass) spectrum of a quantum system.

Discrete Multiplicative

These transformations cannot be constructed from infinitesimal transformations. The most important are parity $\mathcal{P}$ (i.e. the inversion of the coordinate axes), particle–antiparticle conjugation C and time reversal $\mathcal{T}$. The eigenvalues of $\mathcal{P}$ and C are among the quantum numbers of the particles, while $\mathcal{T}$ is not.

Symmetry Breaking

Several symmetries are 'broken', meaning that they are approximate. This can happen in two ways.

- **Explicit breaking** occurs when some interaction does not respect the symmetry or does it only approximately. In this case, only the interactions that do not break the symmetry conserve the corresponding quantum numbers. Only experiments can decide whether or not a certain quantum number is conserved in a given interaction.
- **Spontaneous breaking** occurs when the interaction does respect the symmetry; mathematically, the Lagrangian of the system is invariant under the corresponding group, but its ground state is not.

At the macroscopic level, spontaneous breaking happens for states composed of many identical elements, such as atoms or molecules. Consider, as an analogy, a shoal of fish. Suppose both the surface and the bottom of the sea are far away and the weight is perfectly balanced by the buoyancy. Under these conditions, all directions are equivalent and the system is symmetric under rotations. However, one fish may suddenly decide to change the direction of its motion, and the entire shoal will follow. The symmetry thus broke spontaneously.

Now consider three examples of macroscopic physical systems.

The first example is mechanical instability. Consider a rectangular perfectly symmetric metal plate. Let us place it vertically with its shorter side on a horizontal plane and let us apply an exactly vertical downwards force in the centre of the other short side. The state is symmetric under the exchange of the left and right faces of the plate. If we now gradually increase the intensity of the force, we observe that, at a definite value (which can be calculated from the mechanical characteristics of the system), the plate bows with curvature to the left or to the right. The original symmetry is lost; it has been spontaneously broken.

As a second example, consider a drop of water floating in a space station. In this 'absence of weight' situation, it is perfectly symmetrical for rotations around any axis through its centre. If the temperature is now lowered below 0 °C, the ice that is formed is a crystal with molecules aligned along certain directions. The rotational symmetry is lost. The breakdown is spontaneous because we cannot foresee the directions of the crystal axes by observing the initial, liquid, state.

Spontaneous magnetization is similar. Consider a piece of iron (or any ferromagnetic material) above its Curie temperature. The atomic magnetic moments are randomly oriented. Consider a microcrystal and its spontaneous magnetization axis; for the interaction responsible for the ferromagnetism, the two directions parallel and antiparallel to the axis are completely equivalent, namely the system is symmetric under their exchange. We now lower the temperature below the Curie point. The Weiss domains take form in the crystal. In each of them the magnetic moments have chosen one of the two directions. Again the symmetry has been spontaneously broken.

In these systems, the state of minimum energy, which is called 'the vacuum', is clearly non-symmetric.

Spontaneous breaking is present also at a fundamental level, as discovered by Nambu and collaborators in 1960 (Nambu 1960; Nambu & Jona-Lasinio 1961). The basic reason is that, in relativistic quantum mechanics, the vacuum state is not at all void, but is a very dynamical and vivid state, as we will see. Goldstone then showed (Goldstone 1961) that the spontaneous breaking of perfect symmetry gives rise to a number (depending on the symmetry) of massless bosons. These are called Goldstone bosons. This also happens if the symmetry is not only spontaneously, but also explicitly, broken. However, in this case the bosons acquire mass, which is larger for larger explicit breakings. They are called pseudo-Goldstone bosons. This is a case of the chiral symmetry that we shall study in Section 6.9.

3.2 Parity

The parity operation $\mathcal{P}$ is the inversion of the three spatial coordinate axes. Note that, in two dimensions, the inversion of the axes is equivalent to a rotation, whereas this is not true in three dimensions. The inversion of three axes is equivalent to the inversion of one, followed by a 180° rotation. An object and its mirror image are connected by a parity operation.

The following scheme will be useful. The $\mathcal{P}$ operation (1) inverts the space coordinates $\mathbf{r} \Rightarrow -\mathbf{r}$, (2) does not change the time $t \Rightarrow t$; consequently, (3) it inverts momenta $\mathbf{p} \Rightarrow -\mathbf{p}$, (4) does not change angular momenta $\mathbf{r} \times \mathbf{p} \Rightarrow \mathbf{r} \times \mathbf{p}$, (5) including spin $\mathbf{s} \Rightarrow \mathbf{s}$. More generally, scalar quantities remain unchanged, pseudoscalar ones change their sign, vectors change sign and axial vectors do not.

We can talk of the parity of a state only if it is an eigenstate of $\mathcal{P}$. A vacuum is such a state and its parity is set positive by definition.

A single particle can be, but is not necessarily, in an eigenstate of $\mathcal{P}$ only if it is at rest. The eigenvalue P of $\mathcal{P}$ in this frame is called **intrinsic parity** (or simply parity). The situation is different for bosons on one side and fermions on the other. If we invert the spatial coordinates twice, we obtain the original ones. Then, the operator $\mathcal{P}^2$ is the identity operator, but may be also a rotation of 2π around an axis. Now, for scalars, vectors and tensors, a rotation of 2π is equivalent to no rotation and hence $P^2 = 1$, or $P = \pm 1$. A fermion is described by a spinor ψ. A rotation of 2π transforms ψ into $-\psi$. Consequently, to return to ψ we need to apply $\mathcal{P}$ four times. Hence $P^4 = 1$, or $P = \sqrt[4]{1} = \pm 1, \pm i$. P can be real or imaginary.

The absolute parity of bosons can be defined without ambiguity. We shall see in Section 3.5 how it is measured for the pion. Fermions have half-integer spins, and angular momentum conservation requires them to be produced in pairs. Therefore, only relative parities can be defined. In quantum field theory it is found that the parity of bosons and of their antiparticle is the same, while for fermions

$$P_f P_{\bar{f}} = -1. \tag{3.1}$$

As a consequence, fermion and antifermion have opposite parity if P is real, and equal parity if P is imaginary. It turns out that the intrinsic parity of a fermion obeying the Dirac equation is completely arbitrary. It is chosen to be real. On the contrary, the square of the parity of a fermion obeying the Majorana equation must be -1, and consequently its parity must be imaginary.

By convention, the proton parity is assumed positive, and therefore the parity of the antiproton is negative. The parities of the other non-strange baryons are given relative to the proton. There is no universal convention for the parity of charged leptons. We define as positive the parity of the leptons (as opposed to the antileptons). The parity of neutrinos cannot be defined, because they have only weak interactions that, as we shall see, do not conserve parity.

Strange hyperons are produced in pairs together with another strange particle. This prevents the measurement of both parities. One might expect to be able to choose one hyperon (e.g. the Λ that is the lightest one) and to refer its parity to that of the proton using a decay, for example $\Lambda \to p\pi^-$. This does not work because the decays are weak processes, and weak interactions, as just said, violate parity conservation. By convention, we then take $P(\Lambda) = +1$.

Strange hyperons differ from non-strange ones because of the presence of a strange quark. More hadrons were discovered containing other quark types. The general rule at the quark level is that, by definition, all quarks have positive parity, and antiquarks have negative parity.

The Parity of the Photon

The photon is the quantum equivalent of the classical vector potential $\mathbf{A}$. Therefore, its spin and parity, with a notation that we shall always employ, are $J^P = 1^-$.

The Parity of a Two-Particle System

A system of two particles of intrinsic parities, say, P_1 and P_2, can be a parity eigenstate only in the CM system. In this frame, let us call $\mathbf{p}$ the momentum and θ, ϕ the angles for one particle, and $-\mathbf{p}$ the momentum of the other. We shall write these states as $|p,\theta,\phi\rangle$ or also as $|\mathbf{p},-\mathbf{p}\rangle$. Call $|p,l,m\rangle$ the state with orbital angular momentum l and third component m. The relationship between the two bases is

$$
\begin{aligned}
|p,l,m\rangle &= \sum_{\theta,\phi} |p,\theta,\phi\rangle\langle p,\theta,\phi|p,l,m\rangle \\
&= \sum_{\theta,\phi} Y_l^{*m}(\theta,\phi)|\mathbf{p},-\mathbf{p}\rangle.
\end{aligned}
\tag{3.2}
$$

The inversion of the axes in polar coordinates is $r \Rightarrow r$, $\theta \Rightarrow \pi - \theta$ and $\phi \Rightarrow \pi + \phi$. Spherical harmonics transform as

$$
Y_l^{*m}(\theta,\phi) \Rightarrow Y_l^{*m}(\pi-\theta,\pi+\phi) = (-1)^l Y_l^{*m}(\theta,\phi).
$$

Consequently,

$$
\begin{aligned}
\mathcal{P}|p,l,m\rangle &= P_1 P_2 \sum_{\theta,\phi} Y_l^{*m}(\pi-\theta,\phi+\pi)|-\mathbf{p},\mathbf{p}\rangle \\
&= P_1 P_2 (-1)^l \sum_{\theta,\phi} Y_l^{*m}(\theta,\phi)|\mathbf{p},-\mathbf{p}\rangle \\
&= P_1 P_2 (-1)^l |p,l,m\rangle.
\end{aligned}
\tag{3.3}
$$

In conclusion, the parity of the system of two particles with orbital angular momentum l is

$$
P = P_1 P_2 (-1)^l.
\tag{3.4}
$$

Let us see some important cases.

Two Mesons with the Same Intrinsic Parity (For Example, Two π)

Calling them m_1 and m_2, Eq. (3.4) simply gives

$$
P(m_1,m_2) = (-1)^l.
\tag{3.5}
$$

For particles without spin, such as pions, the orbital angular momentum is equal to the total momentum, $J = l$.

The possible values of parity and angular momentum are: $J^P = 0^+$, 1^-, 2^+, ..., provided the two pions are different.

If the two pions are equal, their status must be symmetrical, as demanded by Bose statistics. Therefore, l, and hence J, must be even. The possible values are $J^P = 0^+$, 2^+,....

Fermion–Antifermion Pair (For Example, Proton–Antiproton)

The two intrinsic parities are opposite in this case. Therefore, if l is again the orbital angular momentum, we have

$$
P(f\bar{f}) = (-1)^{l+1}.
\tag{3.6}
$$

Example 3.1   Find the possible values of J^P for a spin 1/2 particle and its antiparticle if they are in an S wave state, or in a P wave state. (Such is the positronium, which is an e^+e^- atomic system.)

The total spin can be 0 (singlet) or 1 (triplet). In an S wave, the orbital momentum is $l = 0$ and the total angular momentum can be $J = 0$ (in spectroscopic notation $^1S_0$) or $J = 1$ $\left(^3S_1\right)$. Parity is negative in both cases. In conclusion, $^1S_0$ has $J^P = 0^-$, $^3S_1$ has $J^P = 1^-$. The P wave has $l = 1$ hence positive parity. The possible states are $^1P_1$ with $J^P = 1^+$, $^3P_0(J^P = 0^+)$, $^3P_1(J^P = 1^+)$ and $^3P_2(J^P = 2^+)$. $\square$

Parity conservation is not a universal law of physics. Strong and electromagnetic interactions conserve parity, but weak interactions do not. We shall study parity violation in Chapter 7. The most sensitive tests for parity conservation in strong interactions are based on the search for reactions that can proceed only through parity violation.

Experimentally, we can detect parity violation effects if the matrix element is the sum of a scalar and a pseudoscalar term. Actually, if only one of them is present, the transition probability that is proportional to its square modulus is in any case a scalar, meaning that it is invariant under the parity operation. However, if both terms are present, the transition probability is the sum of the two square moduli, which are invariant under parity, and of their double-product, which changes sign. Let us then assume a matrix element of the type

$$M = M_S + M_{PS}. \tag{3.7}$$

A process that violates parity is the decay of an axial vector state into two scalars $1^+ \rightarrow 0^+ + 0^+$. An example is the $J^P = 1^+$ Ne excited ^{20}Ne* $(Q = 13.2$ MeV). If it decays into ^{16}O $\left(J^P = 0^+\right)$ and an α particle $\left(J^P = 0^+\right)$, parity is violated. To search for this decay we look for the corresponding resonance in the process

$$p + ^{19}\text{F} \rightarrow [^{20}\text{Ne}^*] \rightarrow ^{16}\text{O} + \alpha.$$

The resonance was not found (Tonner 1957), a fact that sets the limit for strong interactions

$$|M_S / M_{PS}|^2 \leq 10^{-8}. \tag{3.8}$$

An experimental test of Eq. (3.1) was done by C. S. Wu and J. Shaknov (Wu & Shaknov 1950) on the positronium. This annihilates soon after being formed into two photons from the $^1S_0$ state. Wu and Shaknov found that the correlation between the polarizations of the two photons was characteristic of the decay from an initial $J = 0$, *odd parity*, state, in agreement with Eq. (3.1).

3.3 Particle–Antiparticle Conjugation

The particle–antiparticle conjugation operator C acting on one particle state changes the particle into its antiparticle, leaving space coordinates, time and spin unchanged.

Therefore, the signs of all the additive quantum numbers, electric charge, baryon number and lepton flavour are changed. It is useful to think that if a particle and its antiparticle annihilate then the final state is the vacuum, in which all 'charges' are zero. We shall call this operator the 'charge conjugation', as is often done for brevity, even if the term is somewhat imprecise.

Let us consider a state with momentum $\mathbf{p}$, spin $\mathbf{s}$ and 'charges' $\{Q\}$. Then

$$C|\mathbf{p}, \mathbf{s}, \{Q\}\rangle = |\mathbf{p}, \mathbf{s}, \{-Q\}\rangle. \tag{3.9}$$

Since applying twice the charge conjugation C leads to the original state, the possible eigenvalues are $C = \pm 1$.

Only 'completely' neutral particles, namely particles for which $\{Q\} = \{-Q\} = \{0\}$, can be eigenstates of C. In this case, the particle coincides with its antiparticle. We already know two cases, the photon and the π^0; we shall meet two more, the η and η' mesons. The eigenvalue C for such particles is called their intrinsic charge conjugation, or simply charge conjugation.

The Charge Conjugation of the Photon

Let us consider again the correspondence between the photon and the macroscopic vector potential $\mathbf{A}$. If all the particle sources of the field are changed into their antiparticles, all the electric charges change sign and therefore $\mathbf{A}$ changes its sign. Consequently, the charge conjugation of the photon is negative

$$C|\gamma\rangle = -|\gamma\rangle. \tag{3.10}$$

A state of n photons is an eigenstate of C. Since C is a multiplicative operator

$$C|n\gamma\rangle = (-1)^n |n\gamma\rangle. \tag{3.11}$$

Similarly, the charge conjugation of a state of n neutral mesons is the product of their intrinsic charge conjugations.

The Charge Conjugation of the π^0

The π^0 decays into two photons by electromagnetic interaction, which conserves C, hence

$$C|\pi^0\rangle = +|\pi^0\rangle. \tag{3.12}$$

Charged pions are not C eigenstates; rather, we have

$$C|\pi^+\rangle = +|\pi^-\rangle \qquad C|\pi^-\rangle = +|\pi^+\rangle. \tag{3.13}$$

The Charge Conjugation of the η Meson

The η too decays into two photons, and consequently

$$C|\eta^0\rangle = +|\eta^0\rangle. \tag{3.14}$$

The tests of C conservation are based on searches for C-violating processes. Two examples for the electromagnetic interaction are the experimental limits for the π^0 from McDonough *et al.* (1988) and for the η from Nefkens *et al.* (2005)

$$\Gamma\left(\pi^0 \to 3\gamma\right)/\Gamma_{\text{tot}} \le 3.1 \times 10^{-8} \quad \Gamma\left(\eta \to 3\gamma\right)/\Gamma_{\text{tot}} \le 4 \times 10^{-5}. \tag{3.15}$$

We shall see in Chapter 7 that weak interactions violate C conservation.

Particle–Antiparticle Pair

A system of a particle and its antiparticle is an eigenstate of the particle–antiparticle conjugation in its CM frame. Let us examine the various cases, calling l the orbital angular momentum.

Meson and its Antiparticle (m^+, m^-) with Zero Spin (For Example, π^+ and π^-)

The net effect of C is the exchange of the two mesons; as such it is identical to that of P. Hence

$$C\left|m^+, m^-\right\rangle = (-1)^l \left|m^+, m^-\right\rangle. \tag{3.16}$$

Meson and Antiparticle (M^+, M^-) with Non-Zero Spin $s \ne 0$

The effect of C is again the exchange of the mesons, but now it is not the same as that of $\mathcal{P}$, because C exchanges not only the positions but also the spins. Let us see what happens.

The wave function can be symmetric or antisymmetric under the exchange of the spins. Let us consider the example of two spin 1 particles. The total spin can have the values $s = 0$, 1 or 2. It is easy to check that the states of total spin $s = 0$ and $s = 2$ are symmetric, while the state with $s = 1$ is antisymmetric. Therefore, the spin exchange gives a factor $(-1)^s$. This conclusion is general, as one can show. Therefore, we have

$$C\left|M^+, M^-\right\rangle = (-1)^{l+s} \left|M^+, M^-\right\rangle. \tag{3.17}$$

Fermion and Antifermion ($f\bar{f}$)

Let us start again with an example, namely two spin 1/2 particles. The total spin can be $s = 0$ or 1. This time, the state with total spin $s = 1$ is symmetric, and the state with $s = 0$ is antisymmetric. Therefore, the factor due to the exchange of the spin is $(-1)^{s+1}$. This result too is general.

Fermions and antifermions obeying the Dirac equation have opposite intrinsic charge conjugations, hence a factor of -1. In conclusion,

$$C\left|f\bar{f}\right\rangle = (-1)^{l+s} \left|f\bar{f}\right\rangle. \tag{3.18}$$

Table 3.1. J^{PC} for the spin 1/2 particle–antiparticle systems.

	${}^1S_0$	${}^3S_1$	${}^1P_1$	${}^3P_0$	${}^3P_1$	${}^3P_2$
J^{PC}	$0^{-\,+}$	$1^{-\,-}$	$1^{+\,-}$	$0^{+\,+}$	$1^{+\,+}$	$2^{+\,+}$

The final result is identical to that of the mesons.

We call the reader's attention to the fact that the sum $l + s$ in the above expressions is the sum of two numbers, not the composition of the corresponding angular momenta, that is, it is not in general the total angular momentum of the system.

Example 3.2 Find the eigenvalues of C for the system of a spin 1/2 particle and its antiparticle when they are in an S wave and when they are in a P wave.

The singlets have $S = 0$, hence ${}^1S_0$ has $C = +$, ${}^1P_1$ has $C = -$; the triplets have $S = 1$, hence ${}^3S_1$ has $C = -$, ${}^3P_0, {}^3P_1$ and ${}^3P_2$ have all $C = +$. □

From the results obtained in Examples 3.1 and 3.2, we list in Table 3.1 the J^{PC} values for a fermion–antifermion pair. Notice that not all values are possible. For example, the states with $J^{PC} = 0^{+\,-}$, $0^{-\,-}$, $1^{-\,+}$ cannot be composed of a fermion and its antifermion with spin 1/2.

As we have seen in Section 2.9, a particle obeying the Majorana equation is a completely neutral fermion, f_M, antiparticle of itself. The charge conjugation is the same as for Dirac particles, namely $C = \pm 1$. However, unlike the Dirac case, the charge conjugation of a pair of the same Majorana fermions is, as for photons, simply the product of their intrinsic parities, hence it is positive, independently of l and s, $C|f_m f_m\rangle = +1|f_m f_m\rangle$.

Question 3.1 Write down Table 3.1 for Majorana fermions. □

3.4 Time Reversal and CPT

The time reversal operator T inverts time, leaving the space coordinates unchanged. We shall not discuss it in any detail. We shall only mention that, unlike parity and particle–antiparticle conjugation, there is no time reversal quantum number, because the operator T does not transform as an observable under unitary transformations.

We state here an extremely important property, the Lüders theorem, also called the CPT theorem. If a theory of interacting fields is invariant under the proper Lorentz group (i.e. Lorentz transformations plus rotations), it will also be invariant under the combination of the successive application, in any order, of particle–antiparticle conjugation, space inversion and time reversal.

A consequence of this is that the mass and lifetime of a particle and its antiparticle must be identical, and charges opposite, as already mentioned. Strong limits on CPT violation are provided by the measurements of the masses and of the electric charges of electrons and positrons and of the charge to mass ratios of proton and antiproton. In the proton–antiproton case, for example, S. Ulmer *et al.* (2015), with a high-precision cyclotron frequency comparison of a single antiproton and a negatively charged hydrogen ion carried out in a Penning trap (see Section 5.9) system, provide the limit

$$\frac{|q_{\bar{p}}|}{m_{\bar{p}}} / \frac{|q_p|}{m_p} - 1 = (0.1 \pm 6.9) \times 10^{-11} \tag{3.19}$$

3.5 The Parity of the Pions

The parity of the π^- is determined by observing its capture at rest by deuterium nuclei, a process that is allowed only if the pion parity is negative, as we shall prove. The process is

$$\pi^- + d \rightarrow n + n. \tag{3.20}$$

In practice, one brings a π^- beam of low energy into a liquid deuterium target. The energy is so low that large fractions of the pions come to rest in the liquid after having suffered ionization energy loss.

Once a π^- is at rest, the following processes take place. Since they are negative, the pions are captured, within a time lag of a few picoseconds, in an atomic orbit, replacing an electron. The system is called a 'mesic atom'. The initial orbit has high values of both n and l quantum numbers, but again very quickly (≈ 1 ps), the pion reaches a principal quantum number n of about 7. At these values of n, the wave function of those pions that are in an S orbit largely overlaps with the nucleus. In other words, the probability of the π^- being inside the nucleus is large, and they are absorbed.

The pions that initially are not in an S wave reach it anyway by the following process. Actually, the mesic atom is much smaller than a common atom, because $m_\pi \gg m_e$. Being so small, it eventually penetrates another molecule and becomes exposed to the intense and not uniform electric field present near a nucleus. As a consequence, the Stark–Lo Surdo effect mixes the levels, repopulating the S waves. Then, almost immediately, the pion is absorbed. The conclusion is that practically all captures take place from states with $l = 0$.

This theory was developed by T. B. Day, G. A. Snow and J. Sucker in 1960 (Day *et al.* 1960) and experimentally verified by the measurement of the X-rays emitted from the atomic transitions described above.

Therefore, the initial angular momentum of the reaction (3.20) is $J = 1$, because the spins of the deuterium and of the pion are 1 and 0, respectively, and the orbital momentum is $l = 0$. The deuterium nucleus contains two nucleons, of positive intrinsic parity, in an S wave; hence its parity is positive. In conclusion, the initial parity of the system is equal to that of the pion.

The final state consists of two identical fermions and must be antisymmetric under their exchange. If the two neutrons are in a spin singlet state, which is antisymmetric in the spin exchange, the orbital momentum must be even, and vice versa if the neutrons are in a triplet. Writing them explicitly, we have the possibilities $^1S_0, {}^3P_{0,1,2}, {}^1D_2, \ldots$ The angular momentum must be equal to the initial momentum (i.e. $J = 1$). There is only one choice, namely $^3P_1$. Its parity is negative. Therefore, if the reaction takes place, the parity of π^- is negative.

Panofsky and collaborators (Panofsky *et al.* 1951) showed that the reaction (3.20) proceeds and that its cross-section is not suppressed.

As we mentioned in Section 2.3, a meson, of mass very close to that of the charged pion, was observed to decay mainly as

$$\pi^0 \to 2\gamma. \tag{3.21}$$

To establish it as the neutral pion, we must show that it is a pseudoscalar particle like the charged ones. We shall do this by exploiting symmetry, namely the conditions imposed by angular momentum and parity conservation and by Bose statistics. The method is important per se, but also because it is very similar to that used, as we shall see in Section 9.15, for the Higgs boson. Consequently, we shall describe it in some detail.

We start by showing that the spin of a particle decaying into two photons cannot be 1, both $J^P = 1^+$ and $J^P = 1^-$ being forbidden. The demonstration is simple in the CM frame. The matrix element M should be written using the available vector quantities. There are three of them: the CM momentum $\mathbf{q}$ and the transverse polarizations of the photons, $\mathbf{e}_1$ and $\mathbf{e}_2$. With these, we should build a quantity that, combined with the polarization of the π^0, $\mathbf{S}$, makes a scalar, hence a vector $\mathbf{V}$ if $J^P = 1^-$ or an axial vector $\mathbf{A}$ if $J^P = 1^+$. The matrix element will have the form $M \propto \mathbf{V} \cdot \mathbf{S}$ or $M \propto \mathbf{A} \cdot \mathbf{S}$, respectively. In addition M, and consequently $\mathbf{V}$ or $\mathbf{A}$, must be symmetric under the exchange of the photons, which are identical Bose particles, that is, under the interchange

$$\mathbf{e}_1 \leftrightarrow \mathbf{e}_2 \quad \mathbf{q} \leftrightarrow -\mathbf{q}. \tag{3.22}$$

In addition, the photon polarization must be perpendicular to its momentum, a condition equivalent to the electromagnetic waves being transversal

$$\mathbf{e}_1 \cdot \mathbf{q} = 0 \quad \mathbf{e}_2 \cdot \mathbf{q} = 0. \tag{3.23}$$

No vector or axial vector can be built to satisfy these conditions. The π^0 spin cannot be 1.

We limit the further discussion to the simplest possibilities, $J^P = 0^-$ and $J^P = 0^+$, forgetting in principle possible spin 2 or larger. We can now find a scalar and pseudoscalar combination of the vector and axial vectors of the problem, namely

$$M = a_S \mathbf{e}_1 \cdot \mathbf{e}_2 \qquad \text{scalar,}$$
$$M = a_P \mathbf{e}_1 \times \mathbf{e}_2 \cdot \mathbf{q} \quad \text{pseudoscalar.} \tag{3.24}$$

Question 3.2 Verify that the conditions (3.22) and (3.23) are satisfied. □

In expressions (3.24), a_S and a_P are scalar functions of the kinematic variables called form factors. We shall come back to these soon, but we can already notice how the polarizations of the two photons tend to be parallel, where the dot product is maximum, in the scalar case, and perpendicular, where the cross product is maximum, in the pseudoscalar case. Even if the polarizations cannot be directly measured, we fortunately have an indirect way. Rarely, with a branching ratio of 3.4×10^{-5}, the π^0 decays into two 'virtual' photons, which we call $\gamma *$. We shall learn the meaning of the adjective in Chapter 5. It suffices here to know that a virtual photon can have a mass different from zero with any value, but very small in this case, and it immediately transforms into an electron–positron pair, whose invariant mass is the mass of the virtual photon. The process, called double internal conversion, is

$$\pi^0 \rightarrow \gamma * + \gamma * \rightarrow \left(e^+ e^-\right)_1 + \left(e^+ e^-\right)_2. \tag{3.25}$$

It can be shown that the normal to the plane defined by the momenta of a pair tends to have the same direction of polarization as the photon that gave origin to it. Consequently, the angle ϕ between the two planes tends to be 0 or $\pi / 2$ for a scalar and for a pseudoscalar, respectively.

The form factors are functions of the invariant masses of the two pairs, say m_1 and m_2. Kroll & Wada (1955) have shown that the form factors a_S and a_P differ only by the sign, being positive for the scalar case and that ϕ distributions are given by

$$\frac{dN}{d\phi \, dm_1 \, dm_2} \propto a(m_1, m_2) \cos 2\phi. \tag{3.26}$$

The experiment was done originally in a hydrogen bubble chamber by Samios *et al.* (1962), collecting 112 examples of reaction (3.25). For each of them, the two masses m_1 and m_2 were measured and the value of the form factor was obtained, in both the scalar and pseudoscalar hypotheses. Finally, the ratio of the likelihood functions for the total sample was computed. This statistical analysis favoured the pseudoscalar hypothesis by 3.3 standard deviations. The result can be expressed by stating that the measured weighted average of the form factor is

$$\langle a \rangle = -0.41 \pm 0.24, \tag{3.27}$$

compared with the theoretical expectation of $\langle a \rangle = +0.47$ for the scalar case and $\langle a \rangle = -0.47$ for the pseudoscalar one.

More recently the KTeV experiment (Abouzaid *et al.* 2008) at FNAL collected 30 511 candidates of the (3.25) decay, confirming with this large number the negative π^0 parity. Fig. 3.1 shows the distribution of the ϕ angle together with the expectations for $J^P = 0^+$ and $J^P = 0^-$.

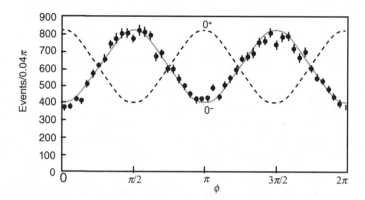

Fig. 3.1. Distribution of the angle ϕ between the planes of the two $e^+ \cdot e^-$ pairs. Continuous curve for $J^P = 0^-$, dashed curve for $J^P = 0^+$.

3.6 Quark Flavours and Baryonic Numbers

The baryon number of a state is defined as the number of baryons minus the number of antibaryons

$$\mathcal{B} = N\left(\text{baryons}\right) - N\left(\text{antibaryons}\right) \tag{3.28}$$

Within the limits of experiments, all known interactions conserve the baryon number. The best limits come from the search for proton decay. In practice, one seeks a specific hypothetical decay channel and finds a lower bound for that channel. We shall consider the most plausible decay, namely

$$p \rightarrow e^+ + \pi^0. \tag{3.29}$$

Notice that this decay also violates the lepton number (see Section 3.7 for the definition) but conserves the difference $\mathcal{B} - \mathcal{L}$.

The present bound is huge, more than 10^{34} years, 10^{24} times the age of the Universe. To reach such levels of sensitivity, one needs to control of the order of 10^{34} protons for several years, ready to detect the decay of a single one, if it should happen.

The main problem when searching for rare phenomena, as in this case, is the identification and the drastic reduction, hopefully the elimination, of the 'background', namely of those natural phenomena that can simulate the events being sought (the 'signal'). The principal background sources are cosmic rays and nuclear radioactivity. In the case of proton decay, the energy of the decay products is of the order of a GeV. Therefore, nuclear radioactivity is irrelevant, because its energy spectra end at 10–15 MeV. Shielding from cosmic rays is obtained by working in deep underground laboratories.

The sensitivity of an experiment grows with its 'exposure', the product of the sensitive mass and of the time for which data are taken.

The most sensitive detector is currently Super-Kamiokande which, as we have seen in Example 1.12, uses the Cherenkov water technique. It is in the Kamioka Observatory at about 1000 m below the Japanese Alps. The total water mass is 50 000 t. Its central part, in which all backgrounds are reduced, is defined as the 'fiducial mass' and amounts to 22 500 t. Let us calculate how many protons it contains. In H_2O the protons are 10/18 of all the nucleons, and we obtain

$$N_p = M \times 10^3 \times N_A (10/18) = 2.25 \times 10^7 \times 10^3 \times 6 \times 10^{23} (10/18) = 7.5 \times 10^{33}.$$

After several years, the exposure reached was $M\Delta t = 450$ kt yr, corresponding to $N_p \Delta t = 1.5 \times 10^{35}$ proton-years.

The irreducible background is due to neutrinos produced by cosmic rays in the atmosphere that penetrate underground. Their interactions must be identified and distinguished from the possible proton decay events. If an event is a proton decay (3.29), the electron gives a Cherenkov ring. The photons from the π^0 decay produce lower-energy electrons that are also detected as rings. The geometrical aspect of an event, the number of rings, their type, etc., is called the event 'topology'. The first step in the analysis is the selection of the events, with a topology compatible with proton decay. This sample contains, of course, background events.

Super-Kamiokande measures the velocity of a charged particle from the position of its centre and from the radius of its Cherenkov ring. Its energy can be inferred from the total number of photons. If the process is the one given in (3.29), then the particles that should be the daughters of the π^0 must have the right invariant mass, and the total energy of the event must be equal to the proton mass. No event was found satisfying these conditions.

We must still consider another experimental parameter: the detection efficiency. Actually, not every proton decay can be detected. The main reason is that the majority of the protons are inside an oxygen nucleus. Therefore, the π^0 from the decay of one of them can interact with another nucleon. If this interaction is accompanied by charge exchange, a process that happens quite often, in the final state we have a π^+ or a π^- and the π^0 is lost. Taking this and other less important effects into account, the calculated efficiency is 44%. The partial decay lifetime in this channel is at the 90% confidence level, from the Super-Kamiokande experiment (Abe *et al.* 2017; Takenaka 2020)

$$\tau \geq B\left(p \to e^+ \pi^0\right) \times 2.4 \times 10^{34} \, \text{yr} \tag{3.30}$$

where B is the unknown branching ratio. Somewhat smaller limits have been obtained for other decay channels, including $\mu^+ \pi^0$ and $K^+ \nu$.

Let us now consider the quarks. Since baryons contain three quarks, the baryon number of the quarks is $\mathcal{B} = 1/3$.

A correlated concept is the 'flavour': the quantum number that characterizes the type of quark. We define the 'down quark number' N_d as the number of down quarks minus the number of anti-down quarks, and similarly for the other flavours. Notice that the strangeness S of a system and the 'strange quark number' are exactly the same quantity. Three other quarks exist, each with a different flavour, called charm

C, beauty B and top T. By definition, the strangeness of the s quark is negative and similarly the beauty of the b quark. Charm of the c and topness of the t are positive. For historical reasons the flavours of the constituents of normal matter, the up and down quarks, do not have a name:

$$
\begin{aligned}
N_d &= N(d) - N(\bar{d}); & N_u &= N(u) - N(\bar{u}) \\
-S = N_s &= N(s) - N(\bar{s}); & C = N_c &= N(c) - N(\bar{c}) \\
-B = N_b &= N(b) - N(\bar{b}); & T = N_t &= N(t) - N(\bar{t}).
\end{aligned}
\tag{3.31}
$$

Strong and electromagnetic interactions conserve all the flavour numbers while weak interactions violate them.

3.7 Leptonic Flavours and Lepton Numbers

The (total) lepton number is defined as the number of leptons minus the number of antileptons.

$$
\mathcal{L} = N(\text{leptons}) - N(\text{antileptons}).
\tag{3.32}
$$

Let us also define the partial lepton numbers or, rather, the lepton flavour numbers: the electronic number (or flavour), the muonic number (or flavour) and the tauonic number (or flavour):

$$
\mathcal{L}_e = N(e^- + \nu_e) - N(e^+ + \bar{\nu}_e)
\tag{3.33}
$$

$$
\mathcal{L}_\mu = N\left(\mu^- + \nu_\mu\right) - N\left(\mu^+ + \bar{\nu}_\mu\right)
\tag{3.34}
$$

$$
\mathcal{L}_\tau = N\left(\tau^- + \nu_\tau\right) - N\left(\tau^+ + \bar{\nu}_\tau\right).
\tag{3.35}
$$

Obviously, the total lepton number is the sum of these three:

$$
\mathcal{L} = \mathcal{L}_e + \mathcal{L}_\mu + \mathcal{L}_\tau.
\tag{3.36}
$$

All known interactions conserve the total lepton number.

The lepton flavours are conserved in all the observed collision and decay processes. The most sensitive tests are based, as usual, on the search for forbidden decays. The best limits (Workman *et al.* 2022) are

$$
\frac{\Gamma\left(\mu^- \to e^- + \gamma\right)}{\Gamma_{\text{tot}}} \le 4.2 \times 10^{-13}, \quad \frac{\Gamma\left(\mu^- \to e^- + e^- + e^+\right)}{\Gamma_{\text{tot}}} \le 1.0 \times 10^{-12},
\tag{3.37}
$$

which are very small indeed.

On this basis, the Standard Model does not allow any violation of the lepton flavour number. However, it has been experimentally observed that neutrinos produced with a certain flavour may later be observed to have a different flavour.

This has been historically discovered in two phenomena, different from collisions and decays:

- The ν_μ flux produced by cosmic radiation in the atmosphere reduces to about 50% over distances of several thousand kilometres, namely crossing part of the Earth. This cannot be due to absorption because cross-sections are too small. Rather, the fraction that has disappeared is transformed into another neutrino flavour, mainly ν_τ.
- The thermonuclear reactions in the centre of the Sun produce ν_e; only about one half of these (or even less, depending on their energy) leave the surface as such. The electron neutrinos, coherently interacting with the electrons of the dense solar matter, transform, partially, in a quantum superposition of ν_μ and ν_τ.

These are the only phenomena so far observed in contradiction of the Standard Model. We shall come back to this in Chapter 10.

On the other hand, there is no evidence, up to now, of violations of the **total** lepton number. The most sensitive test is the $\mu^- \to e^+$ conversion in a muonic atom, namely an atom having a muon in place of an electron (Kaulard 1998)

$$\frac{\sigma\left(\mu^- \mathrm{Tl} \to e^+ \mathrm{Ca}\right)}{\sigma\left(\mu^- \mathrm{Tl} \to \mathrm{all}\right)} < 3.6 \times 10^{-11}$$

However, as we shall see in Chapter 7, the non-observation of the μ-to-e conversion can be guaranteed also by the V–A structure of the charged current weak interactions and by the smallness of the neutrino masses. The most comprehensive approach for probing lepton number conservation is the search for neutrino-less double beta decay of nuclei, $(Z, A) \to (Z+2, A) + 2e^-$, which will be studied in Chapter 10. The current best limits are for ^{136}Xe and for ^{76}Ge

$$T_{1/2}^{0\nu 2\beta}\left(^{136}\mathrm{Xe}\right) \geq 2.3 \times 10^{26} \text{ yr}, \quad T_{1/2}^{0\nu 2\beta}\left(^{76}\mathrm{Ge}\right) \geq 1.8 \times 10^{26} \text{ yr}.$$

3.8 Isospin

The nuclear forces are approximately independent of electric charge. For example, the binding energies of ^{3}H and ^{3}He are very similar, the small difference being due to the electric repulsion between protons. However, this charge-independence property is not simply an invariance under the exchange of a proton with a neutron. Rather it is the invariance under the isotopic spin or, for brevity, isospin, as proposed by W. Heisenberg in 1932 (Heisenberg 1932). The proton and neutron should be considered two states of the same particle, the nucleon, which has isospin $I = 1/2$. The states which correspond to the two values of the third component are the proton with $I_z = +1/2$ and the neutron with $I_z = +1/2$.

The situation is formally equal to that of the angular momentum. The transformations in 'isotopic space' are analogous to the rotations in usual space. The charge

Table 3.2. The lowest isospins and the dimensions of the corresponding representations.

Dimension	1	2	3	4	5	...
I	0	1/2	1	3/2	2	...

independence of nuclear forces corresponds to their invariance under rotations in isotopic space.

The different values of the angular momentum (J) correspond to different representations of the group of the rotations in the normal space. The dimensionality $2J+1$ of the representation is the number of states with different values of the third component of their angular momentum. In the case of the isospin I, the dimensionality $2I+1$ is the number of different particles, or nuclear levels, that can be thought of as different charge states of the same particle, or nuclear state. They differ by the third component I_z. The group is called an isotopic multiplet. Clearly, all the members of a multiplet must have the same mass, spin and parity.

Table 3.2 shows the simplest representations.

There are several isospin multiplets in nuclear physics. We consider the example of the energy levels of the triplet of nuclei: ^{12}B (made of $5p+7n$), ^{12}C ($6p+6n$) and ^{12}N ($7p+5n$). The ground states of ^{12}B and ^{12}N and one excited level of ^{12}C have $J^P = 1^+$. We lodge them in an $I=1$ multiplet with $I_z = -1$, 0 and $+1$ respectively. All of them decay to the ^{12}C ground state: ^{12}B by β^- decay with 13.37 MeV, the excited ^{12}C level by γ decay with 15.11 MeV, and ^{12}N by β^+ decay with 16.43 MeV. If the isotopic symmetry were exact, namely if isospin was perfectly conserved, the energies would have been identical. The symmetry is 'broken' as shown by the small differences, of the order of a MeV. There are two reasons for this. Firstly, as we shall explain below, the symmetry is broken by the electromagnetic interaction, which does not conserve isospin, even if it does conserve its third component. Secondly, the masses of the proton and of the neutron are not identical, but $m_n - m_p \approx 1.3$ MeV. At the quark level, the mass of the d quark is a few MeV larger than that of the u, contributing to making the neutron, which is ddu, heavier than the proton, uud.

In subnuclear physics, it is convenient to describe the isospin invariance with the group $SU(2)$ – comprising all two-dimensional unitary matrices with unit determinant – in place of that of the three-dimensional rotations. The two are equivalent, but $SU(2)$ will make the extension to $SU(3)$ (three-dimensional unitary matrices with unit determinant) easier, as we shall discuss in the next chapter.

Just like nuclear levels, the hadrons are grouped in $SU(2)$ (or isospin) multiplets. This is not possible for non-strongly interacting particles, such as the photon and the leptons. Another useful quantum number defined for the hadrons is the flavour hypercharge (or simply hypercharge), which is defined as the sum of baryon number and strangeness

$$Y = \mathcal{B} + S. \tag{3.38}$$

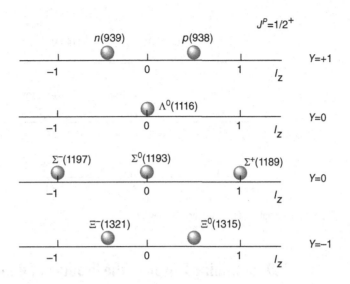

The isospin multiplets of the $J^P = 1/2^+$ baryons.

Since the baryonic number is conserved by all interactions, hypercharge is conserved in the same cases in which strangeness is conserved. For mesons, the hypercharge is simply their strangeness. Here we are limiting our discussion to the hadrons made of the quarks u, d and s only. The particles in the same multiplet are distinguished by the third component of the isospin, which is defined by the Gell-Mann and Nishijima relationship as

$$I_z = Q - Y/2 = Q - (\mathcal{B} + S)/2. \tag{3.39}$$

We can see now that, since electromagnetic interactions conserve electric charge, baryon number and strangeness, they conserve I_z too. Let us see how the hadrons that we have already met are classified in isospin multiplets.

All the baryons we discussed have $J^P = 1/2^+$. They are grouped in the isospin multiplets shown in Fig. 3.2. The approximate values of the mass in MeV are shown next to each particle. The masses within each multiplet are almost, but not exactly, equal. There are small differences for the same reasons that there are small differences for the nucleons. All the members of a multiplet have the same hypercharge, which is shown in the figure next to every multiplet. We shall see more baryons in the next chapter.

For every baryon, there is an antibaryon with identical mass. The multiplets are the same, with opposite charge, strangeness, hypercharge and I_z.

Question 3.3 Draw the figure corresponding to Fig. 3.2 for its antibaryons. □

All the mesons we have met have $J^P = 0^-$ and are grouped in the multiplets shown in Fig. 3.3. The π^- and the π^+ are the antiparticles of each other and are members of the same multiplet. The π^0 in the same multiplet is its own antiparticle. The situation is different for the kaons, which form two doublets containing the particles and their antiparticles respectively. We shall see more mesons in the next chapter.

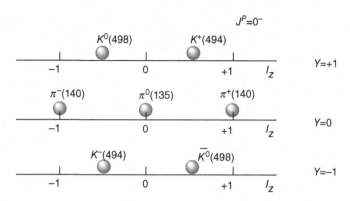

The isospin multiplets of the pseudoscalar mesons.

3.9 Summing Isospins; the Product of Representations

The isospin concept is useful not only for classifying the hadrons, but also for constraining their dynamics in scattering and decay processes. If these proceed through strong interactions, both the total isospin and its third component are conserved; if they proceed through electromagnetic interactions, only the third component is conserved; and if they proceed through weak interactions, neither is conserved.

Isospin conservation implies definite relationships between the cross-sections or the decay probabilities of different strong processes. Consider, for example, a reaction with two hadrons in the final state, and two in the initial one. The two initial hadrons belong to two isospin multiplets, and similarly the final ones. Changing the particles in each of these multiplets we have different reactions, with cross-sections related by isospin conservation. We shall see some examples soon.

In the first step of the isospin analysis, one writes both the initial and final states as a superposition of states of total isospin. The reaction can proceed strongly only if there is at least one common value of the total isospin. In this case, we define a transition amplitude for each isospin value present in both the initial and the final states. The transition amplitude of each process of the set is a linear combination of the isospin amplitudes. We shall now see how.

The rules for isospin composition are the same as for angular momentum. After having recalled them, we shall introduce an alternative notation, which will be useful when dealing with the $SU(3)$ extension of the $SU(2)$ symmetry.

To be specific, let us consider a system of two particles, one of isospin 1 (for example a pion) and one of isospin $1/2$ (for example a nucleon). The total isospin can be $1/2$ or $3/2$. We write this statement as $\mathbf{1} \otimes \mathbf{1/2} = \mathbf{1/2} \oplus \mathbf{3/2}$. This means that the product of the representation of $SU(2)$ corresponding to isospin $\mathbf{1}$ and the representation corresponding to isospin $\mathbf{1/2}$ is the sum of the representations corresponding to isospins $\mathbf{1/2}$ and $\mathbf{3/2}$.

The alternative is to label the representation with the number of its states $(2I+1)$, instead of with its isospin (I). The above written relationship becomes $3 \otimes 2 = 2 \oplus 4$.

Let us start with a few important examples.

Example 3.3   Verify the conserved quantities in the reaction $\pi^- + p \rightarrow \pi^0 + n$. Is the process allowed?

The isospin decomposition of the initial state is $1 \otimes 1/2 = 1/2 \oplus 3/2$; that of the final state is, again, $1 \otimes 1/2 = 1/2 \oplus 3/2$. There are two common values of the total isospin, $1/2$ and $3/2$; hence the isospin can be conserved. For the third component, we initially have $I_z = -1 + 1/2 = -1/2$, and finally $I_z = 0 - 1/2 = -1/2$. The third component is conserved. The interaction can proceed strongly. $\square$

Example 3.4   Does the reaction $d + d \rightarrow^4 \text{He} + \pi^0$ conserve isospin? In the initial state, the total isospin is given by $0 \otimes 0 = 0$. In the final state, it is given by $0 \otimes 1 = 1$. The reaction violates isospin conservation. Experimentally this reaction is not observed, with a limit on its cross-section $< 10^{-2}$ of the value computed in the assumption of isospin violation by the strong interaction. $\square$

Example 3.5   Compute the isospin balance for $\Sigma^0 \rightarrow \Lambda + \gamma$. The γ does not have isospin. Thus the total isospin changes from the initial to the final state from $1 \rightarrow 0$, while the third component is 0 in both the initial and the final states. The decay proceeds via electromagnetic interaction. $\square$

Example 3.6   Compute the isospin balance for $\Lambda \rightarrow p + \pi^-$. The initial isospin is 0. In the final state, the isospin is $1/2 \otimes 1 = 1/2 \oplus 3/2$. The isospin cannot be conserved. The third component is initially $I_z = 0$ and finally $I_z = 1/2 - 1 = -1/2$; it is not conserved. Even if there are only hadrons in the process, it is a weak decay as shown by the violation of I and of I_z. $\square$

We consider now an example of isospin relationships amongst cross-sections. Consider the four reactions: (1) $\pi^+ + p \rightarrow \pi^+ + p$, (2) $\pi^- + p \rightarrow \pi^0 + n$, (3) $\pi^- + p \rightarrow \pi^- + p$, (4) $\pi^- + n \rightarrow \pi^- + n$.

The isospin composition in the initial state is $1 \otimes 1/2 = 1/2 \oplus 3/2$. In this case, but not always, the composition in the final states is the same. In conclusion, the transition probabilities of the four processes are linear combinations of two isospin amplitudes $A_{1/2}$ and $A_{3/2}$. These are complex functions of the kinematic variables.

We shall now see the general rules for finding these linear combinations.

We have two bases. In the first base, the isospins and their third components of each particle are defined. We call these states $|I_1, I_{z1}; I_2, I_{z2}\rangle$. In the second base, the total isospin (I) and its third component (I_z) are defined, as are I_1 and I_2. We call these states $|I, I_z; I_1, I_2\rangle$. The relationship between the two bases is

$$|I, I_z; I_1, I_2\rangle = \sum_{I_{z1}, I_{z2}} |I_1, I_{z1}; I_2, I_{z2}\rangle \langle I_1, I_{z1}; I_2, I_{z2} | I, I_z; I_1, I_2\rangle. \quad (3.40)$$

The quantities $\langle I_1, I_{z1}; I_2, I_{z2} | I, I_z; I_1, I_2\rangle$ are called Clebsch–Gordan coefficients and they can be found in Appendix 4.

Example 3.7 Find the expressions of the cross-sections of the four reactions discussed above in terms of the two isospin amplitudes. We write for simplicity the kets in the second members as $|I, I_z\rangle$. Using the Clebsch–Gordan tables, we find

$$
\begin{aligned}
&|\pi^+ p\rangle = \left|\frac{3}{2}, +\frac{3}{2}\right\rangle; \quad |\pi^- p\rangle = \sqrt{\frac{1}{3}}\left|\frac{3}{2}, -\frac{1}{2}\right\rangle - \sqrt{\frac{2}{3}}\left|\frac{1}{2}, -\frac{1}{2}\right\rangle \\
&|\pi^0 n\rangle = \sqrt{\frac{2}{3}}\left|\frac{3}{2}, -\frac{1}{2}\right\rangle + \sqrt{\frac{1}{3}}\left|\frac{1}{2}, -\frac{1}{2}\right\rangle; \quad |\pi^- n\rangle = \left|\frac{3}{2}, -\frac{3}{2}\right\rangle.
\end{aligned}
\quad (3.41)
$$

With a proportionality constant K equal for all, we obtain

$$\sigma\left(\pi^+ p \to \pi^+ p\right) = K|A_{3/2}|^2 \quad (3.42)$$

$$\sigma\left(\pi^- p \to \pi^0 n\right) = K\left|\frac{\sqrt{2}}{3} A_{3/2} - \frac{\sqrt{2}}{3} A_{1/2}\right|^2 \quad (3.43)$$

$$\sigma\left(\pi^- p \to \pi^- p\right) = K\left|\frac{1}{3} A_{3/2} + \frac{2}{3} A_{1/2}\right|^2 \quad (3.44)$$

$$\sigma\left(\pi^- n \to \pi^- n\right) = K|A_{3/2}|^2. \quad (3.45)$$

In particular we arrive at the prediction $\sigma(\pi^+ p \to \pi^+ p) = \sigma(\pi^- n \to \pi^- n)$ for the same energy. It is experimentally well verified. From these cross-sections we know $K|A_{3/2}|$.

The other cross-sections, and those of other processes such as $\pi^+ + n \to \pi^+ + n$ and $\pi^+ + n \to \pi^0 + p$, depend on two unknowns, $|A_{1/2}|$ and $\arg\left(A_{3/2}^* A_{1/2}\right)$.

At low energies, all of these cross-sections show a large resonance, which was discovered by Fermi (Anderson *et al.* 1952). It is called $\Delta(1236)$, and has a maximum at $\sqrt{s} = 1236$ MeV (see Section 4.2). We know that its isospin is $I = 3/2$ by observing that the cross-section is dominated by $|A_{3/2}|$. Actually, in this case we obtain, from the above expressions,

$$\sigma\left(\pi^+ p \to \pi^+ p\right) : \sigma\left(\pi^- p \to \pi^- p\right) : \sigma\left(\pi^- p \to \pi^0 n\right) = 9 : 1 : 2, \quad (3.46)$$

and the experimental values of the cross-sections in mb are $195 : 22 : 45$. □

3.10 G-parity

G-parity is not a fundamental quantum number; however, it is convenient when dealing with non-strange states with zero baryonic number. These states, typically the pion systems, are eigenstates of G.

The π^0 is an eigenstate of the charge conjugation C. The charged pions (see (3.13)) transform as

$$C|\pi^+\rangle = +|\pi^-\rangle \qquad C|\pi^-\rangle = +|\pi^+\rangle. \tag{3.47}$$

G is defined as C followed by a 180° rotation around the y-axis in isotopic space, namely

$$G \equiv \exp(-i\pi I_y)C. \tag{3.48}$$

The three-π states are the components of a vector in isotopic space (iso-vector). The relationships between Cartesian components $|\pi_x\rangle, |\pi_y\rangle$ and $|\pi_z\rangle$ and charge states are

$$|\pi^+\rangle = \frac{1}{\sqrt{2}}\left(|\pi_x\rangle + i|\pi_y\rangle\right); |\pi^0\rangle = |\pi_z\rangle; |\pi^-\rangle = \frac{1}{\sqrt{2}}\left(|\pi_x\rangle - i|\pi_y\rangle\right). \tag{3.49}$$

Let us apply C and then the rotation to these expressions:

$$|\pi^+\rangle = \frac{1}{\sqrt{2}}\left(|\pi_x\rangle + i|\pi_y\rangle\right) \quad |\pi^-\rangle = \frac{1}{\sqrt{2}}\left(|\pi_x\rangle - i|\pi_y\rangle\right) \quad \frac{1}{\sqrt{2}}\left(-|\pi_x\rangle - i|\pi_y\rangle\right) = -|\pi^+\rangle$$

$$|\pi^0\rangle = |\pi_z\rangle \quad\quad C \Rightarrow |\pi^0\rangle = |\pi_z\rangle \quad\quad e^{i\pi I_y} \Rightarrow -|\pi_z\rangle = -|\pi^0\rangle$$

$$|\pi^-\rangle = \frac{1}{\sqrt{2}}\left(|\pi_x\rangle - i|\pi_y\rangle\right) \quad |\pi^-\rangle = \frac{1}{\sqrt{2}}\left(|\pi_x\rangle + i|\pi_y\rangle\right) \quad \frac{1}{\sqrt{2}}\left(-|\pi_x\rangle + i|\pi_y\rangle\right) = -|\pi^-\rangle.$$

We see that all the charge states are eigenstates with negative eigenvalue

$$G|\pi\rangle = -|\pi\rangle. \tag{3.50}$$

For a system of n_π pions we have

$$G = (-1)^{n_\pi}. \tag{3.51}$$

It is easy to prove that all non-strange non-baryonic states are eigenstates of G. If their isospin is $I = 1$, the situation is identical to that of the pions. The neutral state has $I_z = 0$ and $G = -C$. If $I = 0$, obviously $G = C$.

Only the strong interaction conserves the G-parity because the electromagnetic and weak interactions violate the isospin (and the latter also C).

Problems

3.1   For each interaction type, strong (S), electromagnetic (EM) and weak (W), insert a Y or N in the cell of every quantum number, depending on whether or not it is conserved (I = isospin, I_z its third component, S strangeness, $\mathcal{B}$ baryon number, $\mathcal{L}$ lepton number, $\mathcal{T}$ time reversal, C particle–antiparticle conjugation, $\mathcal{P}$ parity, J angular momentum, J_z its third component).

	I	I_z	S	$\mathcal{B}$	$\mathcal{L}$	$\mathcal{T}$	C	$\mathcal{P}$	J	J_3
S										
EM										
W										

3.2 Consider a π^- beam impinging on a liquid hydrogen target. Find the threshold energy for K^- production.

3.3 The existence of the anti-hyperons was proven by the discovery of an anti-lambda by M. Baldo-Ceolin and D. J. Prowse in 1958. A beam of negative pions with energy $E_\pi = 4.6$ GeV hit an emulsion stack. What is the final state containing an $\bar\Lambda$ that can be produced in a $\pi^- p$ collision at minimum energy? Find the threshold energy if the target protons are free and, approximately, if they are bound inside nuclei with a Fermi momentum $p_f = 150$ MeV. Assuming that the pion beam was produced at a distance $l = 8$ m upstream of the emulsion and that the number of produced pions was $N_0 = 10^6 \, \mathrm{cm}^{-2}$, how many pions per cm^2 reached the emulsion?

3.4 For each of the following reactions, (a) establish whether it is allowed or not, (b) if it is not, give the reason(s) why (there may be more than one) and (c) give the types of interaction that allow it.

    (1) $\pi^- p \to \pi^0 + n$

    (2) $\pi^+ \to \mu^+ + \nu_\mu$

    (3) $\pi^+ \to \mu^+ \bar\nu_\mu$

    (4) $\pi^0 \to 2\gamma$

    (5) $\pi^0 \to 3\gamma$

    (6) $e^+ + e^- \to \gamma$

    (7) $p + \bar p \to \Lambda + \Lambda$

    (8) $p + p \to \Sigma^+ + \pi^+$

    (9) $n \to p + e^-$

    (10) $n \to p + \pi^-$.

3.5 For each of the following reactions, establish whether or not it is allowed; if it is not, give the reasons:

    (1) $\mu^+ \to e^+ + \gamma$

    (2) $e^- \to \nu_e + \gamma$

    (3) $p + p \to \Sigma^+ K^+$

    (4) $p + p \to p + \Sigma^+ + K^-$

    (5) $p \to e^+ + \nu_e$

    (6) $p + p \to \Lambda + \Sigma^+$

    (7) $p + n \to \Lambda + \Sigma^+$

    (8) $p + n \to \Xi^0 + p$

    (9) $p \to n + e^+ + \nu_e$

    (10) $n \to p + e^- + \nu_e$.

3.6 Give the reasons forbidding each of the following decays: (a) $n \to p + e^-$; (b) $n \to \pi^+ + e^-$; (c) $n \to p + \pi^-$; (d) $n \to p + \gamma$.

3.7 Which of the following processes is allowed and which is forbidden by strangeness conservation?

(a) $\pi^- + p \to K^- + p$

(b) $\pi^- + p \to K^+ + \Sigma^-$

(c) $K^- + p \to K^+ + \Xi^0 + \pi^-$

(d) $K^+ + p \to K^- + \Xi^0 + \pi^-$.

3.8 For each of the following reactions, establish whether or not it is allowed; if it is not, give the reasons why.

(a) $p \to n + e^+$

(b) $\mu^+ \to \nu_\mu + e^+$

(c) $e^+ + e^- \to \nu_\mu + \bar\nu_\mu$

(d) $\nu_\mu + p \to \mu^+ + n$

(e) $\nu_\mu + n \to \mu^- + p$

(f) $\nu_\mu + n \to e^- + p$

(g) $e^+ + n \to p + \nu_e$

(h) $e^- + p \to n + \nu_e$.

3.9 Evaluate the ratios between the cross-sections of the following reactions at the same energy, assuming (unrealistically) that they proceed only through the $I = 3/2$ channel:

$$\pi^- p \to K^0 \Sigma^0; \quad \pi^- p \to K^+ \Sigma^-; \quad \pi^+ p \to K^+ \Sigma^+.$$

3.10 Evaluate the ratios between the cross-sections, at the same energy, of (1) $\pi^- p \to K^0 \Sigma^0$, (2) $\pi^- p \to K^+ \Sigma^-$ and (3) $\pi^+ p \to K^+ \Sigma^+$, taking into account the contributions of both isospin amplitudes $A_{1/2}$ and $A_{3/2}$.

3.11 Evaluate the ratio between the cross-sections of the reactions $\pi^- p \to \Lambda K^0$ and $\pi^+ n \to \Lambda K^+$, at the same energy.

3.12 Evaluate the ratio of the cross-sections of the processes $p + d \to {}^3\mathrm{He} + \pi^0$ and $p + d \to {}^3\mathrm{H} + \pi^+$ at the same value of the CM energy $\sqrt{s}$ (${}^3\mathrm{He}$ and ${}^3\mathrm{H}$ are an isospin doublet).

3.13 Evaluate the ratio of cross-sections $\sigma(pp \to d\pi^+)/\sigma(pn \to d\pi^0)$ at the same energy.

3.14 Evaluate the ratio of cross-sections $\sigma(K^- + {}^4\mathrm{He} \to \Sigma^0 + {}^3\mathrm{H})/\sigma(K^- + {}^4\mathrm{He} \to \Sigma^- + {}^3\mathrm{He})$ at the same energy.

3.15 Express the ratios between the cross-sections of (1) $K^- p \to \pi^+ \Sigma^-$, (2) $K^- p \to \pi^0 \Sigma^0$ and (3) $K^- p \to \pi^- \Sigma^+$ in terms of the isospin amplitudes A_0, A_1 and A_2.

3.16 Express the ratio of cross-sections of the elastic $\pi^- p \to \pi^- p$ and the charge exchange $\pi^- p \to \pi^0 n$ scatterings in terms of the isospin amplitudes $A_{1/2}$ and $A_{3/2}$.

3.17 A π^- is captured by a deuteron $d(J^P = 1^+)$ and produces the reaction $\pi^- + d \to n + n$. (a) If the capture is from an S wave, what is the total spin of the two neutrons and what is their orbital momentum? (b) Show that if the capture is from a P state, the neutrons are in a singlet.

3.18 The positronium is an atomic system made by an e^- and an e^+ bound by the electromagnetic force.

(1) Determine the relationship that this condition imposes between the orbital momentum l, the total spin s and the charge conjugation C.

(2) Determine the relationship between l, s and n which allows the reaction $e^-e^+ \to n\gamma$ to occur without violating C.

(3) What is the minimum number of photons in which the ortho-positronium $\left(^3S_1\right)$ and the para-positronium $\left(^1S_0\right)$ can annihilate, respectively?

3.19 Establish from which initial states of the $\bar{p}p$ system amongst $^1S_0, {}^3S_1, {}^1P_1, {}^3P_0, {}^3P_1, {}^3P_2, {}^1D_2, {}^3D_1, {}^3D_2$ and $^3D_3$ the reaction $\bar{p}p \to n\pi^0$ can proceed with parity conservation: (1) for any n; (2) for $n = 2$.

3.20 Consider the strong processes $\bar{K}K \to \pi^+\pi^-$ (where $\bar{K}K$ means both K^+K^- and $\bar{K}^0K^0$). (1) What are the possible angular momentum values if the initial total isospin is $I = 0$? (2) What are they if $I = 1$?

3.21 Consider the following $\bar{p}p$ initial states: $^1S_0, {}^3S_1, {}^1P_1, {}^3P_0, {}^3P_1, {}^3P_2, {}^1D_2, {}^3D_1, {}^3D_2$ and $^3D_3$. Establish from which of these the reaction $\bar{p}p \to \pi^+\pi^-$ can proceed if the two πs are: (1) in an S wave, (2) in a P wave and (3) in a D wave?

3.22 The quark contents of the following charmed particles are as follows: the hyperon Λ_c is udc, the D^+ meson is $c\bar{d}$ and the D^- meson is $\bar{c}d$. Which of the following reactions are allowed: (a) $\pi^+p \to D^+p$, (b) $\pi^+p \to D^-\Lambda_c\pi^+\pi^+$, (c) $\pi^+p \to D^+\Lambda_c$, (d) $\pi^+p \to D^-\Lambda_c$?

3.23 The quark contents of the following particles are: the beauty hyperon $\Lambda_b = dub$, the charmed meson $D^0 = c\bar{u}$, the beauty mesons $B^+ = u\bar{b}$, $B^- = \bar{u}b$ and $B^0 = d\bar{b}$. Which of the following reactions are allowed: (a) $\pi^-p \to D^0\Lambda_b$, (b) $\pi^-p \to B^0\Lambda_b$, (c) $\pi^-p \to B^+\Lambda_b\pi^-$, (d) $\pi^-p \to B^-\Lambda_b\pi^+$, (e) $\pi^-p \to B^-B^+$?

3.24 An η meson decays into 2γ while moving in the x direction with energy $E_\eta = 5$ GeV. (1) If the two γs are emitted in the $+x$ and $-x$ directions, what are their energies? (2) If the two γs are emitted at equal and opposite angles $\pm\theta$ with x, what is the angle between the two.

3.25 The state $\Delta(1232)$ has isospin $I = 3/2$. (1) What is the ratio between the decay rates $\Delta^0 \to p\pi^-$ and $\Delta^0 \to n\pi^0$? (2) What would it have been if $I = 1/2$?

3.26 Consider the search for proton decay in a water Cherenkov detector. The refraction index is $n = 1.33$.

(1) Find the minimum velocity of a charged particle for emitting radiation.

(2) Find the minimum kinetic energy for an electron and a K meson to do so.

(3) Determine whether the charged particle in the decays $p \to e^+ + \pi^0$ and $p \to K^+ + \nu$ is above the Cherenkov threshold.

3.27 Which of the following reactions are forbidden/allowed? Justify your statements.

(a) $\mu^- \to e^- + \gamma$

(b) $\pi^+ \to \mu^+ + \nu_\mu + \bar{\nu}_\mu$

(c) $\Sigma^0 \to \Lambda + \gamma$

(d) $\eta \to \gamma + \gamma + \gamma$

(e) $\gamma + p \to \pi^0 + p$

(f) $p \to \pi^0 + e^+$

(g) $\pi^- \to \mu^- + \gamma$.

3.28 Establish which particle 'X' is in the final states of the following reactions: (a) $\pi^- + p \to \sum^0 + X$, (b) $e^+ + n \to p + X$, (c) $\Xi^0 \to \Lambda + X$?

3.29 Consider the Ξ^0 hyperon produced in the reaction $\pi^+ + p \to K^+ + K^+ + \Xi^0$. Knowing the isospin and its third component for all the other particles, establish I and I_z of the Ξ^0.

3.30 Consider the π^- capture at rest, giving the reaction $\pi^- + p \to n + \gamma$. Calculate the energy of the photon and the kinetic energy of the neutron.

3.31 A π^+ beam of 12 GeV momentum is sent into a bubble chamber. An event is observed with two positive tracks originating from the primary vertex and two V^0s pointing to that vertex. The measurements of the momenta of the two tracks of the first V^0 and of the angle between them give $p_{1+} = 0.4$ GeV, $p_{1-} = 1.9$ GeV and $\theta_1 = 24.5°$, and for the second one give $p_{2+} = 0.75$ GeV, $p_{2-} = 0.25$ GeV and $\theta_2 = 22°$. Assume a relative accuracy of 5%. Determine the nature of the two particles from which the V^0s originate, and guess the complete reaction.

Summary

In this chapter we have studied a fundamental aspect of particle physics, namely symmetry properties and symmetry breakdown, and the corresponding conservation of quantum numbers and their violations. In particular we have seen:

- the different types of symmetries
- the parity, the particle–antiparticle conjugation and the time reversal operations, and when they correspond to quantum numbers and when not
- the CPT invariance
- the parity of the pions and the decay of the charged pion
- the quark flavour numbers and the lepton flavour number and their conservation
- the $SU(2)$ symmetry and the rules for summing spins.

Further Reading

Iachello, F. (2001) Symmetry: the search for order in Nature. *J. Phys. Conf. Ser.* 284 012002

Weyl, H. (1952) *Symmetry*. Princeton University Press

Wigner, E. P. (1963) Nobel Lecture; *Events, Laws of Nature, and Invariance Principles* http://nobelprize.org/nobel–prizes/physics/laureates/1963/wigner-lecture.pdf

Wigner, E. P. (1964) The role of invariance principles in natural philosophy. *Proceedings of the International School of Physics 'Enrico Fermi'* 29 p. 40. Academic Press.

Wigner, E. P. (1965) Violation of symmetry in physics. *Sci. Am.* 213 (6) 28

4

Hadrons

The unstable particles we have discussed up to now decay by weak or electromagnetic interactions. The distance between their production and decay points may be long enough to be observable. If the mass of a hadron is large enough, final states that can be reached by strong interaction (i.e. without violating any selection rule) become accessible to its decay. Therefore, the lifetime is extremely short, of the order of a yoctosecond $\left(10^{-24}\,\mathrm{s}\right)$. These hadrons decay practically where they were born. To fix the orders of magnitude, consider such a particle produced in the laboratory reference frame with a Lorentz factor as large as $\gamma = 300$. In a lifetime, it will travel 100 fm.

These extremely unstable hadrons can be observed as 'resonances' in two basic ways: in the process of 'formation', in the energy dependence of total and partial amplitudes, or in the 'production' process, as a maximum in the invariant mass distribution of a subset of particles in the final states of a reaction. We shall discuss this, after having recalled the features of the similar resonance phenomenon present in classical mechanics and electromagnetism.

Such hadrons, both baryons and mesons, were discovered in rapidly increasing numbers in the 1950s and early 1960s in experiments at proton accelerators, mainly by using bubble chambers as detectors. Their quantum numbers, spin, parity and isotopic spin were measured. After a period of confusion, gradually it became clear that hadrons with the same spin and parity, which we indicate with J^P, could be grouped in multiplets not only of $SU(2)$, as we saw in Chapter 2 for the pseudoscalar mesons and for the $J^P = 1/2^+$ baryons, but also of the larger $SU(3)$ symmetry. We shall first discuss the multiplet of $J^P = 3/2^+$ baryons, which has ten members.

After that, we shall introduce the tools used to determine the spin and parity of mesons decaying into three particles, which is often the case for the mesons: these are the Dalitz plot and its analysis.

We shall see how an octet of $J^P = 1/2^+$ and the aforementioned decuplet of $J^P = 3/2^+$ baryons and two meson nonets, one pseudoscalar and one vector, were established. More multiplets exist, but those are sufficient to establish the $SU(3)$ symmetry of the hadron spectrum. The structure of the experimentally observed multiplets of hadrons is such because these hadrons are made of quarks, of three different types. The quark model was proposed in 1964, but it took several years and much work to firmly establish its validity. We now know that three isotopic doublets of quarks, with charges $-1/3$ and $+2/3$, exist. Each quark doublet is lodged, together with a charged lepton and its neutrino, in one 'family'. The elementary particle families are also called 'generations'. Three families of identical structure, but different masses of the components, exist. The three quarks in the hadrons discovered in the 1960s are the first doublet and the strange quark, which is a member of the second doublet. Its 'brother', called charm, was discovered in 1971, after having been theoretically foreseen, as we shall see in Chapter 7. We shall then discuss the third quark doublet, the beauty and the top, that were discovered in 1977 and in 1995 respectively. These discoveries happened so late because these quarks are quite heavy; the mass of the former is about five times that of the proton, and the mass of the latter is about 200 times that of the proton. Accelerators and colliders of sufficient energy and luminosity had to be built and operated. Being so heavy, the top quark decays before having the time to form hadrons. The decay produces 'jets' of hadrons. A complete discussion of this process requires elements that will be developed in the next few chapters and will be done in Section 9.11.

Having now met all the components of the three families, which are the fundamental fermions, we end the chapter anticipating the different structures of their interactions, the strong, the electromagnetic and the weak ones, which will be discussed later in the book.

4.1 Resonances

Resonant phenomena are ubiquitous in physics, at both the macroscopic and microscopic levels. Even in very different physical situations, ranging from mechanics to electrodynamics, from acoustics to optics, from atomic to nuclear physics, etc., the fundamental characteristics of the phenomenon are the same. Actually, the classical and quantum formalisms are also very similar. Resonances have an extremely important role in hadron spectroscopy.

To recall the fundamental concepts, let us start from a naïve model of an atom. We shall imagine the atom to be made up of a massive central charge surrounded by a cloud of equal and opposite charges. Initially we assume that the system can be described by classical physics.

The system, like an elastic ball, has a triple infinite set of normal modes, each with a proper frequency and a proper width. Let us concentrate on one of these and call ω_0 the proper angular frequency and Γ the proper width. At $t = 0$, we set the system in this mode and we then leave it to evolve freely in time. Let Ψ be a coordinate measuring the displacement of the system from equilibrium. The time dependence of Ψ is an exponentially damped sinusoidal function:

$$\Psi(t) = \Psi_0 \exp\left(-\frac{\Gamma t}{2}\right)\cos\omega_0 t = \Psi_0 \exp\left(-\frac{t}{2\tau}\right)\cos\omega_0 t, \quad \tau \equiv 1/\Gamma. \tag{4.1}$$

Notice that the time constant is denoted by 2τ. This choice gives τ the meaning of the time constant of the intensity, which is the square of the amplitude. We have defined τ as the reciprocal of the width.

Let us now consider the forced oscillations of the system. We act on the oscillator with a periodic force, we slowly vary its angular frequency ω and, waiting for the system to reach the stationary regime at every change, we measure the oscillation amplitude. We obtain the well-known resonance curve. In the neighbourhood of the maximum, the response function is, with a generally good approximation,

$$R(\omega) = \frac{\Gamma^2 \omega^2}{\left(\omega_0^2 - \omega^2\right)^2 + \omega^2 \Gamma^2}. \tag{4.2}$$

Comparing the two expressions, (4.1) and (4.2), we conclude that the width of the resonance curve of the forced oscillator is equal to the reciprocal of the lifetime of its free oscillations.

We also recall that the square of the Fourier transform of the decaying oscillations (4.1) is proportional to the response function (4.2).

If the width is small, $\Gamma \ll \omega_0$, (4.2) can be approximated in the neighbourhood of the peak with a simpler expression, called the Breit–Wigner function. Close to the resonance, the fastest varying factor is the denominator

$$\left(\omega_0^2 - \omega^2\right)^2 + \omega^2 \Gamma^2 = \left(\omega_0 - \omega\right)^2 \left(\omega_0 + \omega\right)^2 + \omega^2 \Gamma^2.$$

Here we see that the variation is mainly due to $\omega_0 - \omega$. We replace ω by ω_0 everywhere but in this term, obtaining

$$\left(\omega_0 - \omega\right)^2 \left(\omega_0 + \omega\right)^2 + \omega^2 \Gamma^2 = 4\omega_0^2 \left[\left(\omega_0 - \omega\right)^2 + \left(\Gamma/2\right)^2\right].$$

Substituting this into (4.2), we finally obtain the Breit–Wigner shape function

$$R(\omega) \approx L(\omega) = \frac{\left(\Gamma/2\right)^2}{\left(\omega_0 - \omega\right)^2 + \left(\Gamma/2\right)^2}. \tag{4.3}$$

Actually, the atom is a quantum, not a classical, object, but its correct quantum description brings us to the same conclusion with the difference, as far as we are concerned here, that the atom has a series of excited levels corresponding to the classical normal modes. Each of them has a proper energy and a proper lifetime.

Continuing with our analogy, we consider the fundamental level of the atom which we call A, and its excited levels which we call $A_i *$, as different particles. Actually, this is a totally correct point of view. Each of these has a certain energy, a certain lifetime – infinite for the fundamental level A, which is stationary, finite for the $A_i *$ – and a certain angular momentum.

Let us suppose that we know of the existence of A and we decide to set up an experiment to search for possible unstable particles $A_i *$. We design a **resonance formation** experiment.

We prepare a transparent container that we fill with the A atoms and build a device capable of producing a collimated and monochromatic light beam, for which we can vary the frequency. We send the beam through the container and measure the intensity of the scattered light (at a certain angle). If we do this, and vary the frequency, we shall find a resonance peak for each of the $A_i *$. The line shapes are described by (4.3) and so we can determine the proper frequencies and the widths of the $A_i *$.

Note that two processes take place at the microscopic level: the formation of the $A_i *$ particle and its decay

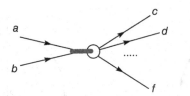 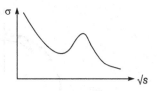

Fig. 4.1. Schematic of a resonance formation study.

$$\gamma + A \rightarrow A_i^* \rightarrow \gamma + A.$$

The time between the two processes may be too short to be measurable, but we can infer the lifetime of A_i^* from the width of the resonance.

The particle A_i^* has a well-defined angular momentum and parity, J^P. These can be determined, as can be easily understood, by measuring, in resonance, the scattered intensity as a function of the angle.

Similarly, in subnuclear physics, we search for very unstable particles by measuring, as a function of the energy, the cross-section of processes of the following type:

$$a + b \rightarrow c + d + \cdots + f. \tag{4.4}$$

Clearly, in this process we can find particles with quantum numbers that are compatible with those of the initial and final states. This is shown schematically in Fig. 4.1.

Example 4.1 The πp resonance $\Delta(1236)$ has the width $\Gamma \approx 120$ MeV. What is its lifetime?

$$\tau = \frac{1}{\Gamma} = \frac{1}{120 \text{ MeV}} = \frac{1}{120 (\text{MeV}) \times 1.52 (\text{s}^{-1}\text{MeV}^{-1}) \times 10^{21}} = 5.4 \times 10^{-24} \text{ s} \quad \square$$

In a quantum scattering process, the relevant functions of energy are the scattering amplitude, which is a complex quantity, and the cross-section, its absolute square, which is the observable we measure. The latter, near to a resonance, is approximately proportional to the shape function (4.3). We shall now look more deeply into the issue with the example of the pion nucleon resonances. Consider the elastic scattering of charged pions on protons (4.5a, b) together with the similar process called charge exchange (4.5c), which will be further discussed in the next section:

$$\pi^+ + p \rightarrow \pi^+ + p \tag{4.5a}$$
$$\pi^- + p \rightarrow \pi^- + p \tag{4.5b}$$
$$\pi^- + p \rightarrow \pi^0 + n. \tag{4.5c}$$

The differential cross-sections of the three processes were measured, in the 1960s, at small steps of beam energies, with accurate control of all the systematic uncertainties. The corresponding scattering amplitudes can be expressed as the sums of terms each having definite orbital momentum L, total angular momentum J, parity P and isotopic spin I. This process is called 'partial wave analysis'. The partial wave amplitudes

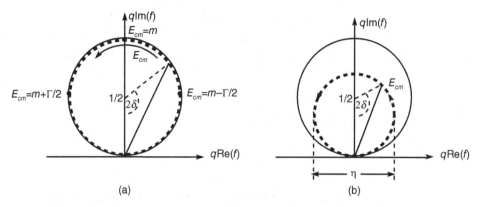

Fig. 4.2. (a) Argand diagram for a Breit–Wigner elastic scattering amplitude; (b) the same for elasticity $\eta < 1$.

are usually denoted as $L_{2I,\ 2J}$ (in practice S_{11}, S_{31} for S waves, P_{11}, P_{13}, P_{31}, P_{33} for P waves, etc.). Calling q the centre of mass momentum, the partial wave amplitude is

$$L_{2I,2J} = \frac{1}{2iq}\left(\eta_{L,I,J}\exp\left(i2\delta_{L,J,I}\right)-1\right), \tag{4.6}$$

where η is the elasticity, namely the ratio of the elastic and total amplitudes, and δ is the phase shift of the partial wave, both functions of energy. Fig. 4.2(a) shows this function graphically in the simplest hypothesis of a purely Breit–Wigner amplitude, namely

$$\frac{\Gamma_{el}/2}{m-E_{cm}-i\Gamma/2}, \tag{4.7}$$

where Γ is the total width, Γ_{el} is the elastic width, m is the resonance mass and E_{cm} is the centre of mass energy. Its absolute square has the behaviour of the Breit–Wigner shape function (4.3).

In the diagrams of Fig. 4.2, called Argand diagrams, the point representing Eq. (4.6) moves on the dashed circle anticlockwise when E_{cm} increases along with the phase-shift δ. The external circle gives the unitarity limit, corresponding to the fact that the scattering probability cannot exceed 1 (unitarity). The diagrams represent the ideal case of purely Breit–Wigner resonances. In Fig. 4.2(a), the scattering is elastic ($\eta = 1$), namely the only open channel is the initial one, and the trajectory of the amplitude follows the unitary circle. In Fig. 4.2(b), more channels are open and the elastic amplitude is a fraction η of the total.

In both cases the following observations are relevant. The phase δ goes rapidly through 90° at the resonance, namely when $E_{cm} = m$, where m is the mass of the resonance. In the elastic case, $\delta = 0$ when the centre of mass energy is half a width below the resonance, that is, $\delta = m - \Gamma/2$, and $\delta = 180°$ when $\delta = m + \Gamma/2$. We shall see in the next section that, in practice, the situation is often more complicated.

The second class of experiments on unstable hadrons is based on **resonance production**.

Assume that we are searching for particles decaying into the stable or metastable particles c and d. Call M_R the mass and Γ the width of such a particle R. We select

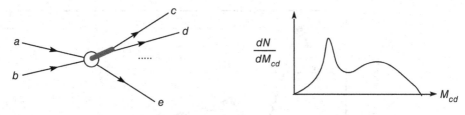

Fig. 4.3.   Schematic of a resonance production study.

a process with a final state that contains those two particles (c and d) and, at least, another one, for example

$$a + b \rightarrow c + d + e. \tag{4.8}$$

If the unstable particle R decaying as $R \rightarrow c + d$ exists, the above reaction proceeds, at least in a fraction of cases, through an intermediate state containing R, namely

$$a + b \rightarrow R + e \rightarrow c + d + e. \tag{4.9}$$

In these cases, which are examples of the 'resonant process', the mass of the system $c + d$, call it M_{cd}, is expected to be equal to M_R or, better, to have a Breit–Wigner distribution peaked at M_R with width Γ. If, on the other hand, the reaction goes directly to the final state (non-resonant process), M_{cd} can have any value within the constraints of energy and momentum conservation. Its distribution is smooth, without peaks, given substantially by the phase-space factor.

We then measure for each event (4.7) the energies and the momenta of the final particles and compute

$$M_{cd} = \sqrt{\left(E_c + E_d\right)^2 - \left(\mathbf{p}_c + \mathbf{p}_d\right)^2}. \tag{4.10}$$

The resonance appears, as sketched in Fig. 4.3, as a peak on a smooth background in the M_{cd} distribution.

Obviously, by the same method one can search for resonances decaying in more than two particles, computing the mass of such systems. Notice, however, that the simple observation of a peak is not enough to establish a resonance. Much more detailed study is necessary.

4.2 The 3 / 2$^+$ Baryons

Up to now we have encountered eight baryons, all with spin 1/2 and positive parity. They are stable, the proton, or metastable because their masses are large enough to allow weak or electromagnetic decays, but not strong decays. Many other baryons exist, both strange and non-strange, with both positive and negative parity and different spin values. These have larger masses and decay strongly. We shall consider only the ground-level baryons that have spin parity 1/2$^+$ or 3/2$^+$.

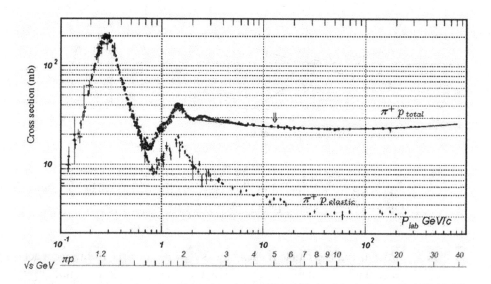

Fig. 4.4. The π^+p total and elastic cross-sections as a function of beam momentum and centre of mass energy. Adapted by permission of Particle Data Group, Lawrence Berkeley National Lab.

The search for strongly decaying baryons follows the principles described in the previous section. Considering first the formation experiments, let us see the nature of the possible targets and beams. To form a baryon, we need a baryon target and a meson beam.

The target must be an elementary particle. In practice, we can have free protons using hydrogen, in the liquid phase for adequate luminosity. We cannot have free neutrons and we must use the simplest nucleus containing neutrons (i.e. deuterium).

Fig. 4.4 shows the total π^+p cross-section and the elastic cross-section, namely that of

$$\pi^+ + p \to \pi^+ + p.$$

We see that, at low energy, the two cross-sections are equal, and the process is purely elastic. This is because no other channel is open. At higher energies, gradually, more channels open up: $\pi^+ + p \to \pi^+ + \pi^0 + p$, $\pi^+ + p \to \pi^+ + \pi^+ + \pi^- + p$, $\pi^+ + p \to \pi^+ + K^+ + K^- + p$, etc., and the elastic cross-section is only a fraction of the total.

A meson beam can be built only if the meson lifetime is long enough. There are not many of them, pions and K-mesons. With charged pions, both positive and negative, the formation of $S = 0$ baryons has been systematically studied, with K^- beams the formation of the $S = -1$ baryons and finally with K^+ beams that of the $S = +1$ baryons.

Systematic investigations carried out with several experiments in all the accelerator laboratories in the 1960s led to the following well-established conclusions. Several dozen resonances exist in the pion–nucleon and K^-–nucleon systems. Their isospins have been measured by comparing the cross-sections in the different charge states, as in Example 3.7. Their spins and parities have been determined by measuring the angular differential cross-sections and performing partial wave analysis.

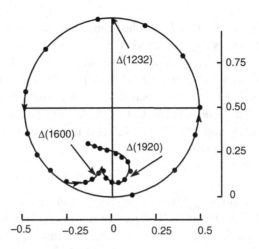

Argand diagram for the $\pi^+ p$ elastic cross-sections.

On the contrary, no resonance exists in the K^+–nucleon system, namely no positively strange $(S = +1)$ baryon $(B = 1)$ exists in nature.

Let us consider the isospin representations now. The pions are an isospin triplet, namely they are in the representation **3**, and the nucleons are a doublet, in the representation **2**. The composition rule gives $\mathbf{3} \otimes \mathbf{2} = \mathbf{2} \oplus \mathbf{4}$. Therefore, the pion–nucleon resonances can be doublets or quartets or, in other words, can have isospin $1/2$ or $3/2$. By convention the $I = 1/2$ states are called $N(\text{xxxx})$, and the $I = 3/2$ ones are called $\Delta(\text{xxxx})$ where xxxx is the mass in MeV.

The first resonance was the discovery of Fermi and collaborators (Anderson *et al.* 1952) working with positive and negative pion beams from the Chicago cyclotron. It is the huge peak in the elastic region, the $\Delta(1232)$ in Fig. 4.4. As we have already discussed, its isospin is $I = 3/2$. Its four charge states are: Δ^-, Δ^0, Δ^+ and Δ^{++}. The partial wave analysis of the angular differential cross-sections for the elastic scattering leads to the Argand diagram shown in Fig. 4.5.

The $\Delta(1232)$ appears as an elastic pure Breit–Wigner resonance going through the unitary circle. This analysis establishes that the orbital momentum of the pion–nucleon system is $L = 1$ and that the total angular momentum is $3/2$. The parity is $(-1)^L$ times the product of the pion and nucleon intrinsic parity, which is -1. In conclusion, $J^P = 3/2^+$.

Fig. 4.4 shows other peaks corresponding to more resonances. However, the vast majority of the resonances cannot be seen 'by eye'; rather, they are found in the Argand diagram. Resonances are marked by a rapid transition of the phase through an extreme, which might even be different from $\pi/2$, possibly superposed on a non-resonant contribution. Two examples, $\Delta(1600)$ and $\Delta(1920)$, both with $J^P = 3/2^+$, are shown in Fig. 4.5.

As a final observation that will be useful in the following, note that as the energy increases the resonances disappear and the cross-section reaches a value of about $\sigma_{\pi+p} \approx 25$ mb and is slowly increasing.

We now consider the hyperons with $S = -1$. They can be formed from, or decay into, a $\bar{K}N$ system (where N stands for nucleon). The kaon and the nucleon are isospin doublets. Following the combination rule $\mathbf{2} \otimes \mathbf{2} = \mathbf{1} \oplus \mathbf{3}$, they can form hyperons with isospin $I = 0$ or $I = 1$. The former are called $\Lambda(\text{xxxx})$, the latter are called $\Sigma(\text{xxxx})$, where xxxx is the mass in MeV.

An $S = -1$ hyperon with mass smaller than the sum of the K meson and proton masses $(494 + 938 = 1432 \text{ MeV})$ cannot be observed as a resonance in the $K^- + p \to K^- + p$ cross-section. Actually, the lowest mass $S = -1$ baryonic system is $\Lambda \pi^{\pm} (m_{\Lambda} + m_{\pi} = 1115 + 140 = 1255 \text{ MeV})$. To search in the mass range between 1255 MeV and 1432 MeV, we must use the production method, searching for $\Lambda \pi^{\pm}$ resonances in the final state of

$$K^- + p \to \Lambda + \pi^+ + \pi^-. \tag{4.11}$$

This was done by Alvarez *et al.* (1963) using a K^- beam with 1.5 GeV momentum produced by the Bevatron. The detector was the 72″ bubble chamber discussed in Section 1.13.

Notice that there are two $\Lambda \pi$ charge states in (4.11). Therefore, there are two possible intermediate states, which are called Σ^*, leading to the same final state

$$\begin{aligned} K^- + p &\to \Sigma^{*+} + \pi^- \to \left(\Lambda + \pi^+\right) + \pi^- \\ K^- + p &\to \Sigma^{*-} + \pi^+ \to \left(\Lambda + \pi^-\right) + \pi^+. \end{aligned} \tag{4.12}$$

We must then consider both masses $\Lambda \pi^+$ and $\Lambda \pi^-$. In the plot in Fig. 4.6, every point is an event. The scales of the axes are the squares of the masses, not the masses, because, as we shall see in Section 4.3, the phase-space volume is uniform in those variables. Therefore, any non-uniformity corresponds to a dynamic feature. The contour of the plot is given by the energy and momentum conservation.

Looking at Fig. 4.6, we clearly see two perpendicular high-density bands. Each of the projections on the axes would show a peak corresponding to the band perpendicular to that axis, similar to that sketched in Fig. 4.3. The bands appear at the same value of the mass, $M = 1385 \text{ MeV}$, and have the same width, $\Gamma = 35 \text{ MeV}$. Finally, being $\Lambda \pi^+$ and $\Lambda \pi^-$ pure $I = 1$ states, the isospin of the observed hyperon is one. It is called $\Sigma(1385)$. The analysis of the angular distributions of its daughters shows that $J^P = 3/2^+$.

Hyperons with isospin $I = 0$ and strangeness $S = -1$ (such as the Λ) and $J^P = 3/2^+$ have been sought, but do not exist. Again, this fact is well explained by the quark model.

The search for $S = -2$ hyperons can be done only in production, looking for possible resonances in the $\Xi \pi$ systems in different charge states. They are generically called $\Xi(\text{xxxx})$. The lowest mass state, $\Xi(1530)$ or Ξ^*, was found using a K^- beam with 1.8 GeV momentum from the Bevatron and the 72″ hydrogen bubble chamber (Pjerrou *et al.* 1962; Schlein *et al.* 1963).

The Ξ^* was observed in the two charge states Ξ^{*0} and Ξ^{*-}. We focus on the former, which was observed in the two reactions

$$K^- + p \to \Xi^{*0} + K^0 \to \left(\Xi^- + \pi^+\right) + K^0 \tag{4.13}$$

$$K^- + p \to \Xi^{*0} + K^0 \to \left(\Xi^0 + \pi^0\right) + K^0. \tag{4.14}$$

Fig. 4.7 shows the $\Xi^- \pi^+$ mass distribution. In this case, the reaction is completely dominated by the resonance. The dashed line is the estimate of the non-resonant background.

The continuous curve is a Breit–Wigner shape obtained as the best fit to the data leaving the mass and the width as free parameters. The result is

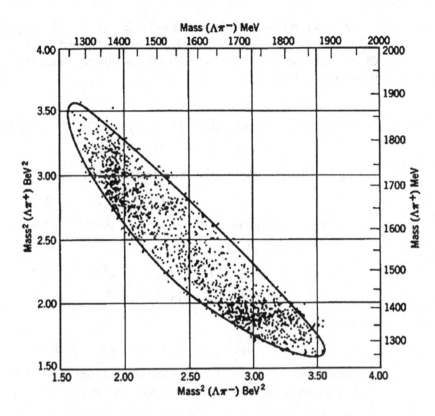

Fig. 4.6. $\Lambda\pi^+$ square mass vs. $\Lambda\pi^-$ square mass for reaction (4.12) (Figure reprinted from Alvarez *et al. Phys. Rev. Lett.* 10 184 1963 Fig. 1. © (1963) by American Physical Society. https://doi.org/10.1103/PhysRevLett.10.184).

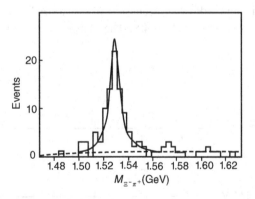

Fig. 4.7.   The $\Xi^-\pi^+$ mass distribution from the 72″ bubble chamber at the Bevatron.

$$m = 1529 \pm 5 \text{ MeV}; \quad \Gamma = 7.2 \pm 2 \text{ MeV} \tag{4.15}$$

The $\Xi^0\pi^0$ mass distribution from (4.14) shows a similar peak, about $1/2$ in height. This result determines the isospin. Let us see how.

Since the Ξ^* decays into $\Xi\pi$, it can have isospin $1/2$ or $3/2$.

The third component of the isospin of the Ξ^{*0} is $+1/2$. It decays with isospin conservation into a Ξ, which has isospin $1/2$, and a π, which has isospin 1. We use the Clebsch–Gordan coefficients to find the weights of the charge states. If the isospin of the Ξ^* is $3/2$, we have

$$
\begin{aligned}
\left|\Xi^{*0}\right\rangle &= \left|3/2,+1/2\right\rangle \\
&= \sqrt{1/3}\left|1/2,-1/2\right\rangle\left|1,+1\right\rangle + \sqrt{2/3}\left|1/2,+1/2\right\rangle\left|1,0\right\rangle \quad\quad (4.16)\\
&= \sqrt{1/3}\left|\Xi^-\pi^+\right\rangle + \sqrt{2/3}\left|\Xi^0\pi^0\right\rangle
\end{aligned}
$$

Taking the squares of the amplitudes, we immediately find

$$
\Gamma\left(\Xi^{*0}\rightarrow\Xi^0\pi^0\right)/\Gamma\left(\Xi^{*0}\rightarrow\Xi^-\pi^+\right) = 2,
$$

in contradiction to experiment. If the isospin of the Ξ^* is $1/2$, we have

$$
\begin{aligned}
\left|\Xi^{*0}\right\rangle &= \left|1/2,+1/2\right\rangle \\
&= \sqrt{2/3}\left|1/2,-1/2\right\rangle\left|1,+1\right\rangle + \sqrt{1/3}\left|1/2,+1/2\right\rangle\left|1,0\right\rangle \quad\quad (4.17)\\
&= \sqrt{2/3}\left|\Xi^-\pi^+\right\rangle + \sqrt{1/3}\left|\Xi^0\pi^0\right\rangle
\end{aligned}
$$

Hence

$$
\Gamma\left(\Xi^{*0}\rightarrow\Xi^0\pi^0\right)/\Gamma\left(\Xi^{*0}\rightarrow\Xi^-\pi^+\right) = 1/2,
$$

in agreement with experiment. In conclusion, the isospin of the Ξ^* is $1/2$.

The analysis of the angular distributions determines the spin parity as $J^P = 3/2^+$.

Question 4.1 For each value of the isospin and for every value of its third component, list the charge states of the $\Xi\pi$ system. Use the Gell–Mann and Nishijima formula (3.39). □

This concludes the discussion of the $J^P = 3/2^+$ strongly decaying baryons, which are summarized in Fig. 4.8. The comparison with the $J^P = 1/2^+$ baryons (Fig. 3.1)

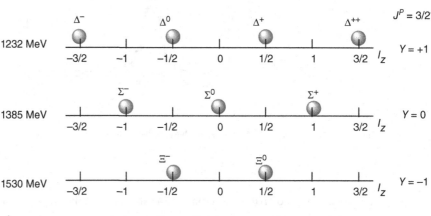

Fig. 4.8. The $J^P = 3/2^+$ baryons with strong decay.

shows the deep difference between the two cases, another feature explained by the quark model.

4.3 The Dalitz Plot

In Section 1.8 we calculated the phase-space volume for a two-particle final state. The next, more complicated, case is the three-particle state. Several important processes, both decays and collisions, have such final states. We shall now discuss the kinematics and the phase-space of systems of three particles of masses m_1, m_2 and m_3. Whether it is the final state of a decay or of a collision is obviously irrelevant.

We shall treat the problem in the CM reference frame, where things are simpler. We shall call the CM energy M, the mass of the mother particle in the case of a decay, and we shall call $\mathbf{p}_1$, $\mathbf{p}_2$ and $\mathbf{p}_3$ the momenta and E_1, E_2 and E_3 the energies of the three particles. The constraints between these variables are the following. Since the masses are given, the energies are determined by the momenta. The nine components of the momenta must satisfy three conditions for momentum conservation, $\mathbf{p}_1 + \mathbf{p}_2 + \mathbf{p}_3 = \mathbf{0}$ and one condition for energy conservation, $E_1 + E_2 + E_3 = M$. We are left with five independent variables.

Since the three momenta add up to the null vector, they are coplanar. Let $\mathbf{n}$ be the unit vector normal to this plane. We choose two of the independent variables as the two angles that define the direction of $\mathbf{n}$. The triangle defined by the three momenta can rotate rigidly in the plane; we take the angle that defines the orientation of the triangle as the third variable.

The last two variables define the shape and the size of the triangle. If we are not interested in the polarization of the initial state, the case to which we limit our discussion, the dependence on the angles of the matrix element is irrelevant and we can describe the final state with the last two variables only. There are a few equivalent choices: two energies, say E_1 and E_2, two kinetic energies, T_1 and T_2, and the masses squared of two pairs. These variables are linked by linear relationships. Take, for example, m_{23}^2:

$$m_{23}^2 = \left(E_2 + E_3\right)^2 - \left(\mathbf{p}_2 + \mathbf{p}_3\right)^2 = \left(M - E_1\right)^2 - \mathbf{p}_1^2 = M^2 + m_1^2 - 2ME_1. \quad (4.18)$$

In conclusion, a configuration of our three-particle system can be represented in a plane defined by, say m_{13}^2 vs. m_{12}^2, as in Fig. 4.9. The loci of the configurations with $m_{23}^2 = $ constant are straight lines such as the one shown in the Fig. 4.9. The above considered diagram in Fig. 4.6 is a Dalitz plot.

The closed line, the contour of the plot, delimits the region allowed by energy and momentum conservation. The diagram goes by the name of R. H. Dalitz (Dalitz 1956), who pointed out that the elements of area in this plane are proportional to the phase-space volume. We have already mentioned this property and we shall now prove it.

From (1.70), the phase-space volume for a three-body system, ignoring constant factors that are irrelevant here, is

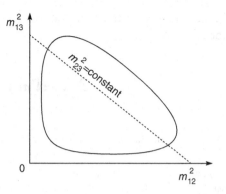

Fig. 4.9.   The Dalitz plot.

$$R_3 \propto \int \frac{d^3 p_1}{E_1} \frac{d^3 p_2}{E_2} \frac{d^3 p_3}{E_3} \delta\left(E_1 + E_2 + E_3 - M\right)\delta^3\left(\mathbf{p}_1 + \mathbf{p}_2 + \mathbf{p}_3\right). \tag{4.19}$$

Integrating on $\mathbf{p}_3$, we obtain

$$R_3 \propto \int \frac{1}{E_1 E_2 E_3} d^3 p_1 d\Omega_1 d^3 p_2 d\Omega_2 \delta\left(E_1 + E_2 + E_3 - M\right). \tag{4.20}$$

We must now integrate over the angles made by the two vectors $\mathbf{p}_1$ and $\mathbf{p}_2$. The following choice is convenient. We fix the angle θ_{12} between $\mathbf{p}_1$ and $\mathbf{p}_2$ and integrate over θ_1, ϕ_1 and ϕ_2. We have

$$R_3 \propto \int \frac{1}{E_1 E_2 E_3} 4\pi p_1^2 dp_1 2\pi p_2^2 dp_2 d\left(\cos\theta_{12}\right)\delta\left(E_1 + E_2 + E_3 - M\right). \tag{4.21}$$

We now use the momentum conservation $\mathbf{p}_3 = -\mathbf{p}_1 - \mathbf{p}_2$, which gives $p_3^2 = p_1^2 + p_2^2 + 2p_1 p_2 \cos\theta_{12}$. Differentiating this expression, keeping p_1 and p_2 constant, we obtain $2 p_3 dp_3 = 2 p_1 p_2 d(\cos\theta_{12})$. By substituting into (4.21), we have

$$R_3 \propto \int \frac{p_1 dp_1 p_2 dp_2 p_3 dp_3}{E_1 E_2 E_3} \delta\left(E_1 + E_2 + E_3 - M\right). \tag{4.22}$$

Differentiating the relationships $E_i^2 = p_i^2 + m_i^2$ we have $p_i dp_i = E_i dE_i$, hence

$$R_3 \propto \int dE_1 dE_2 dE_3 \delta\left(E_1 + E_2 + E_3 - M\right). \tag{4.23}$$

Finally, using the remaining δ-function, we arrive at the conclusion

$$R_3 \propto \int dE_1 dE_2 \propto \int dm_{23}^2 dm_{13}^2 \propto \int dT_1 dT_2. \tag{4.24}$$

This is what we had to prove. The expressions in the last two members are obvious consequences of the linear relationships between all these pairs of variables.

In the next section, we shall use the Dalitz plot for the spin and parity analysis of three-pion systems. Since the three particles are equal, the plot is geometrically symmetrical. Actually, R. H. Dalitz invented the plot for his analysis of the decay of the K-meson into 3π. Let us see his reasoning.

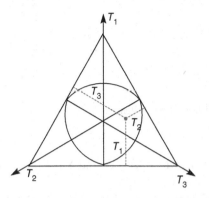

Fig. 4.10.   The 3π Dalitz plot.

The sum of the three CM kinetic energies $T_1 + T_2 + T_3$ is the same for every 3π configuration. Now let us consider a triangle, which we take to be equilateral because the three particles have the same mass. The sum of the distances from the sides is the same for every point inside the triangle. Therefore, if the kinetic energies are measured by these distances, energy conservation is automatically satisfied. Momentum conservation limits the allowed region inside a closed curve, which is tangent to the three sides. The diagram, shown in Fig. 4.10, is equivalent to that in Fig. 4.9. The former explicitly shows the symmetry of the problem.

This diagram is extremely useful in the study of the quantum numbers of the mesons that decay into 3π. Actually, the dependence of the decay matrix element on the position of the representative point in the graph is determined by the angular momentum, the parity and the isospin of the system. In practice, it is necessary to collect a sizeable number of decays of the meson to be studied. Each event is represented as a dot on the Dalitz plot. Provided there are sufficient statistics, every non-uniformity in the point density has to be ascribed to a corresponding variation of the matrix element and the quantum numbers can be determined.

4.4 Spin, Parity and Isospin Analysis of Three-Pion Systems

Three important mesons decay into 3π: the K, which we have already met, the η and the ω, which we shall study in the following sections. In each case, the 3π final system is in a well-defined spin parity and isospin state. The corresponding symmetry conditions of the decay matrix element lead to observable characteristics of the event distributions in the Dalitz plot and hence to the determination of the quantum numbers of the final state. It is important to note that these are not necessarily those of the parent mesons. If the decay is strong, as for the ω, all quantum numbers are conserved; if it is electromagnetic, the isospin is not conserved, as in the case of the η; if it is weak, parity is also violated, as in the case of the K-meson. Historically, these were the

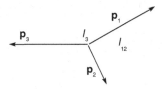

Fig. 4.11. Kinematic variables in a 3π system.

considerations which led Lee and Yang to the hypothesis of parity non-conservation in 1956 (Lee & Yang 1956).

A complete treatment of the methods of analysis for obtaining the quantum numbers of a 3π system was made by C. Zemach (Zemach 1964). Here, we shall limit our discussion to the simplest spin parity assignments. We shall focus on the most conspicuous features, namely the regions of the Dalitz plot where this assignment predicts vanishing density.

We shall now see how to construct a general matrix element $\mathcal{M}$ for the 3π decay of a meson of mass M, taking into account the constraints imposed by isospin, spin and parity and by the Bose statistics. The matrix element $\mathcal{M}$ is Lorentz-invariant, but it can be shown that it reduces to a three-dimensional invariant in the CM system. Therefore, we shall work in the CM reference frame.

Let us begin by dealing with the isospin. The isospin of a 3π system can be any integer number between 0 and 3. As mesons with isospin larger than 1 have not been found, we shall limit our discussion to two possibilities, 0 and 1.

The three pions can have several charge states. We shall consider only those in which all pions are charged or two are charged and one is neutral. Actually, these are the states that can be observed in a hydrogen bubble chamber, the instrument that made the greatest contribution to hadron spectroscopy. There are three such charge combinations: $\pi^+\pi^+\pi^-$ and its charge conjugate $\pi^-\pi^-\pi^+$, which have isospin 1, and $\pi^+\pi^-\pi^0$, which may have isospin 0 or 1.

We now list the elements at our disposal to build $\mathcal{M}$. These must be covariant quantities in three dimensions (scalars, pseudoscalars, vectors or axial vectors).

We start by choosing an order π_1, π_2 and π_3 of the pions and call their momenta in the CM reference frame $\mathbf{p}_1$, $\mathbf{p}_2$ and $\mathbf{p}_3$. These give us two independent vectors, taking into account the relationship

$$\mathbf{p}_1 + \mathbf{p}_2 + \mathbf{p}_3 = 0. \tag{4.25}$$

Fig. 4.11 shows the variables we shall use, in the plane of the three momenta.

There is one axial vector, which is normal to the plane

$$\mathbf{q} = \mathbf{p}_1 \times \mathbf{p}_2 = \mathbf{p}_2 \times \mathbf{p}_3 = \mathbf{p}_3 \times \mathbf{p}_1. \tag{4.26}$$

There are four scalar quantities: one is simply a constant, the other are the energies E_1, E_2 and E_3, which are linked by the relationship

$$E_1 + E_2 + E_3 = M. \tag{4.27}$$

We choose, arbitrarily, two pions, π_1 and π_2 which we call the 'dipion'. We call I_{12} and I_3 the isospin of the dipion and of π_3 respectively. I_{12} can be 0, 1 or 2, while obviously $I_3 = 1$. The total isospin is

$$\mathbf{I} = \mathbf{I}_{12} \otimes \mathbf{I}_3. \tag{4.28}$$

Similarly, let l_{12} be the orbital angular momentum of π_1 and π_2 (in their centre of mass frame). This is also the total angular momentum of the dipion, because the pions have no spin. Let l_3 be the orbital angular momentum between π_3 and the dipion. The total angular momentum is

$$\mathbf{J} = \mathbf{l}_{12} \otimes \mathbf{l}_3. \tag{4.29}$$

Now, let us prove that the spin parity of a 3π system cannot be $J^P = 0^+$. Actually, $J = 0$ implies $l_{12} = l_3$. The total parity is the product of the three intrinsic parities and of the two orbital parities, namely $P = (-1)^3 (-1)^{l_{12}} (-1)^{l_3} = -1$. Hence $J^P = 0^+$ is impossible.

We shall limit this discussion to the three simplest spin parity assignments: $J^P = 0^-$, 1^- and 1^+. This makes six cases in total, taking into account the two possible isospins $I = 0$ and $I = 1$.

For an assumed spin parity J^P of the three-pion system, taking into account that their intrinsic parity is -1, the space part of the amplitude, which we must construct, transforms as J^{-P}. For each isospin choice, we must impose the corresponding symmetry properties under exchanges of two pions.

Firstly, let us consider $I = 0$. In this case, it must be $I_{12} = 1$. This is true for every choice of π_3. Consequently, the matrix element is antisymmetric in the exchange of every pair (completely antisymmetric). Consider now the three J^P cases.

$$J^P = 0^-$$

We must construct a completely antisymmetric scalar quantity. We use the energies. To obtain $1 \leftrightarrow 2$ antisymmetry, we write $E_1 - E_2$. Then we antisymmetrise completely

$$\mathcal{M} \propto (E_1 - E_2)(E_2 - E_3)(E_3 - E_1) \tag{4.30}$$

The vanishing density regions, where M is zero, are all the diagonals as shown in Fig. 4.12(a).

$$J^P = 1^-$$

We need an axial vector. We have only one of them, which, as required, is already antisymmetric. Hence

$$\mathcal{M} \propto \mathbf{q} \tag{4.31}$$

At the periphery of the Dalitz plot (i.e. the kinematic limit), two momenta are parallel and $\mathbf{q} = 0$. The situation is shown in Fig. 4.12(b).

$$J^P = 1^+$$

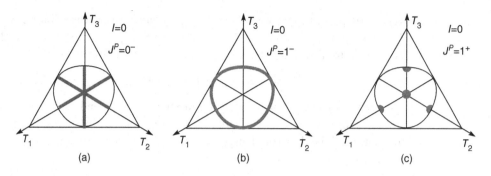

Vanishing density regions for $\pi^0\pi^+\pi^-$ with $I=0$.

We construct a completely antisymmetric vector. We take one of the vectors, $\mathbf{p}_1$, and make it antisymmetric in $2 \leftrightarrow 3$ multiplying by $E_2 - E_3$. We then antisymmetrise completely

$$\mathcal{M} \propto \mathbf{p}_1\left(E_2 - E_3\right) + \mathbf{p}_2\left(E_3 - E_1\right) + \mathbf{p}_3\left(E_1 - E_2\right) \tag{4.32}$$

The three energies are equal at the centre of the plot, hence $\mathcal{M} = 0$. Consider the vertex of a diagonal, for example that corresponding to T_3. Here $\mathbf{p}_1 = \mathbf{p}_2 = -\mathbf{p}_3/2$, hence $E_1 = E_2$ and $\mathcal{M} = 0$. The result is shown in Fig. 4.12(c).

Let us now proceed to $I = 1$. We have two charge states to consider.

$\pi^+\pi^+\pi^-$ (and its charge conjugate)

Let us take as π_3 the different one, $\pi_3 = \pi^-$. Since the other two are identical, the amplitude is symmetric under the exchange $1 \leftrightarrow 2$

$$\pi^0\pi^+\pi^-.$$

We take $\pi_3 = \pi^0$. First we show that $I_{12} = 1$ is forbidden. If this were the case, the isospin of the dipion would be $\left|I_{12}, I_{12,z}\right\rangle = \left|1,0\right\rangle$ and for π_3, obviously, $\left|I_3, I_{3,z}\right\rangle = \left|1,0\right\rangle$; these should total $\left|I, I_z\right\rangle = \left|1,0\right\rangle$. However, this cannot be, since the corresponding Clebsch–Gordan coefficient is zero. We are left with $I_{12} = 0$ or $I_{12} = 2$. In both cases the state is symmetric under the exchange $1 \leftrightarrow 2$.

To sum up, the amplitude must be symmetric under the exchange $1 \leftrightarrow 2$ in both cases. Now, let us move on to the three spin parities.

$J^P = 0^-$

We need a scalar, symmetric in $1 \leftrightarrow 2$. The simplest are E_3 and a constant.

$$\mathcal{M} \propto \text{constant}, \quad \mathcal{M} \propto E_3. \tag{4.33}$$

In both cases, there are no vanishing density regions. See Fig. 4.13(a).

$J^P = 1^-$

We need an axial vector, symmetric in $1 \leftrightarrow 2$. The only axial vector that we have, $\mathbf{q}$, is antisymmetric. We make it symmetric by multiplying it by an antisymmetric (in $1 \leftrightarrow 2$) scalar quantity

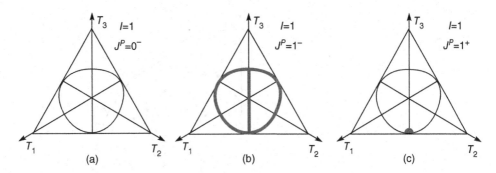

Fig. 4.13. Vanishing density regions for the states $\pi^+\pi^+\pi^-$ and $\pi^0\pi^+\pi^-$ with $I = 1$.

$$\mathcal{M} \propto \mathbf{q}(E_1 - E_2) \tag{4.34}$$

The vanishing points are on the periphery ($\mathbf{q}$) and on the vertical diagonal $(E_1 - E_2)$. See Fig. 4.13(b).

$$J^P = 1^+$$

We need a vector, symmetric in 1,2. We take

$$\mathcal{M} \propto \mathbf{p}_1 + \mathbf{p}_2 = -\mathbf{p}_3 \tag{4.35}$$

It is zero at the foot of the vertical diagonal where $T_3 = 0$, hence $\mathbf{p}_3 = 0$. See Fig. 4.13(c).

4.5 Pseudoscalar and Vector Mesons

Fig. 3.3 showed the pseudoscalar mesons (i.e. those with spin parity 0^-) that we have discussed so far. We have also discussed the measurement of the spin and parity of the charged and neutral pion. We shall now discuss the spin and parity of the charged K^- meson. Two more pseudoscalar mesons exist, called η and η', that we shall discuss soon. Counting all their charge states, the pseudoscalar mesons are nine in number.

There are as many vector mesons, namely with spin parity 1^-, forming isospin multiplets identical to those of the pseudoscalar mesons. We shall deal only with one of these, the ω, and simply list the quantum numbers of the others.

Let us go back, historically, to the early 1950s, when cosmic ray experiments had shown several decay topologies of strange particles of similar masses and lifetimes. This was initially interpreted as evidence of three particles. One of them, decaying into 2π, was called θ; the second, decaying into 3π, was called τ; and the third, with a number of different decays, was called K. The situation was clarified by the G-stack (great-stack) experiment, proposed by M. Merlin in 1953. A 15 l emulsion (63 kg) stack was flown at 27 000 m height on an aerostatic balloon over the Po valley in 1954. The large volume of the emulsion allowed, in several cases, for the complete containment of the tracks of the decays. This, in turn, made it possible to accurately

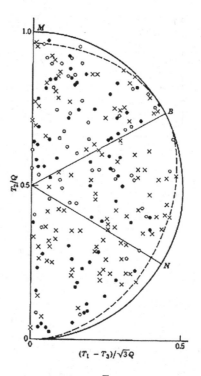

Fig. 4.14. Dalitz plot for 'τ events'. The kinetic energies are divided by $\sqrt{3}$ times the Q value of the decay, namely the difference between initial mass and sum of the final ones. The diagram is folded about the vertical axis, because of two equal pions (Reprinted figure with permission from J. Orear, G. Harris and S. Taylor *Spin and Parity Analysis of Bevatron τ Mesons. Phys. Rev.* 102, 1676, Fig. 2 ©(1956) by American Physical Society DOI https://doi.org/10.1103/PhysRev.102.1676).

measure their energies from the range, and determine with accuracy the mass of the mother particle. A huge analysis and measurement effort followed, and the results were rewarding. The different decay modes were clearly identified and the masses accurately measured. It became clear that different decay modes of the same particle were being observed (Davies *et al.* 1955).

Around the same time, the Bevatron became operational at Berkeley, providing the first K^+ beams. Emulsion exposures allowed accurate measurements of the masses of the strange mesons. The masses of the θ and τ mesons were equal, to within a few per thousand. It was impossible to escape the conclusion that θ and τ were the same particle, now called the K-meson.

Fig. 4.14 shows the Dalitz plot of 220 τ events (i.e. $K^\pm \rightarrow \pi^\pm + \pi^+ + \pi^-$) collected from different experiments. The spin parity analysis was originally done by R. H. Dalitz (Dalitz 1956). We use the conclusions of Section 4.4, in particular with regard to Figs. 4.12 and 4.13, and compare the vanishing density regions shown in those figures with the data in Fig. 4.14. The distribution of the data does not show any depletion, and we conclude that the 3π state has $I = 1$ and $J^P = 0^-$.

However, the K^+ (the θ events) also decays into two pions, $K^+ \rightarrow \pi^+ + \pi^0$. In this case, if $J = 0$, the parity must be positive. The problem became known as the

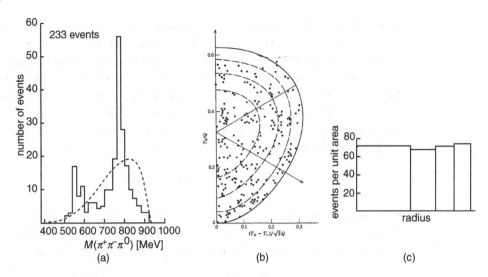

Fig. 4.15. (a) The $\pi^+\pi^-\pi^0$ mass distribution. The dashed curve is the phase-space (Figure reprinted from Pevsner *et al. Phys. Rev. Lett.* 7 421 1961 Fig. 2. © (1961) by American Physical Society, https://doi.org/10.1103/PhysRevLett.7.421). (b) The Dalitz plot of the decay $\eta \to \pi^+\pi^-\pi^0$. (c) Events in equal area zones (Figure reprinted from Alff *et al. Phys. Rev. Lett.* 9 325 1962 Fig. 3. ©(1962) by American Physical Society, https://doi.org/10.1103/PhysRevLett.9.325).

$\theta - \tau$ puzzle. The puzzle was solved, upon suggestion of M. Bloch, by T. D. Lee and C. N. Yang (Lee & Yang 1956) who developed the revolutionary Bloch's hypothesis that the weak interactions violate parity conservation. The hypothesis was confirmed experimentally in the same year, as we shall discuss in Chapter 7.

We conclude by observing that the K^+ decay also violates the isospin conservation, by $\Delta I = 1/2$.

We now turn to the η meson. It was discovered by M. Bloch, A. Pevsner and collaborators (Pevsner *et al.* 1961) in the reaction

$$\pi^+ + d \to \pi^+ + \pi^- + \pi^0 + p + p \tag{4.36}$$

with the 72″ Alvarez bubble chamber filled with liquid deuterium and exposed to a π^+ beam with 1.23 GeV momentum. Fig. 4.15(a) shows the $\pi^+\pi^-\pi^0$ mass distribution. Two resonances are clearly seen: the η at a lower mass, and the ω, which we shall discuss shortly, at a higher mass.

No charged state of the η has ever been observed, hence its isospin $I = 0$. The mass and the width are

$$m_\eta = 548 \text{ MeV} \qquad \Gamma_\eta = 1.3 \text{ MeV}. \tag{4.37}$$

Very soon, the typically electromagnetic decay $\eta \to 2\gamma$ was observed to happen with a probability similar to that of the decay into $\pi^+\pi^-\pi^0$. The present values of the branching ratios are

$$\Gamma\left(\eta \to \pi^+ + \pi^- + \pi^0\right)/\Gamma_{\text{tot}} = 28\% \qquad \Gamma\left(\eta \to 2\gamma\right)/\Gamma_{\text{tot}} = 39.4\%. \tag{4.38}$$

The 2γ decay establishes that the charge conjugation of the η is $C = +1$. Since $I = 0$, the G-parity is $G = C = +1$. However, the 3π final state has $G = -1$, hence $I = 1$. Therefore, this decay violates the isospin and cannot be strong. It is electromagnetic.

Turning to the spin parity, Fig. 4.15(b) shows the Dalitz plot. We compare it with Fig. 4.13, which is relevant for $I = 1$. Again, the only case that has no zeros, in agreement with the uniform experimental distribution, is $J^P = 0^-$. We understand why the G-conserving decay into 2π is forbidden, namely because 2π cannot have $J^P = 0^-$. On the other hand, the decay into 4π is forbidden by energy conservation.

The η' Meson

Without entering into any detail, we simply state that another pseudoscalar, zero-isospin, meson exists (Goldberg *et al.* 1964; Kalbfleish *et al.* 1964). It is called η' and has the same quantum numbers as the η. Its mass and its width are

$$m_{\eta'} = 958 \text{ MeV} \qquad \Gamma_{\eta'} = 0.2 \text{ MeV}. \tag{4.39}$$

The η' meson has, like the η, important electromagnetic decays ($2\gamma, \omega\gamma, \rho\gamma$) and a small width. Surprisingly, its mass is enormous, when compared with pions.

Question 4.2 Given the information on their main decay channels, could the spin of the η be 1? And that of the η'?

The ω Meson

This was discovered in 1961 in the reaction

$$\bar{p} + p \rightarrow \pi^+ + \pi^+ + \pi^0 + \pi^- + \pi^- \tag{4.40}$$

in the 72″ Alvarez hydrogen bubble chamber exposed to the antiproton beam of the Bevatron (Maglić *et al.* 1961; Alff *et al.* 1962).

Note that, for every event, four pion triplets with zero charge exist, four with unit (both signs) charge and two of double charge. Fig. 4.16(a) shows the distributions of these three charge state triplets. A narrow resonance clearly appears in the neutral combination, not in the charged ones. This fixes the isospin, $I = 0$. The ω mass and width are

$$m_\omega = 782 \text{ MeV} \qquad \Gamma_\omega = 8 \text{ MeV}. \tag{4.41}$$

Fig. 4.16(b) shows the Dalitz plot of the $\pi^+\pi^-\pi^0$ combinations, chosen to have a mass in the peak region. Notice that this sample includes not only ω decays, but also background events. The same figure shows the radial distribution of the events, in radial zones of equal areas. The curve is the square of the (4.31) matrix element summed to a background, assumed to be constant. The agreement establishes $J^P = 1^-$.

We shall now consider the other vector mesons $\left(J^P = 1^-\right)$ without entering into any detail, but only summarizing their properties. The isospin multiplets of the vector mesons are equal to those of the pseudoscalar mesons.

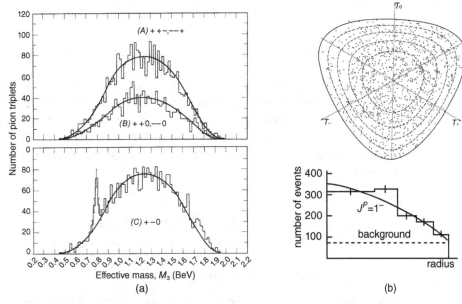

Fig. 4.16. (a) Mass distributions of the triplets with charge 0, 1 and 2 (L. Alvarez, © The Nobel Foundation 1968 Fig. 19 in this link www.nobelprize.org/uploads/2018/06/alvarez-lecture.pdf). (b) Dalitz plot of the decay $\omega \to \pi^+ \pi^- \pi^0$ and numbers of events in equal area zones (Figure reprinted from Alff *et al. Phys. Rev. Lett.* 9 325 1962 Fig. 1. ©(1962) by American Physical Society, https://doi.org/10.1103/PhysRevLett.9.325).

The ρ Meson

This has three charge states, ρ^+, ρ^0 and ρ^-, and isospin $I = 1$. It decays mainly into 2π, hence $G = +1$. It follows that the neutral state ρ^0 has $C = -1$. Its mass is $m \approx 770$ MeV and its width is $\Gamma \approx 150$ MeV (Erwin *et al.* 1961).

The K^* and $\bar{K}^*$ Mesons

The charge states of the former are $+1$ and 0; those of the latter are -1 and 0. Therefore, both have isospin $I = 1/2$. They decay into $K\pi$ and $\bar{K}\pi$ respectively. The mass of both is $m \approx 892$ MeV and the width is $\Gamma \approx 51$ MeV (Alston *et al.* 1961).

The ϕ Meson

This has the same quantum numbers as the ω meson, $J^{PC} = 1^{--}$, $I^G = 0^-$. Its mass is $m = 1019.5$ MeV. Its width is very small, considering its large mass, $\Gamma = 4.25$ MeV (Connolly *et al.* 1963).

An important related feature is as follows. Clearly, the energetically favoured decay channel is the same as that of the ω, namely $\phi \to 3\pi$. However, in this case the branching ratio is small

$$\Gamma(\phi \rightarrow 3\pi)/\Gamma_{\text{tot}} = 15.6\% \tag{4.42}$$

while

$$\Gamma(\phi \rightarrow K\bar{K})/\Gamma_{\text{tot}} = 83\%. \tag{4.43}$$

We experimentally observe that the ϕ meson 'prefers' to decay into K-mesons, even if its mass is barely large enough to allow these decays, $Q = 32$ MeV. This fact hinders the decay and makes the ϕ lifetime long by strong interaction standards. There are two reasons for this behaviour: the quark content of the ϕ, which is mainly $s\bar{s}$, and a dynamical quantum chromodynamics (QCD) property, as we shall see in Chapter 6.

4.6 The Quark Model

We shall now summarize the hadronic states we have met.

- Nine pseudoscalar mesons in two $SU(2)$ singlets, two doublets, one triplet.
- Nine vector mesons in the same multiplets.
- Eight $J^P = 1/2^+$ baryons (and as many antibaryons) in two doublets, with hypercharge $Y = +1$ and $Y = -1$ respectively and a singlet and a triplet with $Y = 0$.
- Nine $J^P = 3/2^+$ baryons in the following multiplets: a quartet with $Y = +1$, a triplet with $Y = 0$ and a doublet with $Y = -1$. As we shall see immediately, we are still missing one hyperon, a singlet with $Y = -2$.

In 1964, G. Zweig (Zweig 1964) and M. Gell-Mann (Gell-Mann 1964) independently proposed that the hadrons are made up of constituents, called quarks by Gell-Mann (and aces by Zweig). Baryons are made of three quarks, the mesons of a quark and an antiquark. It should be noted that the internal structure of the hadrons, as seen by a certain class of experiments, depends on the resolving power. Working at higher momentum transfers, more structures are observed: gluons and additional quark–antiquark pairs. The quark model is a good approximation of the hadrons explored at low resolving power, corresponding to momentum transfers of the order of the GeV. The quarks of the simple structures so observed are called 'valence quarks'. In Chapter 6 we shall discuss the more complicated structure visible at higher resolution.

This scheme extends the isospin internal symmetry, which is based on the group $SU(2)$ to $SU(3)$, a larger unitary group. We immediately stress that the $SU(3)$ symmetry has two very different roles in subnuclear physics: (1) the classification of the hadrons, or rather the hadrons with up (u), down (d) and strange (s) quarks, which we are considering here; and (2) the symmetry of the charges of one of the fundamental forces, the strong one, as we shall discuss in Chapter 6. The two roles are completely different and, to avoid confusion, we shall call the former $SU(3)_f$, where the suffix stands for 'flavour', even if, mathematically, the group is the same in both cases.

Table 4.1. Quantum numbers and masses (from the Particle Data Group) of the three lowest-mass quarks.

	Q	I	I_z	S	C	B	T	$\mathcal{B}$	Y	Mass
d	$-1/3$	$1/2$	$-1/2$	0	0	0	0	$1/3$	$1/3$	$4.67^{+0.48}_{-0.17}$ MeV
u	$+2/3$	$1/2$	$+1/2$	0	0	0	0	$1/3$	$1/3$	$2.16^{+0.49}_{-0.26}$ MeV
s	$-1/3$	0	0	-1	0	0	0	$1/3$	$-2/3$	$93.4^{+8.6}_{-3.4}$ MeV

We now know six different quarks, each with a different flavour. The two quarks present in normal matter, u and d, are an isospin doublet, the former with third component $I_z = +1/2$, the latter with $I_z = -1/2$. We can say that I_z is the flavour of each of them. The flavour of s is the strangeness, which is negative, with the value $S = -1$. The other three quarks that we shall meet in the following sections are the charm quark (c), with flavour which is also called charm, with $C = +1$; the beauty quark (b), with flavour called beauty, $B = -1$; and the top quark (t) with flavour $T = +1$. The sign convention is that the sign of the quark flavour is the same as that of its electric charge.

Table 4.1 gives the quantum numbers of the first three quarks and their masses. Their antiquarks have opposite values of all the quantum numbers, but the isospin I.

The characteristics of the quarks are surprising. Their electric charges and their baryon numbers are fractional. More surprisingly, nobody has ever succeeded in observing a free quark. They have been sought in violent collisions of every type of beam, hadrons, muons, neutrinos, photons, on different targets, they have been looked for in cosmic rays, they were searched for with Millikan-type experiments on ordinary matter and even on rocks brought by the astronauts from the Moon. No quark was ever found. We know today that the reason for that is in the very nature of the colour force that binds the hadrons, as we shall see in Chapter 6. As a quark is never free, its mass cannot be directly measured, but can be indirectly determined, as will be discussed in Chapter 6.

The $SU(3)_f$ representations, namely the multiplets in which the hadrons are grouped, are more complicated than in $SU(2)$. The latter are represented in one dimension, labelling the members by the third component of the isospin. See, for example, the Figs. 3.2, 3.3 and 4.8. Note that $SU(2)$ has as a subgroup $U(1)$, to which the charges of the particles correspond.

Two variables are needed to represent an $SU(3)_f$ multiplet. Therefore, we draw them in a plane taking as axes the third isospin component and the hypercharge. $SU(3)$ has $SU(2)$ as a subgroup and the $SU(3)_f$ multiplets have a substructure made up of $SU(2)$ multiplets.

There are two different fundamental representations, both with dimensions equal to three. They are called **3** and **$\overline{3}$**. We use them to classify the quarks and the antiquarks respectively, as shown in Fig. 4.17. Both **3** and **$\overline{3}$** contain an $SU(2)$ doublet and an $SU(2)$ singlet.

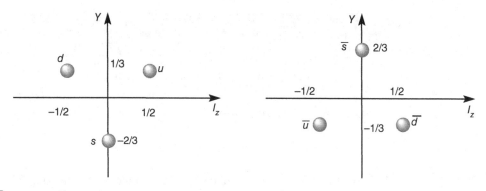

Fig. 4.17.   The 3 and $\overline{3}$ representations.

4.7 Mesons

Mesons are made of a quark and an antiquark. Therefore, they must be members of the multiplets belonging to the product of **3** by $\overline{\mathbf{3}}$. Group theory tells us that

$$\mathbf{3} \otimes \overline{\mathbf{3}} = \mathbf{1} \oplus \mathbf{8}, \tag{4.44}$$

which gives us nine places, exactly the number we need. Let us look at the isospin and hypercharge structure, shown in Fig. 4.18. Notice that in the 'centre' of the octet there are two states with $I_z = Y = 0$, one with $I = 1$ and one with $I = 0$. We see that the octet and the singlet provide exactly the isospin multiplets we need to classify the pseudoscalar and the vector mesons. If the $SU(3)_f$ were exact, all particles in a multiplet would have the same mass. In practice, there are sizeable differences. Unlike the $SU(2)$, which is good for the strong interactions and broken by the electromagnetic interactions, $SU(3)_f$ is already broken by the former.

Now consider the spin parity of the mesons. We expect the quark–antiquark pair to be, for the ground-level mesons we are considering, in the S wave, as is usual for ground states. In this hypothesis the configurations should be $^1S_0$ with $J^{PC} = 0^{-+}$ and $^3S_1$ with $J^{PC} = 1^{--}$. This is just what we observe.

Starting with the pseudoscalar mesons we now must lodge the nine mesons in the octet and in the singlet. We immediately recognize that $\pi^+ = u\overline{d}$, $\pi^- = d\overline{u}$, $K^+ = u\overline{s}$, $K^0 = d\overline{s}$, $K^- = s\overline{u}$ and $\overline{K}^0 = s\overline{d}$, as shown in Fig. 4.19. We still have the three neutral mesons: π^0, η and η'. Since the π^0 is in an isospin triplet, it must be in the octet. As such, it does not contain $\overline{s}s$, which has zero isospin. Moreover, it is easy to show that the antiquark pair that behaves under rotations around the z-axis as an isospin doublet is $(\overline{d}, -\overline{u})$. Using the Clebsch–Gordan coefficients and labelling the state with its $SU(3)_f$ and $SU(2)$ representations, **8** and **3**, we have

$$\left| \pi^0 \right\rangle \equiv \left| 8, 3 \right\rangle = \frac{1}{\sqrt{2}} \left| -\overline{u}u + \overline{d}d \right\rangle. \tag{4.45}$$

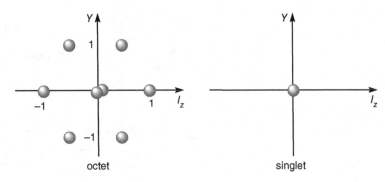

Fig. 4.18.   The octet and the singlet.

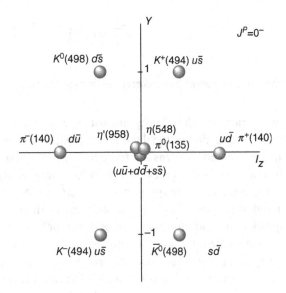

Fig. 4.19.   The pseudoscalar mesons nonet; the approximate values of their masses are in MeV.

The $SU(3)_f$ singlet, which is also an isotopic singlet, is the following state

$$|\eta_1\rangle \equiv |1,1\rangle = \frac{1}{\sqrt{3}}\left|\bar{u}u + d\bar{d} + s\bar{s}\right\rangle, \tag{4.46}$$

which we have not yet identified with a meson. The third combination, namely the iso-singlet of the octet, must be orthogonal to the other two. Imposing this condition, one finds

$$|\eta_8\rangle \equiv |8,1\rangle = \frac{1}{\sqrt{6}}\left|\bar{u}u + \bar{d}d - 2\bar{s}s\right\rangle. \tag{4.47}$$

We have two iso-singlets, η_1 and η_8, and two physical states, states η and η'. However, we cannot identify the latter with the former. Because $SU(3)_f$ is already broken by

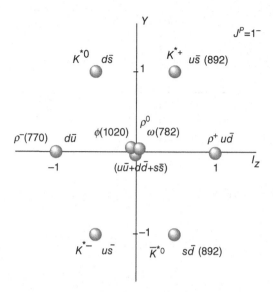

Fig. 4.20.   The vector mesons; the approximate values of the masses are in MeV.

strong interactions, the η and η' are not pure octet and pure singlet. Indeed, the apparently simple question of the flavour composition of the pseudoscalar completely neutral mesons, η and η', turns out to be highly non-trivial.

The reason is that, for these states, the colour interaction induces continuous transitions between quark–antiquark pairs $\left(u\bar{u}, d\bar{d}, s\bar{s} \right)$. A consequence is the large values of the η and η' masses. The correct description of these complicated mixtures of quark–antiquark pairs and gluons can be achieved only within the QCD theory and cannot be given at elementary level. We notice, however, that the experimental and theoretical work on this issue was extremely important in the historical development of QCD. As we shall see in Section 6.8, the QCD vacuum is indeed a very active medium, in which energy fluctuations continuously take place. Pseudoscalar fluctuations couple with η and η', strongly contributing to their mass.

The structure of the vector mesons is similar and is shown in Fig. 4.20. Here, too, the symmetry is broken, as seen from the relevant differences between the masses. The two physical isospin singlets, the ω and the ϕ, are two orthogonal linear superpositions of the $SU(3)_f$ octet and singlet states. These superpositions are such that one of the mesons, the ω, almost does not contain valence s quark, while the other, the ϕ, is made up almost only of strange quarks, that is

$$|\omega\rangle = \frac{1}{\sqrt{2}}\left| u\bar{u} + d\bar{d} \right\rangle$$
$$|\phi\rangle = |s\bar{s}\rangle. \tag{4.48}$$

This explains, firstly, why the masses of the ρ and of the ω are almost equal, and secondly why the 'hidden strangeness' ϕ meson decays preferentially into $K\bar{K}$, that is, into final states in which the s and $\bar{s}$ are still present. If the final state is non-strange, s and $\bar{s}$ must first annihilate into pure energy of the colour field; subsequently, this energy must give origin to non-strange quark–antiquark pairs to produce, in turn, the observed hadronic final state, 3π for example. As we shall discuss in Chapter 6, QCD foresees that this process is severely suppressed.

Many other mesons exist beyond the ground-level pseudoscalar and vector mesons; their quantum numbers are compatible with those of a spin $1/2$ particle–antiparticle pair in excited states with non-zero orbital momenta. In particular their spin parities are amongst those in Table 3.1 but not, for example, $J^{PC} = 0^{+-}, 0^{--}, 1^{-+}$.

4.8 Baryons

Baryons are made of three quarks. Therefore, their classification in $SU(3)_f$ multiplets is less simple than that of mesons. The correspondence between a baryon and the quarks has no ambiguity if three quarks are identical, for example, uuu can be only Δ^{++} and ddd can be only Δ^{-}. However, uud can be p or Δ^{+}, uds can be Σ^{0}, Λ^{0}, $\Sigma^{0}(1385)$ or $\Lambda(1405)$. The physical states correspond to different exchange symmetries of the corresponding three quarks.

Let us start by dealing with three equal quarks. We are still missing a case, namely the baryon with three strange quarks sss. According to the $SU(3)$ symmetry, a hyperon, called Ω^{-}, with strangeness $S = -3$ must exist. Since the isospin of s is zero, the Ω^{-} must be an isospin singlet. We can predict its mass by looking at Fig. 4.8. We see that the $SU(3)_f$ breaking is such that the masses of the $3/2^{+}$ resonances increase by about 145 MeV for a decrease of strangeness of one unit. Therefore, the Ω^{-} mass should be about 1675 MeV. But, if this is true, it cannot decay strongly! Actually the lowest mass final state with $S = -3$ and $\mathcal{B} = 1$ is $\Xi^{0}K^{-}$, which has a mass of 1809 MeV. In conclusion, $SU(3)_f$ was making a precise prediction: a metastable hyperon with the previously mentioned quantum numbers and mass should exist and should have escaped detection.

The decuplet of $3/2^{+}$ baryons was called by Gell-Mann 'decimet' in analogy with the ensemble of ten musical instruments. In 1962, during the general conference on particle physics at CERN, he urged the experimentalists to look for the Ω^{-}. At lunch, with N. Samios and J. Leitner, he sketched out on a napkin how to search for the particle by indicating the cascade of particles into which it would decay. The two experimentalists, who were based at the Brookhaven National Laboratory, took the napkin with them, showed it to the lab's director and convinced him to give them priority for running time on the accelerator. By 1964 the particle was discovered, with exactly the predicted properties, in a bubble chamber experiment on a K^{-} beam (Barnes *et al.* 1964).

The production reaction was

$$K^{-} + p \to \Omega^{-} + K^{+} + K^{0}, \tag{4.49}$$

which conserves the strangeness. The information provided by the bubble chamber is so complete that a single event was sufficient for the discovery. It is shown in Fig. 4.21, which we now analyse.

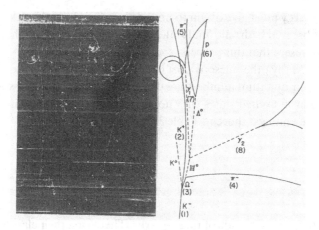

The first Ω^- event (Barnes *et al.* 1964, courtesy Brookhaven National Laboratory).

The track of the Ω^- terminates where the hyperon decays as

$$\Omega^- \to \Xi^0 + \pi^-. \tag{4.50}$$

This weak decay violates the strangeness by one unit, $\Delta S = -1$. Since the Ξ^0 is neutral, it does not leave a track, and decays, again weakly with $\Delta S = -1$, as

$$\Xi^0 \to \Lambda + \pi^0. \tag{4.51}$$

Two other decays follow: a weak one, again with $\Delta S = -1$

$$\Lambda \to p + \pi^-. \tag{4.52}$$

and an electromagnetic one

$$\pi^0 \to 2\gamma. \tag{4.53}$$

Finally, both γs materialize into two electron–positron pairs (a rather lucky circumstance in a hydrogen bubble chamber). Having determined the momenta of all the charged particles by measuring their tracks, a kinematic fitting procedure is performed by imposing the energy–momentum conservation at each decay vertex. In this way, the event is completely reconstructed. The resulting mass and lifetime are

$$m = 1672 \text{ MeV} \qquad \tau = 82 \text{ ps}, \tag{4.54}$$

in perfect agreement with the prediction.

The discovery of the Ω^- marked a triumph of the $SU(3)$ symmetry, but at the same time aggravated a problem that already existed in the quark model. Three baryons, Δ^{++}, Δ^- and Ω^- are composed of three identical quarks. Since they are the ground states, the orbital momenta of the quarks must be zero. Therefore, the spatial part of their wave function is symmetric. Their spin wave function is also symmetric, the total spin being $3/2$. In conclusion, we have states of three equal fermions that are completely symmetric, in contradiction with the Pauli principle. The solution of this puzzle led to the discovery of colour: there are three quarks for each flavour, each with a different colour charge, called red, green and blue.

Table 4.2. The symmetries of the states resulting from the combination of three spin 1/2.

$J = 3/2$	S	$\uparrow\uparrow\uparrow$	$\frac{1}{\sqrt{3}}(\uparrow\uparrow\downarrow + \uparrow\downarrow\uparrow + \downarrow\uparrow\uparrow)$	$\frac{1}{\sqrt{3}}(\downarrow\downarrow\uparrow + \downarrow\uparrow\downarrow + \uparrow\downarrow\downarrow)$	$\downarrow\downarrow\downarrow$
$J = 1/2$	M,A		$\frac{1}{\sqrt{2}}(\uparrow\downarrow - \downarrow\uparrow)\uparrow$	$\frac{1}{\sqrt{2}}(\uparrow\downarrow - \downarrow\uparrow)\downarrow$	
$J = 1/2$	M,S		$\frac{1}{\sqrt{6}}(\uparrow\downarrow + \downarrow\uparrow)\downarrow - \sqrt{\frac{2}{3}}\,\uparrow\uparrow\downarrow$	$-\frac{1}{\sqrt{6}}(\uparrow\downarrow + \downarrow\uparrow)\uparrow - \sqrt{\frac{2}{3}}\,\downarrow\downarrow\uparrow$	
J_z		$+\frac{3}{2}$	$+\frac{1}{2}$	$-\frac{1}{2}$	$-\frac{3}{2}$

With the colour degree of freedom, the wave function is the product of four factors

$$\Psi = \psi_{\text{space}}\psi_{\text{spin}}\psi_{\text{SU3}_f}\psi_{\text{colour}}. \tag{4.55}$$

The Pauli principle requires this product to be antisymmetric. We must now antici-pate that the baryon colour wave function ψ_{colour} is antisymmetric in the exchange of every quark pair. As we have already said, the baryons we are considering, both those with $1/2^+$ and with $3/2^+$, are the ground states and their quarks are in S wave. Therefore, ψ_{space} is symmetric. It follows that

$$\psi_{\text{spin}}\psi_{\text{SU3}_f} = \text{symmetric.} \tag{4.56}$$

This not only solves the paradox of both ψ_{spin} and ψ_{SU3_f} being symmetric for three equal quarks, but also tells us much more. Let us see.

We start by examining the symmetries of the combinations of three $SU(2)$ dou-blets, **2** (namely spin $1/2$). We first take the product of two of them

$$\mathbf{2} \otimes \mathbf{2} = \mathbf{1}_A \oplus \mathbf{3}_S, \tag{4.57}$$

where the suffixes indicate the symmetry. In other words, combining two $1/2$ spins, we obtain a state of total spin 0 that is antisymmetric, and one of total spin 1 that is symmetric. We proceed by taking the product of the result with the third doublet

$$\mathbf{2} \otimes \mathbf{2} \otimes \mathbf{2} = \left(\mathbf{1}_A \otimes \mathbf{2}\right) \oplus \left(\mathbf{3}_S \otimes \mathbf{2}\right) = \mathbf{2}_{M,A} \oplus \mathbf{2}_{M,S} \oplus \mathbf{4}_S. \tag{4.58}$$

Here **M, A** means mixed-antisymmetric, namely antisymmetric in the exchange of two quarks, and similarly, **M, S** stands for mixed-symmetric, which is symmetric in a two-quark exchange. Again, in other words, the spin 0 combined with the third spin $1/2$ gives a spin $1/2$ that is antisymmetric in the exchange of the first two, (**M, A**). The spin 1 with the spin $1/2$ gives a spin $1/2$ that is symmetric in the exchange of the first two, (**M,S**), and a spin $3/2$ (**S**). The situation is spelled out in Table 4.2.

We should now proceed in a similar manner to combine three $SU(3)_f$ triplets, namely three representations **3**. We do not assume the reader to have enough knowl-edge of the group, and give only the result

$$3 \otimes 3 \otimes 3 = 10_S \oplus 8_{M,S} \oplus 8_{M,A} \oplus 1_A. \qquad (4.59)$$

We can understand the structure by the following simple arguments.

- A three-quark system can always be made symmetric in the exchange of each pair (completely symmetric) (**S**). There are three different possibilities with equal quarks (ddd, uuu, sss), six with two equal quarks and one different $(ddu, dds, uud, uus, ssd, ssu)$ and one with all the quarks different (dus); in total **10** states.
- Both the mixed-symmetric (**M,S**) and mixed-antisymmetric (**M,A**) combinations are possible only if at least one of the quarks is different. There are six combinations with two equal quarks and one different. One of these, **M,S**, is for example $(ud+du)u / \sqrt{2}$ and one **M,A** is $(ud-du)u / \sqrt{2}$. There are six different combinations of three all-different quarks (u,d,s). We already used one of these combinations for the **S** states and we shall use a second one for the **A** states immediately. With the remaining four, two **M,S** and two **M,A** combinations can be arranged. Summing up, we have in total **8 M,S** states and **8 M,A** states.
- The three-quark system can be made completely antisymmetric (**A**) only if all of them are different; **1** state $[(sdu-sud+dus-uds+usd-dsu) / \sqrt{6}]$.

We now use the symmetry properties we have found and combine the $SU(3)_f$ and spin multiplets $(\mathbf{SU3, Spin})$ in order to fulfil the Pauli principle, namely Eq. (4.56). We have only two possibilities:

$$\text{one spin } 3/2 \text{ decimet} \qquad \psi^S_{SU3} \psi^S_{spin} \equiv (\mathbf{10}_S, \mathbf{4}_S),$$

$$\text{one spin } 1/2 \text{ octet} \frac{1}{\sqrt{2}} \left(\psi^{M,S}_{SU3} \psi^{M,S}_{spin} + \psi^{M,A}_{SU3} \psi^{M,A}_{spin} \right) \equiv (\mathbf{8}_{M,S}, \mathbf{2}_{M,S}) + (\mathbf{8}_{M,A}, \mathbf{2}_{M,A}).$$

These are precisely the multiplets observed in nature! This result is far from being insignificant and goes beyond the quark model. Actually, the quark model alone would foresee the existence of all the multiplets (4.59). The observed restriction to an octet and a decimet is a consequence of the dynamics, namely the antisymmetry of ψ_{colour}, a fundamental characteristic of QCD. Notice, in particular, that an $SU(3)_f$ hyperon singlet, say a Λ_1, does not exist. The Λ is pure octet.

The octet of $1/2^+$ baryons is shown in Fig. 4.22, the $3/2^+$ decimet in Fig. 4.23.

We conclude this section with an important observation, looking back at the last column of Table 4.1 with the quark masses. As the quarks are never free, we cannot define their mass precisely, but we need to extend the concept of mass itself. This is possible only within a well-defined theoretical scheme. The SM provides this scheme, as we shall see in Chapter 6.

Actually, one issue is already clear. For composite systems like atoms and nuclei the difference between the mass of the system and the sum of the masses of its constituents is small. On the contrary, the mass of the hadrons is much larger than the sum of the masses of their quarks. Notably, u and d have extremely small masses in comparison with the nucleon mass. The mass of the non-strange hadrons is predominantly energy of the colour field. As we shall see in Chapter 6, the colour force is independent of the flavour. This explains why, even if the d quark mass is about twice as large as

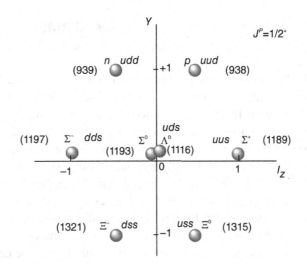

Fig. 4.22.   The baryon octet; in parentheses the masses in MeV.

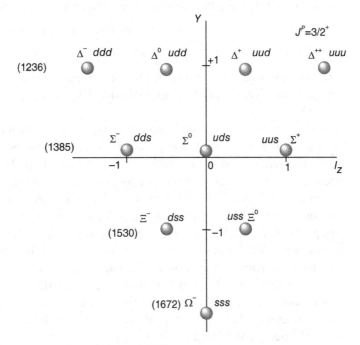

Fig. 4.23.   The baryon decimet; in parentheses the masses in MeV.

that of the u, isospin is a good symmetry. The $SU(3)_f$ case is similar, considering that the s-quark mass is small but not negligible in comparison to the hadron masses. Therefore, the latter symmetry is already broken at the level of the strong interactions. In conclusion, both isospin and $SU(3)_f$ symmetries, when considered at a fundamental level, appear to be accidental. This does not mean that they are less important.

4.9 Charm

Unlike strangeness, the existence of hadrons with a fourth flavour, called 'charm' was theoretically predicted. It was established by the observation of totally unexpected phenomena. In 1970, S. Glashow, I. Iliopoulos and L. Maiani introduced (Glashow *et al.* 1970) a theoretical mechanism, which became known as GIM, to explain the experimentally observed suppression of the 'neutral current' weak processes between quarks of different flavour that should otherwise have been faster by several orders of magnitude. We shall discuss the issue in Chapter 7. Here we consider the essential aspect of the problem in the following example. The decay $K^+ \to \pi^0 + \nu_e + e^+$ is a weak process, as is evident from the presence of a neutrino, and is called a 'charged current' process because the weak interaction, transforming a positive hadron into a neutral one, carries electrical charge. On the other hand, the decay $K^+ \to \pi^+ + \nu + \bar{\nu}$ is, for similar reasons, a weak 'neutral current' process. The two processes are very similar and should proceed with similar partial amplitudes. On the contrary, the latter decay is strongly suppressed, the measured branching ratios (Workman *et al.* 2022) being

$$\begin{aligned}
\mathrm{BR}\left(K^+ \to \pi^+ \nu \bar{\nu}\right) &= \left(1.14^{+0.40}_{-0.33}\right) \times 10^{-10} \\
\mathrm{BR}\left(K^+ \to \pi^0 e^+ \nu_e\right) &> 3 \times 10^{-3}.
\end{aligned} \tag{4.60}$$

The GIM mechanism accounted for the suppression of such neutral current processes, introducing a fourth quark c carrying a new flavour, charm. The electric charge of c is $+2/3$.

A rough evaluation of the masses of the hypothetical charmed hadrons led to values around 2 GeV. Like strangeness, charm is conserved by the strong and electromagnetic interactions and is violated by the weak interaction. The lowest-mass charmed mesons, the ground levels, have spin parity 0^-. There are three such states: two non-strange mesons called D^0 and D^+ with valence quarks $\bar{u}c$ and $\bar{d}c$ respectively, and one strange meson called D_s^+, which contains $\bar{s}c$. They have positive charm, $C = +1$. Their antiparticles, $D^-(d\bar{c})$, $\bar{D}^0(u\bar{c})$ and $D_s^-(s\bar{c})$ have negative charm. These mesons decay weakly with lifetimes of the order of a picosecond. Their lifetimes are roughly an order of magnitude shorter than those of the K-mesons, because of their larger mass.

As we shall see in Section 7.9, another clear prediction of the four-quark GIM model was that the mesons with positive charm would decay preferentially into negative strangeness final states (i.e. at the quark level) by $c \to s + \cdots$. In particular, the favourite decays of D^0 and D^+ are

$$D^0 \to K^- \pi^+, \quad D^+ \to K^- \pi^+ \pi^+ \tag{4.61}$$

and similarly for their charge conjugated, while

$$D^0 \to K^+ \pi^-, \quad D^0 \to \pi^+ \pi^-, \quad D^+ \to K^+ \pi^+ \pi^-, \quad D^+ \to \pi^+ \pi^+ \pi^- \tag{4.62}$$

are strongly suppressed.

In 1974, S. Ting and collaborators at the AGS proton accelerator at Brookhaven (Aubert *et al.* 1974) built a spectrometer designed for the search for heavy particles

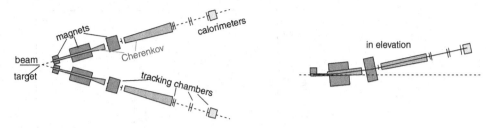

Fig. 4.24.   The Brookhaven double arm spectrometer.

with the same quantum numbers as the photon, $J^{PC} = 1^{--}$. They were not looking for charm. To search for particles of unknown mass one needs to explore a mass spectrum that is as wide as possible. With the assumed quantum numbers, the particle, which was to be called J, would decay into e^+ and e^-. The idea was to search for this decay in hadronic collisions. The overall reaction to search for, calling the non-detected part of the final state X, is

$$p + N \rightarrow J + X \rightarrow e^+ + e^- + X. \qquad (4.63)$$

The spectrometer must then detect two particles of opposite sign. The main difficulty is that the two charged particles are almost always pions. Only once in a million or so might electrons be produced. Consequently, the spectrometer must provide a rejection power against hadrons of at least 10^8. The spectrometer must measure the two momenta with high accuracy, both in absolute value $(p_1$ and $p_2)$ and in direction $(\theta_1$ and $\theta_2)$. The energies are then known, assuming the particles to be electrons, and the mass of the e^+e^- system is calculated:

$$m(e^+e^-) = \sqrt{2m_e^2 + 2E_1 E_2 - 2p_1 p_2 \cos(\theta_1 + \theta_2)}.$$

The particles sought should appear as peaks in this mass distribution.

Fig. 4.24 shows a schematic of the spectrometer, which has two arms, one for the positive and one for the negative particles. In each arm, the measurement of the angle is decoupled from that of the momentum by having the dipole magnets bend in the vertical plane. The range in m can be varied, changing the current intensity in the magnets and thus varying the acceptance for p_1 and p_2.

The necessary rejection power against pions $(>10^8)$ is obtained with the following elements, in each arm: (1) two threshold Cherenkov counters, designed to see the electrons but not pions and heavier hadrons; and (2) a calorimeter that measures the longitudinal profile of the shower, thus distinguishing electrons from hadrons. The spectrometer was designed to be able to handle very high particle fluxes, up to 10^{12} protons hitting the target per second.

Fig. 4.25 shows the $m(e^+e^-)$ distribution, showing a spectacular resonance. Its mass is $m_J = 3100$ MeV and has the outstanding feature of being extremely narrow. Actually, its width is smaller than the experimental resolution and only an upper limit could be determined, $\Gamma < 5$ MeV.

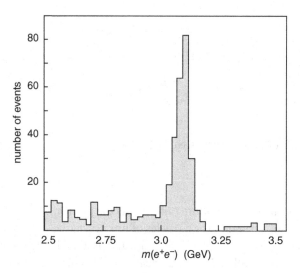

The J particle peak in the proton–antiproton mass distribution.

Around the same time, the SPEAR e^+e^- collider, built by B. Richter and collaborators, was in operation at the SLAC Laboratory. Its maximum energy was $\sqrt{s} = 8$ GeV. A general-purpose detector, called Mark I, had been completed with tracking chambers in a magnetic field and shower counters. After a period of collecting data at high energy, the decision was made to return to $\sqrt{s} \approx 3$ GeV, to check some anomalies that had previously been observed. The energy of the collider was varied in small steps, while measuring the cross-sections of different processes. A huge resonance appeared with all the cross-sections jumping up by more than two orders of magnitude (Augustin *et al.* 1974).

Fig. 4.26 (adapted from Boyarski 1975) shows the cross-sections of the processes

$$e^+e^- \to \text{hadrons}, e^+e^- \to \mu^+\mu^-, e^+e^- \to e^+e^- \tag{4.64}$$

as functions of the centre of mass energy. Note the logarithmic vertical scales and the much expanded energy scale. The resonance is very high and extremely narrow. Since the discovery at SLAC was independent, a different name was given to the new particle, ψ. It is now called J/ψ.

The quantum numbers of the J/ψ are expected to be $J^{PC} = 1^{--}$, because it decays into e^+e^- in the Ting experiment and because it is produced in e^+e^- collisions, both processes being mediated by a photon. However, the characteristics of the J/ψ were so surprising that these assignments had to be confirmed.

First, the J/ψ might have $J^P = 1^+$ and decay with parity violation. To test this possibility, consider the electron–positron annihilation into a muon pair. In general the reaction proceeds in two steps, which, outside the resonance, are the annihilation of the initial electron pair into a photon and the materialization of the photon into the final muon pair

$$e^+e^- \to \gamma \to \mu^+\mu^-. \tag{4.65a}$$

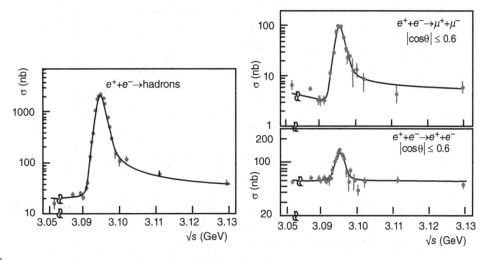

Fig. 4.26.   The ψ particle. Notice the logarithmic scale. (B. Richter, Nobel Lecture 1976, Fig. 5, B. Richter, © The Nobel Foundation 1976 Figure 5 in this link www.nobelprize.org/uploads/2018/06/richter-lecture.pdf).

In resonance an additional contribution is given by the annihilation into a J/ψ that then decays into the final muon pair

$$e^+e^- \to J/\psi \to \mu^+\mu^-. \tag{4.65b}$$

Now, if the J/ψ has $J^P = 1^-$, the two amplitudes interfere because their intermediate states have the same quantum numbers, while, if the J/ψ has $J^P = 1^+$, they do not. Fig. 4.27 shows the ratio between the cross-sections of $e^+e^- \to \mu^+\mu^-$ and $e^+e^- \to e^+e^-$. The expected effect of the interference is a dip below the resonance, shown as a continuous line. The dashed line is the expectation for $J^P = 1^+$. Comparison with the data establishes that $J^P = 1^-$.

Secondly, consider the hadronic decays. Given the very small width of the J/ψ, they might be electromagnetic, as happens for the η meson. By measuring the cross-sections for $e^+e^- \to n\pi$ as a function of energy, the resonance was found for $n = 3$ and $n = 5$, but not for $n = 2$ or $n = 4$. Therefore, the G-parity is conserved, that is, the J/ψ decays strongly. Finally, the isospin of the J/ψ was established to be $I = 0$, observing that its branching ratio into $\rho^0\pi^0$ is equal to that into $\rho^+\pi^-$ (or $\rho^-\pi^+$), while, for $I = 1$ the branching ratio would be zero.

After the discovery of the ψ, a systematic search started at SPEAR scanning in energy at very small steps. Ten days later, a second narrow resonance was found, which was called ψ' (Abrams *et al.* 1974). The quantum numbers of the ψ' were established, in similar manner as for the ψ, to be those of the photon, $J^{PC} = 1^{--}$.

At Frascati, the ADONE e^+e^- collider had been designed with a maximum energy $\sqrt{s} = 3\,\text{GeV}$, just a little too small to detect the resonance. Nevertheless, when S. Ting communicated the discovery, the machine was able to be brought above its nominal maximum energy up to 3100 MeV (Bacci *et al.* 1974). The resonance appeared immediately. Fine-step scanning at lower energies did not show any narrow resonance.

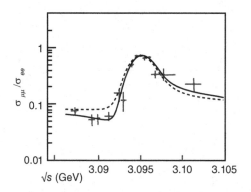

Ratio between $e^+e^- \rightarrow \mu^+\mu^-$ and $e^+e^- \rightarrow e^+e^-$ cross-sections. Notice the logarithmic scale. (B. Richter, Nobel Lecture 1976, Fig. 6, B. Richter, © The Nobel Foundation 1976 Figure 6 in this link www.nobelprize.org/uploads/2018/06/richter-lecture.pdf).

From the line shape of the resonance, one can extract accurate values of the mass and the width. The values obtained at SLAC were

$$m(\psi) = 3097 \text{ MeV} \qquad \Gamma(\psi) = 91 \text{ keV}$$
$$m(\psi') = 3686 \text{ MeV} \qquad \Gamma(\psi') = 281 \text{ keV}. \tag{4.66}$$

Again note the width values, which are surprisingly small for strongly decaying particles. Let us see how the widths are determined. One might think of taking simply the half maximum width. However, this is the convolution of the natural width and of the experimental resolution, and the method does not work if the latter is much wider than the former, as in this case.

We shall now use an expression of the cross-section as a function of the CM energy E, in the neighbourhood of a resonance in ultra-relativistic conditions, namely when the energies are much larger than the masses. Under these conditions energy and momentum of a particle are practically equal. Let M_R be the mass of the resonance, Γ its width and J its spin. We call Γ_i and Γ_f the partial widths of the initial and final states, respectively. Let s_a and s_b be the spins of the particles a and b in the initial state. We state without proof that the following expression is valid in the Breit–Wigner approximation, namely if $\Gamma \ll M_R$. For the cross-sections of the processes (4.65) with $s_a = s_b = 1/2$ and $J = 1$, if $\Gamma_i = \Gamma_e$, the initial state (e^+e^-) partial width, we have

$$\sigma(E) = \frac{(2J+1)}{(2s_a+1)(2s_b+1)} \frac{4\pi}{E^2} \frac{\Gamma_e \Gamma_f}{(E-M_R)^2 + (\Gamma/2)^2}$$
$$= \frac{3\pi}{s} \frac{\Gamma_e \Gamma_f}{(E-M_R)^2 + (\Gamma/2)^2}. \tag{4.67}$$

A quantity that can easily be measured is the 'peak area', which is not altered by the experimental resolution. The integration of (4.67) gives (see Problem 4.24)

$$\int \sigma(E) dE = \frac{6\pi^2}{M^2} \frac{\Gamma_e \Gamma_f}{\Gamma}. \tag{4.68}$$

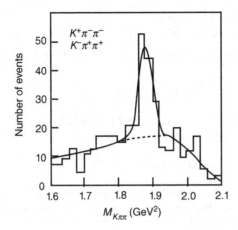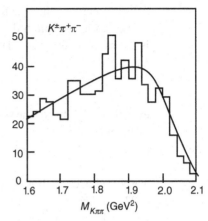

$K\pi\pi$ mass distributions in multiparticle events at $\sqrt{s} = 4.02$ GeV . (B. Richter, Nobel Lecture 1976, Fig. 18, B. Richter, © The Nobel Foundation 1976 Figure 18 in this link www.nobelprize.org/uploads/2018/06/richter-lecture.pdf).

The partial widths that appear in the numerator are determined by the measurements of the ratios between the peak areas in the corresponding channels. Then (4.68) gives the total width.

The extremely small widths show that J/ψ and ψ' are hadronic states of a completely new type, they are 'hidden charm' states. Both are made of a $c\bar{c}$ pair in a $^3S_1$ configuration, as follows from the quantum numbers $J^{PC} = 1^{--}$. The former is the fundamental level (1^3S_1 in spectroscopic notation), the latter is the first radial excited level $\left(2^3S_1\right)$. However, further experimental work was needed before the extremely small values of the widths could be understood.

The search for the charmed pseudoscalar mesons at $\sqrt{s} = 4.02$ GeV started at the Mark I detector in 1976, after having improved its K to π discrimination ability, in the channels

$$e^+ + e^- \to D^0 + \bar{D}^0 + X \qquad e^+ + e^- \to D^+ + D^- + X, \tag{4.69}$$

where X means anything else. The mesons are expected to have very short lifetimes. Consequently, they appear as resonances in the final state. A first resonance was observed in the $K^{\pm}\pi^{\mp}$ mass distributions in multiparticle events, corresponding to the $D^0 \to K^-\pi^+$ and $\bar{D}^0 \to K^+\pi^-$ decays. Its mass was 1865 MeV and its width was smaller than the experimental resolution (Goldhaber *et al.* 1976).

Soon afterwards (Peruzzi *et al.* 1976), as shown in Fig. 4.28, the charged D-mesons were observed at the slightly larger mass of 1875 MeV (now 1869.5 MeV) in the channels

$$D^+ \to K^-\pi^+\pi^+ \qquad D^- \to K^+\pi^-\pi^-. \tag{4.70}$$

However, no resonance was present in the channels

$$D^+ \to K^+\pi^+\pi^- \qquad D^- \to K^-\pi^+\pi^-. \tag{4.71}$$

This is precisely what the four-quark model requires.

We can now explain the reason for the narrow widths of J/ψ and ψ'. The reason is the same as for the ϕ. However, while the ϕ can decay into non-QCD-suppressed channels

$$\phi \to K^0 \bar{K}^0; K^+ K^-,$$

even if slowly, due to the small Q, in the cases of J/ψ and ψ', the corresponding decay channels

$$J/\psi \to D^0 \bar{D}^0; D^+ D^-; \qquad \psi' \to D^0 \bar{D}^0; D^+ D^-, \qquad (4.72)$$

are closed because both $m_{D^+} + m_{D^-}$ and $m_{D^0} + m_{\bar{D}^0}$ are larger than $m_{J/\psi}$ and $m_{\psi'}$.

This conclusion is confirmed by the observation, again at SLAC, of the third $3^3 S_1$ level, called ψ'', with mass and width

$$m(\psi'') = 3770 \text{MeV} \qquad \Gamma(\psi'') = 24 \text{ MeV}. \qquad (4.73)$$

In this case, the decay channels (4.72) are open and, consequently, the width is 'normal'.

Before leaving the subject, let us take another look at Fig. 4.28: the resonance curves are not at all symmetric around the maximum as the Breit–Wigner formula would predict, but their high-energy side is higher than the low-energy one. This feature is general, for resonances in the $e^+ e^-$ colliders. Actually, the energy on the horizontal axis is the energy we know, namely the energy of the colliding beams. This is not always the nominal electron–positron collision energy. It may happen that one or both initial particles radiate a photon (bremsstrahlung) before colliding. In this case, the collision energy is smaller than the collider energy. If the latter is above the resonance energy, the process returns the collision to resonance, increasing the cross-section. On the other hand, if the machine energy is below the resonance, the initial state radiation takes the collision energy even farther from resonance, decreasing the cross-section. This explains the high-energy tails.

A further element in the history of the discovery of charm must be mentioned, namely the precursor observations made in Japan before 1974, which, however, were substantially ignored in the West. In Japan the technology of nuclear emulsions exposed to cosmic rays at high altitudes by airplane and balloon flights had progressed considerably. This was true in particular at Nagoya, where K. Niu and collaborators had developed the 'emulsion chamber'.

An emulsion chamber is made up of two main components. The first is a sandwich of several emulsion sheets; the second is another sandwich of lead plates, about 1 mm thick, alternated with emulsion sheets. The former gives an accurate tracking of the charged particles; the latter provides the gamma conversion (and the detection of the π^0s), the identification of the electrons and the measurement of their energy. The momenta of the charged particles are determined by an accurate measurement of the multiple scattering of the tracks.

The first example of associated production of charm, published in 1971, is shown in Fig. 4.29 (Niu *et al.* 1971). The short horizontal bars in Fig. 4.29 show the points

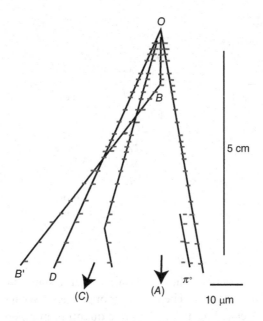

Fig. 4.29.   The first associated production of charm particles. Adapted from Niu *et al. Prog. Theor. Phys.* No. 46 (1971), p. 1644, Fig. 3a with permission of the author and of the Physical Society of Japan. https://doi.org/10.1143/PTP.46.1644.

where the tracks crossed the different emulsion sheets (the measurement accuracy is much better than the length of the bars). The straight lines are the interpolated tracks.

We now analyse the picture, making the following observations. The primary interaction has all the features of a strong interaction. Two particles decay, after 1.38 cm and 3.76 cm, respectively, corresponding to proper times of the order of several 10^{-14} s. Therefore, the two particles are produced in association and decay weakly. The primary particle had energy of several TeV, as evaluated by the measured energies of the secondary particles. Notice that, at the time, no accelerator at this energy scale existed. Tracks OB and BB' and the π^0, shown by the gammas that materialize in the lower part, are coplanar.

Niu dubbed X the particle decaying in the point B and evaluated its mass to be $m_X = 1.8$ GeV if it was a meson, $m_X = 2.9$ GeV if it was a baryon. Consequently, it could not be a strange particle; a new type of hadron had been discovered. In the following few years the Japanese group observed other examples of the X particles, neutral and charged, in emulsion chambers exposed both to cosmic rays and to the proton beam at FNAL. The new particles had all the characteristics foreseen for charmed hadrons.

Many hadrons containing the c quark, the charmed hadrons, are known today. Table 4.3 summarizes the characteristics of the lowest-mass mesons. The charmed vector mesons, not shown, have larger masses and can decay strongly, without violating charm conservation.

Several 'hidden charm' mesons exist. Indeed, we can think of the $c\bar{c}$ system as an 'atom', which is called charmonium, in which quark and antiquark are bound by

Table 4.3. The lowest-mass hidden and open charm mesons. (For more precise values and uncertainties see Appendix 3.)

State	quark	M (MeV)	Γ / τ	J^{PC}	I	Principal decays
$J/\psi\left(1^3S_1\right)$	$c\bar{c}$	3097	93 keV	1^{--}	0	hadrons (88%), e^+e^- (6%), $\mu^+\mu^-$ (6%)
$\psi'\left(2^3S_1\right)$	$c\bar{c}$	3686	294 keV	1^{--}	0	$\psi + 2\pi$ (50%)
$\psi''\left(3^3S_1\right)$	$c\bar{c}$	3774	27 MeV	1^{--}	0	$D\bar{D}$ dominant
η_c	$c\bar{c}$	2984	32 MeV	0^{-+}	0	hadrons
D^+	$c\bar{d}$	1870	1 ps	0^-	1/2	semileptonic, hadrons
D^0	$c\bar{u}$	1865	0.4 ps	0^-	1/2	semileptonic, hadrons
D_s^+	$c\bar{s}$	1968	0.5 ps	0^-	0	hadrons

strong interaction. The charmonium-bound states have the same quantum numbers as the hydrogen atom. The ψs are amongst these. We mention in particular the 1^1S_0 level, which is called η_c. It has the same quantum numbers as the η and the η', $J^{PC} = 0^{-+}$, $I = 0$, and in principle could mix with η and the η'. This does not happen in practice, due to the large mass difference, and η_c is a pure $c\bar{c}$ state.

As expected, the charmed hyperons exist too, containing one, two or three c quarks and consequently with $C = 1$, 2 or 3, with any combination of the other quarks. We shall not enter into any detail here. However, we report the characteristics of the principal charmed hadrons in Appendix 3.

Question 4.3 Write down the possible charm values of a meson and those of a meson with charge $Q = +1$. □

Question 4.4 Why is the radiative decay $\psi' \to \psi + \gamma$ forbidden? Why is $\psi \to \pi^0 + \pi^0$ forbidden? □

4.10 The Third Family

The basic constituents of ordinary matter are electrons and, inside the nucleons, the up and down quarks, with charges $-1/3$ and $2/3$, respectively. In the nuclear beta decays, a fourth particle appears, the electron neutrino. In total, we have two quarks and two leptons.

However, there are more in nature. As we have seen, in cosmic rays a second charged lepton is present, the μ, identical to but heavier than the electron, and its associated neutrino. In cosmic radiation too, hadrons containing two further quarks, s with charge $-1/3$ and c with charge $2/3$, have been discovered. Nature appears to have repeated the same scheme twice. The two groups of elementary particles are called first and second 'families'.

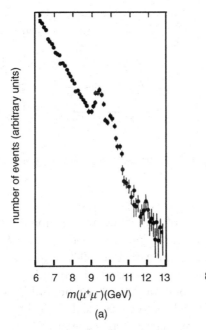
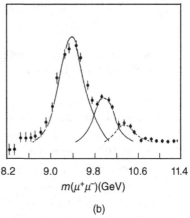

We have already seen that a third lepton and its neutrino exist. This suggests the existence of two more quarks, in three colours each, with charges $-1/3$ and $+2/3$. These are the 'beauty' (also called 'bottom') b and 'top' t quarks. Their flavours are, respectively, $B = -1$ and $T = +1$. As for the other flavours, B and T are conserved by strong and electromagnetic interactions and are violated by weak ones.

In 1977, L. Lederman and collaborators (Herb *et al.* 1977) built a two-arm spectrometer at FNAL designed to study $\mu^+\mu^-$ pairs produced by high-energy hadronic collisions. The reaction studied was

$$p + (\text{Cu}, \text{P}t) \rightarrow \mu^+ + \mu^- + X. \tag{4.74}$$

The 400 GeV proton beam extracted from the Tevatron was aimed at a metal target made of Cu or Pt. The two arms measure the momenta of the positive and negative particles, respectively. Since the events looked for are extremely rare, the spectrometer must accept very intense fluxes and must provide a high rejection power against charged pions and other hadrons. The rejection is obtained by using a sophisticated 'hadron filter' located on the path of the secondary particles, before they enter the arms of the spectrometer. A block of beryllium, eighteen interaction lengths thick, stops the hadrons while letting the muons through. The price to pay is some degradation in the momentum measurement. The corresponding resolution on the mass of the two-muon system was $\Delta m_{\mu\mu} / m_{\mu\mu} \approx 2\%$.

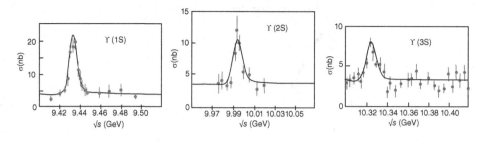

Fig. 4.31. The hadronic cross-section measured by the CLEO experiment at the CESR e^+e^- collider, showing the $\Upsilon\left(1^1S_3\right)$, $\Upsilon\left(2^1S_3\right)$ and $\Upsilon\left(3^1S_3\right)$ states.

To have an idea of the orders of magnitude, we mention that at every extraction, namely at every accelerator cycle, 10^{11} protons hit the target. With a total exposure of 1.6×10^{16} protons on target, a sample of about 9000 $\mu^+\mu^-$ events with $m_{\mu\mu} > 5$ GeV was obtained. The $m(\mu^+\mu^-)$ mass distribution is shown in Fig. 4.30(a) and, after subtracting a non-resonating (i.e. continuum) background, in Fig. 4.30(b). Three barely resolved resonances are visible, which were generically called Υ.

Precision study of the new resonances was made at the e^+e^- colliders at DESY (Hamburg) and at Cornell in the USA. Fig. 4.31, with the data from CLEO at Cornell (Andrews *et al.* 1980), shows that the peaks are extremely narrow. The first measurement of the masses and of the widths of the Υs, made with the method we discussed for ψ, gave the results

$$m\left(1^3S_1\Upsilon\right) = 9460 \text{ MeV} \qquad \Gamma\left(1^3S_1\Upsilon\right) = 53 \text{ keV}$$
$$m\left(2^3S_1\Upsilon\right) = 10\,023 \text{ MeV} \qquad \Gamma\left(2^3S_1\Upsilon\right) = 43 \text{ keV} \qquad (4.75)$$
$$m\left(3^3S_1\Upsilon\right) = 10\,355 \text{ MeV} \qquad \Gamma\left(3^3S_1\Upsilon\right) = 26 \text{ keV}$$

The situation is very similar to that of the ψs, now with three very narrow resonances, all with $J^{PC} = 1^{--}$ and $I = 0$: they are interpreted as the states $^3S_1$ of the $b\bar{b}$ 'atom', the bottomium, with increasing principal quantum number. None of them can decay into hadrons with 'explicit' beauty because their masses are below threshold.

The lowest-mass beauty hadrons are the pseudoscalar mesons with a $\bar{b}$ antiquark and a d, u, s or c quark. Therefore, there are two charged, $B^+ = u\bar{b}$ and $B_c^+ = c\bar{b}$, and two neutral, $B^0 = d\bar{b}$ and $B_s^0 = s\bar{b}$, mesons and their antiparticles. The masses of the B^0 and of the B^+ are practically equal. The mass of the B_s^0 is about 100 MeV higher, owing to the presence of the s, and that of the B_c^+ is about 1000 MeV higher because of the c. Table 4.4 gives a summary of the beauty particles we are discussing.

The pseudoscalar beauty mesons, as the lowest-mass beauty states, must decay weakly. Their lifetimes, shown in Table 4.4, are, surprisingly, of the order of a picosecond, larger than those of the charmed mesons, notwithstanding their much larger masses. As we shall see in Chapter 7, in the weak decay of every quark both the electric charge and the flavour change. In the case of charm, there are two possibilities, $c \to s + \ldots$ and $\to d + \ldots$. The former, as we saw, is favourite; the second is suppressed.

Table 4.4. The principal hidden and open beauty hadrons (for more precise values and uncertainties see Appendix 3).

State	Quark	M (MeV)	Γ / τ	J^{PC}	I
$\Upsilon(1^1S_3)$	$b\bar{b}$	9460	54 keV	1^{--}	0
$\Upsilon(2^1S_3)$	$b\bar{b}$	10 023	32 keV	1^{--}	0
$\Upsilon(3^1S_3)$	$b\bar{b}$	10 355	20 keV	1^{--}	0
$\Upsilon(4^1S_3)$	$b\bar{b}$	10 579	20 MeV	1^{--}	0
B^+	$u\bar{b}$	5279.3	1.6 ps	0^-	1/2
B^0	$d\bar{b}$	5279.6	1.5 ps	0^-	1/2
B^0_s	$s\bar{b}$	5366.9	1.5 ps	0^-	0
B^+_c	$c\bar{b}$	6274.5	0.5 ps	0^-	0

Notice that, in the former case, the initial and final quarks are in the same family; in the latter they are not. In the case of beauty, the 'inside family' decay $b \rightarrow t + \dots$ cannot take place because the t mass is larger than the b mass. The beauty must decay as $b \rightarrow c + \dots$, that is, with change of one family (from the third to the second), or as $b \rightarrow u + \dots$, that is, with change of two families (from the third to the first). We shall come back to this hierarchy in Section 7.9.

The non-QCD-suppressed decays of the Υs are those into a beauty–antibeauty pair. The smaller masses of these pairs are $m_{B^+} + m_{B^-} \approx 2m_{B^0} = 10\,558$ MeV and $2m_{B^0_s} = 10\,740$ MeV. Therefore $\Upsilon(1^1S_3)$, $\Upsilon(2^1S_3)$ and $\Upsilon(3^1S_3)$ are narrow. The next excited level, the $\Upsilon(4^1S_3)$, is noticeable. Since it has a mass of 10 580 MeV, the decay channels

$$\Upsilon(4^3S_1) \rightarrow B^0 + \bar{B}^0; \qquad \rightarrow B^+ + B^- \qquad (4.76)$$

are open. The width of the $\Upsilon(4^1S_3)$ is consequently larger, namely 20 MeV.

The third family still needed an up-type quark, but it took 20 years from the discovery of the τ and 18 years from the discovery of beauty to find it. This was because the top is very heavy, more than 170 GeV in mass. Taking into account that it is produced predominantly via strong interactions in a pair, a very high centre of mass energy is necessary. Finally, in 1995, the Collider Detector Facility (CDF) experiment at the Tevatron collider at FNAL at $\sqrt{s} = 2$ TeV reported 27 top events with an estimated background of 6.7 ± 2.1 events. More statistics were collected over the following years thanks to a substantial increase in the collider luminosity.

Let us see now how the top was discovered. We must anticipate a few concepts that we shall develop in Chapter 6. Consider a quark, or an antiquark, immediately after its production in a hadronic collision. It moves rapidly in a very intense colour field, to which it contributes. The energy density is so high that the field materializes in a number of quark–antiquark pairs. Quarks and antiquarks, including the original one, then join to form hadrons. This process, which traps the quark into a hadron, is called 'hadronization'. In this process, the energy–momentum that initially belonged to the quark is distributed amongst several hadrons. In the reference frame of the quark, their momenta are typically of half a GeV. In the reference frame of the collision, the centre of mass of the group moves with the original quark momentum, which is typically several dozen GeV. Once hadronized, the quark appears to our detectors as a 'jet' of hadrons in a rather narrow cone.

Top is different from the other flavours in that there are no top hadrons. The reason for this is that the lifetime is very short due to the top large mass. In the Standard Model its expected value is $\tau \approx 5 \times 10^{-25}$ s (corresponding to a width of about 1.3 GeV). On the other hand, the time needed for one quark to combine with another in a QCD bound state is of the order of $d/c = 10^{-22}$ s, where $d \approx 1$ fm is the diameter of a hadron. The top quark decays before hadronizing. Unlike the other quarks, the top lives freely, but very briefly. As mentioned above, it was discovered in 1995 by the CDF experiment at Tevatron (Abe *et al.* 1995).

At Tevatron, top production is a very rare event; it happens once in 10^{10} collisions. Experimentally one detects the top by observing its decay products. To distinguish these from the background of non-top events, one must look at the channels in which the top 'signature' is as different from the background as possible. The top decays practically always into final states containing a W boson and a b quark or antiquark. Therefore, one searches for the processes

$$p + \bar{p} \rightarrow t + \bar{t} + X; \quad t \rightarrow W^+ + b; \quad \bar{t} \rightarrow W^- + \bar{b}. \tag{4.77}$$

The W boson, the mediator of the weak interactions, has a mass of about 80 GeV and a very short lifetime. It does not leave an observable track and must be detected by observing its daughters. The W decays most frequently into a quark–antiquark pair, but these decays are difficult to distinguish from much more common events with quarks directly produced by proton–antiproton annihilation. We must search for rare but cleaner cases, such as those in which both Ws in Eq. (4.77) decay into leptons

$$W \rightarrow e\nu_e \text{ or } \rightarrow \mu\nu_\mu. \tag{4.78}$$

Another clean channel occurs when one W decays into a lepton and the other into a quark–antiquark pair, adding the request of presence of a b and a $\bar{b}$ from the t and $\bar{t}$ decays. Namely, one searches for the following sequence of processes:

$$p + \bar{p} \rightarrow t + \bar{t} + X; \quad t \rightarrow W^+ + b \rightarrow W^+ + \text{jet}(b); \quad \bar{t} \rightarrow W^- + \bar{b} \rightarrow W^- + \text{jet}(\bar{b})$$
$$W \rightarrow e\nu_e \text{ or } \rightarrow \mu\nu_\mu \quad \text{and} \quad W \rightarrow q\bar{q}' \rightarrow \text{jet} + \text{jet}. \tag{4.79}$$

The requested 'topology' must have one electron or one muon; one neutrino; and four hadronic jets, two of which contain a beauty particle. Fig. 4.32 shows this topology pictorially. Fig. 4.33 shows one of the first top events observed by CDF in 1995. The right-hand part is an enlarged view of the tracks near the primary vertex showing the presence of two secondary vertices. They flag the decays of two short-lived particles, such as the beauties. The high-resolution picture is obtained thanks to a silicon micro-strip vertex detector (see Section 1.13). The calorimeters of CDF surround the

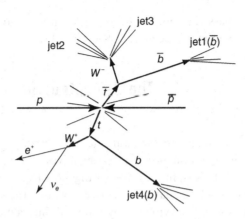

Schematic view of reactions (4.79); the flight lengths of the *W*s and the *t*s are exaggerated.

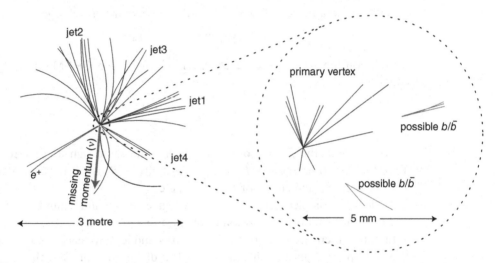

An example of reaction (4.79) from CDF. One observes the four hadronic jets; the track of an electron, certified as such by the calorimeter; and the direction of the reconstructed missing momentum. The enlargement shows the *b* candidates in jets 1 and 4 (Credit Fermi National Accelerator Laboratory).

interaction point in a 4π solid angle, as completely as possible. This makes it possible to check if the sum of the momenta of the detected particles is compatible with zero. If this is not the case, the 'missing momentum' must be the momentum of the unde-tectable particles, the vector sum of the neutrinos' momenta. The missing momentum is also shown in Fig. 4.33.

As the top decays before hadronizing, we can measure its mass from the energies and momenta of its decay products, as for any free particle. The actual value is

$$m_t = 172.69 \pm 0.30 \text{ GeV} \tag{4.80}$$

The top mass is an important quantity in the Standard Model. We shall discuss its measurement more precisely in Section 9.11.

4.11 The Elements of the Standard Model

Let us now summarize the hadronic spectroscopy we have studied. The hadrons have six additive quantum numbers, called flavours, which are two values of the third component of the isospin (I_z), the strangeness (S), the charm (C), the beauty (B) and the top (T). All the flavours are conserved by the strong and by the electromagnetic interactions and are violated by the weak interaction. There is a quark for each flavour. Quarks do not exist as free particles (except for top); rather they live inside the hadrons, to which they give flavour, baryonic number and electric charge. They have spin $1/2$ and, by definition, positive parity.

With a generalization of (3.38), we define as flavour hypercharge

$$Y = \mathcal{B} + S + C + B + T. \tag{4.81}$$

Its relationship to the electric charge is given by the Gell-Mann and Nishijima equation

$$Q = I_z + \frac{Y}{2}. \tag{4.82}$$

By convention, the flavour of a particle has the same sign as its electric charge. Therefore the strangeness of K^+ is +1, the beauty of B^+ is +1, the charm of D^+ is +1, both strangeness and charm of D_s^- are −1, etc.

Table 4.5, a complete version of Table 4.1, gives the quantum numbers of the quarks, and their masses. For a discussion of the meaning of quark masses, see Chapters 6.

In nature there are three families of quarks and leptons, each with the same structure: an up-type quark with charge $+2/3$, a down-type quark with charge $-1/3$, a charged lepton with charge −1 and a neutrino. We shall see an experimental proof of the number of families in Chapter 9.

Table 4.5. Quantum numbers and masses of the quarks.

	Q	I	I_z	S	C	B	T	$\mathcal{B}$	Y	Mass
d	−1/3	1/2	−1/2	0	0	0	0	1/3	1/3	$4.67^{+0.48}_{-0.17}$ MeV
u	+2/3	1/2	+1/2	0	0	0	0	1/3	1/3	$2.16^{+0.49}_{-0.26}$ MeV
s	−1/3	0	0	−1	0	0	0	1/3	−2/3	$93.4^{+8.6}_{-3.4}$ MeV
c	+2/3	0	0	0	+1	0	0	1/3	4/3	1.27 ± 0.002 GeV
b	−1/3	0	0	0	0	−1	0	1/3	−2/3	$4.18^{+0.03}_{-0.02}$ GeV
t	+2/3	0	0	0	0	0	+1	1/3	4/3	172.69 ± 0.30 GeV

In the following chapters we shall study, at an elementary level, the fundamental properties of the interactions between quarks and leptons, namely their dynamics.

Each of the three fundamental interactions differs from gravitation in that there are 'charges', which are the sources and the receptors of the corresponding force, and vector bosons that mediate them. The fundamental characteristics of the charges and of the mediators are very different in the three cases, as we shall study in the following chapters. We anticipate all this with a summary of the main properties.

(1) The electromagnetic interaction has the simplest structure. There is only one charge, the electric charge, with two different types. Charges of the same type repel each other; charges of different types attract each other. The two types are called positive and negative. Note that these are arbitrary names. The mediator is the photon, which is massless and has no electric charge. In Chapter 5, we shall study the fundamental aspects of quantum electrodynamics (QED) and we shall introduce instruments that we shall use for all the interactions.

(2) The strong interaction sources and receptors are the 'colour' charges, where the name colour has nothing to do with everyday colours. The structure of the colour charges is more complex than that of the electric charge, as we shall study in Chapter 6. There are three charges of different colours, instead of the one of QED, called red R, green G and blue B. The quarks have one colour charge and only one; the leptons, which have no strong interaction, have no colour charge. The colour force between quarks is independent of their flavours. For example, the force between a red up quark and a green strange quark is equal to the force between a red down and a green beauty, provided the states are the same. There are 18 quarks in total, with 6 flavours and 3 colours. Similarly to the case of the electric charge, one might define a positive and negative 'redness', a positive and negative 'greenness' and a positive and negative 'blueness'. However, positive and negative colour charges are called 'colour' and 'anticolour' respectively. This is simply a matter of names. The repulsive or attractive character of the colour force between quarks cannot be established simply by looking at the signs of their charges, a fundamental difference compared to the electromagnetic force. The colour force mediators are the gluons, which are massless. The limited range of the strong force is not due to the mass of the mediators, but to a more complex mechanism, as we shall see. There are eight different gluons, which do have colour charges; hence they also interact strongly amongst themselves. We shall study QCD in Chapter 6.

(3) The weak interaction has a still different structure. All the fundamental fermions, quarks, charged leptons and neutrinos have weak charges. The weak charge of a fermion depends on its 'chirality'. As we discussed in Section 2.8, chirality, not to be confused with helicity, is the eigenvalue of the Dirac γ_5 matrix. It can be equal to $+1$ or -1. A state is often called 'right' if its chirality is positive, 'left' if it is negative. Electrons and positrons can have both positive and negative chirality, while, strangely enough, only negative chirality neutrinos exist. The mediators of the weak interaction are three bosons, two charged, W^+, W^-, and one neutral, Z^0.

Table 4.6. The 24 fundamental fermions and their antiparticles. Every column is a family or generation.

	Fermions			Antifermions	
d^R	s^R	b^R	$\bar{d}^{\bar{R}}$	$\bar{s}^{\bar{R}}$	$\bar{b}^{\bar{R}}$
d^G	s^G	b^G	$\bar{d}^{\bar{G}}$	$\bar{s}^{\bar{G}}$	$\bar{b}^{\bar{G}}$
d^B	s^B	b^B	$\bar{d}^{\bar{B}}$	$\bar{s}^{\bar{B}}$	$\bar{b}^{\bar{B}}$
u^R	c^R	t^R	$\bar{u}^{\bar{R}}$	$\bar{c}^{\bar{R}}$	$\bar{t}^{\bar{R}}$
u^G	c^G	t^G	$\bar{u}^{\bar{G}}$	$\bar{c}^{\bar{G}}$	$\bar{t}^{\bar{G}}$
u^B	c^B	t^B	$\bar{u}^{\bar{B}}$	$\bar{c}^{\bar{B}}$	$\bar{t}^{\bar{B}}$
ν_e	ν_μ	ν_τ	$\bar{\nu}_e$	$\bar{\nu}_\mu$	$\bar{\nu}_\tau$
e^-	μ^-	τ^-	e^+	μ^+	τ^+

All of them are massive, the mass of the charged bosons is about 80 GeV and that of the neutral one is about 90 GeV. The mediators have weak charges and, consequently, interact between themselves, as the gluons do. The phenomenology of weak interactions is extremely rich. We shall discuss it in Chapters 7, 8 and 9.

Table 4.6 contains all the known fundamental fermions, particles and antiparticles, with their interaction charges. The colour is the apex at the left of the particle symbol.

A further component of the Standard Model is the Higgs boson, which will be discussed in Chapter 9. Its fundamental role in the SM is to 'give the mass', through a specific symmetry-breaking mechanism, to the W and Z bosons, to the quarks and to the charged leptons, while neutrinos remain, in theory, rigorously massless.

Two observations are in order, both on neutrinos. Neutrinos are the most difficult particles to study, owing to their extremely small interaction probability. They are also amongst the most interesting. Their study has always provided surprises.

- The neutrino states in Table 4.6 are the states of defined lepton flavour. These are the states in which neutrinos are produced by the weak interaction and the states that we can detect, again by weak interaction. Nevertheless, unlike for the other particles in the table, these are not the stationary states. The stationary states, called ν_1, ν_2 and ν_3, are quantum superpositions of ν_e, ν_μ and ν_τ. The stationary states are the states of definite mass, but do not have definite flavour and, therefore, cannot be classified in a family.

- What we have just said implies that the lepton flavour numbers are not conserved. Moreover, even if never observed so far, we cannot completely exclude a very small violation of the total lepton number. Actually, the lepton number is the only quantum number that distinguishes the neutrino from the antineutrino. If it is violated, neutrino and antineutrino may well be two states of the same particle, described by the Majorana equation (2.69). This is not, of course, the assumption of the SM.

Problems

4.1 Consider the following three states: π^0, $\pi^+\pi^+\pi^-$ and ρ^+. Define which of them is a G-parity eigenstate and, in this case, give the eigenvalue.

4.2 Consider the particles: ω, ϕ, K and η. Define which of them is a G-parity eigenstate and, in this case, give the eigenvalue.

4.3 From the observation that the strong decay $\rho^0 \to \pi^+\pi^-$ exists but $\rho^0 \to \pi^0\pi^0$ does not, what information can be extracted about the ρ quantum numbers: J, P, C, G, I?

4.4 Find the distance travelled by a K^* with momentum $p = 90$ GeV in a lifetime.

4.5 In a bubble chamber experiment on a K^- beam, a sample of events of the reaction $K^- + p \to \Lambda^0 + \pi^+ + \pi^-$ is selected. A resonance is detected both in the $\Lambda^0\pi^+$ and $\Lambda^0\pi^-$ mass distributions. In both, the mass of the resonance is $M = 1385$ MeV and its width $\Gamma = 50$ MeV. It is called $\Sigma(1385)$. (a) What are the strangeness, the hypercharge, the isospin and its third component of the resonance $\Lambda^0\pi^+$? (b) If the study of the angular distributions establishes that the orbital angular momentum of the $\Lambda^0\pi$ systems is $L = 1$, what are the possible spin parity values J^P?

4.6 The $\Sigma(1385)$ hyperon is produced in the reaction $K^- + p \to \pi^- + \Sigma^+ (1385)$, but is not observed in $K^+ + p \to \pi^+ + \Sigma^+ (1385)$. Its width is $\Gamma = 50$ MeV; its main decay channel is $\pi^+\Lambda$. (a) Is the decay strong or weak? (b) What are the strangeness and the isospin of the hyperon?

4.7 State the three reasons forbidding the decay $\rho^0 \to \pi^0\pi^0$.

4.8 The ρ^0 has spin 1; the f^0 meson has spin 2. Both decay into $\pi^+\pi^-$. Is the $\pi^0\gamma$ decay forbidden for one of them, for both, or for none?

4.9 Calculate the branching ratio $\Gamma(K^{*+} \to K^0 + \pi^+)/\Gamma(K^{*+} \to K^+ + \pi^0)$ assuming, in turn, that the isospin of the K^* is $I_{K^*} = 1/2$ or $I_{K^*} = 3/2$.

4.10 Calculate the ratios $\Gamma(K^-p)/\Gamma(\overline{K}^0n)$ and $\Gamma(\pi^-\pi^+)/\Gamma(\overline{K}^0n)$ for the $\Sigma(1915)$ that has $I = 1$.

4.11 A low-energy antiproton beam is introduced into a bubble chamber. Two exposures are made, one with the chamber full of liquid hydrogen (to study the interactions on protons) and one with the chamber full of liquid deuterium (to study the interactions on neutrons). The beam energy is such that the antiprotons come to rest in the chamber. We know that the stopped antiprotons are captured in an 'antiproton' atom and, when they reach an S wave, annihilate. The $\overline{p}p$ and $\overline{p}n$ in S wave are, in spectroscopic notation, the triplet $^3S_1$ and the singlet $^1S_0$. List the possible values of the total angular momentum and parity J^P and isospin I. Establish what are the eigenstates of C and those of G and give the eigenvalues. What are the quantum numbers of the possible initial states of the process $\overline{p}\,n \to \pi^-\pi^-\pi^+$? Consider the following three groups of processes. Compute for each the ratios between the processes:
(a) $\overline{p}n \to \rho^0\pi^-$; $\overline{p}n \to \rho^-\pi^0$
(b) $\overline{p}p(I = 1) \to \rho^+\pi^-$; $\overline{p}p(I = 1) \to \rho^0\pi^0$; $\overline{p}p(I = 1) \to \rho^-\pi^+$
(c) $\overline{p}p(I = 0) \to \rho^+\pi^-$; $\overline{p}p(I = 0) \to \rho^0\pi^0$; $\overline{p}p(I = 0) \to \rho^-\pi^+$.

4.12 Establish the possible total isospin values of the $2\pi^0$ system.

4.13   Find the Dalitz plot zeros for the $3\pi^0$ states with $I = 0$ and $J^P = 0^-$, 1^- and 1^+.

4.14   Knowing that the spin and parity of the deuteron are $J^P = 1^+$, give its possible states in spectroscopic notation.

4.15   What are the possible charm (C) values of a baryon, in general? What is it if the charge is $Q = 1$, and what is it if $Q = 0$?

4.16   A particle has baryon number $\mathcal{B} = 1$, charge $Q = +1$, charm $C = 1$, strangeness $S = 0$, beauty $B = 0$, top $T = 0$. Define its valence quark content.

4.17   Consider the following quantum number combinations, with, in every case $\mathcal{B} = 1$ and $T = 0$: $Q,C,S,B = -1,\ 0,\ -3,\ 0$; $Q,C,S,B = 2,\ 1,\ 0,\ 0$; $Q,C,S,B = 1,\ 1,\ -1,\ 0$; $Q,C,S,B = 0,\ 1,\ -2,\ 0$; $Q,C,S,B = 0,\ 0,\ 0,\ -1$. Define the valence quark contents for each one.

4.18   Consider the following quantum number combinations, with, in every case $\mathcal{B} = 0$ and $T = 0$: $Q,S,C,B = 1,\ 0,\ 1,\ 0$; $Q,S,C,B = 0,\ 0,\ -1,\ 0$; $Q,S,C,B = 1,\ 0,\ 0,\ 1$; $Q,S,C,B = 1,\ 0,\ 1,\ 1$. Define the valence quark contents of each one.

4.19   Explain why each of the following particles cannot exist according to the quark model: a positive strangeness and negative charm meson; a spin 0 baryon; an antibaryon with charge +2; a positive meson with strangeness −1.

4.20   Suppose you do not know the electric charges of the quarks. Find them using the other columns of Table 4.5.

4.21   What are the possible electric charges in the quark model of (a) a meson and (b) a baryon?

4.22   The mass of the J/ψ is $m_J = 3.097$ GeV and its width is $\Gamma = 91$ keV. What is its lifetime? If it is produced with $p_J = 5$ GeV in the L reference frame, what is the distance travelled in a lifetime? Consider the case of a symmetric $J/\psi \rightarrow e^+ e^-$ decay, that is, with the electron and the positron at equal and opposite angles $\pm\theta_e$ to the direction of the J/ψ. Find this angle and the electron energy in the L reference frame. Find θ_e if $p_J = 50$ GeV.

4.23   Consider a D^0 meson produced with energy $E = 20$ GeV. We wish to resolve its production and the decay vertices in at least 90% of cases. What spatial resolution will we need? Mention adequate detectors.

4.24   Consider the cross-section of the process $e^+ e^- \rightarrow f^+ f^-$ as a function of the centre of mass energy $\sqrt{s}$ near a resonance of mass M_R and total width Γ. Assuming that the Breit–Wigner formula correctly describes its line shape, calculate its integral over energy (the 'peak area'). Assume $\Gamma / 2 \ll M_R$.

4.25   A 'beauty factory' is (in particle physics) a high-luminosity electron–positron collider dedicated to the study of the $e^+ e^- \rightarrow B^0 \bar{B}^0$ process. Its centre of mass energy is at the $\Upsilon(4^1 S_3)$ resonance, namely at 10 580 MeV. This is only 20 MeV above the sum of the masses of the two Bs. Usually, in a collider, the energies of the two beams are equal. However, in such a configuration the two Bs are produced with very low energies. They travel distances that are too small to be measured. Therefore, the beauty factories are asymmetric. Consider PEP2 at SLAC, where the electron momentum is $p_{e-} = 9$ GeV and the positron momentum is $p_{e+} = 3$ GeV. Consider the case in which the two Bs are produced with the same energy. Find the distance travelled by the Bs in a lifetime and the angles of their directions to the beams.

4.26 A baryon decays strongly into $\Sigma^+\pi^-$ and $\Sigma^-\pi^+$, but not into $\Sigma^0\pi^0$ nor into $\Sigma^+\pi^+$, even if all are energetically possible. (1) What can you tell about its isospin? (2) You should check your conclusion looking at the ratio between the widths in the two observed channels. Neglecting phase-space differences, what is the value you expect?

4.27 Write the scattering amplitudes of the following processes in terms of the total isospin amplitudes: (1) $K^-p \to \pi^-\Sigma^+$, (2) $K^-p \to \pi^0\Sigma^0$, (3) $K^-p \to \pi^+\Sigma^-$, (4) $\bar{K}^0p \to \pi^0\Sigma^+$, (5) $\bar{K}^0p \to \pi^+\Sigma^0$.

4.28 Which of the following reactions is allowed or forbidden by strong interactions? Specify the reason(s) in each case.
(a) $\pi^- + p \to K^- + \Sigma^+$
(b) $\pi^- + p \to K^0 + \Lambda$
(c) $\pi^+ + p \to K^0 + \Sigma^+$
(d) $\Lambda \to \Sigma^- + \pi^+$
(e) $K^- + p \to K^0 + n$
(f) $\Xi^- \to \Lambda + \pi^-$
(g) $\Omega^- \to \Xi^- + \pi^0$.

4.29 Which of the following reactions is allowed or forbidden by strong interactions? Specify the reason(s) in each case.
(a) $\pi^- + p \to \bar{K}^0 + \Sigma^0$
(b) $\pi^- + p \to \Omega^- + K^+ + K^0 + \pi^0$
(c) $\pi^+ + p \to \Lambda + K^+ + \pi^+$
(d) $\pi^- + p \to \Xi^- + K^+ + K^0$
(e) $\pi^- + p \to \Lambda + \pi^-$
(f) $\Xi^0 \to \Sigma^+ + \pi^-$
(g) $\Xi^- \to p + \pi^- + \pi^-$.

4.30 Find the possible values of isospin, parity, charge conjugation, G-parity and total angular momentum J, up to $J = 2$, for a $\rho^0\rho^0$ state. For each value of J, specify also the orbital momentum L and the total spin S.

4.31 Find the possible values of isospin, parity, charge conjugation, G-parity and spin, up to $J = 2$, for a $\rho^0\pi^0$ state.

4.32 Find the threshold energy needed to produce $\Lambda_b\,(udb)$ with a π^- beam on a hydrogen target.

4.33 Find the threshold energy needed to produce $\Sigma_c^{++}\,(uuc)$ with a π^- beam on a hydrogen target.

4.34 Find the threshold energy needed to produce $\Omega_c^0\,(ssc)$ with a π^- or a K^- or a K^+ beam on a hydrogen target.

Summary

In this chapter we have studied the spectroscopy of hadrons. We have learnt that mesons and baryons are not elementary, but are made of quarks. These are bound by the strong colour field, of which the gluons are the quanta. In particular we have seen:

- Many of the hadrons decay via strong interactions and are detected as resonances.
- There are three families, or generations, of quarks, each composed of a doublet with charges $-1/3$ and $+2/3$, called down-type and up-type respectively, after the names of the first family.
- Baryons are made of three (valence) quarks, mesons of a quark and an antiquark.
- Quarks are never free.
- The quarks of the first family, u and d, have very small mass, much smaller than that of the hadrons they belong to. The strange quark s has a somewhat larger mass, but still small. The other three quarks are heavy.
- The $SU(2)$ symmetry, of the isotopic spin, of the hadrons made of the u and d quarks is a consequence of their small mass. Similarly for $SU(3)$ symmetry, which is more broken due to the larger s-quark mass.
- Only some $SU(3)$ multiplets and spin combinations are realized by nature: those corresponding to an antisymmetric colour wave function.

Further Reading

Alvarez, L. (1968) Nobel Lecture; *Recent Developments in Particle Physics*

Fowler, W. B. & Samios, N. P. (1964) The Omega minus experiment. *Sci. Am.* 211 (4) 36 (October 1964)

Lederman, L. M. (1988) Nobel Lecture; *Observations in Particle Physics form Two Neutrinos to the Standard Model*

Richter, B. (1976) Nobel Lecture; *From the Psi to Charm – The Experiments of 1975 and 1976*

Ting, S. B. (1976) Nobel Lecture; *The Discovery of the J Particle: A Personal Recollection*

5 Quantum Electrodynamics

In the previous chapters we learnt the main properties of the elementary particles, including the hadrons and how they are composed of smaller structures, the quarks and the gluons.

Starting with this chapter, we shall discuss the interactions of quarks and leptons, the electromagnetic, the strong and the weak ones, and finally their unification in the SM. Each of the three fundamental interactions is described by a quantum field theory. In this chapter we deal with quantum electrodynamics (QED).

We call the attention of the reader to the following important aspects of the quantum field theories, in the example of QED. As the reader knows, the equations of classical electromagnetism, the Maxwell equations, can be written in terms of both the electric, $\mathbf{E}$, and the magnetic, $\mathbf{B}$, fields and, in an equivalent manner, in terms of the potentials, the scalar potential ϕ and the vector potential $\mathbf{A}$, or, in relativistic notation, the four-potential A_μ. In classical physics, the most important quantities are the fields, which are measured, because the force acting on a charged particle is directly determined by them. The physical meaning of the potentials is linked to the *interaction* between charged particles and the electromagnetic field. The scalar potential ϕ is the *energy per unit charge locally available to the exchange between charged matter and the field*; the vector potential $\mathbf{A}$ *is the momentum per unit charge locally available to the exchange between charged matter and the field*. In quantum mechanics in general, and in quantum field theory in particular, forces are no longer important; the most relevant quantities are energy and momentum. In QED the fundamental field is the four-potential A_μ. Its quanta are the photons. The QED Lagrangian is formally the same as the classical one, but A_μ is an operator, rather than a number.

In quantum field theory, not only are the interactions between the constituents of matter described by a quantized field, but also the constituents themselves, the quarks and the leptons, are quanta of their field. The Lagrangian of the field of spin $1/2$ particles is the Dirac Lagrangian, where ψ is no longer the wave-function, but the field of the particle, an operator rather than a number. Scattering and decay are events in a single space-time point, in which the operator A_μ may 'create' or 'annihilate' photons and the operator ψ may create or annihilate the particles of which it is the field, electrons and positrons for example.

A property of classical electromagnetism is the gauge invariance. As the reader knows, there is no unique choice of potentials to represent a given electromagnetic field ($\mathbf{E}$ and $\mathbf{B}$). The four-divergence of A_μ can be fixed arbitrarily. This operation is called 'gauge fixing' and the corresponding degree-of-freedom property is called 'gauge invariance'. Gauge invariance in classical physics is substantially a mathematical curiosity, whereas in quantum physics it has a fundamental role. Gauge invariance, under a chosen group of transformations, determines the very structure of the interaction Lagrangian, as we shall see in this chapter for QED. In this case, the gauge group is the simplest one, namely $U(1)$. We shall start by recalling the relevant properties of classical electrodynamics and then introduce the QED Lagrangian.

The fundamental experiment showing that the quantization of the electromagnetic field was insufficient to describe nature was done by Lamb and Retherford in 1947. This masterpiece of atomic experimental physics is described in Section 5.2.

We shall introduce the Feynman diagram, whose lines are world lines of the particles in space-time. They are graphic representations of mathematical expressions of scattering or decay amplitudes, which we cannot treat at the level of this book. The diagrams, however, visually suggest the underlying physics.

We shall introduce some important concepts that will be used in the following chapters. The Lagrangians of all the elementary interactions are scalar products of two fermion *currents*. The 'current' is a Dirac covariant (see (2.50)) of those bispinors. Not all the covariants will be needed, but only the vector current in QED and QCD, and the vector current and the axial currents in weak interactions. The world lines of the incoming and leaving particles are called the *external legs* of the diagram. The *vertices* are where particles are created and destroyed by one of the currents. The simplest scattering diagrams have two vertices, one for each current. The *propagator* describes a 'virtual' particle moving in space-time from one vertex to another. This means that the relationship between energy, momentum and mass is not the same as for 'real' particles. Virtual particles may move backward in time. These are virtual antiparticles moving forward in time. The existence of antiparticles is a consequence of the scattering amplitude being an analytical function of the kinematic variables.

We shall discuss the important example of the e^+e^- annihilation into $\mu^+\mu^-$, and then show how the fine-structure constant, which is the dimensionless expression of the electromagnetic charge, is not really constant. Rather it depends on how close we look at it, or, rather, on the momentum transfer between the probe and the target charge in the scattering experiment we are performing. The 'evolution' or 'running' of the coupling constants is a property of all the interactions, as we shall see in the following chapters.

Finally, in Section 5.9, we shall discuss two tests of the SM, obtained with the highest precision measurements and theoretical prediction of the magnetic moments of the electron and of the muon.

5.1 The Lagrangian Density

We start by recalling some relevant aspects of classical electromagnetism, in the relativistic formalism. Taking the metric tensor diagonal elements (the others are zero) as

$$g_{00} = -1, g_{11} = +1, g_{22} = +1, g_{33} = +1, \qquad (5.1)$$

the electromagnetic field in covariant form is the antisymmetric tensor $F^{\mu\nu}$, which, in terms of the four-potential A_μ,

$$A_\mu = (\phi/c, \mathbf{A}),\tag{5.2}$$

is

$$F_{\mu\nu} = \partial_\mu A_\nu - \partial_\nu A_\mu,\tag{5.3}$$

and in terms of the electric and magnetic fields, $\mathbf{E}$ and $\mathbf{B}$, is

$$F_{\mu\nu} = \begin{pmatrix} 0 & E_x/c & E_y/c & E_z/c \\ -E_x/c & 0 & -B_z & B_y \\ -E_y/c & B_z & 0 & -B_x \\ -E_z/c & -B_y & B_x & 0 \end{pmatrix}.\tag{5.4}$$

The four-current is the space-time vector

$$J_\mu = (\rho c, \mathbf{j}),\tag{5.5}$$

where ρ is the charge density and $\mathbf{j}$ the current density. The Lagrangian density, or Lagrangian for short, is

$$\mathcal{L} = \frac{1}{4\mu_0} F_{\mu\nu} F^{\mu\nu} - J^\mu A_\mu\tag{5.6}$$

Let us now see how the Euler–Lagrange equation applied to (5.6) gives the Maxwell equations. We start by expressing $F_{\mu\nu}$ in terms of the four-potential with Eq. (5.4):

$$\mathcal{L} = -\frac{1}{4\mu_0}\left(\partial_\mu A_\nu - \partial_\nu A_\mu\right)\left(\partial^\mu A^\nu - \partial^\nu A^\mu\right) - J^\mu A_\mu$$

$$= \frac{1}{4\mu_0}\left(\partial_\mu A_\nu \partial^\mu A^\nu - \partial_\mu A_\nu \partial^\nu A^\mu - \partial_\nu A_\mu \partial^\mu A^\nu + \partial_\nu A_\mu \partial^\nu A^\mu\right) - J^\mu A_\mu$$

The 1st and 4th terms in parentheses are equal, as are the 2nd and 3rd terms, so we can write

$$\mathcal{L} = -\frac{1}{2\mu_0}\left(\partial_\mu A_\nu \partial^\mu A^\nu - \partial_\nu A_\mu \partial^\mu A^\nu\right) - J^\mu A_\mu.\tag{5.7}$$

The Euler–Lagrange equation(s) for the (components of the) A field is (are)

$$\frac{\partial \mathcal{L}}{\partial A_\nu} - \partial_\mu \frac{\partial \mathcal{L}}{\partial(\partial_\mu A_\nu)} = 0.\tag{5.8}$$

From (5.7), we obtain: $\mathcal{L} = -\partial_\mu \dfrac{1}{2\mu_0}\left(\partial^\mu A^\nu - \partial^\nu A^\mu\right) - J^\mu A_\mu$, which, going back to the electromagnetic field, is

$$\partial^\mu F_{\mu\nu} = \mu_0 J_\nu\tag{5.9}$$

These are the non-homogeneous Maxwell equations.

The homogeneous Maxwell equations are automatically satisfied by defining the tensor $F_{\mu\nu}$ as antisymmetric.

Electric charge conservation is a fundamental law in physics. The best experimental limits are obtained by searching for the decay of the electron, which, since it is the lightest charged particle, can decay only by violating charge conservation. The present most stringent limit comes from the BOREXINO experiment (Agostini *et al.* 2015), which searched for the photon in the hypothetical decay $e^- \rightarrow v_e + \gamma$, without finding it.

$$\tau_e > 6.6 \times 10^{28} \text{ yr.} \tag{5.10}$$

In the theory, as the reader knows, charge conservation is not an independent assumption, but is a consequence of the Maxwell equations. Charge is conserved locally, as expressed by the continuity equation

$$\nabla \cdot \mathbf{j} + \frac{\partial \rho}{\partial t} = 0, \tag{5.11}$$

or, in terms of the four-current J^μ

$$\partial_\mu J^\mu = 0 \tag{5.12}$$

Let us now consider the 'gauge invariance' of electromagnetism, a property of both classic and quantum electromagnetism.

Experimentally, we observe the effects of the Lorentz force, which depends on $\mathbf{E}$ and $\mathbf{B}$. The implication is that we have some freedom in the definition of the potentials. The Maxwell equations are invariant under the gauge transformations of the potentials $\mathbf{A}$ and ϕ

$$\mathbf{A} \Rightarrow \mathbf{A}' = \mathbf{A} + \nabla \chi; \quad \phi \Rightarrow \phi' = \phi - \frac{\partial \chi}{\partial t}, \tag{5.13}$$

where the scalar function $\chi(\mathbf{r}, t)$ is called the 'gauge function'.

The freedom to choose a particular gauge, called 'gauge fixing', is used in practice to simplify the mathematical expressions. In electrodynamics the 'Lorentz gauge' is chosen:

$$\nabla \cdot \mathbf{A} = -\mu_0 \varepsilon_0 \frac{\partial \phi}{\partial t}.$$

In covariant notation, the substitution (5.13) is

$$A_\mu \rightarrow A_\mu + \partial_\mu \chi(x_0, x_1, x_2, x_3). \tag{5.14}$$

The invariance is immediately verified as $F_{\mu\nu} \rightarrow \partial_\mu (A_\nu + \partial_\nu \chi) - \partial_\nu (A_\mu + \partial_\mu \chi) = \partial_\mu A_\nu - \partial_\nu A_\mu = F_{\mu\nu}$.

In covariant notation the Lorentz gauge is

$$\partial^\mu A_\mu = 0. \tag{5.15}$$

Up to now we have discussed the 'free' electromagnetic field. Similarly, the Dirac Lagrangian in Eq. (2.51) governs the wave function of the 'free' spin $1/2$ particles. As already stated, it is also the Lagrangian of the field of these particles. In practice, the electromagnetic and the Dirac fields interact, if the particle is charged. The interaction is ruled by the 'coupling constant', which is the well-known **fine-structure constant**

$$\alpha = \frac{1}{4\pi\varepsilon_0} \frac{q_e^2}{\hbar c} \simeq \frac{1}{137}, \tag{5.16}$$

where q_e is the elementary charge. Note that α is dimensionless; it is a pure number, and is small. It is one of the fundamental constants in physics and one of the most accurately measured, as we shall see.

In QED the electromagnetic current J^μ is due to the field of the charged particle, which we call electron for simplicity. It is given by

$$J^\mu = -q_e\bar{\psi}\gamma^\mu\psi \tag{5.17}$$

The Lagrangian density of the interacting electron and electromagnetic field is

$$\mathcal{L} = i\bar{\psi}\gamma^\mu\partial_\mu\psi - m\bar{\psi}\psi + \frac{1}{4\mu_0}F_{\mu\nu}F^{\mu\nu} - q_e\bar{\psi}\gamma^\mu\psi A_\mu \tag{5.18}$$

where the first two terms describe the free charged particle, the third describes the free photons and the last describes the interaction between them. Let us now check if it is gauge invariant. The substitution of (5.14) in the interaction term gives

$$-q_e\bar{\psi}\gamma^\mu\psi A_\mu \rightarrow -q_e\bar{\psi}\gamma^\mu\psi\left(A_\mu + \partial_\mu\chi\right) = -q_e\bar{\psi}\gamma^\mu\psi A_\mu - q_e\bar{\psi}\gamma^\mu\psi\partial_\mu\chi. \tag{5.19}$$

The second term on the right-hand side violates the gauge invariance.

V. Fock pointed out in 1926 (Fock 1926) that the invariance can be restored if the field of the charged particles is transformed at the same time as the four-potential. The required transformation is

$$\psi \rightarrow e^{-iq_e\chi}\psi, \quad \bar{\psi} \rightarrow e^{+iq_e\chi}\bar{\psi}. \tag{5.20}$$

The rotations of the phase angle, namely the set of the complex numbers of absolute value 1, form a Lie group, the unitary group $U(1)$, which is the gauge symmetry of QED.

Let us verify that. The mass term in the Dirac Lagrangian is clearly invariant. The substitution of (5.19) in the term with the four-gradient gives

$$\bar{\psi}e^{+iq_e\chi}\gamma^\mu i\partial_\mu\left(e^{-iq_e\chi}\psi\right) = \bar{\psi}e^{+iq_e\chi}\gamma^\mu i\left(e^{-iq_e\chi}\partial_\mu\psi - iq_e\psi\partial_\mu\chi\right)$$

$$= i\bar{\psi}\gamma^\mu\partial_\mu\psi + q_e\bar{\psi}\gamma^\mu\psi\partial_\mu\chi.$$

We see that the second term in the final expression exactly cancels the non-invariant term in Eq. (5.19).

Gauge invariance is a basic principle of the SM. All the fundamental interactions, not only the electromagnetic one, are gauge invariant. The gauge transformations of each of the three interactions form a Lie group, $SU(3)$ for QCD and $SU(2)\otimes U(1)$ for the electroweak interaction.

We have already used $SU(2)$ and $SU(3)$ to classify the hadrons and to correlate the cross-sections and the decay rates of different hadronic processes. We have observed that these symmetries are only approximate and are a result of the masses of two of the six quarks being very small, compared with the hadrons, and that the mass of a third is also small, although not as small. We now meet the same symmetry groups, and the symmetries are exact. Their role is now much deeper because they determine the very structure of the fundamental interactions.

We conclude by observing that other 'charges' that might look similar at first sight, namely the baryonic and the leptonic numbers, do not correspond to a gauge invariance. Therefore, from a purely theoretical point of view, their conservation is not as fundamental as that of the gauge charges.

5.2 The Lamb and Retherford Experiment

In 1947, Lamb and Retherford performed a crucial atomic physics experiment on the simplest atom, hydrogen (Lamb & Retherford 1947). The result showed that the motion of the atomic electron could not be described simply by the Dirac equation in an external, classically given field. The theoretical developments that followed led to a novel description of the interaction between charged particles and the electromagnetic field, and to the construction of the first quantum field theory, QED.

Let us start by recalling the aspects of the hydrogen atom relevant for this discussion. We shall use the spectroscopic notation, nL_j, where n is the principal quantum number, L is the orbital angular momentum and j is the total electronic angular momentum (i.e. it does not include the nuclear angular momentum, as we shall not need the hyperfine structure). We have not included the spin multiplicity $2s + 1$ in the notation since this, being $s = 1/2$, is always equal to 2.

Since the spin is $s = 1/2$, there are two values of j for every L, $j = L+1/2$ and $j = L-1/2$, with the exception of the S wave, for which it is only $j = 1/2$.

A consequence of the $-1/r$ dependence of the potential on the radius r is a large degree of degeneracy in the hydrogen levels. In a first approximation, the electron motion is non-relativistic ($\beta \approx 10^{-2}$) and we can describe it by the Schrödinger equation. As is well known, the energy eigenvalues in a $V \propto -1/r$ potential depend only on the principal quantum number

$$E_n = -\frac{Rhc}{n^2} = -\frac{13.6}{n^2} \text{eV},\tag{5.21}$$

where R is the Rydberg constant.

However, high-resolution experimental observation of the spectrum, for example with a Lummer plate or a Fabry–Pérot interferometer, resolves the spectral lines into multiplets. This is called the 'fine structure' of the spectrum.

We are interested here in the $n = 2$ levels. Their energy above the fundamental level is

$$E_2 - E_1 = Rhc\left(1-\frac{1}{4}\right) = \frac{3}{4}Rhc = 10.2 \text{ eV}\tag{5.22}$$

We recall that the fine structure is a relativistic effect. It is theoretically interpreted by describing the electron motion with the Dirac equation. The equation is solved by expanding in a power series of the fine-structure constant, which is much smaller than one. We give the result at order $\alpha^2 \left[= (1/137)^2 \right]$:

$$E_{n,j} = -\frac{Rhc}{n^2}\left[1+\frac{\alpha^2}{n}\left(\frac{1}{j+1/2}-\frac{3}{4n}\right)\right].\tag{5.23}$$

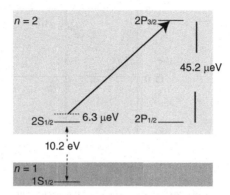

Fig. 5.1. Sketch of the energy levels relevant to the Lamb experiment.

We see that all levels, apart from the S level, split into two. This is the well-known spin–orbit interaction between the orbital and the spin magnetic moments of the electron.

However, the degeneracy is not completely eliminated: states with the same values of the principal quantum number n and of the angular momentum j with a different orbital momentum L have the same energy. In particular, the levels $2S_{1/2}$ and $2P_{1/2}$ are still degenerate. The aim of the Lamb experiment was to check this crucial prediction, namely whether it really is $E\left(2S_{1/2}\right) - E\left(2P_{1/2}\right) = 0$, or, in other words, whether there is a shift between these levels. We can expect this shift, even if it exists, to be small in comparison with the energy splits of the fine structure, which, as shown in Fig. 5.1, are tens of μeV.

The energy of a level cannot be measured in absolute value, but only in relative value. Lamb and Retherford measured the energy differences between three (for redundancy) $2P_{3/2}$ levels, taken as references, and the $2S_{1/2}$ level, searching for a possible shift (now called the Lamb shift) of the latter. The method consisted of forcing transitions between these states with an electromagnetic field and measuring the resonance frequency (order of tens of GHz). One of these transitions is shown as an arrow in Fig. 5.1. Also shown are the levels relevant to the experiment; the solid line for the $2S_{1/2}$ level is drawn according to Eq. (5.23); the dotted line includes the Lamb shift.

Let us assume that $E\left(2S_{1/2}\right) > E\left(2P_{1/2}\right)$. This is the actual case; the discussion for the opposite case would be completely similar, inverting the roles of the levels. In our hypothesis, $2S_{1/2}$ is metastable, meaning that its lifetime is of the order of 100 μs, much longer than the usual atomic lifetimes, which are of the order of 10 ns. Indeed, one of the a priori possible transitions, the $2S_{1/2} \Rightarrow 1S_{1/2}$ is forbidden by the $\Delta l = \pm 1$ selection rule and the second, $2S_{1/2} \Rightarrow 2P_{1/2}$ would be extremely slow, because the transition probability is proportional to the cube of the energy difference.

Now consider the atom in a magnetic field. All the energy levels split depending on the projection of the angular momentum in the direction of **B** (Zeeman effect). Fig. 5.2 gives the energies, in frequency units, of $2S_{1/2}$ and $2P_{1/2}$ as functions of the field. We have let the $2S_{1/2}$ and $2P_{1/2}$ energies be slightly different at zero field, because this possible difference is precisely the sought-after Lamb shift.

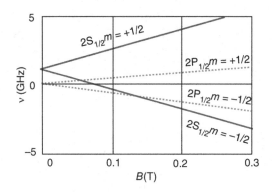

Fig. 5.2. Sketch of the dependence of the energy levels on the magnetic field.

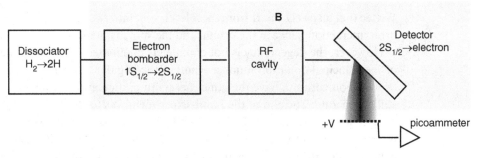

Fig. 5.3. Schematic block diagram of Lamb and Retherford's apparatus. RF, radio-frequency.

Note that, when the field increases, the level $(2S_{1/2}, m = -1/2)$ approaches the $2P_{1/2}$ levels and even crosses some of them. Therefore, it mixes with these levels, loses its metastability and decays in times of the order of 10^{-8} s. On the other hand, the level $(2S_{1/2}, m = +1/2)$ moves farther from the $2P_{1/2}$ levels and remains metastable.

Let us now discuss the logic of the experiment with the help of Fig. 5.3. The principal elements of the apparatus are as follows.

(1) The oven where, at 2500 K, 65% of the H_2 molecules dissociate into atoms. The atoms and the remaining molecules exit from an aperture with a Maxwellian velocity distribution with an average speed $\langle \upsilon \rangle \approx 8000$ m s^{-1}.

(2) The $1S_{1/2}$ to $2S_{1/2}$ excitation stage. This cannot be done with light because the transition is forbidden, as already mentioned. Instead, the atoms are bombarded with electrons with approximately 10 eV of energy. In this way, one succeeds in exciting only a few atoms to the $2S_{1/2}$ level, about one in 10^8.

(3) The separation of the Zeeman levels. The rest of the apparatus is in a magnetic field of adjustable intensity perpendicular to the plane of the figure. The atoms in the metastable level $(2S_{1/2}, m = +1/2)$ fly in a lifetime over distance $d = 10^{-4}\,(\text{s}) \times 8 \times 10^3\,(\text{m s}^{-1}) = 0.8$ m, enough to cross the apparatus. The non-metastable atoms, those in the level $(2S_{1/2}, m = -1/2)$ in particular, can travel only $d \approx 10^{-8}\,(\text{s}) \times 8 \times 10^3\,(\text{m s}^{-1}) = 0.08$ mm.

(4) The pumping stage: the beam, still in the magnetic field, enters a cavity in which the radio-frequency field is produced. Its frequency can be adjusted to induce a transition from the $(2S_{1/2}, m = +1/2)$ level to one of the Zeeman $2P_{3/2}$ levels. There are four of these, but one of them, $(2P_{3/2}, m = -3/2)$, cannot be reached because this would require $\Delta m = -2$. The other three, $(2P_{3/2}, m = -1/2)$, $(2P_{3/2}, m = +1/2)$ and $(2P_{3/2}, m = +3/2)$, however, can be reached. Therefore, for a fixed magnetic field value, there are three resonance frequencies for transitions from $(2S_{1/2}, m = +1/2)$ to a $2P_{3/2}$ level. The atoms pumped into one of these levels, which are unstable, decay immediately. Therefore, the resonance conditions are detected by measuring the disappearance, or a strong decrease, of the intensity of the metastable $(2S_{1/2}, m = +1/2)$ atoms after the cavity.

(5) The excited atom detector: a tungsten electrode. The big problem is that the atoms to be detected (i.e. those in the $(2S_{1/2}, m = +1/2)$ level) are a very small fraction of the total, a few in a billion as we have seen, when they are present. However, they are the only excited ones that reach the detector; the others have already decayed. To build a detector sensitive to the excited atoms only, Lamb used the capability of extracting electrons from a metal, tungsten in this case. The atoms in the $n = 2$ level, which are 10.2 eV above the fundamental level, when in contact with a metal surface de-excite and a conduction electron is freed. This is energetically favoured because the work function of tungsten is $W_W \approx 6$ eV < 10.2 eV. Obviously, atoms in the fundamental level cannot do that.

(6) Electron detection: this operation is relatively easy. An electrode, at a positive potential relative to the tungsten (which is earthed) collects the electron flux, measured as an electric current with a picoammeter.

The results are given in Fig. 5.4. The measuring procedure was as follows: a value of the radio-frequency in the cavity, v, was fixed; the magnetic field intensity was then varied and the detector current measured in search of the resonance conditions, appearing as minima in the current intensity. The points in Fig. 5.4 were obtained.

The resonance frequencies correspond to the energy differences ΔE between the levels according to

$$hv = \Delta E. \tag{5.24}$$

One can see that the experimental points fall into three groups, each with a linear correlation. Clearly each group corresponds to a transition. The three lines extrapolate to a unique value at zero field, as expected, but they are shifted from the positions expected according to Dirac's theory, the dashed lines. The experiment shows that the $S_{1/2}$ level is shifted by about 1 GHz. More precisely, the Lamb shift value as measured in 1952 was

$$\Delta E \left(2S_{1/2} - 2P_{1/2} \right) = 1057.8 \pm 0.1 \, \text{MHz} \tag{5.25}$$

As we shall discuss in Section 5.9, in 1947, P. Kusch and H. M. Foley (Kusch & Foley 1947) accurately measured the electron magnetic moment, finding a difference at the per mille level from the Dirac value. The effect will be called 'anomalous magnetic moment'. The consequence of both observations was the development of QED.

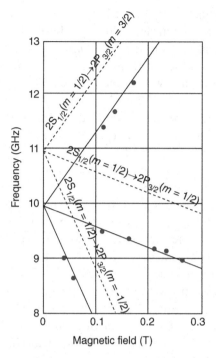

Fig. 5.4. Measured values of the transition frequencies for different magnetic field intensities (dots). Linear interpolation of the data (continuous lines) and expected behaviour in the absence of the shift (dashed lines) (adapted from Lamb & Retherford 1947).

5.3 Quantum Field Theory

The theoretical developments originated by the discoveries in the previous section led to the creation of a fundamental description of the basic forces, the quantum field theories. To interpret the Lamb–Retherford and Kusch–Foley experiments, we must not think of the electric field of the proton seen by the electron as an external field classically given once and for ever, as for example in the Bohr description of the atom. On the contrary, the field itself is a quantum system, made of photons that interact with the charges. We shall proceed in our description by successive approximations.

Let us use for the first time, with the help of intuition, a Feynman diagram. It is shown in Fig. 5.5 and represents an electron interacting with a nucleus. We must think of a time coordinate on a horizontal axis running from left to right and of a vertical axis giving the particle position in space. The thin lines represent the electron, which exchanges a photon, the wavy line, with the nucleus of charge Ze. The nucleus is represented by a line parallel to the time axis because, having a mass much larger than the electron, it does not move during the interaction.

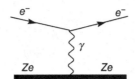

Fig. 5.5.   Diagram of an electron interacting with a nucleus.

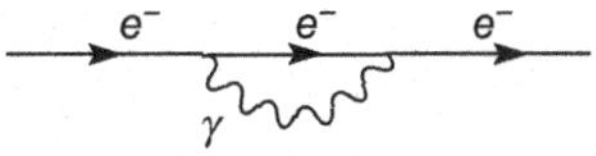

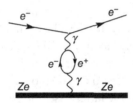

Fig. 5.6.   Diagram of an electron emitting and reabsorbing a photon.

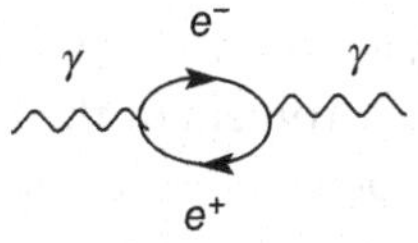

Fig. 5.7.   Vacuum polarization by a photon.

Fig. 5.8.   An electron interacting with a nucleus with vacuum polarization.

The Feynman diagram, and Fig. 5.5 in particular, represents a well-defined physical quantity, the probability amplitude of a process.

Now consider a free electron in a vacuum. The quantum vacuum is not really empty, because processes such as that shown in Fig. 5.6 continually take place. The diagram shows the electron emitting and immediately reabsorbing a photon.

In a similar way, a photon in a vacuum is not simply a photon. Fig. 5.7 shows a photon that materializes into an e^+e^- pair followed by their re-annihilation into a photon. This process is called 'vacuum polarization'.

The e^+e^- pair production and annihilation also occur for the virtual photon mediating the electron–nucleus interaction, as shown by the diagram in Fig. 5.8.

The careful reader will have noticed that the processes we have just described do not conserve the energy. However, to be able to detect an energy violation, say ΔE, we need to measure with an energy resolution better than ΔE. This, according to the uncertainty principle, requires some time. Therefore, if the duration Δt of the violation is very short, namely if

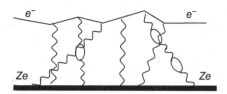

Fig. 5.9.   An electron bound to a nucleus.

$$\Delta E \Delta t \leq \hbar, \tag{5.26}$$

the available time is not sufficient, the violation is not detectable, and may occur.

In conclusion, the atomic electron interacts both with the external field and with its own field. As in classical electromagnetism, this self-interaction implies an infinite value of the electron mass-energy. A fundamental theoretical contribution was given by H. Bethe in 1947, a month after the Lamb and Retherford experiment (Bethe 1947). He observed that the problem of the infinite value of the auto-interaction term could be avoided because such a term is not observable. One could 'renormalize' the mass of the electron by subtracting an infinite term.

After this subtraction, if the electron is in a vacuum, the contribution of the self-interaction is zero (by construction). However, this does not happen for a bound electron. Indeed, we can imagine the electron as moving randomly around its unperturbed position, due to the above-mentioned quantum fluctuations. The electron appears, effectively, as a small charge distribution and, consequently, its binding energy is a little less than that of a point particle. This small increase in total energy is a little larger for the zero orbital momentum states such as $2S_{1/2}$, than for that of $2P_{1/2}$. This is because, in the latter case, the electron has a smaller probability of being close to the nucleus.

Now consider the new interpretation of the Dirac equation mentioned above. If the electron field is not quantized, $|\psi|^2$ is the probability of finding the electron. However, as we have seen, the hydrogen atom does not always contain only one electron. Sometimes two electrons are present, together with a positron, or even three electrons and two positrons. As long as the system is bound, the electrons move in the neighbourhood of the nucleus, continuously exchanging photons, as in the diagram in Fig. 5.9. In QED, the number of particles is not constant. We must describe by a quantum field not only the 'force' – the electromagnetic field – but also the particles, such as the electron, that are the sources and receptors of that field. The electron field contains operators that 'create' and 'destroy' electrons and positrons. Consider the simple diagram of Fig. 5.5. It shows two oriented electron lines, one entering the 'vertex' and one leaving it. The correct meaning of this is that the initial electron disappears at the vertex and is destroyed by an 'annihilation operator'; at the same time, a 'creation operator' creates the final electron. Asking whether the initial and final electrons are the same or different particles is meaningless because all the electrons are identical.

5.4 The Interaction as an Exchange of Quanta

Now consider, in general, a particle a interacting through the field mediated by the boson V. When moving in a vacuum it continually emits and re-absorbs V bosons, as shown in Fig. 5.10(a).

Now suppose that another particle b, with the same interaction as a, comes close to a. Then, sometimes, a mediator V emitted by a can be absorbed not by a but by b, as shown in Fig. 5.10(b).

We say that particles a and b interact by exchanging a field quantum V.

The V boson in general has a mass m different from zero, and, consequently, the emission process $a \rightarrow a+V$ violates energy conservation by $\Delta E = m$. The violation is equal and opposite in the absorption process. The net violation lasts only for a short time, Δt, which must satisfy the relationship $\Delta E \Delta t \leq \hbar$. As the V boson can reach a maximum distance $R = c\Delta t$ in this time, the range of the force is finite

$$R = c\Delta t = \hbar / (mc). \tag{5.27}$$

This is a well-known result: the range of the force is inversely proportional to the mass of its mediator.

The diagram in Fig. 5.10(b) gives the amplitude for the elastic scattering process $a + b \rightarrow a+b$. It contains three factors, namely the probability amplitude for the emission of V, the probability amplitude for the propagation of V from a to b and the probability amplitude for the absorption of V. The internal line is called the 'propagator' of V. We shall now find the mathematical expression of the propagator by using a simple argument.

We start with the non-relativistic scattering of a particle a of mass m from the central potential $\phi(\mathbf{r})$. The potential is due to a centre of forces of mass M, much larger than m. Let g be the 'charge' of a, which therefore has energy $g\phi(\mathbf{r})$, and let g_0 be the charge of the central body. Note that, since it is in a non-relativistic situation, the use of the concepts of potential and potential energy is justified.

The scattering amplitude is given by the diagram in Fig. 5.11, where $\mathbf{p}_1$ and $\mathbf{p}_2$ are the momenta of a before and after the collision. The central body does not move, assuming its mass to be infinite.

The momentum

$$\mathbf{q} = \mathbf{p}_2 - \mathbf{p}_1 \tag{5.28}$$

transferred from the centre to a is called 'three-momentum transfer'. Obviously, a transfers the momentum $-\mathbf{q}$ to the centre of forces.

(a)   (b)

Fig. 5.10. Diagrams showing the world lines of (a) particle a emitting and absorbing a V boson and (b) particles a and b exchanging a V boson.

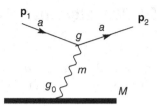

Fig. 5.11.   Diagram of the scattering of particle *a* in the potential of the infinite-mass centre *M*.

The situation is similar to that considered in Section 1.8, with a generic charge g in place of the electron electric charge q_e. Let us calculate the transition matrix element. In the initial and final states the particle a is free, hence its wave functions are plane waves. Neglecting uninteresting constants, the matrix element is given by Eq. (1.84) with g in place of q_e

$$\langle \psi_f | g\phi(r) | \psi_i \rangle \propto g\int \exp(i\,\mathbf{p}_2 \cdot \mathbf{r})\phi(r)\exp(-i\,\mathbf{p}_1 \cdot \mathbf{r})dV = g\int \exp[i\,\mathbf{q}\cdot\mathbf{r}]\phi(r)dV. \quad (5.29)$$

Again, the scattering amplitude depends only on the three-momentum transfer. Calling this amplitude $f(\mathbf{q})$, we have

$$f(\mathbf{q}) \propto \int \exp[i\,\mathbf{q}\cdot\mathbf{r}]\phi(r)dV. \quad (5.30)$$

We now assume the potential corresponding to a meson of mass m to be the Yukawa potential of range $R = 1/m$

$$\phi(r) = \frac{g_0}{4\pi r}\exp\left(-\frac{r}{R}\right) = \frac{g_0}{4\pi r}\exp(-rm). \quad (5.31)$$

Let us calculate the scattering amplitude

$$f(\mathbf{q}) = g\int_{\text{space}}\phi(r)e^{i\mathbf{q}\cdot\mathbf{r}}dV = g\int_{\text{space}}\phi(r)e^{iqr\cos\theta}d\varphi\,\sin\theta\,d\theta\,r^2dr$$

$$= g2\pi\int_0^\infty\phi(r)r^2dr\int_0^\pi e^{iqr\cos\theta}d\cos\theta = g4\pi\int_0^\infty\phi(r)\frac{\sin qr}{qr}r^2dr$$

that, with the potential (5.31) becomes

$$f(\mathbf{q}) = gg_0\int_0^\infty e^{-rm}\frac{\sin qr}{q}dr = g_0g\int_0^\infty e^{-mr}\left(\frac{e^{iqr}-e^{-iqr}}{2iq}\right)dr.$$

Finally, calculating the above integral, we obtain the very important equation

$$f(\mathbf{q}) = \frac{g_0g}{|\mathbf{q}|^2+m^2}. \quad (5.32)$$

As anticipated, the amplitude is the product of the two 'charges' and the propagator, of which we now have the expression.

We now consider the relativistic situation, no longer assuming an infinite mass of the scattering centre. Therefore, the particle a and the particle of mass M exchange both momentum and energy. The kinematic quantities are defined in Fig. 5.12.

The relevant quantity is now the four-momentum transfer. Its norm is

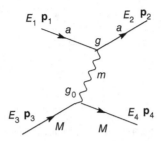

Fig. 5.12.   Kinematic variables for the scattering of two particles.

$$t \equiv \left(E_2 - E_1 \right)^2 - \left(\mathbf{p}_2 - \mathbf{p}_1 \right)^2 = \left(E_4 - E_3 \right)^2 - \left(\mathbf{p}_4 - \mathbf{p}_3 \right)^2 \tag{5.33}$$

which, we recall, is negative or zero.

We noted above that the emission and absorption processes at the vertices do not conserve energy and, we may add, momentum. When using the Feynman diagrams we take a different point of view, assuming that at every vertex energy and momentum are conserved. The price to pay is the following. Since the energy of the exchanged particle is $E_2 - E_1$ and its momentum is $\mathbf{p}_2 - \mathbf{p}_1$, the square of its mass is given by Eq. (5.33). This is not the physical mass of the particle on the propagator. We call it a 'virtual particle'. The following language is also used. When a particle is real, and the relation between its energy, momentum and mass is (1.28), it is said to be on the mass shell; when it is virtual, and that relation does not hold, it is said to be off the mass shell.

We do not calculate, but simply give, the relativistic expression of the scattering amplitude, that is

$$f(t) = \frac{g_0 g}{m^2 - t}, \tag{5.34}$$

which is very similar to (5.29). The 'vertex factors' are the probability amplitudes for emission and absorption of the mediator, that is the charges of the interacting particles. The propagator, namely the probability amplitude for the mediator to move from one particle to the other, is

$$\prod(t) = \frac{1}{m^2 - t}. \tag{5.35}$$

The probabilities of the physical processes, cross-sections or decay rates, are proportional to $\left| \prod(t) \right|^2$, to the coupling constants and to the phase-space volume.

5.5 The Feynman Diagrams

From the historical point of view, QED was the first quantum field theory to be developed. It was created independently by S. I. Tomonaga (Tomonaga 1946), R. Feynman (Feynman 1948) and J. Schwinger (Schwinger 1948). Feynman developed simple rules

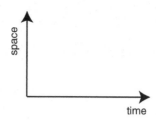

Fig. 5.13. Space-time reference frame used for the Feynman diagrams in the text.

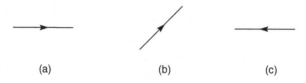

Fig. 5.14. World lines of fermions in the Feynman diagrams.

for evaluating the transition matrix elements. In QED, and in general in all quantum field theories, the probability of a physical process is expressed as a series of diagrams that become more and more complex as the order of the expansion increases. These 'Feynman diagrams' represent mathematical expressions, defined by a set of precise rules, which we shall not discuss here. However, the Feynman diagrams are also pictorial representations that clearly suggest intuitive interaction mechanisms, and we shall use them as such.

Consider the initial and the final states of a scattering or decay process. They are defined by specifying the initial and final particles and the values of the momenta of each of them. We must now consider that there is an infinite number of possibilities for the system to go from the initial to the final state. Each of these has a certain probability amplitude, a complex number with an argument and a phase. The probability amplitude of the process is the sum, or rather the integral, of all these partial amplitudes. The probability of the process, the quantity we measure, is the square modulus of the sum.

The diagrams are drawn on a sheet of paper, on which we imagine two axes, one for time, the other for space (on the paper we have only one dimension for the three spatial dimensions), as in Fig. 5.13. The particles, both real and virtual, are represented by lines, which are their world lines. A solid line with an arrow is a fermion; it does not move in Fig. 5.14(a), and it moves upwards in Fig. 5.14(b). The arrow shows the direction of the flux of the charges relative to time. For example, if the fermion is an electron, its electric charge and electron flavour advance with it in time. In Fig. 5.14(c), all the charges go back in time: it is a positron moving forward in time. We shall soon return to this point.

We shall use the symbols in Fig. 5.15 for the vector mediators of the fundamental interactions, that is, the 'gauge bosons'.

An important element of the diagrams is the vertex, shown in Fig. 5.16 for the electromagnetic interaction. The particles f are fermions, of the same type on the

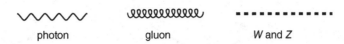

Fig. 5.15. World lines of the vector bosons mediating the interactions in the Feynman diagrams.

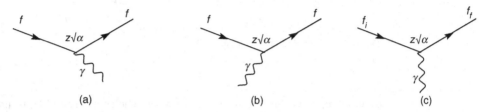

Fig. 5.16. The electromagnetic vertex.

two sides of the vertex of electric charge z. In Fig. 5.16(a) the initial f disappears in the vertex, while two particles appear in the final state: a fermion f and a photon. The initial state in Fig. 5.16(b) contains a fermion f and a photon that disappear at the vertex; in the final state there is only one fermion f. The two cases represent the emission and the absorption of a photon. Actually the mathematical expression of the two diagrams is the same, evaluated at different values of the kinematic variables, namely the four-momentum of the photon. Therefore, we can draw the diagram in a neutral manner, as in Fig. 5.16(c) (where we have explicitly written the indices i and f for 'initial' and 'final').

The vertex corresponds to the interaction Lagrangian

$$z\sqrt{\alpha} A_\mu \overline{\psi}_f \gamma^\mu \psi_f. \tag{5.36}$$

The operators ψ_f and $\overline{\psi}_f$ are Dirac bispinors. The combination $\overline{\psi}_f \gamma^\mu \psi_f$ is called the 'vector current' of the fermion f (electron current, up-quark current, etc.), where 'vector' stands for its properties under Lorentz transformations. The vector current is extremely important; it appears in all the interactions of the fermion f: electromagnetic if it is charged (leptons and quarks), strong if it is a quark, weak for all of them, including neutrinos. In the weak interaction, as we shall see, a further term, the axial current $\overline{\psi}_f \gamma^\mu \gamma_5 \psi_f$ is present. On the other hand, the coupling constant, α, and the field A_μ, the quantum analogue of the classical four-potential, are characteristic of the electromagnetic interaction. The actions of the operators ψ_f and $\overline{\psi}_f$ in the vertex are as follows: ψ_f destroys the initial fermion (f_i in Fig. 5.16), $\overline{\psi}_f$ creates the final fermion (f_f). The four-potential is due to a second charged particle that does not appear in the figure, because the vertex it shows is only a part of the diagram. Fig. 5.17 shows an example of a complete diagram of the elastic scattering

$$e^- + \mu^- \rightarrow e^- + \mu^-. \tag{5.37}$$

It contains two electromagnetic vertices.

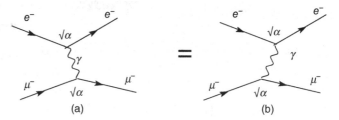

Fig. 5.17. Feynman diagram for the electron–muon scattering.

The lines representing the initial and final particles are called 'external legs'. The four-momenta of the initial and final particles, which are given quantities, define the external legs completely. On the contrary, there are infinite possible values of the virtual photon four-momentum, corresponding to different directions of its line. The scattering amplitude is the sum of these infinite possibilities. The diagram represents this sum. Therefore, we can draw the propagator in any direction. For example, the two parts of Fig. 5.17 are the same diagram whether the photon is emitted by the electron and absorbed by the μ, or vice versa, if it is emitted by the μ and absorbed by the electron.

The probability amplitude is given by the product of two vertex factors (5.36)

$$\left(\sqrt{\alpha} A_\nu \overline{\psi}_e \gamma^\nu \psi_e \right) \left(\sqrt{\alpha} A_\nu \overline{\psi}_\mu \gamma^\nu \psi_\mu \right), \tag{5.38}$$

where ν is the label on which we sum, and e and μ identify the particles. Note that, since the emission and absorption probability amplitudes are proportional to the charge of the particle, namely to $\sqrt{\alpha}$, the scattering amplitude is proportional to $\alpha \, (\sim 1/137)$ and the cross-section to α^2.

Summarizing, the internal lines of a Feynman diagram represent virtual particles, which exist only for short times, since they are emitted and absorbed very soon after. The relationship between their energy and their momentum is not that of real particles. We shall see that, although they live for such a short time, the virtual particles are extremely important.

The amplitudes of the electromagnetic processes, such as those in Eq. (5.37), are calculated by performing an expansion in a series of terms of increasing powers of α, called a perturbative series. The diagram of Fig. 5.17 is the lowest term of the series; this is called 'tree level'. Fig. 5.18 shows two of the next-order diagrams. They contain four virtual particles and are proportional to $\alpha^2 \left(\sim 1/137^2 \right)$. One can understand that the perturbative series rapidly converges, owing to the smallness of the coupling constant. In practice, if a high accuracy is needed, the calculations may require supercomputers, because the number of different diagrams grows enormously with the increasing order. In the higher-order diagrams, closed patterns of virtual particles are always present. They are called 'loops'.

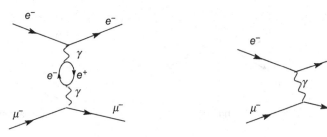

Fig. 5.18.   Two diagrams at the next-to-tree level.

5.6 Analyticity and the Antiparticles

Consider the two-body scattering

$$a + b \rightarrow c + d. \tag{5.39}$$

Let us consider the two invariant quantities: the centre of mass energy squared

$$s = \left(E_a + E_b\right)^2 - \left(\mathbf{p}_a + \mathbf{p}_b\right)^2 = \left(E_c + E_d\right)^2 - \left(\mathbf{p}_c + \mathbf{p}_d\right)^2, \tag{5.40}$$

where the meaning of the variables should be obvious, and the norm of the four-momentum transfer is

$$t = \left(E_b - E_a\right)^2 - \left(\mathbf{p}_b - \mathbf{p}_a\right)^2 = \left(E_d - E_c\right)^2 - \left(\mathbf{p}_d - \mathbf{p}_c\right)^2. \tag{5.41}$$

We recall that $s \geq 0$ and $t \leq 0$.

The amplitude corresponding to a Feynman diagram is an analytical function of these two variables, representing different physical processes for different values of the variables, joined by analytical continuation. Consider, for example, the following processes: the electron–muon scattering and the electron–positron annihilation into a muon pair

$$e^- + \mu^- \rightarrow e^- + \mu^- \quad \text{and} \quad e^- + e^+ \rightarrow \mu^- + \mu^+. \tag{5.42}$$

Fig. 5.19 shows the Feynman diagrams. They are drawn differently, but they represent the same function. They are called 's channel' and 't channel' respectively.

In the special case $a = c$ and $b = d$, the particles in the initial and final states are the same for the two channels. Therefore, as shown in an example in Fig. 5.20, the two channels contribute to the same physical process. Its cross-section is the absolute square of their sum, namely the sum of the two absolute squares and of their cross product, the interference term.

Returning to the general case, we recall that $\sqrt{s}$ and $\sqrt{t}$ are the masses of the virtual particles exchanged in the corresponding channel. In the t channel the mass is imaginary, whereas it is real in the s channel. In the latter, something spectacular may happen. When $\sqrt{s}$ is equal, or nearly equal, to the mass of a real particle, such as J/ψ

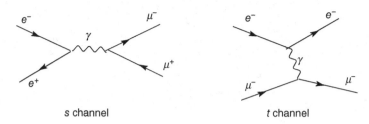

Fig. 5.19.   Photon exchange in s and t channels.

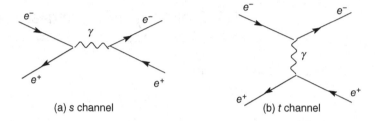

Fig. 5.20.   Feynman diagrams for $e^- + e^+ \rightarrow e^- + e^+$ showing the photon exchange in the s and t channels.

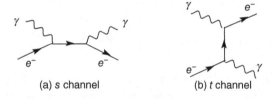

Fig. 5.21.   A fermion propagator. Compton scattering. (a) s channel. (b) t channel.

for example, the cross-section shows a resonance. Notice that the difference between virtual and real particles is quantitative, not qualitative.

Up to now we have discussed boson propagators, but fermion propagators also exist. Fig. 5.21 shows the t channel and the s channel diagrams for Compton scattering.

Let us focus on the t channel in order to make a very important observation. As we know, all the diagrams, differing only by the direction of the propagator, are the same diagram, and represent the same function. In Fig. 5.22(a) the emission of the final photon, event A, happens before the absorption of the initial photon, event B. The shaded area is the light cone of A. In Fig. 5.22(a) the virtual electron line is inside the cone. The AB interval is time-like; the electron speed is less than the speed of light. In Fig. 5.22(b) the AB interval is outside the light cone, it is space-like. We state without any proof that the diagram is not zero in these conditions; in other words, virtual particles can travel faster than light. This is a consequence of the analyticity of the scattering amplitude that follows, in turn, from the uncertainty of the measurement of the speeds intrinsic to quantum mechanics.

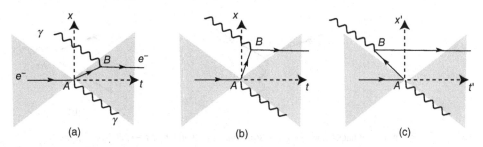

Feynman diagrams for Compton scattering. The grey region is the light cone. (a) The virtual electron world line is inside the cone (time-like). (b) The virtual electron world line is outside the cone (space-like), (c) As in (b), as seen by an observer in motion relative to (a) and (b).

This observation has a very important consequence. If two events, A and B, are separated by a space-like interval, the order of their sequence in time is reference-frame dependent. We can always find a frame in which event B precedes event A, as shown in Fig. 5.22(c). An observer in this frame sees the photon disappearing in B and two electrons appearing, one advancing and one going back in time. The observer interprets the latter as an anti-electron, with positive charge, moving forward in time. Event B is the materialization of a photon in an electron–positron pair. Event A, coming later in time, is the annihilation of the positron of the pair with the initial electron.

We must conclude that the virtual particle of one observer is the virtual antiparticle of the other. However, the sum of all the configurations, which is what the diagram is for, is Lorentz-invariant. Lorentz invariance and quantum mechanics, once joined together, necessarily imply the existence of antiparticles.

Every particle has an amplitude to go back in time, and therefore has an antiparticle. This is true for both fermions and bosons. Consider, for example, Fig. 5.17(b). We can read it thinking that the photon is emitted at the upper vertex, moves backward in time and is absorbed at the lower vertex, or that it is emitted at the lower vertex, moves forward in time and is absorbed at the upper vertex. The two interpretations are equivalent because the photon is completely neutral, that is, photon and antiphoton are the same particle. This is the reason why there is no arrow in the wavy line representing the photon in Fig. 5.15.

An important property of the interaction vertex, which we present here for the electromagnetic interaction but which is true for all of them, is the helicity conservation which is true with a very good approximation at ultra-relativistic energies, namely at energies much higher than the masses of the particles – as is often the case – and rigorously if the masses are zero.

Fig. 5.23(a) shows the electromagnetic vertex as it appears in the scattering process of the fermion f; Fig. 5.23(b) shows it as it appears in the fermion–antifermion annihilation (and in the inverse process of pair creation). As already discussed, it represents the same analytical function in different kinematic regions.

In ultra-relativistic conditions, the vertex couples an electron and a positron if their helicities are opposite, not if they are equal. Let us prove this statement. With refer-

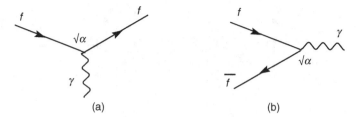

Fig. 5.23. The electromagnetic vertex (a) in a scattering process and (b) in an annihilation process.

ence to Fig. 5.23(b), consider, for example, the case in which both helicities are positive. Since the masses can be neglected, the electron is a positive chirality fermion and the positron is a negative chirality antifermion, namely

$$\psi_R = \frac{1+\gamma_5}{2}\psi; \quad \overline{\psi}_L = \overline{\psi}\frac{1+\gamma_5}{2}.$$

We then write the electromagnetic interaction and, taking into account that $\gamma_5^2 = 1$, we have

$$\overline{\psi}_L\gamma_\mu\psi_R = \overline{\psi}\frac{1+\gamma_5}{2}\gamma_\mu\frac{1+\gamma_5}{2}\psi = \overline{\psi}\gamma_\mu\frac{1+\gamma_5}{2}\frac{1-\gamma_5}{2}\psi = 0.$$

In a similar way one proves that the initial and final fermions in Fig. 5.26 are coupled only if their helicities are equal, provided they can be considered as massless.

We can now understand why the probability of the elastic scattering of an electron by a massive target becomes zero at 180°, as foreseen by the Mott formula (1.95). Indeed, in such conditions the incoming and outgoing electrons would have opposite velocities, but spins in the same direction, hence opposite helicities.

Finally, in Fig. 5.15 we have included for completeness the gauge bosons of the weak interactions. The Z is, like the photon, completely neutral; it is its own antiparticle. On the contrary, W^+ and W^- are the antiparticles of each other. A W^+ moving back in time is a W^-, and vice versa. To be rigorous this would require including an arrow in the graphic symbol of the Ws in Fig. 5.15, but this is not really needed in practice.

The situation is similar for the gluons. There are eight gluons in total, two completely neutral and the other six coming in three particle–antiparticle pairs. We shall study them in Chapter 6.

5.7 Electron–Positron Annihilation into a Muon Pair

When an electron and a positron annihilate, they produce a pure quantum state, with the quantum numbers of the photon, $J^{PC} = 1^{--}$. We have already seen how resonances appear when $\sqrt{s}$ is equal to the mass of a vector meson. Actually, the contributions of the e^+e^- colliders to elementary particle physics were also extremely important outside

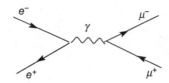

Fig. 5.24.   Lowest-order diagram for $e^+ + e^- \rightarrow \mu^+ + \mu^-$.

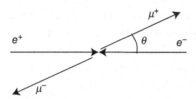

Fig. 5.25.   Initial and final momenta in the process $e^+ + e^- \rightarrow \mu^+ + \mu^-$, and definition of the scattering angle θ.

the resonances. In the next chapter we shall see what they have taught us about strong interaction dynamics, namely, QCD. Now consider the process

$$e^+ + e^- \rightarrow \mu^+ + \mu^-, \tag{5.43}$$

at energies that are high compared with the masses of the particles. This process is easily described by theory, because it involves only leptons that have no strong interactions. It is also easy to measure because the muons can be unambiguously identified.

Fig. 5.24 shows the lowest-order diagram for reaction (5.43), the photon exchange in the s channel. The t channel does not contribute. The differential cross-section of (5.43), for unpolarized initial and final states, is given by Eq. (1.77). Neglecting the electron and muon mass, we have $p_f = p_i$ and

$$\frac{d\sigma}{d\Omega_f} = \frac{1}{(8\pi)^2} \frac{1}{E^2} \frac{p_f}{p_i} \overline{\sum_{\text{initial}} \sum_{\text{final}}} |M_{fi}|^2 = \frac{1}{(8\pi)^2} \frac{1}{s} \frac{1}{4} \sum_{\text{spin}} |M_{fi}|^2 . \tag{5.44}$$

We do not perform the calculation, but we give the result directly. Defining the scattering angle θ as the angle between the μ^- and the e^- (Fig. 5.25), we have

$$\frac{1}{4} \sum_{\text{spin}} |M_{fi}|^2 = (4\pi\alpha)^2 (1 + \cos^2 \theta). \tag{5.45}$$

We observe here that the cross-section in (5.44) is proportional to $1/s$. This important feature is common to the cross-sections of the collisions between point-like particles at energies much larger than all the implied masses, both of the initial and final particles and of the mediator. This can be understood by a simple dimensional argument. The cross-section has the physical dimensions of a surface, or, in NU, of the reciprocal of an energy squared. Under our hypothesis, the only available dimensional quantity is the centre of mass energy. Therefore the cross-section must be inversely proportional

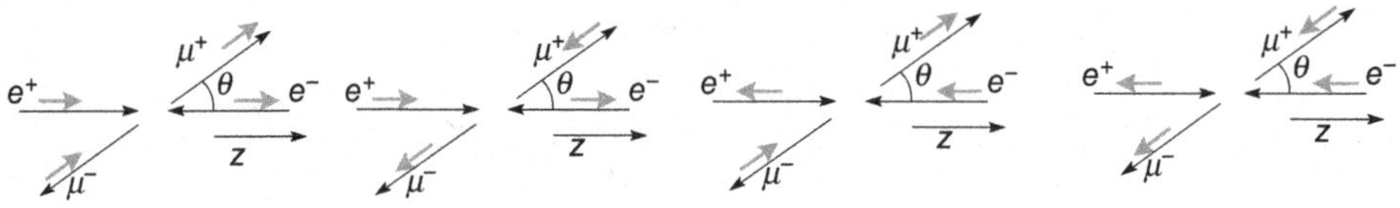

Fig. 5.26. Four polarization states for $e^+ + e^- \rightarrow \mu^+ + \mu^-$.

to its square. This argument fails if the mediator is massive at energies that are not very high compared with its mass. We shall consider this case in Section 7.1.

Let us now discuss the origin of the angular dependence (5.45). Since reaction (5.43) proceeds through a virtual photon, the total angular momentum is defined to be $J = 1$. We take the angular momenta quantization axis z along the positron line of flight. As the helicity is conserved, the third components of the spins of the electron and the positron can both be $+1/2$ or both $-1/2$, but not one $+1/2$ and one $-1/2$.

In the final state we choose as quantization axis z' the line of flight of one of the muons, say the μ^+. The third component of the orbital momentum is zero and therefore the third component of the total angular momentum can be, again, $m' = +1$ or $m' = -1$. The components of the final spins must again be both $+1/2$ or both $-1/2$. In total, we have four cases, as shown in Fig. 5.26.

The matrix element for each $J = 1$, m, m' case is proportional to the rotation matrix from the axis z to the axis z', namely to $d^1_{m,m'}(\theta)$, that is, the four contributions are proportional to

$$d^1_{1,1}(\theta) = d^1_{-1,-1}(\theta) = \frac{1}{2}(1 + \cos\theta); \quad d^1_{1,-1}(\theta) = d^1_{-1,1}(\theta) = \frac{1}{2}(1 - \cos\theta). \quad (5.46)$$

The contributions are distinguishable and we must sum their absolute squares. We obtain the angular dependence $(1 + \cos^2\theta)$ that we see in Eq. (5.45). This result is valid for all the spin $1/2$ particles.

The arguments we have made give the correct dependence on energy and on the angle, but cannot give the proportionality constant. The complete calculation gives, for the total cross-section of $e^+ + e^- \rightarrow \mu^+ + \mu^-$,

$$\sigma = \frac{4}{3}\pi\frac{\alpha^2}{s} = \frac{86.8 \text{ nb}}{s(\text{GeV}^2)}. \quad (5.47)$$

We introduce now a very important quantity, called the 'hadronic cross-section'. It is the sum of the cross-sections of the electron–positron annihilations in all the hadronic final states

$$e^+ + e^- \rightarrow \text{hadrons}. \quad (5.48)$$

Fig. 5.27 shows the hadronic cross-section as a function of $\sqrt{s}$ from a few hundred MeV to about 200 GeV. Notice the logarithmic scales. The dashed line is the 'point-like' cross-section, which does not include resonances. We see a very rich spectrum of resonances, the ω, the ρ (and the ρ', which we have not mentioned), the ϕ, the ψs, the Υs and finally the Z.

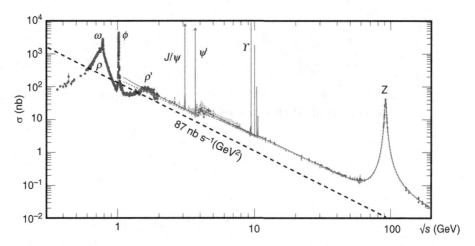

Fig. 5.27. The hadronic cross-section. Adapted by permission of Particle Data Group, Lawrence Berkeley National Lab.

Before leaving this figure, we observe another feature. While the hadronic cross-section generically follows $1/s$ behaviour, it also shows a step every so often. These steps correspond to the thresholds for the production of quark–antiquark pairs of flavours of increasing mass.

5.8 The Evolution of α

We have already mentioned that infinite quantities are met in quantum field theories and that the problem is solved by the theoretical process called 'renormalization'. In QED, two quantities are renormalized: the charge and the mass. We are interested here in the charge, namely the coupling constant. One starts by defining a 'bare' charge that is infinite, and not observable, and an 'effective' charge that we measure. Then one introduces counter terms, which are subtracted cancelling the divergences. The counter terms are infinite.

The situation is illustrated in Fig. 5.28. The coupling constant at each vertex is the bare constant. However all the terms of the series contribute to the measured quantity, reducing the bare charge to the effective charge. Note that the importance of the higher-order terms grows as the energy of the virtual photon increases. Therefore, the effective charge depends on the distance at which we measure it. When we go closer to the charge we must include diagrams of higher order.

We proceed by analogy, considering a small sphere with a negative charge immersed in a dielectric medium. The charge polarizes the molecules of the medium, the dipoles of which tend to become oriented towards the sphere, as shown in Fig. 5.29. This causes the well-known screening action that macroscopically appears as the dielectric constant. Imagine measuring the charge from the deflection of a charged probe

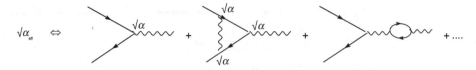

Fig. 5.28. The lowest-order diagrams contributing to the electromagnetic vertex, illustrating the relationship between the 'bare' coupling and the 'effective' (measured) one.

Fig. 5.29. A charge in a dielectric medium.

Fig. 5.30. A charge in a vacuum.

particle. We understand that the charge 'seen' by the probe is smaller and smaller at increasing distances of closest approach.

In quantum physics the vacuum becomes, spontaneously, polarized at the microscopic level. Actually, e^+e^- pairs appear continuously, live for a short time, and recombine. If a charged body is present the pairs become oriented. If its charge is, for example, negative, the positrons tend to be closer to the body, the electrons somewhat farther away, as schematically shown in Fig. 5.30. The virtual particle cloud that forms around the charged body reduces the power of its charge at a distance by its screening action.

If we repeat scattering experiments with the probe particle, we find an effective charge that is larger and larger at smaller and smaller distances.

The fine-structure constant, which we shall call simply α without the suffix 'eff', is not, because of the above discussion, constant; rather, it 'evolves' with the four-momentum transfer or, in other cases, with the centre of mass energy at which we perform the measurement. Let us call Q^2 the relevant Lorentz-invariant variable, namely s or t depending on the situation. The coupling constants of all the fundamental forces are functions of Q^2. These functions are almost completely specified by renormalization theory, which, however, is not able to fix an overall scale constant, which must be determined experimentally.

Suppose for a moment that only one type of charged fermion exists: the electron. Then only e^+e^- pairs fluctuate in the vacuum. The expression of α is

$$\alpha(Q^2) = \frac{\alpha(\mu^2)}{1 - \frac{\alpha(\mu^2)}{3\pi}\ln(|Q|^2/\mu^2)}. \tag{5.49}$$

Once the coupling constant α is known at a certain energy scale μ, this expression gives its value at any other energy. Note also that in (5.49) the dependence is on the absolute value of Q^2, not on its sign.

Equation (5.49) is valid at small values of $|Q|$ when only e^+e^- pairs are effectively excited. At higher values, more and more particle–antiparticle pairs are resolved, $\mu^+\mu^-, \tau^+\tau^-, u\bar{u}, d\bar{d}, \ldots$. Every pair contributes proportionally to the square of its charge. The complete expression is

$$\alpha(Q^2) = \frac{\alpha(\mu^2)}{1 - z_f\frac{\alpha(\mu^2)}{3\pi}\ln(|Q|^2/\mu^2)}, \tag{5.50}$$

where z_f is the sum of the squares of the charges (in units of the electron charge) of the fermions that effectively contribute at the considered value of $|Q|^2$, in practice those with mass $m < |Q|/2$.

For example, in the range 10 GeV $< Q <$ 100 GeV, three charged leptons, two up-type quarks, u and c (charge 2/3), and three down-type quarks, d, s and b (charge 1/3), contribute, and we obtain

$$z_f = 3(\text{leptons}) + 3(\text{colours}) \times \frac{4}{9} \times 2(u,c) + 3 \times \frac{1}{9} \times 3(d,s,b) = 6.67,$$

hence

$$\alpha(Q^2) = \frac{\alpha(\mu^2)}{1 - 6.67\frac{\alpha(\mu^2)}{3\pi}\ln(|Q|^2/\mu^2)} \quad \text{for 10 GeV} <|Q|< 100 \text{ GeV.} \tag{5.51}$$

The dependence on Q^2 of the reciprocal of α is particularly simple, namely

$$\alpha^{-1}(Q^2) = \alpha^{-1}(\mu^2) - \frac{z_f}{3\pi}\ln(|Q|^2/\mu^2). \tag{5.52}$$

We see that α^{-1} is a linear function of $\ln(|Q|^2/\mu^2)$, as long as thresholds for more virtual particles are not crossed. The crossing of thresholds is an important aspect of the evolution of the coupling constants, as we shall see.

The fine-structure constant cannot be measured directly; instead, its value at a certain Q^2 is extracted from a measured quantity, a cross-section or a lifetime. The relationship between the former and the latter is obtained by a theoretical calculation in the framework of QED, or of the entire SM when needed.

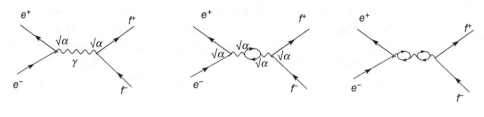

Fig. 5.31. Three diagrams for $e^+ + e^- \rightarrow f^+ + f^-$.

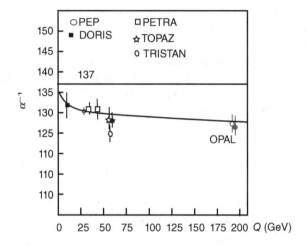

Fig. 5.32. $1/\alpha$ versus energy (Reproduced from G. Abbiendi *et al. Euro. Phys. J.* C33 (2004) 173–212, fig. 17. With kind permission of Springer Science and Business Media. https://doi.org/10.1140/epjc/s2004-01595-9).

The fine-structure constant has been determined (Fan *et al.* 2023) at $Q^2 = 0$ with an accuracy of 0.1 ppb (1 part in 10^{10}), by measuring the electron magnetic moment with an accuracy of 0.13 parts per trillion (ppt) (see Section 5.9) and using the SM for its relationship with α, as we shall discuss in the next section, as

$$\alpha^{-1}(0) = 137.035999166 \pm 0.000000015 \qquad (5.53)$$

The evolution, or 'running', of α has been determined both for $Q^2 > 0$ and for $Q^2 < 0$ at the e^+e^- colliders. Let us see how.

To work at $Q^2 > 0$, we use an s channel process, measuring the cross-section of the electron–positron annihilations into fermion–antifermion pairs (for example $\mu^+\mu^-$):

$$e^+ + e^- \rightarrow f^+ + f^-.$$

Fig. 5.31 shows the first three diagrams of the series contributing to the process.

The measured quantities are the cross-sections as functions of $Q^2 = s$, from which the function $\alpha(s)$ is extracted with a QED calculation. The result is shown in Fig. 5.32 in which $1/\alpha$ is given at different energies. The data show that, indeed, α is not a constant and that its behaviour perfectly agrees with the prediction of quantum field theory.

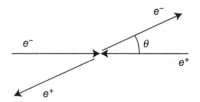

Fig. 5.33.   Bhabha scattering.

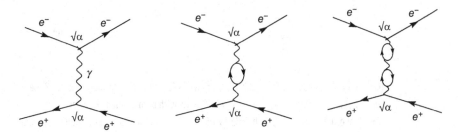

Fig. 5.34.   Three diagrams for the Bhabha scattering.

A high-precision determination of α at the Z mass was made by the LEP experiments, with a combined resolution of 35 parts per million (ppm). The value is

$$\alpha^{-1}\left(M_Z^2\right) = 128.936 \pm 0.046. \tag{5.54}$$

To verify the prediction of the theory for space-like momenta, namely for $Q^2 < 0$, we measure the differential cross-section of the elastic scattering (called Bhabha scattering)

$$e^+ + e^- \rightarrow e^+ + e^-. \tag{5.55}$$

The four-momentum transfer depends on the centre of mass energy and on the scattering angle θ (see Fig. 5.33) according to the relationship

$$|Q|^2 = -t = \frac{s}{2}(1 - \cos\theta). \tag{5.56}$$

Fig. 5.34 shows the lowest-order diagrams contributing to the Bhabha scattering in the t channel. We see that $|Q|^2$ varies from zero in the forward direction $(\theta = 0)$ to s at $\theta = 180°$, and that to have a large $|Q|^2$ range one must work at high energies. Another condition is set by the consideration that we wish to study a t channel process. As a consequence, we should be far from the Z peak where the s channel is dominant. The highest energy reached by LEP, $\sqrt{s} = 198$ GeV satisfies both conditions. The L3 experiment measured the differential cross-section at this energy between almost 0° and 90°, corresponding to 1800 GeV^2 $\langle |Q|^2 < 21600$ GeV^2.

Let $d\sigma^{(0)} / dt$ be the differential cross-section calculated with a constant value of α and let $d\sigma / dt$ be the cross-section calculated with α as in (5.51). The relationship between them is

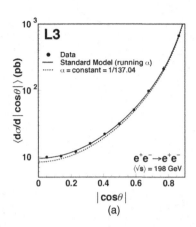
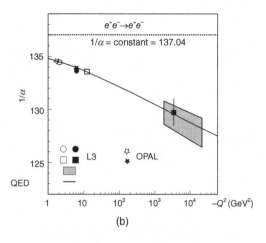

Fig. 5.35.   (a) Differential cross-section of Bhabha scattering at $\sqrt{s} = 198$ GeV as measured by L3. Reprinted from P. Achard *et al. Phys. Lett.* B623 26 (2005) Figs. 3 and 5, © 2005 with permission by Elsevier. https://doi.org/10.1016/j.physletb.2005.07.052. (b) $1/\alpha$ in the space-like region from the OPAL (Abbiendi *et al.* 2006) and L3 (Achard *et al.* 2005) experiments.

$$\frac{d\sigma}{dt} = \frac{d\sigma^{(0)}}{dt} \left[\frac{\alpha(t)}{\alpha(0)} \right]^2. \tag{5.57}$$

To be precise, things are a little more complicated, mainly because of the s channel diagrams. However, these contributions can be calculated and subtracted.

Fig. 5.35(a) shows the measurement of the Bhabha differential cross-section. The dotted curve is $d\sigma^{(0)}/dt$ and is clearly incompatible with the data. The solid curve is $d\sigma/dt$ with $\alpha(t)$ given by Eq. (5.51), in perfect agreement with the data. Fig. 5.35(b) shows a number of measurements of $1/\alpha$ at different values of $-Q^2$. In particular, the trapezoidal band is the result of the measurement just discussed. The solid curve is Eq. (5.51); the dotted line is the constant as measured at $Q^2 = 0$.

Quantum electrodynamics is a beautiful theory that allows us to calculate the observables of the electromagnetic processes. Its predictions have been experimentally tested with enormous accuracy. However, this beautiful construction has a logical pitfall, known as the Landau pole. In 1955, Landau discovered (Landau 1955) that the charge of a point particle increases when we test it at smaller and smaller distances, so much so that it becomes infinite at a certain distance, or equivalently at a certain momentum transfer, which we call Λ_{EM}. This fact can be immediately seen in (5.51); the denominator has a zero. Let us start from the value of α at $\mu = M_Z$ given by (5.54) and, considering that we will deal with very high energies at which the top quark is also active, let us take $z_f = 8$. Solving the equation

$$z_f \frac{\alpha(M_Z^2)}{3\pi} \ln\left(\frac{Q^2}{M_Z^2} \right) = 1 \quad \text{for } Q = \Lambda_{EM}$$

we obtain

$$\Lambda_{EM} \simeq 10^{35}\,\text{GeV}. \tag{5.58}$$

This value is enormous, much larger even than the Planck scale, and the presence of the divergence has no practical consequence, but it shows that the theory is not logically consistent. One might object that (5.50) is the lowest and most important term in a series expansion. However, even including higher-order corrections does not avoid the problem. In Chapter 9 we shall see how the electromagnetic and weak interactions 'unify' at the energy scale of about 100 GeV. The evolution of α changes, but, again, the divergence is moved to even much higher energy, but does not disappear. The bottom line is that the present theory is incomplete and needs to be modified at very high energies.

5.9 Magnetic Moments

We can now come back to the 1947 measurement by Kusch and Foley of the electron magnetic moment (Kusch & Foley 1947), mentioned at the end of Section 5.2. Since then, the magnetic moment of the electron and that of the muon have been measured with increasing precision and are now known with the astonishing accuracy of 1.3 parts in 10^{13} and 4.6 parts in 10^{7}, respectively. In parallel, the theoretical calculations have advanced in precision to match those of the experiments. Here we discuss both the original measurements and then, skipping the history, the most recent ones.

Let us start by recalling that the magnetic moment of the atom has a component due to the nucleus, and two due to its electrons, one corresponding to the orbital motion of the electrons, one to their spin. The ratio of the orbital magnetic moment μ_l to the orbital momentum l is the Bohr magneton μ_B

$$\mu_B = \frac{q_e \hbar}{2m_e}. \tag{5.59}$$

The ratio of the intrinsic magnetic moment μ_s and the spin $s = 1/2$ is twice as large, as foreseen by the Dirac equation. To include both cases, one writes (in absolute values)

$$\mu_s = g\mu_B s, \qquad \mu_l = g\mu_B l, \tag{5.60}$$

where g is called the gyromagnetic ratio. It is equal to 1 for the orbital moment and foreseen by the Dirac equation to be equal to 2 for the spin.

For the proton, the corresponding quantity is the nuclear magneton

$$\mu_N = \frac{q_e \hbar}{2m_p}. \tag{5.61}$$

Notice, in all cases, the inverse proportionality to the mass of the particle.

Question 5.1 Prove that, in non-relativistic classical physics, the ratio of the magnetic moment to the angular momentum of a point charge q of mass m, moving in a circle at constant speed, is also independent of the radius of the orbit and of the velocity and is $q/(2m)$. $\square$

A consequence of the inverse proportionality of the magnetic moment to the mass is that the magnetic moments of the atoms are dominated by the electrons, their total spin S and their total orbital momentum L. Combining the two, we have the total angular momentum J of the atomic electrons.

The absolute determination of the atomic magnetic moment requires a measurement of the Zeeman splitting of the energy levels in an atomic beam in a known magnetic field, a technique developed at Columbia University by Rabi starting in the 1930s.

Soon after, the radio-frequency techniques were strongly developed during World War II for radar. The frequencies of the lines in the Zeeman spectrum, which give the energy differences between atomic levels, could be measured with the techniques available in 1947 to within 1 part in 10 000 or 20 000, but the magnetic field could not be measured with such accuracy.

To overcome the problem, Kusch and Foley noticed that, from measurements of the frequencies of the Zeeman lines in two atomic states arising in either the same or different atoms in exactly the same constant magnetic field, it is possible to deduce the ratio of the atomic gyromagnetic ratios of these states, without any very accurate knowledge of the field. The procedure profits from the fact that the g_J value associated with a state is, as just mentioned, a linear combination of the electronic orbital and spin g values, g_L and g_S. If one chooses states with different combinations, g_S can be evaluated in terms of g_L, provided that the coefficients of the combinations are known with sufficient accuracy. This is indeed the case of atoms with single electrons outside the closed shells (that have zero angular momentum) for which the Russell–Saunders LS coupling is a good approximation and the coefficients that relate the various g values are known.

Kusch and Foley measured, in a weak field of about 0.04 T, the atomic g_J factors for three pairs of states, $^2P_{3/2}$ and $^2P_{1/2}$ of Ga, $^2S_{1/2}$ of Na and $^2P_{3/2}$ of Ga, and $^2S_{1/2}$ of Na and $^2P_{1/2}$ of In, and extracted three values of g_S / g_L for the electron, consistently finding them to be larger than 2. The reported best value for the electron gyromagnetic ratio is

$$g_S / 2 = 1.00119 \pm 0.00005. \tag{5.62}$$

Immediately after this result was made public, Julian Schwinger (Schwinger 1948) showed that it could be explained considering that the electron charge polarizes the vacuum. His calculation led to

$$a_e = \frac{g-2}{2} = \frac{1}{2}\frac{\alpha}{\pi} = 1.00116, \tag{5.63}$$

where a_e is called the (electron) magnetic anomaly. The theory agrees perfectly with the experiment. It was the birth of QED.

At the time of writing (winter 2023/24), the electron magnetic anomaly is the most precisely measured and calculated quantity in physics. In the experiment by the group led by G. Gabrielse (Fan *et al.* 2023), a single electron is suspended in a magnetic field **B**, in the vertical (z) direction, in a 'Penning trap', a device that will be described soon. Fig. 5.36 pictorially shows the two (axial) vectors that characterize the electron, its

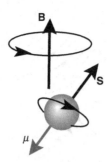

Fig. 5.36. Pictorial view of the electron, represented as a classic top, with its spin **S** and magnetic moment μ, and the rotation of their direction in a magnetic field.

spin **S** and its magnetic moment μ, in opposite directions to one another because the charge of the electron is negative.

A charge in a magnetic field, in a vacuum, can move at uniform speed in a plane perpendicular to the field on a circle, called cyclotron motion, at angular frequency ω_c given by

$$\hbar\omega_c = \hbar\frac{q_e B}{m} = 2\mu_B B \tag{5.64}$$

The spin, and with it the magnetic moment, rotates at a frequency, ω_s, that would be equal to ω_c if it were exactly $g = 2$. The difference between the two $\omega_a = \omega_s - \omega_c$ encodes the magnetic anomaly.

The confinement of a charged particle in space requires a potential minimum in all three directions. This is impossible using only an electrostatic field. The Penning trap uses a superposition of an axial strong uniform magnetic field ($B = 5.3$ T in this experiment) and a weak electrostatic quadrupole field providing axial confinement. Fig. 5.37(a) shows schematically the fields. The trap developed for the experiment is made of five coaxial cylindric electrodes, with appropriately chosen relative dimensions and potentials to produce in the central region the required quadrupole field shown in Fig. 5.37(a). In addition, the trap is also a cylindrical microwave cavity at frequencies of up to 160 GHz.

The quadrupole potential, $V \propto z^2 - \left(x^2 + y^2\right)/2$, has a saddle point in the centre. Its quadratic dependence on the coordinates leads to a linear dependence of the force ($\mathbf{F}_e$ in Fig. 5.37(a)) from the coordinates. In the (vertical) z direction the force is directed to the centre, resulting in harmonic oscillation with angular frequency ω_z. In the radial directions, the electric force is directed away from the centre, namely in the horizontal x–y plane, the electrostatic potential has a maximum on the axis. An electron released on the top of the 'hill' will immediately move to the walls of the trap. This is contrasted by the magnetic force ($q_e \mathbf{v} \times \mathbf{B}$) that is directed towards the axis. The equilibrium orbit is a circle at a certain height around the hilltop, called magnetron motion (from the name of the thermionic valve working on similar principles), at frequency ω_m. The cyclotron motion, which is the interesting one here, is also in the x–y plane

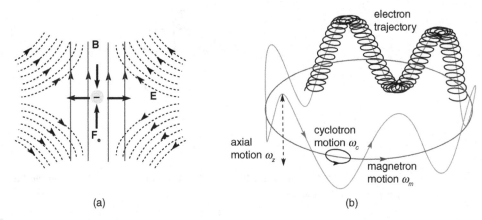

Fig. 5.37. (a) Schematic view of the confining magnetic (continuous lines, not all drawn) and quadrupole electric fields (dashed lines, not all drawn) and of the electric force ($\mathbf{F}_e$). (b) The three periodic component motions of the electro trajectory in the Penning trap.

around a much smaller circle at frequency ω_c. The resulting orbit in the x–y plane is an epicycloid. The three motion components and a segment of the resulting trajectory are shown in Fig. 5.37(b).

The experiment detects the image charges induced by the electron on the upper and lower walls of the trap.

If the Penning trap is cooled at a low enough temperature, the 'artificial atom' made by the trapped electron is a quantum system. The charge distribution, which is a few dozen nanometres across, has discrete energy levels. For the cyclotron frequency, the energy levels of the Fock state $|n\rangle$ are those of the harmonic oscillator, $E_n = (n+1/2)\hbar\omega_c$. For each of them, the electron spin component on the magnetic field can be up, $m_s = +1/2$ or down, $m_s = -1/2$, with an energy difference $m_s\hbar\omega_s$, where ω_s corresponds to the spin precession frequency. The energy levels are

$$E = \hbar\omega_s m_s + \hbar\omega_c \left(n + \frac{1}{2}\right), \tag{5.65}$$

Fig. 5.38 shows the lowest energy levels, the ones used in the experiment. Preliminary research had shown that when the Penning trap is at a temperature of 50 mK, the electron permanently remains in the lowest energy level, the ground state of the system. The state $|n, m_s\rangle = |1, +1/2\rangle$ is also stable over hours at least.

Fig. 5.38 shows the relevant transitions and corresponding frequencies, namely $\hbar\omega_s$ for $|0, -1/2\rangle \rightarrow |0, +1/2\rangle$, $\hbar\omega_c$ for $|0, +1/2\rangle \rightarrow |1, +1/2\rangle$ and $\hbar\omega_a$ for $|1, -1/2\rangle \rightarrow |1, +1/2\rangle$. The frequencies are related to the magnetic moment and its anomaly by

$$-\frac{\mu_e}{\mu_B} = \frac{\omega_s}{\omega_c} = 1 + \frac{\omega_a}{\omega_c}. \tag{5.66}$$

We see that measuring ω_a / ω_c directly gives the anomaly, and so gains three orders of magnitude in precision over measuring ω_s / ω_c, which is larger by about this factor. A

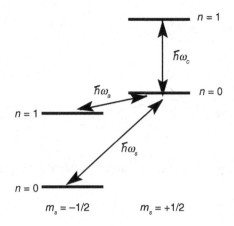

Fig. 5.38. The energy levels of the artificial cyclotron atom and the forced transitions.

precision of 1 part 10^{10} on the measurements of ω_a / ω_c results in the design precision of 1 part in 10^{13}. Notice that frequency is the physical quantity that can be measured with the highest precision, with the exception of the ratio of two frequencies. Indeed, the ratio can be measured with the same clock.

Both the cyclotron $|0,+1/2\rangle \to |1,+1/2\rangle$ and anomaly $|1,-1/2\rangle \to |1,+1/2\rangle$ transitions are forced by injecting an electromagnetic field in the cavity containing the electron; the frequency of the field is varied to scan for ω_a and ω_c. The transitions appear as resonance peaks in the rate of 'quantum jumps between the levels'. The height of each peak is proportional to the transition probability.

The final result is

$$-\frac{\mu_e}{\mu_B} = \frac{g}{2} = 1.00115965218059(13) \ [0.13 \ ppt], \tag{5.67}$$

where the two digits in parentheses are the one standard deviation uncertainty.

To test the validity of the theory, the anomaly $a_e = (g-2)/2$ must be calculated to match the precision of the measurement. This was done by Aoyama *et al.* (2018). The perturbation theory is used, which is a series expansion of powers of $\alpha / \pi \sim 2.3 \times 10^{-3}$. The Feynman diagrams for the magnetic moment of a particle like the electron describe its interaction with an external electromagnetic field. Consequently, they include an incoming and an outgoing electron and a photon. The lowest-order diagram, shown in Fig. 5.39(a), in which all the particles are real, gives the Dirac value, $g=2$. The next order one, shown in Fig. 5.39(b), is the one considered by Schwinger in 1948: the two electrons exchange a virtual photon, and the real photon is coupled to two virtual electrons, namely the photon–electron vertex.

Since $(\alpha / \pi)^5 \sim 0.07 \times 10^{-12}$, in order to match the experimental precision, all the terms up to the 10th order must be evaluated (Fig. 5.40(a) shows an example of the 8th order). In addition, smaller but significant contributions from other sectors of the SM must be included

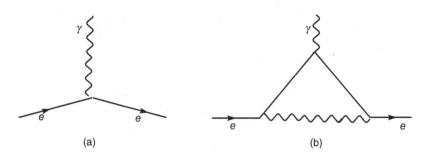

Fig. 5.39. Lowest-order photon–electron vertex diagrams in the electron magnetic moment.

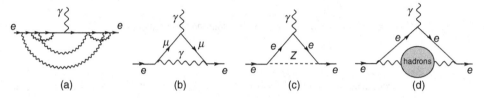

Fig. 5.40. Examples of higher-order Feynman diagrams for the electron magnetic moment. (a) 8th order self-energy. (b) 2nd order with virtual muons. (c) A weak interaction with virtual Z-boson. (d) Hadronic.

$$\frac{g}{2} = 1 + C_2\left(\frac{\alpha}{\pi}\right) + C_4\left(\frac{\alpha}{\pi}\right)^2 + C_6\left(\frac{\alpha}{\pi}\right)^3 + C_8\left(\frac{\alpha}{\pi}\right)^4$$
$$+ C_{10}\left(\frac{\alpha}{\pi}\right)^5 + a_{\mu\tau} + a_{\text{hadronic}} + a_{\text{EW}}. \tag{5.68}$$

The number of diagrams grows enormously with the order: 72 for C_6, 891 for C_8 and 12 672 for C_{10}. We mention only that the four C_i are of the order of unity, ranging from about −2 to +7.

The term $a_{\mu\tau}$ in Eq. (5.68), of which Fig. 5.40(b) shows an example, contains the QED diagrams with virtual muon or tau lines. Their perturbative expansion is considered separately because its evaluation needs experimental input, the ratios of the lepton masses. However, $a_{\mu\tau}$ is small, $a_{\mu\tau} = 2.747\,5719(13)\,10^{-12}$, and the experimental uncertainties on these quantities give negligible contributions to the overall uncertainty.

The term a_{EW}, of which Fig. 5.40(c) shows an example, contains the electroweak contribution. These diagrams contain the exchange of virtual particles that are weakly interacting, such as the W and Z bosons or neutrinos. Consequently, this term is quite small, $a_{\text{EW}} = 0.03053(23)\,10^{-12}$.

Finally, the term a_{hadronic} (Fig. 5.40(d)) contains the exchange of hadrons. Their contribution cannot be calculated with a perturbative expansion of QCD, because at the low energies at which they take place, the coupling constant is large. As we shall see in Chapter 6, calculations on a lattice are done. The result is $a_{\text{hadronic}} = 1.6927(120)\,10^{-12}$.

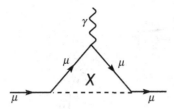

A lowest-order photon–electron vertex diagram including a hypothetical unknown X particle.

Summing up, Aoyama *et al.* (2018) gave the theoretical prediction for the electron magnetic anomaly

$$a_e = 1\,159\,652\,182.032\,(13)(12)(720) \times 10^{-12}, \qquad (5.69)$$

where the first and second uncertainties are due to the QED and hadronic terms respectively. The third and largest uncertainty comes from the uncertainty on the fine-structure constant as determined by independent, and less accurate, experiments. The difference between experiment and theory on $g/2$ is $0.7 10^{-12}$, with the dominant uncertainty arising from the independent experimental value of α. The agreement allows us to establish deviations from the SM. One of them is a limit on the electron radius as less than 3.2×10^{-19} m

If we assume that the theoretical value of a_e is correct, by equating it to the measured value, we obtain the most precise value of the fine-structure constant, Eq. (5.53).

Question 5.2 From the value of α in Eq. (5.53), calculate in SI units the value of the vacuum permittivity ε_0 and its uncertainty. Use a value of π with a sufficient number of digits to make negligible its uncertainty. ☐

We now discuss the muon magnetic anomaly. Today we know that the muon and the electron have exactly the same properties, except for the larger mass of the muon, which makes it unstable. Historically, an important proof of the former statement was found establishing that their magnetic anomalies are equal. On the other hand, the larger muon mass ($m_\mu / m_e \approx 206.8$) and the finite muon lifetime ($\tau_\mu \approx 2.2$ μs) have important practical consequences.

From the theoretical point of view, the contributions to the magnetic anomaly of virtual heavy particles in the relevant Feynman diagrams are more important for the muon than for the electron, in proportion of their mass ratio squared, namely $(m_\mu / m_e)^2 \approx 43000$. This fact, on the one hand, makes the muon anomaly more sensitive to possible physics beyond the SM, as, for example, to the presence of an unknown X particle, as shown in Fig. 5.41. On the other hand, however, the very same fact makes the QCD contributions, which are more difficult to evaluate, more important too.

Muons are artificially produced at an accelerator by sending the accelerated proton beam to a target. The charged pions produced by the collisions will decay via weak interaction as $\pi^+ \to \mu^+ + \nu_\mu$ or $\pi^- \to \mu^- + \bar{\nu}_\mu$, producing muons of one or the other

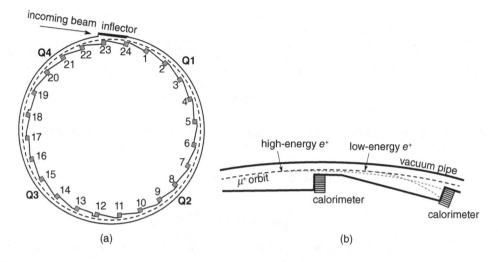

FNAL muon $g-2$ E821 experiment. (a) Schematic view of the storage ring with the 24 calorimeters (squares), the 4 focussing quadrupoles (Qi) and the inflector to put the beam in orbit. (b) Two segmented calorimeters and examples of trajectories.

sign. We anticipate from Chapter 7 that, as a consequence of the parity violation in the weak interaction, muons are born 100% polarized when pions decay at rest. For a beam of pions, the very forward, and very backward, muons have, as a consequence, a strong degree of polarization. And the parity violation of weak interaction helps in measuring the polarization too, because it implies that the direction of highest-energy positrons, or electrons, in the decay $\mu^+ \rightarrow e^+ + \nu_e + \bar{\nu}_\mu$ or $\mu^- \rightarrow e^- + \bar{\nu}_e + \nu_\mu$ is strongly correlated with the muon spin direction.

In 1960, Gawrin et al., using a muon beam produced with the Navis cyclotron, established the existence of a magnetic anomaly of the muon, with the value $g_\mu - 2 = 1.00113^{+0.00016}_{-0.00012}$, which, within its 12% accuracy (see note in proof in the article), is equal to that of the electron. A series of experiments followed, to the present time, constantly increasing the accuracy. Three generations of experiments were performed at CERN from 1961 to 1979. The action than moved to Brookhaven National Laboratory (BNL) from 1989 to 2001, following the basic CERN design, and finally at FNAL with the E821 experiment. The last uses the improved BNL apparatus and has published the first result with an accuracy of 0.46 ppm (Abi et al. 2021).

The experiments employ a muon beam, in general μ^+, produced by a proton accelerator, which is injected into a storage ring, using a magnet called an 'inflector', shown in Fig. 5.42(a) for the E821 experiment. The storage ring is a circular vacuum pipe between the poles of magnets designed to provide an extremely uniform magnetic field **B**. In the field, the muons move similarly to the electrons we have considered above. As in the case of the electron, the angular frequency (of which we now consider the (axial) vector character) of the spin rotation, AB, is different from the cyclotron one ω_c, their difference being ω_a, which is proportional to the magnetic anomaly that we want to measure:

$$\omega_a = \omega_s - \omega_c = -a_\mu \frac{q_e}{m} \mathbf{B}. \tag{5.70}$$

Unlike electrons, muons are unstable. Statistical accuracy requires measuring over as many periods as possible. This is done using relativistic muons, 'dilating' their life, in the laboratory frame, by the Lorentz γ factor. A number of detectors (24 for E821) are arranged around the ring to measure the arrival time and the energy of the decay positrons. The ideal orbit is not stable and, as in the case of the electron, quadrupole electric fields are used to stabilize it vertically (the Qi positions in Fig. 5.42(a)).

The number of decay positrons as a function of time t (with a suitable choice of $t=0$) is given by

$$N(t) = N_0 e^{-\frac{t}{\gamma\tau}} \left[1 + A\cos(\omega_a t) \right], \tag{5.71}$$

where A is the product of the beam degree of polarization and of the weak decay asymmetry and τ the muon lifetime, which are known quantities. A price to pay for having relativistic particles is that the field $\mathbf{E}$ in the laboratory frame Lorentz transforms in the particle frame with a magnetic component, proportional to $\beta \times \mathbf{E}$, where β is the velocity of the muon divided by c. The cyclotron angular frequency is then

$$\omega_c = -\frac{q_e}{m} \left[\frac{\mathbf{B}}{\gamma} - \frac{\gamma}{\gamma^2-1} \frac{1}{c} (\beta \times \mathbf{E}) \right]. \tag{5.72}$$

Considering the motion in a plane perpendicular to the field $\mathbf{B}$, the frequency ω_a to be measured is then

$$\omega_a = -\frac{q_e}{m} \left[a_\mu \mathbf{B} - \left(a_\mu - \frac{\gamma}{\gamma^2-1} \right) \frac{1}{c} (\beta \times \mathbf{E}) \right]. \tag{5.73}$$

The existence of the additional magnetic field appears as a substantial problem for precision, because the quadrupole fields are not homogeneous. Fortunately, however, the minus sign in the parentheses allows a first-order cancellation choosing the 'magic' value of the Lorentz factor $\gamma_m = 29.3$ (corresponding to $a_\mu = 1.166 \times 10^{-3}$). We have

$$\omega_a = -\frac{q_e}{m} a_\mu \mathbf{B}. \tag{5.74}$$

After very careful corrections for all the higher-order effects, the result was

$$a_\mu = 116\,592\,040(54) \times 10^{-11} \quad (0.46\,\text{ppm}), \tag{5.75}$$

which, combined with previous measurements, gives

$$a_\mu = 116\,592\,061(41) \times 10^{-11} \quad (0.35\,\text{ppm}). \tag{5.76}$$

To test the SM, its prediction must be evaluated with an accuracy matching that of the experiment. The principal difficulty is that, relative to the electron case, the hadronic vacuum contribution, of which Fig. 5.39(b) gives an example in the electron case, is about 40 000 times larger and cannot be evaluated perturbatively because, as already mentioned, the QCD coupling constant is large at low energies. As we shall discuss in the next chapter, however, the 'lattice QCD' computing technology has been developed recently to reach the sub-per-cent accuracy needed to match the experimental one. The Budapest–Marseille–Wuppertal collaboration (Borsanyi *et al.* 2021) has found the following value for the dominant term, technically called the leading-order hadronic vacuum polarization (LO-HVP)

$$a_\mu^{\text{LO-HVP}} = 7075(55) \times 10^{-11}, \tag{5.77}$$

which has a 0.77% accuracy. The theoretical muon magnetic anomaly is then

$$a_\mu = 116\,591\,953(58) \times 10^{-11}, \tag{5.78}$$

which agrees with the experimental value (within 1.5 standard deviations).

Problems

5.1 Estimate the speeds of an atomic electron, a proton in a nucleus and a quark in a nucleon.

5.2 Evaluate the order of magnitude of the radius of the hydrogen atom.

5.3 Calculate the energy difference due to the spin–orbit coupling between the levels $P_{3/2}$ and $P_{1/2}$ for $n = 2$ and $n = 3$ for the hydrogen atom $[Rhc = 13.6 \text{ eV}]$.

5.4 Consider the process $e^+ + e^- \to \mu^+ + \mu^-$ at energies much larger than the masses. Evaluate the spatial distance between the two vertices of the diagram Fig. 5.19 (s channel) in the CM reference frame and in the reference frame in which the electron is at rest.

5.5 Draw the tree-level diagrams for the Compton scattering $\gamma + e^- \to \gamma + e^-$.

5.6 Draw the diagrams at the next-to-tree-level order for the Compton scattering (17 in total).

5.7 Give the values that the cross-section of $e^+ e^- \to \mu^+ \mu^-$ would have in the absence of resonance at the ρ, the ψ, the $\varUpsilon$ and the Z. What is the fraction of the angular cross-section $\theta > 90°$?

5.8 Calculate the cross-sections of the processes $e^+ e^- \to \mu^+ \mu^-$ and $e^+ e^- \to$ hadrons at the J/ψ peak $\left(m_\psi = 3.097 \text{ GeV}\right)$ and for the ratio of the former to its value in the absence of resonance. Neglect the masses and use the Breit–Wigner approximation $[\Gamma_e / \Gamma = 5.9\%, \Gamma_h / \Gamma = 87.7\%]$.

5.9 Consider the narrow resonance $\varUpsilon \left(m_\varUpsilon = 9.460 \text{ GeV}\right)$ that was observed at the $e^+ e^-$ colliders in the channels $e^+ e^- \to \mu^+ \mu^-$ and in $e^+ e^- \to$ hadrons. Its width is $\Gamma_\varUpsilon = 54 \text{ keV}$. The measured 'peak areas' are $\int \sigma_{\mu\mu}(E)\,dE = 8 \text{ nb MeV}$ and $\int \sigma_h(E)\,dE = 310 \text{ nb MeV}$. In the Breit–Wigner approximation, calculate the partial widths Γ_μ and Γ_h. Assume all the leptonic widths to be equal.

5.10 Two photons flying in opposite directions collide. Let E_1 and E_2 be their energies. (1) Find the minimum value of E_1 to allow the process $\gamma_1 + \gamma_2 \to e^+ + e^-$ to occur if

$E_2 = 10$ eV. (2) Answer the same question if $E_1 = 2E_2$. (3) Find the centre of mass speed in the latter case. (4) Draw the lowest-order Feynman diagram of the process.

5.11 Calculate the reciprocal of the fine-structure constant at $Q^2 = 1$ TeV^2, knowing that $\alpha^{-1}\left(M_Z^2\right) = 129$ and that $M_Z = 91$ GeV. Assume that no particles beyond the known ones exist.

5.12 If no threshold is crossed, $\alpha^{-1}\left(Q^2\right)$ is a linear function of $\ln\left(|Q|^2/\mu^2\right)$. What is the ratio between the quark and lepton contributions to the slope of this linear dependence for $4 < Q^2 < 10$ GeV^2?

5.13 Calculate the energy threshold (E_γ) for the conversion of a photon into a e^+e^- pair in the electric field of (1) an oxygen nucleus, (2) an electron and (3) for the postproduction of a $\mu^+\mu^-$ pair in the field of a proton. In any case, the photon, which has mass equal to zero, converts into a pair of mass $m_{ee} > 0$. In which configuration is m_{ee} a minimum? Consider in particular electron energies $E_+, E_- \gg m_e$.

5.14 Weakly decaying negative particles may live long enough to come to rest in matter and be captured by a nucleus. Consider the simplest case of the capture by a proton. (a) Evaluate the Bohr radius for the μ^-p system (muonium), the π^-p system (pionium), the K^-p system (kaonium) and the $\bar{p}p$ system (antiprotonium). (b) Calculate the energy released in the capture of an electron at rest into the ground state by a proton.

5.15 Is the decay $\omega \to \pi^+\pi^-$ allowed by strong interactions? Is it allowed by electromagnetic interactions?

5.16 Draw the (lowest-order) Feynman diagrams for (a) $\pi^0 \to \gamma\gamma$, (b) $\Sigma^0 \to \Lambda\gamma$, (c) $e^- + e^- \to e^- + e^-$, (d) $e^+ + e^+ \to e^+ + e^+$, (e) $e^+ + e^- \to e^+ + e^-$, (f) $e^+ + e^- \to \mu^+ + \mu^-$, (g) $e^+ + \mu^+ \to e^+ + \mu^+$.

5.17 Determine the charge conjugation, the lowest value of the orbital momentum and the isospin of the 2π systems in the decays (a) $\eta \to \pi^+\pi^-\gamma$, (b) $\omega \to \pi^+\pi^-\gamma$ and (c) $\rho^0 \to \pi^+\pi^-\gamma$. State also the ΔI in the decay. (d) State whether the following decays are allowed or forbidden and why: $\eta \to \pi^0\pi^0\gamma$, $\omega \to \pi^0\pi^0\gamma$ and $\rho^0 \to \pi^0\pi^0\gamma$.

5.18 Consider the decays of the J/ψ into $\rho\pi$ with their measured branching ratios $\dfrac{J/\psi \to \rho\pi}{all} = (1.69 \pm 0.15) \times 10^{-2}$ and $\dfrac{J/\psi \to \rho^0\pi^0}{all} = (0.56 \pm 0.07) \times 10^{-2}$. Determine the isospin of the J/ψ.

5.19 Consider two counter-rotating beams of electrons and positrons stored in the LEP collider both with energy $E_e = 100$ GeV. The beam intensity slowly decays due to beam–beam interactions and various types of losses. One of the losses (very small indeed) may be the interaction of the electrons with the photons of the cosmic microwave background. Assume the photon energy $E_\gamma = 0.25$ meV and a photon density $\rho = 3 \times 10^8$ m^{-3}. (a) Evaluate the energy for a photon $E_{\gamma FT}$ against an electron at rest to have the same CM energy. (b) Assuming the Thomson cross-section $$\sigma = \left(\frac{\alpha\hbar}{m_ec}\right)^2 = 7.9 \times 10^{-30} \text{ m}^2,$$ evaluate the interaction length and the corresponding time. Considering that $N_e = 1.6 \times 10^{12}$ electrons are stored in each ring, determine the rate of the considered events in the whole ring. Take the average value $\cos\theta = 0$ for the angle between the initial photon and the beam. (c) Determine the energy or the photons that scatter backward at 180° after a head-on collision.

Summary

In this chapter:

- we have studied the QED Lagrangian, the prototype of that of all the interactions
- we have seen how the Lamb and Retherford experiment gave origin to the quantum field theoretical description of nature
- we have seen, in the case of QED, how a local gauge invariance, under a symmetry group, generates the interaction itself
- we have learnt that the Lagrangians of the fundamental interactions are scalar products of currents, of the fermion fields, which may be vector or axial vector
- we have learnt the basic structure of the Feynman diagrams
- we have studied the process $e^+e^- \rightarrow \mu^+\mu^-$
- we have studied the running (i.e. the energy dependence) of the electromagnetic coupling constant
- we have seen how the extremely precise measurements of the electron and muon magnetic moments and the predictions at the same level of accuracy theoretically calculated in the SM provide stringent tests of the theory.

Further Reading

Feynman, R. P. (1985) *QED*. Princeton University Press
Feynman, R. P. (1987) *The Reason for Antiparticles. Elementary Particles and the Laws of Physics*. Cambridge University Press
Jackson J. D. & Okun, L. B. (2001) Historical roots of gauge invariance. *Rev. Mod. Phys.* **73** 663
Kusch, P. (1955) Nobel Lecture; *The Magnetic Moment of the Electron*
Lamb, W. E. (1955) Nobel Lecture; *Fine Structure of the Hydrogen Atom*

In this chapter, we study a second fundamental interaction, the strong one. It binds together the quarks by exchange of gluons. The strong charges, corresponding to the electric charge, are called colours and their theoretical description is called quantum chromodynamics (QCD). There are similarities with QED, but also fundamental differences. One of them, called confinement, is that quarks are never free. Nobody has succeeded in breaking a proton and extracting its quarks, no matter which energy is used. We shall see why.

We begin by showing how the colour charges were experimentally discovered by studying the production of hadrons, mainly pions, in experiments with electron–positron colliders. We shall also see that the underlying process $e^+ + e^- \rightarrow q + \bar{q}$ becomes evident when the CM energies are large enough. The quarks appear as *jets* of hadrons. When one of the quarks radiates a hard gluon, this appears as a third jet. In this way the gluon was shown to exist and its spin was determined by studying its angular distribution.

We saw in Chapter 4 how hadron spectroscopy points to the internal quark composition of hadrons. However, quarks could explain the observed spectroscopy even if they were mathematical rather than physical objects. Gell-Mann (but not the other proponent of the model, Zweig) considered them to be mathematical objects until experiments at SLAC that looked into the proton with high-energy electron beams and high-resolution showed, like the Rutherford experiment on atoms, the presence of an internal structure.

We shall see how these *deep inelastic* experiments, with both electron and neutrino beams, have measured the distributions in the relevant kinematic variables of the nucleon components, called partons, namely the quarks and the gluons. The structure of the nucleon we 'see' depends on the spatial resolution of the probe we employ, namely on the momentum transfer. Nucleons, and in general hadrons, do not contain only the quarks corresponding to the spectroscopy, which are called *valence* quarks, but also a *sea* of quark–antiquark pairs, of all the flavours.

We are now ready to discuss the properties of the colour charges. The symmetry group of the QCD Lagrangian is $SU(3)$ (containing all the three-dimensional unitary matrices with unit determinant), which is more complex than the $U(1)$ of QED, and is non-abelian. The colour charges are three, called red, green and blue. A consequence of the gauge being non-abelian is that the gluons are 'coloured', rather than just the quarks. When a charged particle emits, or absorbs, a photon, its charge does not change. On the contrary, the colour charge of a quark can change when emitting or absorbing a gluon. Consequently, gluons interact with each other. We shall see how colour charges work and how they bind three quarks or a quark–antiquark pair together, forming hadrons that are 'white', namely with zero overall colour charges, and how the multiplets of the hadrons that can be formed in this way are just those that are experimentally observed.

We shall then see how the QCD coupling constant evolves. With increasing momentum transfer, it decreases, rather than increasing as the fine-structure constant does. Quarks become 'free', in the sense that they have vanishing interactions when they are very close to each other, rather than very far apart (asymptotic freedom).

To measure the mass of a particle, it must be free. We can define the mass of a quark only by properly extending the concept of mass. This can be done in the frame of QCD, but the extraction of the mass value of each of the six quarks from the measured observables requires calculations that depend on the adopted theoretical 'subtraction scheme'. In addition, as we shall see, the quark masses depend on the energy scale at which they are determined: as do the coupling constants, the quark masses run. We shall discuss that for the example of the b quark.

In Section 6.7 we shall see the origin of the proton mass and how only a very small fraction of it is due to the quark masses, 99% being the energy of the colour field. This leads us to discuss the QCD vacuum, the status of minimum energy. It is a very active medium indeed.

The observed masses of the pseudoscalar mesons, the pion in particular, are much smaller than those of their scalar partners, a consequence of the spontaneous breaking of the chiral symmetry. This important feature of QCD will be discussed in Section 6.9, together with the relations between quark and meson masses that can be obtained with chiral expansion.

In QED, observables can be calculated perturbatively at any energy because the coupling constant is small. In QCD this is possible only at high enough energies where the coupling constant has evolved to small enough values. At low energies, a powerful alternative exists, by discretizing the space-time in a lattice. In the last section we shall discuss the principles of lattice QCD (LQCD) and how its provides predictions of the hadron spectrum.

When matter first appeared in the Universe, in the first microsecond after the Big Bang, quarks and gluons moved freely in a hot 'soup' called the quark–gluon plasma (QGP). The QGP phase of matter can now be created in the ultra-relativistic heavy ion colliders, such as the LHC at CERN and the Relativistic Heavy Ion Collider (RHIC) at BNL, studied in the laboratory and theoretically analysed with LQCD. A short discussion can be found in Section 6.11.

6.1 Hadron Production at Electron–Positron Colliders

We have already anticipated the importance of the experimental study of the process

$$e^+ + e^- \to \text{hadrons} \tag{6.1}$$

in the electron–positron colliders. We shall now see why this study is important.

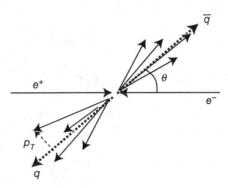

Hadronization of two quarks into jets.

We interpret the process as a sequence of two stages. In the first stage a quark–antiquark pair is produced

$$e^+ + e^- \rightarrow q + \bar{q}. \tag{6.2}$$

Here q and $\bar{q}$ can be any quark above threshold, namely with mass m such that $2m < \sqrt{s}$. The second stage is called hadronization, the process in which the quark and the antiquark produce hadronic jets, as shown in Fig. 6.1.

The energies of the quarks are of the order of $\sqrt{s}$. Their momenta are of the same order of magnitude, at high enough energy that we can neglect their masses, and are directed in equal and opposite directions, because we are in the CM frame. The quark immediately radiates a gluon, similar to an electron radiating a photon, but with a higher probability due to the larger coupling constant. The gluons, in turn, produce quark–antiquark pairs, and quarks and antiquarks radiate more gluons, etc. During this process, quarks and antiquarks join to form hadrons. The radiation is most likely soft, the hadrons having typical momenta of 0.5–1 GeV. In the collider frame, the typical hadron momentum component in the direction of the original quark is a few times smaller than the quark momentum. Its transverse component p_T is between about 0.5 and 1 GeV.

Therefore, the opening angle of the group of hadrons is of the order

$$\frac{p_T}{p} \approx \frac{0.5}{\sqrt{s}/2} = \frac{1}{\sqrt{s}}, \tag{6.3}$$

with $\sqrt{s}$ in GeV. If, for example, $\sqrt{s} = 30$ GeV the group opening angle is of several degrees and it appears as a rather narrow 'jet'. On the contrary, if the energy is low, the opening angle is so wide that the jets overlap and are not distinguishable.

Fig. 6.2 shows the transverse (to the beams) projection of a typical hadronic event at the JADE detector at the PETRA collider at the DESY laboratory at Hamburg, with CM energy $\sqrt{s} = 30$ GeV. The final-state quark pairs appear clearly as two back-to-back jets.

Nobody has ever seen a quark by trying to extract it from a proton. To 'see' the quarks we must change our point of view, as we have just done, and focus our attention on the energy and momentum flow, rather than on the single hadrons. The quark then appears as a flow in a narrow solid angle with the shape of a jet.

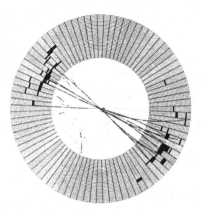

Fig. 6.2.   Two jet events at the JADE detector at the PETRA collider at DESY. Reprinted from B. Naroska, *Physics Reports* 148 (1987) 67–215, Fig. 5.1(a), © 1987 with permission by Elsevier. https://doi.org/10.1016/0370-1573(87)90031-7.

The total hadronic cross-section (6.1) can be measured both at high energies when the quarks appear as well-separated jets and at lower energies where the hadrons are distributed over all the solid angle and the jets cannot be identified. It is useful to express this cross-section in units of the point-like cross-section, that is, the one into $\mu^+\mu^-$ that we studied in Section 5.7, namely

$$R = \frac{\sigma\left(e^+ + e^- \to \text{hadrons}\right)}{\sigma\left(e^+ + e^- \to \mu^+ + \mu^-\right)}. \tag{6.4}$$

If the quarks are point-like, without any structure, this ratio is simply given by the ratio of the sum of the squared electric charges

$$R = \sum_i q_i^2 / 1, \tag{6.5}$$

where the sum is over the quark flavours with production above threshold.

In 1969 the experiments at ADONE first observed that the hadronic production was substantially larger than expected. However, at the time quarks had not yet been accepted as physical entities and a correct theoretical interpretation was not given. In retrospect, since the u, d and s quarks are produced at the ADONE energies ($1.6 < \sqrt{s} < 3$ GeV), we expect $R = 2/3$, whilst the experiments indicated values of between 1 and 3. This was the first, not understood, evidence for colour.

Actually, the quarks of every flavour come in three types, each with a different colour. Consequently, R is three times larger

$$R = 3 \sum_{\text{flavour}} q_i^2. \tag{6.6}$$

Fig. 6.3 shows the R measurements in the range 10 GeV $< \sqrt{s} < 40$ GeV. In the energy region 2 GeV $< \sqrt{s} < 3$ GeV, quark–antiquark pairs of the three flavours, u, d and s, can be produced; between 5 GeV and 10 GeV, $c\bar{c}$ pairs are also produced; and, finally, above 20 GeV, $b\bar{b}$ pairs are also produced. In each case R is about three times

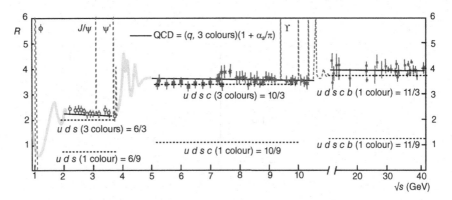

Fig. 6.3. Ratio R of hadronic to point-like cross-section in e^+e^- annihilation as a function of $\sqrt{s}$. Adapted by permission of Particle Data Group, Lawrence Berkeley National Lab.

larger than foreseen in the absence of colour. To be precise, QCD also interprets well the small residual difference above the prediction of Eq. (6.6). This is because of the gluons, which themselves have colour charges. QCD predicts that (6.6) must be multiplied by the factor $(1+\alpha_s/\pi)$, where α_s is the QCD coupling constant, corresponding to the QED α, as we shall see shortly.

Question 6.1 Evaluate α_s at $\sqrt{s}=40$ GeV from Fig. 6.3. Compare your result with Fig. 6.23. □

In Section 5.7 we studied the differential cross-section for the electron–positron annihilation into two point-like particles of spin 1/2. If the spin of the quarks is 1/2, the cross-section of the process

$$e^+ + e^- \rightarrow q + \bar{q} \rightarrow \text{jet} + \text{jet} \qquad (6.7)$$

should be

$$\frac{d\sigma}{d\Omega} = \frac{z^2\alpha^2}{4s}\left(1+\cos^2\theta\right), \qquad (6.8)$$

where z is the quark charge in elementary charge units.

The scattering angle θ is the angle between, say, the electron and the quark. As we cannot measure the direction of the quarks, we take the common direction of the total momenta of the two jets. We know only the absolute value $|\cos\theta|$ because we cannot tell the quark from the antiquark jet. Fig. 6.4 shows the angular cross-section of (6.7) at measured at $36.8<\sqrt{s}<46.8$ GeV by the CELLO experiment at DESY (Beherend *et al.* 1987). It shows that the quark spin is 1/2.

We have seen that the soft gluon radiation by the quarks gives rise to the hadronization process. More rarely the quark radiates 'hard' gluons, meaning with a large relative momentum. The hard gluons hadronize, much like the quarks do, becoming visible as a hadronic jet. At typical collider energies $\sqrt{s}=30-100$ GeV, a third jet appears in the detector about 10% of the time. In this way, the gluon was discovered at

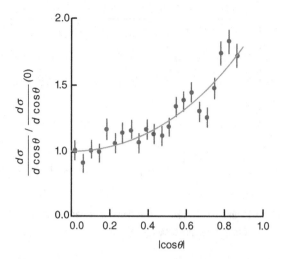

Fig. 6.4.   Two-jet differential cross-section from data of the CELLO experiment at DESY. The curve is $1 + \cos^2\theta$.

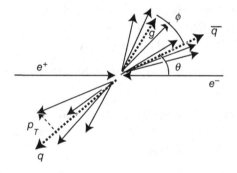

Fig. 6.5.   Sketch of the gluon radiation.

the PETRA electron positron collider at DESY, the first collider capable of reaching an energy high enough to do so, namely $\sqrt{s} = 30$ GeV. Fig. 6.5 shows the schematics, and Fig. 6.6 shows a three-jet event observed by the JADE detector at PETRA. In conclusion, at large enough centre of mass energy, the gluons are clearly detectable as hadronic jets.

The physical characteristics of the gluon jets are very similar to those of the quark jets. Therefore, it is possible to establish which is the gluon jet only on a statistical basis. The following criterion is adequate: in every three-jet event, we classify the jets in decreasing CM energy order, E_1, E_2, E_3, and we define jet 3 as the gluon. We then transform to the jet1 − jet2 CM frame and calculate the angle ϕ between the common direction of the pair and jet 3.

The distribution of ϕ depends on the gluon spin. Fig. 6.7 shows the measurement of the $\cos\phi$ distribution made by the TASSO experiment at PETRA (Brandelik *et al.* 1980). The two curves are calculated assuming the spin parity of the gluon to be 0^+ and 1^-. The data clearly show that the gluon is a vector particle.

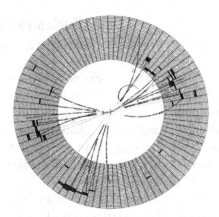

Fig. 6.6.   Three-jet event at the JADE detector at the PETRA collider at DESY. Reprinted from B. Naroska, *Physics Reports* 148, 67–215 (1987) Fig. 5.1(b), © 1987 with permission by Elsevier. https://doi.org/10.1016/0370-1573(87)90031-7.

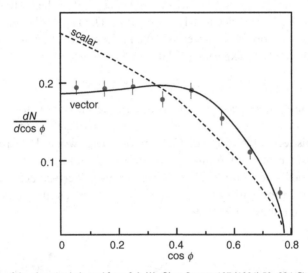

Fig. 6.7.   Angular distribution of the gluon jet (adapted from S. L. Wu *Phys. Reports* 107 (1984) 59–324, Fig. 3.23. © 2005 with permission by Elsevier (https://doi.org/10.1016/0370-1573(84)90033-4).

In conclusion, we have seen that the experiments at the e^+e^- colliders at centre of mass energies of several tens of GeV have given the following fundamental pieces of information.

(1) The hadronic cross-section cannot be understood without the colour charges.
(2) The quarks are observed as hadronic jets with the angular distribution of the elementary spin 1/2 particles.
(3) The gluons are seen as a third jet. Their angular distribution relative to the quark jets is that foreseen for a vector particle. Sometimes more gluons are radiated and are detected as further jets.
(4) The R value shows that not only the quarks, but also the gluons, are coloured.

6.2 Nucleon Structure

In the 1960s, a two mile long linear electron accelerator (LINAC) was built at Stanford in California. Its maximum energy was 20 GeV. The laboratory, after that, was called the Stanford Linear Accelerator Center (SLAC).

J. Friedman, H. Kendall and collaborators at the Massachusetts Institute of Technology (MIT), and R. Taylor and collaborators at SLAC (Taylor 1991), designed and built two electron spectrometers, with maximum energies of 8 GeV and 20 GeV, respectively. These instruments were set up to study the internal structure of the proton and of the neutron. The layout and a picture are shown in Fig. 6.8.

The electron beam extracted from the LINAC is brought onto a target, which is liquid hydrogen when the proton is being studied or liquid deuterium when it is the neutron. The beam is collimated and approximately monochromatic with known energy E. The spectrometers measure the energy E' of the scattered electron and the scattering angle θ, as accurately as possible. The rest of the event, namely what happens to the nucleon, is not observed. We indicate it by X. This type of measurement is called an 'inclusive' experiment. The reaction is

$$e^- + p \rightarrow e^- + X. \tag{6.9}$$

The 8 GeV spectrometer decouples the measurement of the angle from that of the momentum by using bending magnets that deflect in the vertical plane. Scaling up this technique to the 20 GeV spectrometer would have required a very large vertical displacement. A brilliant solution was found by K. Brown and B. Richter, who proposed a novel optics arrangement allowing vertical bending while keeping the vertical dimension within bounds. The first experimental results of a 17 GeV energy beam were published in 1969.

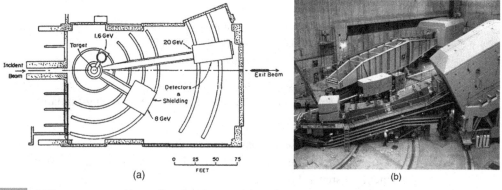

(a)   (b)

Fig. 6.8. (a) The spectrometers ride on rails and can be rotated about the target to change the angle of the detected electrons (R. E. Taylor, Nobel Lecture 1990, Fig. 14: R. E. Taylor, © The Nobel Foundation 1990 Figure 14 in this link www.nobelprize.org/uploads/2018/06/taylor-lecture.pdf). (b) The detectors are inside the heavy shielding structures visible at the ends of the spectrometers (courtesy of SLAC).

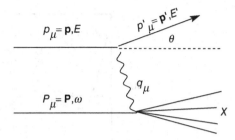

Fig. 6.9. Sketch of the deep inelastic scattering.

Fig. 6.9 shows the kinematics and defines the relevant variables. To detect small structures inside the nucleon we must hit them violently, breaking the nucleon. The process is called 'deep inelastic scattering' (DIS).

We must now define the kinematic variables that we shall use. The first is the four-momentum transfer to the nucleon, t, which is negative. To follow the convention, we also define its opposite Q^2. With reference to Fig. 6.9 we have

$$
\begin{aligned}
-Q^2 &\equiv t = q^\mu q_\mu = (E' - E)^2 - (\mathbf{p}' - \mathbf{p})^2 \\
&= 2m_e^2 - 2(EE' - pp'\cos\theta) \approx -2EE'(1 - \cos\theta),
\end{aligned}
\tag{6.10}
$$

where we have neglected the electron mass because the energy is high enough. Hence we have

$$
Q^2 = 2EE'\sin^2\frac{\theta}{2}.
\tag{6.11}
$$

We see that to know Q^2 we must measure the scattered electron energy E' and its direction θ. Another invariant quantity that can be measured is the square of the mass of the hadronic system W^2

$$
\begin{aligned}
W^2 &= (P_\mu + q_\mu)(P^\mu + q^\mu) \\
&= m_p^2 + 2P_\mu q^\mu - Q^2 = m_p^2 + 2m_p\nu - Q^2.
\end{aligned}
\tag{6.12}
$$

In the last expression of (6.12) we have introduced a further Lorentz-invariant quantity

$$
\nu \equiv P_\mu q^\mu / m_p.
\tag{6.13}
$$

To see its physical meaning, we look at its expression in the laboratory frame, where $P_\mu = (m_p, \mathbf{0})$ and $q_\mu = (E - E', \mathbf{q})$. Therefore

$$
\nu = E - E'.
\tag{6.14}
$$

We see that ν is the energy transferred to the target in the laboratory frame. We determine it by measuring E' and knowing the incident energy. We shall then use the two variables ν and Q^2 that are measured as just specified.

In Section 1.8 we gave the Mott cross-section, which is valid in conditions similar to those we are considering now for point-like targets. One can show that the

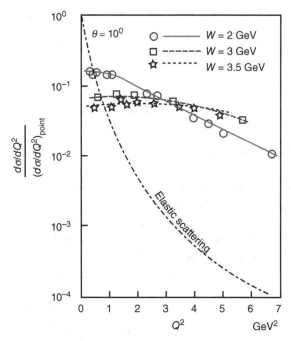

Fig. 6.10. Deep inelastic scattering cross-section measured at SLAC (J. Friedman, Nobel Lecture 1990, Fig. 1: J. Friedman, © The Nobel Foundation 1990 Figure 1 in this link www.nobelprize.org/uploads/2018/06/friedman-lecture.pdf).

scattering cross-section from a target with an internal structure can be expressed in terms of two 'structure functions' $W_1\left(Q^2,v\right)$ and $W_2\left(Q^2,v\right)$. The former describes the interaction between the electron and nucleon magnetic moments, and as such is sensitive to the current density distribution in the nucleon; the latter describes the interaction between the charges and is sensitive to the charge distribution. In the kinematic conditions of the experiments that we shall consider, the contribution of W_1 is negligible, in a first approximation, and we can write

$$\frac{d\sigma}{d\Omega dE'} = \left(\frac{d\sigma}{d\Omega}\right)_{\text{point}} W_2\left(Q^2,v\right). \tag{6.15}$$

To determine the function $W_2\left(Q^2,v\right)$ experimentally, one measures the deep inelastic differential cross-section at several values of Q^2 and v, or, in practice, for different beam energies and scattering angles.

The main results of the measurements made at SLAC with the above-described spectrometers are shown in Fig. 6.10. Three sets of data points are shown, each for a different fixed value of the hadronic mass W. The points are the measured cross-section values divided by the calculated point-like cross-section, namely $W_2\left(Q^2,v\right)$, as functions of Q^2. Surprisingly, we see that in the deep inelastic region, namely for large enough values of W, the function W_2 does not vary, or varies just a little, with Q^2 and, moreover, is independent of W. It is difficult to avoid the conclusion that the nucleon

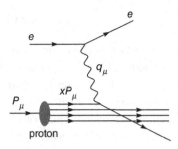

Fig. 6.11. Proton structure and kinematics of DIS in the infinite-momentum frame.

contains point-like objects, as for the nucleus in the Geiger–Marsden experiment. Notice, for comparison, the steep decrease of the elastic cross-section.

The deep physical implications of the experimental data were identified by R. Feynman (Feynman 1969). Initially he dubbed the hard objects inside the nucleons 'partons'; these were later identified as the quarks. We follow his argument and consider the scattering process in a frame in which the proton moves with a very large four-momentum, P_μ. In this frame we can neglect the transverse momenta of the partons and consider them all to be moving in the same direction with very large, but not necessarily equal, momenta, as schematically shown in Fig. 6.11.

Let us indicate by x the fraction of four-momentum of a given parton. Therefore its four-momentum is xP_μ.

Feynman put forward the hypothesis that the electron–parton collision can be considered as taking place on a free parton. We shall justify this 'impulse approximation' in Section 6.5.

Let q_μ be the four-momentum transferred from the electron to the parton. Let us assume the mass m of the parton to be negligible and write

$$m^2 = \left(xP_\mu + q_\mu\right)\left(xP^\mu + q^\mu\right) \approx 0,$$

that is, $x^2 m_p^2 - Q^2 + 2xP_\mu q^\mu = 0$, and, if $Q^2 \gg x^2 m_p^2$,

$$x = \frac{Q^2}{2P_\mu q^\mu} = \frac{Q^2}{2\nu m_p}. \tag{6.16}$$

If this model is correct, the dependence of the structure function on Q^2 for a fixed x is the Fourier transform of the charge distribution in the parton that is found at x. If the parton is point-like, then the transform is a constant, independent of Q^2. Moreover, the structure function should depend only on x. In other words, the function should not vary where ν or Q^2 vary, provided their ratio is kept constant. This property is known as the Bjorken 'scaling law' after its discoverer (Bjorken 1969).

Let us now move on to another pair of kinematic variables, x and Q^2, and let us also define the dimensionless structure function F_2 (while W_2 is dimensionally an inverse energy)

$$F_2\left(x,Q^2\right) \equiv vW_2\left(Q^2,v\right).\tag{6.17}$$

The scaling law predicts that the values of F_2 measured for different values of Q^2 must be equal if x is the same. This is just what is shown by the data, as we shall see immediately, confirming that the scattering centres inside the nucleon are point-like and hard. Therefore, the quarks, which were introduced to explain hadron spectroscopy, are physical, not purely mathematical, objects.

Moreover, the nucleons, and in general the hadrons, contain much more than what is shown by spectroscopy. In summary, the high-resolution probes, principally the electron and neutrino beams, have shown that the nucleons contain the following components.

- The three quarks that determine the spectroscopy, called 'valence' quarks.
- The gluons that are the quanta of the colour field.
- The quark and antiquark of the 'sea'. Actually the following processes continually happen in the intense colour field: a gluon materializes in a quark–antiquark pair, which soon annihilates, two gluons fuse into one, etc. The sea contains quark–antiquark pairs of all flavours, with decreasing probability for increasing quark masses. Therefore, there are many $u\bar{u}$ and $d\bar{d}$ pairs, fewer $s\bar{s}$ pairs and just a few $c\bar{c}$ pairs.

We define $f(x)$ as the distribution of momentum fraction for the quark of f flavour; consequently, $f(x)dx$ is the probability that this quark carries a momentum fraction of between x and $x+dx$, and $xf(x)dx$ is the corresponding amount of momentum fraction.

We also call $\bar{f}(x)$ the analogous function for the antiquark of f flavour and $g(x)$ that of the gluons. Having no electric and no weak charges, the gluons are not seen either by electrons or by neutrinos. These functions are called parton distribution functions (PDFs).

Since the charm contribution is small, we shall neglect it for simplicity. We have 12 functions of x to determine experimentally: the distribution functions of the up, down and strange quarks and of their antiquarks in the proton and in the neutron. However, not all of these functions are independent.

The isospin invariance gives the following relationships between proton and neutron distribution functions

$$\begin{aligned} u_p\left(x\right) &= d_n\left(x\right); & d_p\left(x\right) &= u_n\left(x\right); \\ \bar{d}_p\left(x\right) &= \bar{u}_n\left(x\right); & \bar{u}_p\left(x\right) &= \bar{d}_n\left(x\right) \end{aligned}\tag{6.18}$$

and

$$s_p\left(x\right) = s_n\left(x\right); \quad \bar{s}_p\left(x\right) = \bar{s}_n\left(x\right).\tag{6.19}$$

Finally, the sea quarks have the same distributions as the antiquarks of the same flavour and, letting $s(x) \equiv s_p(x) = s_n(x)$, we have

$$s(x) = \bar{s}(x).\tag{6.20}$$

We are left with five independent functions. We call $u(x)$ the distribution of the u quark in the proton and of the d quark in the neutron, $u(x) \equiv u_p(x) = d_n(x)$, and similarly $d(x) \equiv d_p(x) = u_n(x)$, etc.

Notice that the u and d distribution functions contain the contributions both of the valence $(u_v$ and $d_v)$ and of the sea quarks $(u_s$ and $d_s)$. The sea contribution is equal to the distribution function of its antiquark

$$\bar{u}(x) = u_s(x), \quad \bar{d}(x) = d_s(x). \tag{6.21}$$

Experimentally, the distribution functions are obtained from the measurement of the deep inelastic differential cross-sections of electrons, neutrinos and antineutrinos.

The electrons 'see' the quark charge, which, in units of the elementary charge, we call z_f for the quarks and $-z_f$ for the antiquarks. The measured structure function is the sum of the contributions of all the q and $\bar{q}$, weighted with the square of the charge z_f^2. Therefore the electrons do not distinguish quarks from antiquarks. We have

$$F_2(x) = x\sum_f z_f^2 \left[f(x) + \bar{f}(x) \right]. \tag{6.22}$$

With proton and neutron targets, we have for the electron–proton scattering

$$\frac{F_2^{ep}(x)}{x} = \frac{4}{9}\left[u(x) + \bar{u}(x)\right] + \frac{1}{9}\left[d(x) + \bar{d}(x) + s(x) + \bar{s}(x)\right]. \tag{6.23a}$$

and for the electron–neutron scattering

$$\frac{F_2^{en}(x)}{x} = \frac{4}{9}\left[d(x) + \bar{d}(x)\right] + \frac{1}{9}\left[u(x) + \bar{u}(x) + s(x) + \bar{s}(x)\right]. \tag{6.23b}$$

Both muon neutrino and antineutrino beams can be built at a proton accelerator as we have seen in Section 2.4. These are very powerful probes because they see different quarks. The reactions

$$\begin{aligned} \nu_\mu + d \to \mu^- + u; \quad \nu_\mu + \bar{u} \to \mu^- + \bar{d}; \\ \bar{\nu}_\mu + u \to \mu^+ + d; \quad \bar{\nu}_\mu + \bar{d} \to \mu^+ + \bar{u} \end{aligned} \tag{6.24}$$

are allowed, while

$$\begin{aligned} \nu_\mu + u \to \mu^+ + d; \quad \nu_\mu + \bar{d} \to \mu^+ + \bar{u}; \\ \bar{\nu}_\mu + d \to \mu^- + u; \quad \bar{\nu}_\mu + \bar{u} \to \mu^- + \bar{d} \end{aligned} \tag{6.25}$$

violate the lepton number and are forbidden. One might expect to have four independent processes, neutrino and antineutrino beams on proton and neutron targets, but only two of them are such, as can be easily seen. We then consider the proton targets only. By measuring the cross-sections of the DISs

$$\nu_\mu + p \to \mu^- + X; \quad \bar{\nu}_\mu + p \to \mu^+ + X \tag{6.26}$$

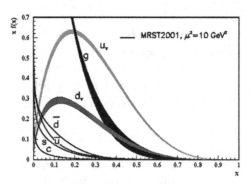

Fig. 6.12. The parton distribution functions. By permission of Particle Data Group.

one extracts the structure functions

$$\nu_\mu p \qquad \frac{F_2^{\nu_\mu p}(x)}{x} = 2[d(x) + \bar{u}(x)] \qquad (6.27a)$$

$$\bar{\nu}_\mu p \qquad \frac{F_2^{\bar{\nu}_\mu p}(x)}{x} = 2\left[u(x) + \bar{d}(x)\right], \qquad (6.27b)$$

where the factor 2 comes from the V–A structure of the weak interaction that we shall study in Chapter 7. Actually, there are other pieces of experimental information that we shall not discuss. Without entering into details, we give the results in Fig. 6.12. The bands are the uncertainties on the corresponding function. The gluon structure function will be discussed below.

We make the following observations: the valence quarks have a broad maximum in the range $x = 0.15 - 0.3x$, and go to zero both for $x \to 0$ and for $x \to 1$. The probability for a valence quark to have more than, say, 70% of the momentum is rather small. The sea quarks, on the contrary, have high probabilities at very low momentum fractions, less than about $x \approx 0.3$.

Question 6.2 Show that the cross-sections $\nu_\mu n$ and $\bar{\nu}_\mu n$ give the same relationships as (6.27). □

One would think that the sum of the momenta carried by all the quarks and antiquarks is the nucleon momentum, but it is not so. Indeed, by integrating the measured distribution functions, one obtains

$$\int_0^1 x\left[u(x) + d(x) + \bar{u}(x) + \bar{d}(x) + s(x) + \bar{s}(x)\right]dx \approx 0.50. \qquad (6.28)$$

Half of the momentum is missing! We conclude that 50% of the nucleon momentum is carried by partons that have neither electric nor weak charges. These are the gluons.

The HERA electron–proton collider was built at the DESY laboratory with the main aim of studying the proton structure functions in a wide range of kinematic variables with a high resolving power. The colliding beams were electrons at 30 GeV and protons at 800 GeV. The two experiments ZEUS and H1 measured

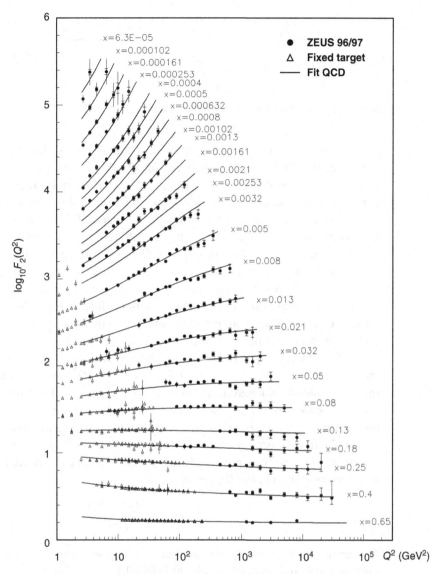

Fig. 6.13. The F_2 parton structure function as a function of Q^2 at different values of x as measured by the ZEUS experiment at HERA and by several fixed-target experiments. Lines are the Dokshitzer–Gribov–Lipatov–Altarelli–Parisi (DGLAP) theoretical predictions (adapted from S. Chekanov *et al. Eur. Phys. J.* C21 (2001) 443, Fig. 13. Reproduced with kind permission of Springer Science and Business Media. https://doi.org/10.1007/s100520100749).

the structure function F_2 with high accuracy in the momentum transfer range $2.7 < Q^2 < 30\,000$ GeV^2. Fig. 6.13 shows the results at different x values for $6 \times 10^{-5} < x < 0.65$. Data from fixed-target experiments are also included.

We see that, for the main range of x values, say for $x > 0.1$, the structure function F_2 is substantially independent of Q^2. The scaling law, as we had anticipated, is experimentally verified.

(a)   (b)

Fig. 6.14.   Looking at partons with smaller and larger resolving powers.

However, the data show that, at small x values, the scaling law is falsified: the structure function increases with increasing Q^2, that is, when we look into the proton with increasing resolving power. The scaling law violation bring us beyond the naïve parton model, which is only a first approximation. Indeed, they had been theoretically predicted by Yu. L. Dokshitzer (Dokshitzer 1977), V. N. Gribov and L. N. Lipatov (Gribov & Lipatov 1972) and G. Altarelli and G. Parisi (Altarelli & Parisi 1977) between 1972 and 1977 (DGLAP). The theoretical predictions are the curves in Fig. 6.13 and, as we can see, they are in perfect agreement with the data, proof of the validity of the theory. Consequently, F_2 is a function of x and Q^2, $F_2(x, Q^2)$.

We try to understand the phenomenon with the help of Fig. 6.14. The quarks in the nucleon emit and absorb gluons, with higher probability at lower x. Consider a quark with momentum fraction x emitting a gluon, which takes the momentum fraction $x - x'$. Therefore, the quark momentum fraction becomes x', which is less than before the emission. If the resolving power is not enough (Q^2 not large enough), one sees the quark and the gluon as a single object and measures x. If Q^2 is large enough, one resolves the two objects and measures the quark momentum fraction to be x'. Therefore, at small x, the distribution functions increase with increasing resolving power Q^2.

The gluon structure function, shown in Fig. 6.13, is larger where F_2 varies faster with increasing resolving power. More precisely, it can be shown that, at each value of x, it is proportional to the Q^2 derivative of F_2

$$G(x, Q^2) \propto \frac{\partial F_2(x, Q^2)}{\partial Q^2}. \tag{6.29}$$

As seen in Fig. 6.12, the gluon contribution is important at low momentum fractions, say below 0.2, and becomes dominant below 0.1.

The scaling law violations depend, as one can understand, on the coupling α_s (and on its Q^2 dependence). The curves in Fig. 6.13 have been calculated with α_s left as a free parameter, determined by the best fit of the curves to the data. This is one of the ways in which α_s is determined.

6.3 The QCD Lagrangian and the Colour Charges

The gauge symmetry of the strong interaction is $SU(3)$. It is an exact symmetry or, in other words, the colour charges are absolutely conserved. Since this group is more complex than the $U(1)$ group of QED, the colour charge structure is more complex than that of the electric charge. The Lagrangian density is both analogous to and substantially different from QED. While entering into any detail would need a knowledge of the Lie algebra of $SU(3)$, which is beyond the introductory level of this textbook, we give some hints. In QED, the gauge potential A_μ is a real-valued field; in QCD, the gauge potential G_μ is a 3×3 Hermitian field. The 'free' gluon field term, or 'kinetic' term, similar to $F^{\mu\nu} F_{\mu\nu}$ (where $F_{\mu\nu} = \partial_\mu A_\nu - \partial_\nu A_\mu$) in QED, is $G^{\mu\nu} G_{\mu\nu}$, with $G_{\mu\nu} = \partial_\mu G_\nu - \partial_\nu G_\mu + i g_s [G_\mu, G_\nu]$. The term in the square parentheses is the commutator, which is different from zero and is given by the algebraic structure of the non-abelian group $SU(3)$. The constant g_s is equivalent, in QCD, to the elementary charge in electromagnetism. Its appearance in the Lagrangian of the gluon-only term also means that the gluons, and not only quarks, have colour charges, and they interact among themselves. Gluons are also sources of the chromodynamic field, unlike in QED. As a matter of fact, the theory foresees the existence of gluon-only bound states, dubbed 'glueballs'. However, because they have the same quantum numbers as some quark–antiquark mesons, they get mixed with the latter and cannot be firmly established experimentally at present.

The interaction of the field ψ_f of the quark of flavour f appears in the Lagrangian as $\bar\psi_f \left[i\gamma_\mu \left(\partial_\mu + q_f A_\mu + i g_s G_\mu \right) - m_f \right] \psi_f$. The second term in the round parentheses gives the QED interaction of the quark, with q_f being its electric charge. The third term is the QCD interaction. Similar to the fine-structure constant α, in QCD the 'alpha strong' constant α_s is defined as

$$\alpha_s = \frac{1}{4\pi} g_s^2. \tag{6.30}$$

In both QED and QCD, the charges belong to a fundamental representation of the group. In the former case, $U(1)$ is simply a singlet. The structure of $SU(3)$ is richer; the group has two fundamental representations, **3** and **3̄** Correspondingly, there are three different charges, called red, green and blue (R, G, B). Each of them can have two values, say + and −. The former are in the **3**, the latter in the **3̄**. The quarks have colour charges +, the antiquarks have −. By convention, instead of speaking of positive and negative colour, one speaks of colour and anticolour; for example, a negative red charge is called antired. We shall use this convention but one can easily think of charges of both signs for every colour, as the reader prefers. The strong force depends only on the colour and is independent of the flavour and of the electric charge.

The gluons belong to the octet that is obtained by 'combining' a colour and an anticolour

$$\mathbf{3} \otimes \bar{\mathbf{3}} = \mathbf{8} \oplus \mathbf{1}. \tag{6.31}$$

We see that the situation is similar to that we have met in the quark model. Indeed, we are dealing with the same symmetry group, that is, $SU(3)$. We can then profit from the analogy, but keep in mind that it is only formal. In this analogy the colour triplet **3** corresponds to the flavour quark triplet d, u, s and the anticolour antitriplet $\overline{3}$ to the antiquark antitriplet. Note, however, that there is no analogy of the isospin.

Recalling Eq. (4.46), the singlet is

$$g_0 = \frac{1}{\sqrt{3}}\left(R\overline{R} + B\overline{B} + G\overline{G}\right), \tag{6.32}$$

Which is completely symmetric. In the singlet, the colour charges neutralize each other. As a result it does not interact with the quarks. Consequently, there is no singlet gluon.

Similarly with the meson octet, the eight gluons are

$$g_1 = R\overline{G}, \, g_2 = R\overline{B}, \, g_3 = G\overline{R}, \, g_4 = G\overline{B}, \, g_5 = B\overline{R},$$
$$g_6 = B\overline{G}, \, g_7 = \frac{1}{\sqrt{2}}\left(R\overline{R} - G\overline{G}\right), \, g_8 = \frac{1}{\sqrt{6}}\left(R\overline{R} + G\overline{G} - 2B\overline{B}\right) \tag{6.33}$$

The meson octet contains three meson–antimeson pairs: π^+ and π^-, K^+ and K^-, and K^0 and $\overline{K}^0$. Similarly, six gluons have a colour and a different anticolour and make up three particle–antiparticle pairs: g_1 and g_3, g_2 and g_5, and g_4 and g_6. The other two are antiparticles of themselves (completely neutral). Notice that g_7, analogous to π^0, has two colours and two anticolours and that g_8, analogous to η_8, has three colours and three anticolours. There is no octet–singlet mixing because the $SU(3)$ symmetry is unbroken.

Fig. 6.15(a) shows, for comparison, the vertex of the electromagnetic interaction. The ingoing and outgoing particles are equal; their charge is z_1 (elementary charges). The overall interaction amplitude between two charges z_1 and z_2 is proportional to $(z_1 z_2 \alpha)$.

The chromodynamic vertex is similar, but more complex, as shown in Fig. 6.16. The first difference with the QED vertex is that the incoming fermion q and the outgoing one q' may be different, for example two quarks of the same flavour and different colours. In this case, the gluon has the colour of one of them and the opposite of the colour of the other one and no flavour. Secondly, the vertex contains not only the coupling $\sqrt{\alpha_s}$ but also a 'colour factor' $\kappa_\lambda^{c_i \overline{c}_j}$, where c_i and c_j are the colours of the two quarks and λ is the gluon type. Finally, by convention, there is a factor $1/\sqrt{2}$.

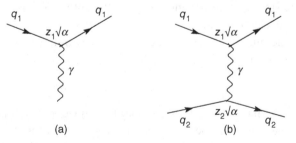

(a)

(b)

Fig. 6.15. The electromagnetic vertex and scattering.

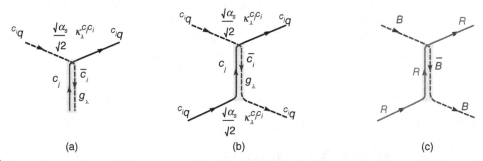

Fig. 6.16. (a) General quark–gluon vertex, showing the colour lines. (b) Quark–quark scattering with gluon exchange. (c) A blue quark–red quark scattering.

The gluon is usually represented by a helix, without a direction. However, it should have one for every colour it carries. For example, g_1 moving backward in time is g_3 moving forward. In this section we shall highlight this by drawing the gluon as a grey band in which the colour flows are represented by arrows, as in Fig. 6.16.

Fig. 6.16(c) shows an example of colour interaction, a diagram contributing to the scattering of a B (blue) quark and an R (red) quark. A blue quark changes to red by emitting a red–antiblue gluon that is absorbed by a red quark changing to blue. The same process can be seen also as a red quark changing to blue by emitting a blue–antired gluon that is absorbed by a blue quark changing to red. The general rule is that the colour lines are continuous through the diagram. However, a quantitative evaluation requires not only following the colour but also including the appropriate colour factors, which are simply the numerical factors appearing in (6.31), namely

$$\kappa_1^{R\bar{G}} = 1; \ \kappa_2^{R\bar{B}} = 1; \ \kappa_3^{G\bar{R}} = 1; \ \kappa_4^{G\bar{B}} = 1; \ \kappa_5^{B\bar{R}} = 1; \ \kappa_6^{B\bar{G}} = 1;$$

$$\kappa_7^{R\bar{R}} = \frac{1}{\sqrt{2}}; \ \kappa_7^{G\bar{G}} = -\frac{1}{\sqrt{2}}; \ \kappa_8^{R\bar{R}} = \frac{1}{\sqrt{6}}; \ \kappa_8^{G\bar{G}} = \frac{1}{\sqrt{6}}; \ \kappa_8^{B\bar{B}} = -\frac{2}{\sqrt{6}}. \tag{6.34}$$

The colour factors of antiquarks are the opposite of those of quarks.

We now look at two examples. Let us start with the interaction between two quarks of the same colour, BB for example

$$^{B}q + {}^{B}q \rightarrow {}^{B}q + {}^{B}q \tag{6.35}$$

shown in Fig. 6.17. From Eq. (6.34) we see that only one gluon can mediate this interaction, g_8. We have

$$\frac{1}{\sqrt{2}}\kappa_8^{B\bar{B}} \frac{1}{\sqrt{2}}\kappa_8^{B\bar{B}} = \frac{1}{2}\left(\frac{-2}{\sqrt{6}}\right)\left(\frac{-2}{\sqrt{6}}\right) = \frac{1}{3}.$$

Now consider RR (Fig. 6.18)

$$^{R}q + {}^{R}q \rightarrow {}^{R}q + {}^{R}q. \tag{6.36}$$

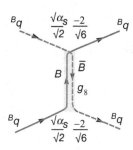

Fig. 6.17. Interaction between two blue quarks.

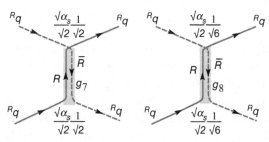

Fig. 6.18. Interaction between two red quarks.

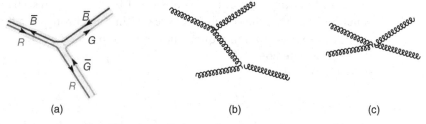

(a)   (b)   (c)

Fig. 6.19. (a) Three-gluon vertex: a red–antiblue gluon splits into a red–antigreen and a green–antiblue gluon. (b) Gluon–gluon scattering with gluon exchange. (c) Direct four-gluon scattering.

We have two contributions, g_7 and g_8. We sum them

$$\frac{1}{\sqrt{2}}\kappa_7^{R\bar{R}}\frac{1}{\sqrt{2}}\kappa_7^{R\bar{R}} + \frac{1}{\sqrt{2}}\kappa_8^{R\bar{R}}\frac{1}{\sqrt{2}}\kappa_8^{R\bar{R}} = \frac{1}{2}\left(\frac{1}{\sqrt{2}}\right)\left(\frac{1}{\sqrt{2}}\right) + \frac{1}{2}\left(\frac{1}{\sqrt{6}}\right)\left(\frac{1}{\sqrt{6}}\right) = \frac{1}{3}.$$

As expected from the symmetry, the force between R and R is the same as between B and B. The positive sign means that the force is repulsive. As in electrostatics, same sign colour charges repel each other.

Question 6.3 Verify the intensity of the force between G and G. ☐

Since gluons are coloured, they can interact coupled by continuous colour lines, as shown in Fig. 6.19(a). In this example, a red–antiblue gluon 'splits' into a green–antiblue gluon and a red–antigreen gluon. Gluon–gluon scattering can happen by exchanging another gluon as shown in Fig. 6.19(b), which includes two vertices of the type in Fig. 6.19(a). A further contribution to gluon–gluon scattering is the four-gluon coupling shown in Fig. 6.19(c).

6.4 Colour-Bound States

The hadrons do not have any net colour charge, but are made up of coloured quarks. It follows that the colour charges of these quarks must form a 'neutral' combination. An electromagnetic analogue is the atom, which is neutral because it contains as many positive charges as negative ones. In QCD the neutrality is the colour singlet state. Let us see how this happens for mesons and for baryons.

We start with the mesons, the simpler case. They are bound quark–antiquark states. The colour of the quark is in the $\mathbf{3}$ representation, the colour of the antiquark in $\bar{\mathbf{3}}$. They bind because their product contains the singlet

$$\mathbf{3} \oplus \bar{\mathbf{3}} = \mathbf{8} \oplus \mathbf{1}. \tag{6.37}$$

The singlet state is

$$\left(q\bar{q}\right)_{\text{singlet}} = \frac{1}{\sqrt{3}}\left(\, ^{B}q\, ^{\bar{B}}\bar{q} + \, ^{R}q\, ^{\bar{R}}\bar{q} + \, ^{G}q\, ^{\bar{G}}\bar{q}\right). \tag{6.38}$$

We notice that, by symmetry, the interactions between the three pairs in this expression are equal. So it is enough to compute one of them, say $^{B}q\, ^{B}\bar{q}$ and multiply by 3. In the calculation we must take all the possibilities into account; the initial state is $^{B}q\, ^{B}\bar{q}$, but the final state can be any quark–antiquark pair. Consequently, we have the diagrams of Fig. 6.20. Recalling that the antiquark colour factors are the opposite to those of the quarks, and including the normalization factor $\left(1/\sqrt{3}\right)^{2}$, we find the total colour factor for the (6.38) interaction

$$3\left(\frac{1}{\sqrt{3}}\right)^{2}\frac{1}{2}\left[\kappa_{8}^{B\bar{B}}\kappa_{8}^{B\bar{B}} + \kappa_{2}^{R\bar{B}}\kappa_{2}^{\bar{R}B} + \kappa_{4}^{G\bar{B}}\kappa_{4}^{\bar{G}B}\right]\alpha_{s} = \frac{1}{2}\left(-\frac{4}{6}-1-1\right)\alpha_{s} = -\frac{4}{3}\alpha_{s}. \tag{6.39}$$

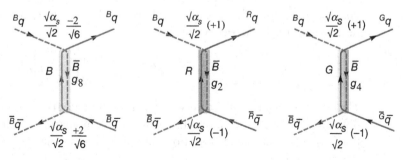

Fig. 6.20. Diagrams for a blue quark–antiquark interaction.

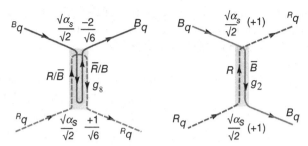

Diagrams for a blue quark red–quark interaction.

Notice in particular the negative sign. As in electrostatics, two opposite charges attract each other.

Now let us consider the baryons, which contain three quarks. Their colours are in the $\mathbf{3}$ representation. As the product $\mathbf{3} \otimes \mathbf{3} \otimes \mathbf{3} = \mathbf{10} \otimes \mathbf{8} \otimes \mathbf{8} \otimes \mathbf{1}$ contains a singlet (the neutral colour combination), three quarks can bind together. Let us see the structure in detail, starting from the first product

$$\mathbf{3} \oplus \mathbf{3} = \mathbf{6} \oplus \bar{\mathbf{3}}. \tag{6.40}$$

The $\mathbf{6}$ is symmetric; the $\bar{\mathbf{3}}$ is antisymmetric. Taking the second product we have

$$(\mathbf{3} \otimes \mathbf{3}) \otimes \mathbf{3} = \mathbf{6} \otimes \mathbf{3} \otimes \bar{\mathbf{3}} \otimes \mathbf{3} \tag{6.41}$$

There is no singlet in the product $\mathbf{6} \otimes \mathbf{3}$; the only one is in $\bar{\mathbf{3}} \otimes \mathbf{3}$, as shown by Eq. (6.37). In conclusion, every quark pair inside a baryon is in the antisymmetric colour $\bar{\mathbf{3}}$ and couples with the third quark to form the singlet. Recalling the discussion in Section 4.8, we have

$$(qqq)_{\text{singlet}} = \frac{1}{\sqrt{6}} \Big[({}^{R}q\,{}^{B}q - {}^{B}q\,{}^{R}q){}^{G}q + ({}^{G}q\,{}^{R}q - {}^{R}q\,{}^{G}q){}^{B}q + ({}^{B}q\,{}^{G}q - {}^{G}q\,{}^{B}q){}^{R}q \Big]. \tag{6.42}$$

It is easy to show that the colour factors of the three addenda are equal. We then calculate one of them, say the first, and multiply by 6. The two contributing diagrams are shown in Fig. 6.21, corresponding to ${}^{R}q + {}^{B}q \to {}^{R}q + {}^{B}q$ and ${}^{R}q + {}^{B}q \to {}^{B}q + {}^{R}q$.

We must be careful with the signs. The contribution of the second process must be taken with a minus sign because the final quarks are inverted and because the wave function is antisymmetric. We obtain

$$6 \left(\frac{1}{\sqrt{6}} \right)^{2} \frac{1}{2} \Big[\kappa_{8}^{R\bar{R}} \kappa_{8}^{B\bar{B}} - \kappa_{2}^{R\bar{B}} \kappa_{2}^{R\bar{B}} \Big] \alpha_{s} = \frac{1}{2} \Big[-\frac{2}{6} - 1 \Big] \alpha_{s} = -\frac{2}{3} \alpha_{s}. \tag{6.43}$$

The negative result implies a very important difference from the electric charges. Two different colour charges in an antisymmetric combination attract the charge of the third colour of the same sign. For example, the combination $(RB - BR)$ attracts G exactly as $\bar{G}$ does.

Question 6.4 Calculate the contribution of the third addendum in (6.42). □

Question 6.5 Given three objects R, G and B, how many symmetric combinations of the three, equal or different, can be made? How many antisymmetric combinations can be made? ☐

The characteristics of the colour charges that we have seen are demonstrated by the 'hyperfine structure' of the meson and baryon spectra. We start by recalling the hyperfine structure of the hydrogen atom, which is made of two opposite charge spin 1/2 particles, bound by the electromagnetic interaction. Since the photon is a vector particle, the hyperfine structure term appears as the interaction between the magnetic moments of the proton and the electron

$$\Delta E \propto -\mu_e \cdot \mu_p \propto -q_e q_p \mathbf{s}_1 \cdot \mathbf{s}_2, \tag{6.44}$$

where q_p and q_e are their equal and opposite electric charges. Consider in particular the S states. The two spins can be parallel $\left({}^3S_1, J=1\right)$ or antiparallel $\left({}^1S_0, J=0\right)$. The energy difference $E\left({}^3S_1\right) - E\left({}^1S_0\right)$ is very small and positive.

The mesons are also bound states of two spin 1/2 opposite colour particles. The ${}^3S_1$ states are the vector mesons, the states ${}^1S_0$ the pseudoscalar mesons. The differences are now large, and positive, for example $m\left(K^*\right) - m\left(K\right) = 395$ MeV. Now consider a baryon and take two of its quarks. If their total spin is 1, the baryon is in the decimet; if it is 0, it is in the octet. The difference between the levels is again large and again positive, for example $m(\Delta) - m\left(p\right) = 293$ MeV.

The interaction responsible for the difference between the levels is mediated for QCD, as for QED, by massless vector bosons and appears as an interaction between 'colour magnetic moments'. These have the direction of the spins, but the charges to be considered are the colour charges. Therefore, we have

$$\Delta E \propto \mu_e \mu_p \propto -\kappa_1 \kappa_2 \, \mathbf{s}_1 \cdot \mathbf{s}_2 \tag{6.45}$$

where κ_1 and κ_2 are the colour factors. There are two important differences with respect to electrodynamics. First, the level separation is much larger, because the colour coupling is bigger. The second difference requires more discussion.

Let us start with the mesons. Recalling that if $\mathbf{J}$ is the sum of $\mathbf{s}_1$ and $\mathbf{s}_2$, from

$$\langle \mathbf{J}^2 \rangle = \langle \mathbf{s}_1^2 \rangle + \langle \mathbf{s}_2^2 \rangle + 2 \langle \mathbf{s}_1 \cdot \mathbf{s}_2 \rangle$$

we have

$$2 \langle \mathbf{s}_1 \cdot \mathbf{s}_2 \rangle = J\left(J+1\right) - s_1\left(s_1+1\right) - s_2\left(s_2+1\right) = J\left(J+1\right) - \frac{3}{2}.$$

Hence

$$2 \langle \mathbf{s}_1 \cdot \mathbf{s}_2 \rangle = -\frac{3}{2} \quad \text{for } J=0; \quad 2 \langle \mathbf{s}_1 \cdot \mathbf{s}_2 \rangle = +\frac{1}{2} \quad \text{for } J=1 \tag{6.46}$$

We shall now see that, if the colour charges were to behave like the electric charges, the resulting hyperfine structure would be wrong. In this hypothesis the product of the charges would be $\kappa_1 \kappa_2 = -1$ and

$$\Delta E \propto -\kappa_1 \kappa_2 \, \mathbf{s}_1 \cdot \mathbf{s}_2 = +\mathbf{s}_1 \cdot \mathbf{s}_2 \propto -\frac{3}{2} \quad \text{for } J=0;$$

$$\Delta E \propto +\frac{1}{2} \quad \text{for } J=1 \tag{6.47}$$

Calling K a positive proportionality constant, we have

$$m\left(^3S_1\right) - m\left(^1S_0\right) = +2K.\tag{6.48}$$

In the case of the baryons we must consider the contributions of all the quark pairs and sum them. Let us start with the sum of the dot products

$$\begin{aligned}\Sigma &\equiv \langle 2\left(\mathbf{s}_1 \cdot \mathbf{s}_2 + \mathbf{s}_2 \cdot \mathbf{s}_3 + \mathbf{s}_3 \cdot \mathbf{s}_1\right)\rangle \\ &= \langle\left(s_1 + s_2 + s_3\right) - s_1\left(s_1 + 1\right) - s_2\left(s_2 + 1\right) - s_3\left(s_3 + 1\right)\rangle \\ &= J\left(J+1\right) - \frac{9}{4}\end{aligned}$$

and

$$\Sigma = -\frac{3}{2} \text{ for } J = \frac{1}{2}; \quad \Sigma = +\frac{3}{2} \text{ for } J = \frac{3}{2}.\tag{6.49}$$

If it were similar to the electric charges it would be $\kappa_1\kappa_2 = +1$ and, consequently,

$$\Delta E \propto -\kappa_1\kappa_2\, \mathbf{s}_1 \cdot \mathbf{s}_2 \propto +\frac{3}{2} \text{ for } J = \frac{1}{2}; \propto -\frac{3}{2} \text{ for } J = \frac{3}{2}$$

and

$$m\left(\mathbf{10}\right) - m\left(\mathbf{8}\right) = -3K.\tag{6.50}$$

In conclusion, if the colour charges were to behave like the electric charges, the vector meson masses should be larger than the pseudoscalar meson masses, which is right, and the masses of the decimet should be smaller than those of the octet, which is wrong. Moreover, in absolute value, the hyperfine structure of the baryons would be one and a half times larger than that of the mesons; instead it is somewhat smaller.

However, the colour force structure is given by $SU\left(3\right)$, not by $U\left(1\right)$, and the colour factors given by (6.39) and (6.43) must be considered. We have

$$\begin{aligned}m(^3S_1) - m(^1S_0) &= -2K \times \left(-4/3\right)\alpha_s = +\alpha_s K \times 8/3 \\ m\left(\mathbf{10}\right) - m\left(\mathbf{8}\right) &= -3K \times \left(-2/3\right)\alpha_s = +\alpha_s K \times 2.\end{aligned}\tag{6.51}$$

The predicted mass splittings have the same sign and the splitting of the mesons is larger than that of the baryons, as experimentally observed.

To complete the picture, we mention another difference between electric and colour charges, which is a consequence of the abelian and non-abelian characters of QED and QCD respectively. Both theories are gauge invariant, under the transformations of $U\left(1\right)$ and $SU\left(3\right)$, respectively. In both cases, the calculation of a physical observable, such as a cross-section, requires us to fix the gauge (see Section 5.1). Fixing one gauge, instead of another, does not alter the values of the observables. They are gauge invariant.

The difference we are discussing is that the electric charge is gauge invariant, the colour charges are not. A quark that is red in one gauge can be blue in another. However, a system that is colour-singlet in a gauge is the same in all the gauges. Consequently, the gauge non-invariance of colour has no physically observable consequences, because only colour-singlets are observable.

Let us, however, try to understand the reason for the difference. Consider a small volume surrounding an electron in one case, a quark in another one. The electric and the colour charges in that volume are, in both cases, the integral of the corresponding charge density over that volume. In both cases the volume contains the particle and the quanta of its field, photons and gluons, respectively. In the first case only, the charge of the electron contributes, because the photon is neutral; in the second, that of the quark and those of the gluons contribute. Moreover, it happens that the latter contribution is gauge dependent.

6.5 The Evolution of α_s

The strong interaction coupling constant α_s, which is dimensionless, as is α, is renormalized in a similar manner, but with one fundamental difference. As shown in Fig. 6.22, we must include in the vertex expansion not only fermion loops, but also gluon loops, because the gluons carry colour charges. The theory shows that bosonic and fermionic loop contributions have opposite signs.

The effect of vacuum polarization due to the quarks is like that which we have seen in electrodynamics, with the colour charges in place of the electric charge. The quark–antiquark pairs popping out of the vacuum shield the colour charge, reducing its value for increasing distance, or for decreasing momentum transfer in the measuring process.

However, the action of gluons is a smearing of the colour charge, which results in an effect of the opposite sign from that of quarks, called 'antiscreening'. The net result is that the colour charges decrease with decreasing distance. D. Politzer (Politzer 1973) and D. Gross and F. Wilczek (Gross & Wilczek 1973) discovered this property theoretically in 1973. G. 't Hooft had already presented, but not published, this conclusion in Marseille in 1972. The expression they found for the evolution of α_s is

$$\alpha_s\left(|Q|^2\right) = \frac{\alpha_s\left(\mu^2\right)}{1 + \dfrac{\alpha_s\left(\mu^2\right)}{12\pi}\left(33 - 2n_f\right)\ln\left(|Q|^2/\mu^2\right)}. \tag{6.52}$$

As in QED, once the coupling constant α_s is known at a certain energy scale μ, this expression gives its value at any other energy. The quantity n_f is the number of quark flavours effectively contributing to the loops, namely those with mass about $m_f < |Q|$. We see that the coupling constant decreases when $|Q|^2$ increases because

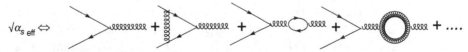

Fig. 6.22.   The lowest-order diagrams contributing to the QCD vertex, illustrating the relationship between the 'bare' coupling and the 'effective' (measured) coupling.

$(33 - 2n_f)$ is always positive since n_f is never larger than 6. As we saw for α, the dependence of the reciprocal of α_s on $\ln\left(|Q|^2 / \mu^2\right)$ is linear, but for the important effect of the opening of thresholds, it is given by

$$\alpha_s^{-1}\left(|Q|^2\right) = \alpha_s^{-1}\left(\mu^2\right) + \frac{33 - 2n_f}{12\pi}\ln\left(|Q|^2 / \mu^2\right). \tag{6.53}$$

Equation (6.52) can be usefully written in an equivalent form, defining a scale 'lambda-QCD' Λ_{QCD}, with the dimension of a mass, as the free parameter in lieu of μ

$$\Lambda_{QCD}^2\left(n_f\right) \equiv \mu^2 \exp\left[-\frac{12\pi}{\left(33 - 2n_f\right)\alpha_s\left(\mu^2\right)}\right]. \tag{6.54}$$

With this definition we have

$$\alpha_s\left(Q^2\right) = \frac{12\pi}{\left(33 - 2n_f\right)\ln\left(|Q|^2 / \Lambda_{QCD}^2\right)}. \tag{6.55}$$

Equation (6.55) shows that $\alpha_s\left(Q^2\right)$ diverges for $Q^2 = \Lambda_{QCD}^2$. The situation is similar to that we met at the end of Chapter 5 in QED, but for QCD the divergence is not at very high energies, but at quite low ones. As in QED, introducing higher-order corrections does not solve the problem. At high enough energies (see Fig. 6.23(b)), α_s is small, but with decreasing energy we approach Λ_{QCD}; the interaction becomes very strong and the theory does not admit a perturbative expansion. In the region of 'non-perturbative QCD', however, computing techniques 'on the lattice' have been developed (LQCD), which we shall deal with in Section 6.10.

Like α, α_s cannot be measured directly, but must be extracted from measured quantities with a theoretical calculation. Values of α_s have been extracted in a coherent way from observables measured in a wealth of processes at different Q^2 values. We quote, for example, (1) the probability of observing a third jet in e^+e^- hadronic processes, which is proportional to α_s, (2) the excess of hadronic production noticed with reference to Fig. 6.3 and (3) the scaling law violations in DIS. The most precise determination has been done at LEP at the Z mass. Consequently, it is convenient to choose the scale $\mu = M_Z$. The value is

$$\alpha_s\left(M_z^2\right) = 0.1179 \pm 0.0009$$

At this energy, the number of 'active' quarks is $n_f = 5$. Consider now the behaviour of Eq. (6.52) for decreasing values of Q^2. When we reach the mass of the bottom quark $Q^2 = m_b^2$, the factor $(33 - n_f)$ is discontinuous. We can avoid a discontinuity in $\alpha_s\left(Q^2\right)$ defining different values of Λ_{QCD} for different values of n_f. We determine the $\Lambda_{QCD}\left(n_f\right)$ values by imposing the continuity condition in Eq. (6.55), $\alpha_s\left(n_f = 5, Q^2 = m_b^2\right) = \alpha_s\left(n_f = 4, Q^2 = m_b^2\right)$ and similarly at the mass of the charm quark $\alpha_s\left(n_f = 4, Q^2 = m_c^2\right) = \alpha_s\left(n_f = 3, Q^2 = m_c^2\right)$. We cannot go below the charm mass because our expressions are no longer valid. In this way (and introducing higher-order corrections in (6.52)) one obtains, with uncertainties of about 10 MeV,

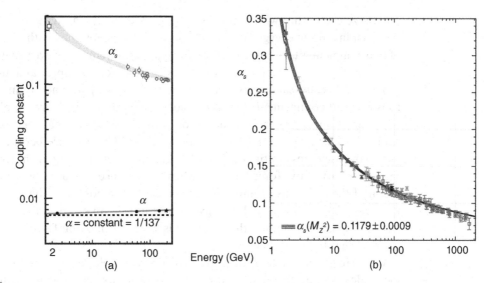

Fig. 6.23. (a) Comparison of the evolution of α and α_s (from S. Mele, personal communication 2005). (b) α_s from 2 GeV to 2 TeV (from Particle Data Group, Lawrence Berkeley National Lab).

$$\Lambda_{QCD}(5) \approx 215\,\text{MeV}; \quad \Lambda_{QCD}(4) \approx 300\,\text{MeV}; \quad \Lambda_{QCD}(3) \approx 340\,\text{MeV}. \quad (6.56)$$

Notice that Λ_{QCD} depends exponentially on α_s, the quantity more directly linked to the measurements. Consequently, even a small uncertainty in the latter results in a large variation in the former.

Fig. 6.23(a) shows a few determinations of α_s and of α for comparison, showing the decrease of the former and the (slower) increase of the latter with energy. Fig. 6.23(b) shows α_s up to the maximum energy at which it has been determined. The curve, and its width, show the theoretical prediction, and uncertainty, obtained with perturbation theory at higher energies and with LQCD at lower energies. The value of α_s at a few hundred MeV (not shown in the figure) is around 10; at 2 TeV it is about 0.03. Opposite electric charges attract one another at finite distances, becoming asymptotic free when the distance increases to infinity. Opposite colour quark and antiquark attract one another at finite distance too, but they become free when the distance tends to zero, or when the energy scale at which we look at them increases to infinity. This is called 'asymptotic freedom', somewhat the opposite of the confinement

The antiscreening action of the coloured gluons deserves further discussion, which we shall have by following the arguments of Wilczek. Consider a free quark, with its colour charge. In its neighbourhood the quantum vacuum pulsates; quark–antiquark pairs form and immediately disappear, gluons appear from nothing and fade away. This cloud of virtual particles antiscreens the central quark, making the colour charge grow indefinitely with increasing distance from the quark. However, this would require an infinite energy, which is impossible. This catastrophic growth can be avoided if an antiquark is present near the quark, because their clouds neutralize each other where they overlap. Consequently, a quark and its antiquark can exist in a finite energy system. The same result is obtained with a pair of quarks of two complementary colours, in an antisymmetric state. However, neither a quark, nor an antiquark, nor a quark pair can exist alone for an appreciable time.

The mechanism that keeps quarks and antiquarks permanently inside the hadrons is called confinement. Let us consider the mesons, which are simpler. In a first approximation, a meson is made up of a quark–antiquark pair and of the colour field, with all its virtual particles, between them. The distance between quark and antiquark oscillates continuously with a maximum elongation of the order of 1 fm. Indeed, the attractive force increases when the distance increases, because the cancellation of the two antiscreening clouds decreases. Suppose now that we try to break the meson by sending a high-energy particle into it, an electron for example. If the electron hits, for example, the quark, this will start moving further from the antiquark. What happens then?

We try to give a simplified description of a very complex phenomenon. We start with the analogy of the electrostatic force. Fig. 6.24 shows the electrostatic field between two equal and opposite charges. When the distance increases, the energy density of the field decreases.

The behaviour of the colour field is different, for reasons we cannot explain here. Fig. 6.25 shows the colour field lines between a quark and an antiquark. At distances of about 1 fm, the colour field is concentrated in a narrow 'tube'. When the separation between quark and antiquark increases, the length of the tube increases, but its diameter remains approximately constant. Therefore the field energy density remains constant and the total energy in the tube increases proportionally to its length. When the energy in the tube is large enough, it becomes energetically favourable to break the tube, producing a new quark–antiquark pair at the two new ends, as in Fig. 6.25(c). We have now a second meson, which is colour neutral. The process continues and more hadrons are created out of the colour field energy. It is the hadronization process.

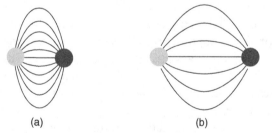

(a)   (b)

Fig. 6.24.   The electrostatic field lines between two equal and opposite charges. The lines going to infinity are not shown, for simplicity.

(a)   (b)   (c)

Fig. 6.25.   Sketch of the colour field lines between a quark and an antiquark.

The situation is similar for the quark confinement in a baryon, in which there are three colour tubes.

We can now answer another question. In the SLAC deep inelastic experiments of Section 6.2, a quark is hit by an electron and is suddenly accelerated. Why does it not radiate? Why, in other words, is the impulse approximation a good one? The explanation is, again, the antiscreen. At small distances from the quark, the quark charge is small and therefore the virtual particle cloud is only feebly attached to the quark. The hit quark darts away leaving its cloud behind, almost as if it had no charge. Later on, when the virtual particles respond to the change, a new cloud forms around the quark and moves with it. However, this last process does not imply a significant momentum and energy radiation. This is why, in the inclusive experiments, which measure only energy and momentum flows, the quarks behave as free, even if they are confined in a nucleon.

We also understand now the claim we made in Section 6.1, when we said that soft radiation by a quark is frequent, whereas hard radiation is rare. Indeed, at small momentum transfer, the interaction constant is large, but it is small at large momentum transfer.

We saw in Chapter 4 that the decays of particles of hidden flavour such as the $\phi\left(s\bar{s}\right)$, the $J/\psi\left(c\bar{c}\right)$ and the $\Upsilon\left(b\bar{b}\right)$ into final states not containing the 'hidden' quark are suppressed. This property was noticed by several authors and became known as the OZI rule after the names of three of them (Okubo, Zweig and Iizuka). The rule remained on a purely heuristic ground until QCD gave the reason for it.

Fig. 6.26 shows, as an example, the case of J/ψ. In the process shown in Fig. 6.26(a), a soft gluon radiated by one of the quarks materializes in a quark–antiquark pair. This process is favoured by QCD, but forbidden by energy conservation. Therefore, the charm–anticharm pair must annihilate into gluons. How many? The original pair, being colourless, cannot annihilate into one gluon, which is coloured; it cannot annihilate into two gluons, because, as shown by Landau (Landau 1948), invariance under rotations and inversion forbids a vector particle to decay into two massless vectors. The minimum number of gluons is three, as in Fig. 6.26(b). The norm of their four-momentum is the square of the mass of the decaying meson, m_V. Since this is rather large, the gluons are hard. The decay probability is proportional to $\alpha_s^3\left(m_V^2\right)$, where V indicates a generic vector boson. This is a small quantity. We have

$$\alpha_s^3\left(m_\phi^2\right) \approx 0.5^3 = 0.13; \; \alpha_s^3\left(m_{J/\psi}^2\right) \approx 0.3^3 = 0.03; \; \alpha_s^3\left(m_\Upsilon^2\right) \approx 0.2^3 = 0.008. \quad (6.57)$$

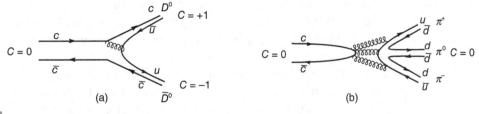

Fig. 6.26.   Two diagrams for the J/ψ decay.

6.6 The Running of the Quark Masses

In Table 4.5 we gave a summary of the quark quantum numbers, including the values of their masses with their uncertainties. We have already emphasized that quark masses are not measurable, because the quarks are never free. The top decays before hadronizing and we determine its mass by measuring the jets and the leptons originating from its decay. Still, a dependence on theory remains, because the top is coloured, but the observed final state is not. The 'masses' of the other quarks can be determined only indirectly through their influence on the properties of the hadrons. The definition of quark mass has a quantitative meaning only within a specific theoretical framework. The mass of a quark is defined as the parameter that appears in the mass term of the Lagrangian

$$\mathcal{L} = \cdots m\overline{\psi}\psi, \tag{6.58}$$

where ψ is the quark field. In QED the lepton masses are observable and can be identified in the theory with the pole of the propagator, remember Eq. (5.35), without ambiguity. In QCD, it turns out that, despite the impossibility of observing the quarks free, quark masses can be consistently treated in perturbation theory, as can the coupling constants. Renormalization is the procedure leading to finite values of the scattering and decay amplitudes. It uses a subtraction scheme to render the scattering amplitudes finite. As we have seen when discussing the coupling constant, this requires the introduction of a scale parameter μ, having the dimension of energy. Moreover, there are different possible choices of renormalization scheme. The calculated observables must be independent of the renormalization scheme and scale parameter chosen.

The mass parameters in the mass term of the Lagrangian (6.58), which are not observable, depend both on the renormalization scheme used and on the 'normalization scale' μ. The latter property is called the 'running' of the quark masses. The most used scheme in QCD is the 'modified minimal subtraction scheme', the $\overline{MS}$ scheme.

We can now specify that the quark masses in Tables 4.1 and 4.5 are in the $\overline{MS}$ scheme. The masses of the light quarks u, d and s are given for the normalization scale $\mu = 2$ GeV. The running masses of the heavy quarks c and b are given at the scale equal to the quark mass, namely $m(\mu)$ at $\mu = m_c$ and $\mu = m_b$, respectively.

Three quarks, u, d and s, have masses substantially smaller than Λ_{QCD} (light quarks), the other three, c, b and t, are much larger than Λ_{QCD} (heavy quarks). Only the masses of the latter are in the perturbative regime.

Consider first the light quarks, of which the most common hadrons are composed. The light quark masses can be extracted from the measured values of the hadrons' mass spectrum, if we can develop a theory connecting the former with the latter. The masses of the hadrons made of u, d and s quarks are in the region where the coupling constant α_s is large and a perturbative development is not possible. Lattice QCD must be used, as we shall see in Section 6.10. A complementary approach, able to give information on the ratios of the masses, is the exploitation of a symmetry of QCD, the chiral symmetry, as we will discuss in Section 6.9. For the heavy quarks, c and b, perturbative methods can be used. The masses and decay rates of hadrons containing the heavy quarks, such as the D and B mesons, are calculated and the quark masses extracted. The mass of the top quark will be discussed in Section 9.11.

We now discuss the experimental evidence for the running of the mass of the b quark at LEP and LHC.

Measured observables are the cross-sections of physical processes and the decay widths. The effects of the running of the quark mass are foreseen to be proportional to m^2 / Q^2, where Q is the energy scale of the measurement. On the other hand, in order to have good control on the QCD calculations, one needs to work at $Q \gg \Lambda_{QCD}$. This is the case at LEP on the Z peak, namely $Q = M_Z$, and of LHC in Higgs boson decays, namely $Q = M_H$. However, one pays a price for that, namely that $m^2 / Q^2 \ll 1$; consequently, the effects are small. This is why running is best known for the heavy b quark, for which $m_b^2 / M_Z^2 \sim 0.003$.

Let us consider first the LEP data. The probability for a quark produced in the $e^+ e^-$ annihilation to radiate a gluon is a function of its mass. Let us then consider the two processes represented in Fig. 6.27: the gluon radiation by a light quark (represented by a d in Fig. 6.27(a)) and by a b quark, and the ratio between their cross-sections. Perturbative QCD calculations, at the first order in α_s^2, in the $\overline{MS}$ scheme, give the following expression for the running b mass, valid for $\mu < 2M_w$,

$$\frac{m_b\left(\mu^2\right)}{m_b\left(\mu^2 = 4.2^2 \text{ GeV}^2\right)} = \left[\frac{\alpha_s\left(\mu^2\right)}{\alpha_s\left(\mu^2 = 4.2^2 \text{ GeV}^2\right)}\right]^{-\frac{12}{23}} \tag{6.59}$$

Experimentally, the events corresponding to the diagrams of Fig. 6.27 are three-jet events. If one of the jets originates by a b quark, it can be identified by observing the B decay vertex with the vertex detectors, as we have discussed for the top quark discovery. In this way, the LEP experiments measured the ratio of the radiation probability by the b quark and by the u, d, s and c quarks

$$R_3^{b,udsc} \equiv \frac{\Gamma_{3j}^b / \Gamma^b}{\Gamma_{3j}^{udsc} / \Gamma^{udsc}}. \tag{6.60}$$

Three experiments, ALEPH (Abdallah $et\ al.$ 2006), DELPHI (Abreu $et\ al.$ 1998) and OPAL (Abbiendi $et\ al.$ 2001), have consistently found values for $R_3^{b,udsc}$ of around 0.97.

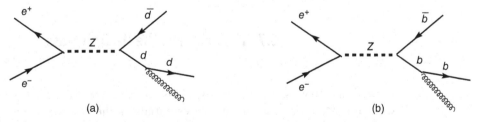

Fig. 6.27. Tree-level diagrams for single gluon radiation from (a) a light quark, generically labelled as d, and (b) a b quark.

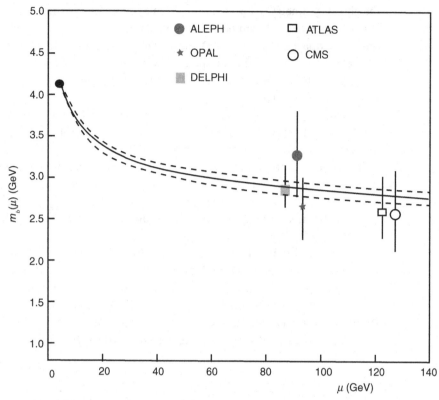

Fig. 6.28.　The running b quark mass as predicted by QCD (continuous line) with uncertainties (dashed lines) and the values extracted from the measurements of the LEP (ALEPH, OPAL, DELPHI) and LHC (ATLAS, CMS) experiments.

The effect is really small. Fig. 6.28 shows the b quark mass at the renormalization scale m_b (black dot) and the prediction of QCD with its uncertainty. The values of $m_b(M_Z)$ extracted by the three experiments are also shown (plotted displaced for clarity).

At LHC, the measurement of the partial width of the Higgs boson in beauty quarks, namely $H \to b\bar{b}$, provides a determination of the b quark mass at the renormalization scale of the Higgs boson mass (about 125 GeV). The values obtained by ATLAS (ATLAS Collaboration 2020) and CMS (CMS Collaboration 2019) are also shown in Fig. 6.28.

As can be seen, the values of $m_b(\mu)$ at $\mu = M_Z$ and $\mu = M_H$ are substantially smaller than at $\mu = m_b$, in agreement with the QCD predictions.

6.7　The Origin of the Hadron Mass

In Section 6.10 we shall see how LQCD is able to correctly predict the hadron mass spectrum. Here we show the physics essentials on the basis of a very simple model. Consider the proton that has a mass, about 1 GeV, much larger than the sum of the masses of the component u and d quarks, namely, about 7 MeV. What then is the origin of the proton mass? The proton mass being the largest fraction of the matter we know.

Let us start with a well-known problem: the mass of the hydrogen atom. Its size, the average distance a between electron and proton, is dictated by the uncertainty principle. The electron potential energy is negative and decreases with decreasing distance between electron and proton. But the electron cannot approach the proton too much because the localization of the wave function has an energy cost. The smaller the uncertainty of the electron position, the greater the uncertainty of its momentum. As a consequence, the average momentum itself is larger and, finally, the average kinetic energy is larger as well. The atomic radius is the distance at which the sum of potential and kinetic energies is at a minimum. This fact, known from atomic physics, is recalled in Exercise 5.2. If $E(a)$ is the total (kinetic plus potential) electron energy at a distance a, the atom mass is

$$m_H = m_p + m_e + E(a) = m_p + m_e - 13.6 \, \text{eV}. \tag{6.61}$$

We can phrase this as follows: the mass of the hydrogen atom is the sum of the masses of its constituents and of the work that must be done on the system to move the constituents in a configuration in which their interaction energy is vanishing. This configuration, for the atom, is when the constituents are far apart. The work is negative and small in comparison with the masses of its constituents.

Having recalled a familiar case, let us go back to the proton. The QCD interaction amongst the three valence constituent quarks is strong at a distance of the order of the proton radius (a little less than 1 fm). On the other hand, if the three quarks were located at the same point, they would not interact because the three antiscreening clouds would cancel each other out exactly (in the $SU(3)$ singlet configuration in which they are). This cannot happen precisely because of the energy cost of the localization of the wave functions. The three quarks adjust their positions at the average distance to minimize the energy, as in the case of the atom. We can take this distance as the proton radius r_p.

We start with the evaluation of the proton mass. Again this is the sum of the masses of the constituent quarks (a small fraction of the total) and of the work that must be done on the system to bring the constituents into a configuration in which the interaction strength vanishes; this is now where the quarks are **very close** to each other, namely the asymptotic freedom. The work is positive because it corresponds to the extraction of energy from the system (the 'spring' is contracting) and is by far the largest contribution to the proton mass.

The scale of the energy difference between an intense and a negligible interaction is, of course, Λ_{QCD}. In order of magnitude, the work to bring one quark into a non-interacting configuration is $\Lambda_{\text{QCD}} \approx 340$ MeV. In total (three quarks), we have $m_p \approx 3 \, \Lambda_{\text{QCD}} \approx 1$ GeV.

Having obtained a reasonable value for the proton mass, let us now check if we find a reasonable value for the proton radius. Following the arguments of Section 6.5, we assume that the energy of the colour field increases proportionally to the average distance x between the quarks, say as $k\,x$, where k is the 'spring' constant. The quark velocities are close to the speed of light and we can assume their kinetic energy to be equal to their momentum p. In conclusion, the energy of the three quarks is $E = 3p + kx$. The uncertainty principle now gives $p\,x \approx 1$ and we have

$$E = \frac{3}{x} + kx. \tag{6.62}$$

We now find a relation between the unknown constant k and the equilibrium inter-quark distance x_p by imposing the energy to be minimum

$$\left(\frac{dE}{dx}\right)_{x_p} = 0 = -\frac{3}{x_p^2} + k. \tag{6.63}$$

We obtain k by stating that the minimum energy must be equal to the proton mass. From (6.62)

$$m_p = E\left(x_p\right) = 6/x_p. \tag{6.64}$$

For $m_p \approx 1\,\text{GeV}$, we obtain $x_p \approx 1.2\,\text{fm}$. This is the average distance between two quarks. If the three quarks are at the vertices of an equilateral triangle, as they should be on average, the radius of the proton is

$$r_p = \frac{1}{\sqrt{3}} x_p \approx 0.7\text{fm}, \tag{6.65}$$

which is the correct value. This result is even too good for the very rough calculation we made.

In conclusion, the proton (the nucleon) mass would be very small if the three constituent quarks were in exactly the same position, because the antiscreen clouds would cancel each other out. The distance between the quarks is imposed by the energetic cost of the localization and, in turn, determines, due to the incomplete overlap of the clouds, the proton (nucleon) mass. The largest fraction, 99%, of the proton mass, the largest fraction of the mass of the matter we know, is the energy of the colour field.

The situation is similar for all hadrons that contain only u and d as valence quarks; the mass of the quarks make an appreciable contribution in the case of s, a large contribution in the case of c and a dominant contribution in the case of b.

6.8 The Quantum Vacuum

We have already discussed, even if only qualitatively, the 'vacuum polarization' phenomenon in the vicinity of a particle with electric charge (in Chapter 5) and of a particle with colour charge (in this chapter). However, this phenomenon occurs even if no particle is present, as we shall now see.

Diagrams of vacuum polarization: (a) a positron–electron loop and (b) a quark–antiquark loop.

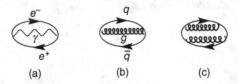

Higher-order diagrams of the vacuum polarization by positron–electron and quark–antiquark pairs.

We say that a region of space is empty, on large scales, if it does not contain particles or fields. Macroscopically, the electromagnetic field is zero and so is the colour field. The latter condition is obvious because this field exists only inside hadrons that are absent in a vacuum. Quantum mechanics teaches us that the vacuum is not empty at all, but, on small scales, contains virtual particles, their antiparticles and the quanta of their interactions.

The positron–electron pairs that we have met in our discussion of the evolution of α and drawn in Fig. 6.29 are also present in the absence of the central particle of that figure, even in a vacuum. Fig. 6.30(a) shows a positron–electron pair popping out of the vacuum. It recombines after a time Δt short enough to allow energy conservation to be compatible with the uncertainty of the measurement process, namely

$$\Delta t \leq \frac{1}{2m}. \tag{6.66}$$

Similar processes happen for every fermion–antifermion pair ($\mu^+\mu^-$, $\tau^+\tau^-$ and quark–antiquark), as exemplified in Fig. 6.30(b). In general, the mass m in Eq. (6.66) is the mass of the fermion.

To fix the scale, a positron–electron pair with $2m \approx 1\,\text{MeV}$ typically lives $\Delta t \approx 6.6 \times 10^{-22}\,\text{s}$, hence in a region smaller than $c\Delta t \approx 200\,\text{fm}$, while $d\bar{d}$ pairs, with masses of an order of magnitude larger, live 10 orders of magnitude less, in volumes within a 10 fm radius maximum.

During the short life of the couple, one or more photons or one or more gluons may be present, as in Fig. 6.31. There is more to it than that: as gluons carry colour charge themselves, gluon-only processes can occur. Since the gluon mass is zero, these processes take place at all energy scales.

In conclusion, the vacuum, when seen at the scale of 1 fm or smaller, is alive. It contains mass and energy fluctuations that grow larger at decreasing time and space scales. The fluctuations can be calculated with LQCD by using powerful parallel

Chromodynamics

Fig. 6.31. Diagram of the vacuum polarization by gluons.

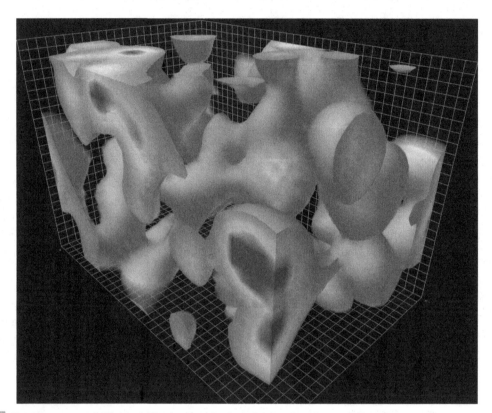

Fig. 6.32. The quantum vacuum. The length of each side is 2 fm. Credit D. Leinweber, University of Adelaide. www.physics .adelaide.edu.au/theory/staff/leinweber/VisualQCD/Nobel/ActionX08T30blackLattHR4096x3072.png.

computers. Fig. 6.32 shows an example of such calculations made by D. Leinweber, smoothing over the very small scales. It is a snapshot of the energy fluctuations in a volume of about 2 fm per side (more precisely of the action density, which is used in LQCD rather than energy, as we shall see in Section 6.10). In the time-dependent simulation, the energy 'lumps' evolve, changing shape, merging, disappearing and reappearing, but keeping the same general appearance.

The quantum vacuum is an extremely dynamic medium; its properties determine, to a large extent, the properties of matter itself.

The vacuum contains energy, in the same way a hadron does. But the presence of the quarks, real not virtual, in the hadron fosters the materialization of energy as mass. As we have seen, this is 99% of the mass of the matter we know.

6.9 The Chiral Symmetry of QCD and the Mass of the Pion

As we have seen in Chapter 4, the strong interactions are, to a good approximation, invariant under the isotopic spin symmetry, the flavour $SU(2)$. The symmetry is explicitly broken by the electromagnetic interaction and by the mass difference between the d and u quarks. The strong interaction is invariant, to a worse approximation, under $SU_f(3)$, which is broken by the larger mass of the s quark.

As the light quark masses are very small, let us consider the limit in which they are zero. In this limit, chirality coincides with helicity and, as we have seen in Section 2.8, the SM Lagrangians can be written as the sum of two similar terms, one involving only left fields and one involving only right fields. Both terms are invariant under $SU(3)$. We call $SU_L(3)$ and $SU_R(3)$ the two symmetries. The flavour symmetry for massless quarks is then larger than $SU(3)$: it is $SU_L(3) \otimes SU_R(3)$. It is called 'chiral' symmetry.

If the chiral symmetry were exact, the meson spectrum would show both pseudoscalar and scalar states with the same masses, and similarly for the higher spins. But this is not the case.

We start, for simplicity, with the non-strange mesons, which contain the u and d quarks. The chiral symmetry is reduced to $SU_L(2) \otimes SU_R(2)$. It is broken in two ways: explicitly, as we already mentioned, and spontaneously. The reason for the spontaneous breaking is that the physical states are the hadrons, not simply the quarks, and that the lowest-energy hadronic state, the hadronic vacuum, is not symmetric under the chiral symmetry. It can be shown that the corresponding Goldstone bosons form a pseudoscalar isospin triplet. These mesons have non-zero masses because the symmetry is also explicitly broken, by the non-zero values of the quark masses, and are called pseudo-Goldstone bosons. More precisely, it can be seen that the squares of the meson masses are proportional to the quark masses. Furthermore, the difference between the u and d masses produces a square-mass difference between the neutral and charged members of the triplet, as we shall see below. The isospin triplet is easily identified in the pions. Indeed, the pion mass squared is so small because it is a pseudo-Goldstone boson and because the u and d quark masses are very small.

However, a pseudoscalar isospin triplet is not a complete multiplet of $SU_L(2) \otimes SU_R(2)$. The simplest such multiplet contains four members, the fourth being a scalar–isoscalar meson. Scalar mesons exist but are very difficult to identify experimentally because they have large widths and, consequently, important overlap between resonances and with non-resonating background. Their classification in an $SU(2)$ multiplet is also difficult due to the possible existence of glueball states. Historically, evidence for the first $J^{PC} = 0^{++}$ state, an isoscalar, dates back to 1966, when it was observed in the $\rho^0 \rho^0$ system in the reaction $\bar{p}n \to \rho^0 \rho^0 \pi^- \to (\pi^+ \pi^-)(\pi^+ \pi^-)\pi^-$ (Bettini *et al.* 1966). The state was later observed to decay also in $\pi\pi$ and in $K\bar{K}$, but it took several years to establish its bona fide resonance nature (see Peláez *et al.* 2023). The experiment was done in a bubble chamber filled with deuterium, in order to have the simplest possible neutron target, and with an antiproton beam. The beam energy was low enough to allow the antiprotons to come to rest in the chamber, and then annihilate. This procedure was followed because at rest the antiproton becomes bound to a deuterium nucleus, forming an 'antiprotonic atom' and, soon after, the $\bar{p}n$ annihilation takes place mainly from the S wave of this 'atom'. The process is described by the Day–Snow–Sucker theory discussed in a similar case in Section 3.5. Therefore, the five-pion final state we are considering can have its origin only from the singlet $^{1}S_{0}$, the triplet $^{3}S_{1}$ being forbidden by G-parity. Consequently, the total quantum numbers of the process are known: $I = 1$, $J^{P} = 0^{-}$. We readily see then that the total spin of the $\rho\rho$ system, say $S_{\rho\rho}$, must be equal to its orbital momentum $L_{\rho\rho}$ and its parity $P_{\rho\rho} = (-1)^{L_{\rho\rho}} = (-1)^{S_{\rho\rho}}$, that is $J^{P}_{\rho\rho} = 0^{+}, 1^{-}, 2^{+}, \ldots$. The study of the angular distributions allowed us to unambiguously choose $J^{P}_{\rho\rho} = 0^{+}$. The resonance is called $f_{0}(1370)$, where the number in parentheses is the mass in MeV. We see that the square mass of the scalar is as large as about 2 GeV^{2}, two orders of magnitude larger than that of the pion. The chiral symmetry is largely broken, the pion is a pseudo-Goldstone boson, the $f_{0}(1370)$ is not.

The spontaneous chiral symmetry breaking can be seen with the following argument, due to Banks and Susskind, quoted by 't Hooft (2000). The valence quark and the antiquark in a pseudoscalar meson are in an S wave. Consequently, their movement inside the meson is essentially an oscillation along a diameter. However, they cannot invert their motion at the end of the oscillation, because this would imply a change of helicity, which is impossible for a massless quark and almost so for the actual light quark. It is more likely that the quark continues in its direction and disappears into the 'sea' of quark–antiquark pairs present inside the meson and another one of opposite helicity jumps out of the sea. This hints at the importance of the connection between the $q\bar{q}$ pseudoscalar states and the QCD vacuum. The situation for the scalar mesons is different because the valence quark and antiquark in $J^{PC} = 0^{++}$ are in the $^{3}P_{0}$ configuration, namely with orbital momentum $l = 1$, a state in which velocities have non-zero non-radial components.

Consider now also the mesons containing the s quark. The situation is similar, with the symmetry $SU_{L}(3) \otimes SU_{R}(3)$. The pseudo-Goldstone bosons form an $SU(3)$ octet of pseudoscalar mesons, which we identify with the octet we discussed in Chapter 4. The masses can be calculated numerically with LQCD, as we shall see in the next section, but some results can be more easily reached on the basis of the chiral symmetry, by expanding in series of powers of the quark masses. As anticipated, it is found that, to the first order in this chiral expansion, the squares of the meson masses are proportional to the quark masses. The following relations are found between the former and the latter, with two unknowns, B, having the dimensions of a mass, and the electromagnetic square-mass difference Δ_{em}:

$$
\begin{aligned}
m^{2}_{\pi^{0}} &= B(m_{u} + m_{d}) \\
m^{2}_{\pi^{\pm}} &= B(m_{u} + m_{d}) + \Delta_{em} \\
m^{2}_{K^{0}} &= B(m_{d} + m_{s}) \\
m^{2}_{K^{\pm}} &= B(m_{u} + m_{s}) + \Delta_{em} \\
m^{2}_{\eta} &= \frac{1}{3} B(m_{u} + m_{d} + 4m_{s}).
\end{aligned}
\tag{6.67}
$$

Eliminating the unknowns, we obtain two mass ratios

$$\frac{m_u}{m_d} = \frac{2m_{\pi^0}^2 - m_{\pi^\pm}^2 + m_{K^\pm}^2 - m_{K^0}^2}{m_{K^0}^2 - m_{K^\pm}^2 + m_{\pi^\pm}^2} = 0.56, \tag{6.68}$$

and

$$\frac{m_s}{m_d} = \frac{m_{K^\pm}^2 + m_{K^0}^2 - m_{\pi^\pm}^2}{m_{K^0}^2 - m_{K^\pm}^2 + m_{\pi^\pm}^2} = 20.2. \tag{6.69}$$

The squared mass of the ninth pseudoscalar, η', is much larger than the squared masses of the octet, about 50 times larger than that of the pion. The understanding of this problem was of fundamental importance in the understanding of the QCD hadronic vacuum. The bottom line is that the η' is not a pseudo-Goldstone boson, hence its squared mass does not need to be small. Basically, for the same reason, its mixing with the η is small.

Before concluding, we make the following important observation. We saw in Section 6.7 that the mass of the nucleons is predominantly due to the confinement energy. The small u and d quark masses give only a 1% contribution. However, those tiny masses determine the size of the nuclei, namely the range of the nuclear forces, which is inversely proportional to the mass of the pion. The square of the latter, as we have just seen, is proportional to the quark masses, as a consequence of the breakdown of the chiral symmetry.

6.10 QCD on the Lattice

The idea of calculating QCD processes on a lattice (LQCD) was advanced by K. Wilson in 1974 for a theoretical interpretation, from first principles, of the confinement and asymptotic freedom. The idea, taken from statistical physics of condensed matter, was to replace the infinite and continuous four-dimensional space-time with a finite hypercubic lattice. The following developments of the algorithms, together with the availability of constantly increasing computer power, mean it is now possible to calculate accurate QCD predictions with numerical methods. Depending on the problem, an integral of the 'density' of the relevant physical quantity must be performed on the space-time. The integral is reduced to a discrete sum over the lattice cells. The measure of the integral becomes the sum of products of a very large, but finite, number of differentials. The theory is non-perturbative, being the sums over all the relevant Feynman diagrams at once.

Fig. 6.33 shows three dimensions, two spatial and one temporal, of the lattice hypercube. The lattice spacing is a and side length L. To perform the calculations, the fermionic variables, that is the fermion fields, are placed on the lattice sites, and the gauge fields, which are 3×3 matrices, are placed on the links connecting these sites. Mathematically, this LQCD is like a classical four-dimensional statistical system.

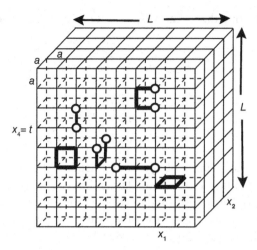

The lattice hypercube. Dots represent the spinor fields; heavy lines represent QCD gluon fields in interaction both with spinors and with themselves.

The physical result is obtained going to the infinite size limit (volume $V = L^4 \to \infty$), numerically or analytically, and to the continuum limit ($a \to 0$). To obtain the latter, one needs to repeat the calculations for several decreasing values of a, say one half each time. Doing that, the number of operations at each step increases by $2^4 = 16$. The more steps, the greater the final accuracy. For a sub-per-cent accuracy, these are typically of the order of 10^{10}. This is today possible with the availability of petaflop (10^{15} floating point operations per second) supercomputers, five orders greater than the power of a laptop.

As we discussed already, QCD, and in general quantum field theories, are not mathematically well behaved at the high energy limit. They suffer from infinities that are controlled with renormalization procedures. By putting the theory on a discrete lattice, we automatically introduce a high-energy cut-off, corresponding to the lattice spacing a. However, removing the discretization effects when taking the limit $a \to 0$ is not at all a trivial task.

Another major difficulty that has been solved in the last decade is to maintain the chiral property of the continuous function in its discretization, something that is mandatory for the light quarks.

Lattice QCD is a fundamental approach to compute from first principles many physical quantities. A few examples are the magnetic anomaly discussed in Section 5.9, the strong coupling constant (Section 6.5), the quark masses (Section 6.6) and the QCD vacuum (Section 6.8). In Section 6.7 we gave a rough evaluation of the proton mass based on basic physics arguments; LQCD is able to compute with precision the hadron spectrum, as we shall soon show. Lattice QCD is also mandatory to compute hadronic contributions to weak processes, such as the hadron decays we shall see in Chapter 7 and the quark mixing and meson oscillations that we shall study in Chapter 8.

We come now to LQCD prediction on the masses of the lighter hadron multiplets, specifically the non-singlet pseudoscalar and vector meson octets and the baryon octet and decimet. The full LQCD calculation was performed *ab initio* by the Budapest–Marseille–Wuppertal (BMW) Collaboration (Dürr *et al.* 2008) with only three experimental input parameters. The measured masses of the pion and of the K meson are used to fix the isospin averaged mass of the u, d light doublet and of the s quark. QCD does not predict the hadron masses in physical units; only dimensionless combinations, such as mass ratios, can be calculated. The dimensional quantity chosen to fix the overall physical scale is the mass of the Ξ baryon. This is chosen because it belongs to the octet, for which the calculation is more precise than for the decimet, and, in the octet, it is the particle with maximum content of s quarks, which makes its mass less sensitive to the chiral behaviour of the lighter quarks.

Substantial progress has been achieved with the inclusion of the effects of the light sea quarks, which dramatically improve both the systematic uncertainties and the agreement of QCD with experiment. As for the volume of the hypercube, the authors use a side $L \gtrsim 4 / m_\pi = 6.3\,\text{fm}$, which results in masses that practically coincide with the infinite volume results (a look at Fig. 6.32 shows that the typical size of the vacuum fluctuations is one order of magnitude smaller). As for the spacing, it was found that the calculated hadron masses deviate from their continuum $a \to 0$ limit by less than approximately 1% for $a \cong 0.125\,\text{fm}$. The comparison of the calculated QCD light hadron spectrum and the measured one is shown in Fig. 6.34.

An observation is in order to conclude. At high energies, where the coupling constant is sufficiently small, the predictions of QCD can be calculated perturbatively at high precision to be tested against experiments. At low energies, where the coupling constant is large and, consequently, perturbative calculations are impossible, the lattice calculations have reached the stage where all the systematic errors can be completely controlled, leading to sub-per-cent accuracy.

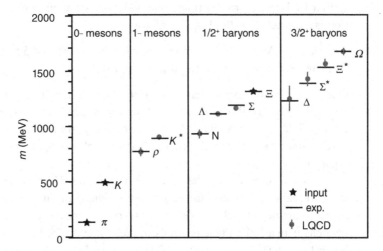

Fig. 6.34. The light hadron spectrum of QCD. The horizontal bars are the experimental values. The solid circles are the QCD predictions, for which vertical error bars represent the combined statistical and systematic error estimates. π, K and Ξ have no error bars, because they are used to set the light quark mass, the strange quark mass and the overall scale, respectively.

6.11 Deconfinement and the Quark–Gluon Plasma

When matter first appeared in the Universe, in the first microsecond after the Big Bang, the basic constituents of the matter we know, quarks and gluons, moved freely in a hot 'soup' called, in analogy to the plasma of electrons and nuclei, quark–gluon plasma (QGP). Cosmologists also tell us that this very first phase of matter had the characteristics of a liquid, possibly with quite small viscosity. As the Universe quickly expanded, and cooled, protons and neutrons and other hadrons 'froze out'. Quarks became confined, bound together by exchanging gluons.

The existence of the QGP phase of matter and a number of its physical properties are predicted by QCD. As the QGP energy scale is small relative to Λ_{QCD}, numerical predictions of the relevant quantities are obtained via LQCD. Experimentally, QGP is studied in ultra-relativistic heavy ion collisions. The heavy ions are accelerated at high energies and stored as two counter-rotating beams in a collider. In the most central collisions, if the energy density and the temperature are high enough, the QGP may be formed in the initial instants. Then the initial fireball rapidly expands, cooling down, and quarks hadronize, become bound in hadrons, hadrons that, finally, we can detect.

After pioneering studies at lower energies, centre of mass energies sufficient for detailed studies were reached with the RHIC built on purpose in the BNL, which started operation in the year 2000. Centre of mass energy per colliding nucleon pair reached $\sqrt{s_{NN}} = 200$ GeV in Au + Au collisions. The LHC project at CERN was designed to include the options of Pb + Pb, Xe + Xe (or other heavy nuclei) and p + Pb (or other nuclei) collisions. Note that, in the laboratory, the nuclei appear similar to flat discs, rather than spherical, because their diameter in the beams' direction is contracted by the Lorentz factor γ.

Question 6.6 Assuming the Pb nucleus to be a sphere of 5.5 fm radius, compute its diameter in the beam direction if its energy is 2.5 TeV. □

One of the LHC experiments, ALICE, was designed specifically to study heavy ion collisions, QGP and more. At the end of 2010, the first Pb + Pb collisions at $\sqrt{s_{NN}} = 2.76$ TeV took place. In the following years the energy per nucleon reached 5 TeV. Since then, the QGP and the hadronic phases have been studied with increasing accuracy.

Let us start considering the orders of magnitude. The atomic nuclei quarks and the sea antiquarks are bound in the nucleons, whose radius is about 0.88 fm. In nuclei, nucleons are not densely packed; there is some space between them. The nucleon number density in a ^{204}Pb nucleus, for example, is 0.29 fm^{-3}. This is 14 orders of magnitude larger than the density of water. If we successfully reach substantially larger densities in high-energy nuclear collisions, the distance between quarks decreases accordingly. The coupling constant decreases as a consequence, and, if the temperature is above the freeze-out value, the quarks are no longer confined but can move over distances much larger than the size of the nucleon. This state of matter is the QGP. However, the structure, and the phases, of strongly interacting matter under extreme conditions are much more complex than a single phase and their study is at the frontier of modern research. Here we discuss only a very small sector.

In the laboratory, the collisions between high-energy nuclei take place at all possible impact parameters. The central, head-on collisions are experimentally selected on the basis of the 'multiplicity', namely the number of detected particles produced in the collision. Events with multiplicity larger than a suitably defined value give the sample of 'central' events. Another important quantity is the 'pseudorapidity', defined for each produced particle as

$$\eta = -\ln\left[\tan\left(\frac{\theta}{2}\right)\right],\tag{6.70}$$

where θ is the angle of the particle with the beam's direction. Pseudorapidity is used in place of simply θ because its distribution is much flatter in the central region. QGP is studied selecting events at central rapidity, say $|\eta| < 1.4$.

In this kinematical region, at LHC energies, hot and dense matter that contains approximately equal numbers of quarks and antiquarks is produced, for a short duration of a few 10^{-23} s. Initially the system is in its deconfined QGP phase. Its temperature, measured in units $k = 1$ (where k is the Boltzmann constant) is above 300 MeV. For comparison, remember that the 300 K room temperature is 26 meV in this units. The QGP at LHC is 10^{10} times hotter, as we shall now see. Quickly, the QGP cools down, expands and undergoes the transition to a fireball of hadron gas. The temperature at which the inelastic interactions between particles cease is called chemical freeze-out temperature T_{CF}, where 'chemical' is in analogy with macroscopic thermal systems.

Analogously, a macroscopic fluid such as water is in its deconfined phase, vapour, at high temperatures. It undergoes the transition to the confined, liquid, phase when the temperature decreases below the boiling temperature at that pressure. Unlike the QGP, however, the density of the confined phase is greater than that of the deconfined one. While the electromagnetic interaction – the cohesion force is such – vanishes with increasing distance, the chromodynamic one vanishes with decreasing distance between quarks, as we have already discussed. Another important difference will be discussed soon.

Indeed, the system we are discussing is, even if very small, a thermodynamic system and, as such, is theoretically described by statistical analysis, called the statistical hadronization model. Here we limit the discussion to the deconfining transition. This can be most properly represented by plotting the two phases' coexistence temperature T_{CF} versus the baryon chemical potential μ_B. The latter is defined as the change in the internal energy U by a change in the baryon number N_B, namely $\mu_B = \partial U / \partial N_B$. Notice that N_B is the difference between the number of baryons and the number of antibaryons, namely the baryon–antibaryon unbalance. The larger the chemical potential, the larger the unbalance. Let us see how these quantities are determined.

Experimental access to the hot initially created plasma is given by measuring the yield of 'thermal' direct photons (the term will be defined soon). As the photon mean free path in the plasma is much larger than the size of the fireball, these photons reach our detectors unaffected, delivering information on QGP at early times. Other photons are produced by electromagnetic interactions at later stages. Direct photons can be distinguished from the other photons because they dominate the lower transverse momentum region $\left(p_T \lesssim 3\,\mathrm{GeV} \right)$ of the spectrum, which, in this region shows an exponential behaviour. This is the Boltzmann factor expected for any system in thermal equilibrium (hence the adjective above). From the expression

$$\frac{d^2 N_{\gamma_{\mathrm{dir}}}}{p_T\, dp_T\, dy} \propto e^{-p_T/T_{\mathrm{eff}}}, \qquad (6.71)$$

the 'effective' temperature of the initial hot matter T_{eff} is obtained. At Pb + Pb central collisions at $\sqrt{s_{NN}} = 2.76\,\mathrm{TeV}$, ALICE finds $T_{\mathrm{eff}} = \left(304 \pm 41 \right)\,\mathrm{MeV}$. This is larger than the critical deconfinement temperature expected from LQCD, showing that the QGP phase has been produced.

The statistical hadronization model predicts the yields of different particles. ALICE exploited its superior features of particle identification to measure, in Pb + Pb collisions at $\sqrt{s_{NN}} = 2.76\,\mathrm{TeV}$, in the central rapidity region, the yields of a number of hadrons: $\pi^+, \pi^-,\ K^+, K^-, K^0,\ p, \bar{p},\ \Lambda, \bar{\Lambda}, \phi,\ \Xi^-, \bar{\Xi}^+,\ \Omega^-, \bar{\Omega}^+,\ d, \bar{d}$ and of light nuclei, of a lambda hypernucleus and of their antinuclei: $d, \bar{d},\ ^3\mathrm{He}, ^3\overline{\mathrm{He}},\ ^3_\Lambda\mathrm{H}, ^3_{\bar{\Lambda}}\overline{\Lambda},\ ^4\mathrm{He}, ^4\overline{\mathrm{He}}$. The yields span 9 orders of magnitude from the pion (yield per unity rapidity interval 6×10^2) to the helium (yield per unity rapidity interval 7×10^{-7}). The statistical hadronization model predicts correctly the observed values (see Andronic *et al.* 2018), within the experimental uncertainties, by optimizing the values of the thermal quantities. Amongst these, the baryon chemical potential turns out to be small, $\mu_B = 0.7 \pm 3.8\,\mathrm{MeV}$, consistent with zero, as expected by the observation of equal production of matter and antimatter at LHC.

The fitted chemical freeze-out temperature is $T_{CF} = 165.5 \pm 1.5\,\mathrm{MeV}$. Remarkably, this value agrees within the error with the prediction of LQCD of the critical temperature of the deconfinement transition at zero chemical potential, $T_c = 154 \pm 9\,\mathrm{MeV}$. More precisely, this temperature is called 'pseudocritical' because the phase transition predicted by LQCD is of the crossover type, namely a smooth but rapid variation of the thermodynamic quantities in a narrow region around T_c, different from the first-order phase transition as in the above-considered example of water vapour.

Finally, Fig. 6.35 shows a schematic representation of the QCD phase diagram, adapted from the ALICE Collaboration (2022). We can see the crossover band predicted by LQCD, in which the transition from QGP to the confined hadrons phase takes place, with a rapid but not discontinuous variation of the temperature. The pseudocritical temperature curve is also shown. The measurements agree with the theory. As mentioned above, the measured initial effective temperature is about twice as large as the transition temperature. The status of the Universe when it was about 1 μs old was similar. The baryon chemical potential was zero as the matter and antimatter contents were equal. Nuclear matter is much colder and, as it contains only baryons, the chemical potential is large.

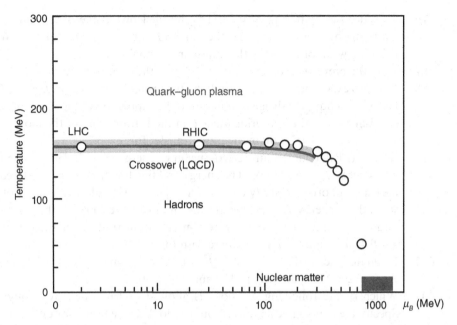

Fig. 6.35. LQCD predictions are shown as the line for the pseudocritical temperature, and the band for the half-width of the crossover transition (i.e. the temperatures where the QGP and hadrons can co-exist). The open points show experimental results. The location of atomic nuclei is also approximately shown. Adapted from the ALICE Collaboration (2022).

Problems

6.1 How many gluons exist? Give the electric charges of each of them. Give the values of their strangeness, charm and beauty. What is the gluon spin? How many different quarks exist for every flavour? What are their charges? Does QCD define the number of families?

6.2 Evaluate $R \equiv \sigma\left(e^+e^- \to \text{hadrons}\right)/\sigma\left(e^+e^- \to \mu^+\mu^-\right)$ at $\sqrt{s} = 2.5$ GeV and at $\sqrt{s} = 4$ GeV.

6.3 Consider the reaction $e^+ + e^- \to q + \bar{q}$ at a collider with CM energy $\sqrt{s} = 20$ GeV. Give a typical value of the hadronic jet opening angle in a two-jet event. If θ is the angle of the common jet direction with the beams, what is the ratio between the counting rates at $\theta = 90°$ and $\theta = 30°$?

6.4 In a DIS experiment aimed at studying the proton structure, an $E = 100$ GeV electron beam hits a liquid hydrogen target. The energy E' and the direction of the scattered electron are measured. If x and Q^2 are, respectively, the momentum fraction and the four-momentum transfer, find E' for $Q^2 = 25$ GeV^2 and for $x = 0.2$.

6.5 What is the value of the x variable in elastic scattering? Find the expression (1.83) $E' = E / \left[1 + \dfrac{E}{m}\left(1 - \cos\theta\right)\right]$, taking the elastic cross-section as the limit of the inelastic cross-section.

6.6 Consider the scattering of ν_μ and $\bar{\nu}_\mu$ by nucleons in the quark model, in terms of scattering by quarks. Consider the d, u and s quarks and antiquarks. Write the contributing weak processes with a muon in the final state.

6.7 As in the previous problem, but considering the quarks c and $\bar{c}$.

6.8 In a DIS experiment aimed at studying the proton structure, an $E = 100$ GeV electron beam hits a liquid hydrogen target. Find the expression of the momentum transfer Q^2 as a function of the scattering angle θ in the L frame and of the momentum fraction x. What is the maximum momentum transfer for $x = 0.2$.

6.9 In the HERA collider, an electron beam of energy $E_e = 30$ GeV hits a proton beam with energy $E_p = 820$ GeV. The energy and the direction of the scattered electron are measured in order to study the proton structure. Calculate the centre of mass energy $\sqrt{s}$ and the energy $E_{e,f}$ an electron beam must have to reach the same $\sqrt{s}$ at a fixed target. Calculate the maximum four-momentum transfer of the electron Q^2_{max} for $x = 0.4$, 0.01 and 0.0001. Compare with Fig. 6.13.

6.10 Evaluate the ratio α / α_s at $Q^2 = (10 \text{ GeV})^2$ and at $Q^2 = (100 \text{ GeV})^2$. Take $\Lambda_{QCD} = 200$ MeV, $\alpha^{-1}(m_Z^2) = 129$ and $M_Z = 91$ GeV.

6.11 Which of the following reactions is allowed or forbidden by strong interactions? Specify the reason in each case. You may look at the tables for establishing the quark compositions: (a) $\pi^- + p \rightarrow \Lambda_c^+ + \pi^-$, (b) $\pi^- + p \rightarrow \Lambda_c^+ + D^-$, (c) $\pi^- + p \rightarrow \Lambda_c^+ + D^0$ and (d) $\pi^- + p \rightarrow \Lambda_c^+ + D_s^-$.

6.12 A non-charmed baryon has strangeness $S = -2$ and electric charge $Q = 0$. What are the possible values of its isospin I and of its third component I_z? What is it usually called if $I = 1/2$?

6.13 The proton has uud as valence quarks. Write down the triplet wave function in its spin, isospin and colour factors, taking into account that all the orbital momenta are zero.

6.14 As in the previous problem, but for the Λ hyperon.

6.15 Consider the processes (1) $e^+ + e^- \rightarrow \mu^+ + \mu^-$ and (2) $e^+ + e^- \rightarrow$ hadrons at the two centre of mass energies $\sqrt{s} = 2$ GeV and $\sqrt{s} = 20$ GeV. Calculate the ratio of the cross-section of process (1) at the two energies. Calculate (approximately) the ratios of the cross-sections of the two processes at each of the two energies. What is the ratio of the cross-section of process (2) at the two energies?

6.16 We observe the elastic scattering of $E = 5$ GeV electrons from protons $[m_p = 938 \text{ MeV}]$ at the angle $\theta = 8°$ and we measure their energy. (1) What is its expected value (neglecting the electron mass)? (2) What is the scattered electron energy in the CM frame?

6.17 A beam of α particles of kinetic energy $E = 10$ MeV and intensity $I = 1 \mu A$ hits a lead target $[A = 207, Z = 82, \rho = 1.14 \times 10^4 \text{kg m}^{-3}]$ of thickness $t = 0.2$ mm. We locate a detector of area $S = 1$ cm^2 at a distance of $l = 0.5$ m beyond the target at the angle $\theta = 40°$. Neglecting, when necessary, the variation of the angle on the detector, find: (a) the number of incident particles per second R_i, (b) the solid angle $\Delta\Omega$ under which the target sees the detector, (c) the differential cross-section at the detector and (d) how many hits the detector counts per second.

6.18   Which of the following reactions is allowed or forbidden by strong interactions? Specify the reason in each case. You may look at the tables for establishing the quark compositions: (a) $\pi^- + p \to \Lambda_b^0 + K^0$, (b) $\pi^- + p \to \Lambda_b^0 + D^0$, (c) $\pi^- + p \to \Lambda_b^0 + B^0$, (d) $\pi^- + p \to \Sigma_b^- + B^+$ and (e) $\pi^- + p \to \Sigma_b^+ + B^-$.

6.19   A beam of electrons of energy $E_e = 1$ GeV hits a liquid hydrogen target. A calorimeter measures the energy of the scattered electrons at the angle $\theta = 20°$. Calculate the energy in the case of elastic scattering. Similarly with a liquid He target. Similarly with an iron target $(A = 56)$. (Use approximations $p = E$, $p' = E'$ and $m_e^2 = 0$.)

6.20   Consider an electron–proton collision at HERA. The energy of the electron beam is $E_e = 28$ GeV and that of the proton beam is $E_p = 820$ GeV. One electron is observed to scatter at the angle $\theta = 120°$ and its energy is measured to be $E_e' = 223$ GeV. Calculate the centre of mass energy and the kinematical variables Q^2, x, ν and W.

6.21   A monochromatic photon beam of unknown energy but of known direction scatters on a liquid hydrogen target. The energy of the photons scattered at 20° is measured, finding $E' = 12$ GeV. What is the energy of the beam?

6.22   Consider the elastic scattering of an electron of energy $E = 15$ GeV on a proton. Calling E' the energy of the scattered electron: (a) find the maximum four-momentum transfer Q^2 and the corresponding recoil kinetic energy of the proton, and (b) the same questions for $E = 20$ MeV and a ^{56}Fe nucleus as a target. How can the expression be simplified?

6.23   Draw the Feynman diagrams for the gluon exchanges for the following couples of vertices: (a) $^R q \to {}^B q$, $^B q \to {}^R q$; (b) $^R q \to {}^B q$, $^R q \to {}^{\bar{B}} q$; (c) $^R q \to {}^R q$, $^R q \to {}^{\bar{R}} q$. Specify which gluon(s) is (are) exchanged and the colour charges at both vertices. (d) Explain the meanings of the signs of the results.

6.24   Draw the Feynman diagrams for the gluon exchanges for the following couples of vertices: (a) $^B q \to {}^G q$, $^G q \to {}^B q$; (b) $^G q \to {}^G q$, $^R q \to {}^{\bar{R}} q$; (c) $^G q \to {}^G q$, $^{\bar{G}} q \to {}^{\bar{G}} q$. Specify which gluon(s) is (are) exchanged and the colour charges at both vertices. (d) Explain the meanings of the signs of the results.

6.25   Draw the (lowest-order) Feynman diagrams for (a) $\phi \to K^+ K^-$, (b) $\phi \to K^0 \bar{K}^0$, (c) $\phi \to \pi^+ \pi^- \pi^0$.

6.26   Consider a two-gluon system in a colour singlet state, in which the colour wave function is symmetrical. Let S be the total spin, L the total orbital momentum, J the total angular momentum, P the parity and C the charge conjugation. Write down the possible values of all these quantities up to $L = 1$.

6.27   The SPEAR $e^+ e^-$ collider at Stanford observed two narrow ψ resonances. Consider the second one the $\psi(3686)$, where the number in parentheses is the mass in MeV. Bound P wave $c\bar{c}$ states called χ_c were also discovered. Consider in particular the reaction $e^+ e^- \to \psi(2S) \to \chi_c + \gamma$, followed by $\chi_c \to \pi^+ \pi^-$ in which the photon and the two pions are detected. The energy of the gamma is $E_\gamma = 0.26$ GeV. Determine the energy of the χ_c and the possible values of its spin, parity, charge conjugation and isospin. Assume the radioactive decay to be E1 (i.e. the photon takes away overall one unit of angular momentum and negative parity).

6.28   Consider the P wave $c\bar{c}$ states, which are called χ_c. Establish their number and the possible values of J^{PC}. Which of them can decay into two gluons (and then into hadrons)?

Summary

In this chapter we have studied the strong interaction, QCD and the colour charges. In particular we have seen the following:

- the evidence from experiments at e^+e^- collisions in hadronic final states for colour, quark jets and gluon jets
- how deep inelastic experiments show the internal structure of the nucleon, the distribution functions of its constituents (valence and sea quarks and gluons), the scaling laws and their violations
- the QCD Lagrangian and how colour forces act in binding quarks and in producing the hadrons and the colours of the gluons
- why quarks cannot exist free and that hadrons are always 'white'
- the evolution of the strong coupling constant
- the meaning of quark masses and their evolution
- how the mass of the proton is mainly the energy of the colour field
- the properties of the QCD vacuum
- the chiral symmetry of QCD and its spontaneous breaking
- how, with LQCD, observables in the non-perturbative regime can be calculated
- a few basic features of the deconfinement transition to the QGP.

Further Reading

Friedman, J. I. (1990) Nobel Lecture; *Deep Inelastic Scattering: Comparison with the Quark Model*

Roberts, B. L. (2019) The history of muon (g−2) experiments. *SciPost Phys.* 1 (032)

Taylor, R. E. (1990) Nobel Lecture; *Deep Inelastic Scattering: The Early Years*

Wilczek, F. A. (2004) Nobel Lecture; *Asymptotic Freedom: from Paradox to Paradigm*

7

Weak Interactions

In this chapter we begin the study of weak interactions, which will continue in Chapters 8 and 9, in which we shall see how electromagnetic and weak interactions are unified. Two types of fermionic currents exist in weak interactions: charged currents (CCs), mediated by the charged vector bosons W^+ and W^-, and neutral currents (NCs) mediated by Z^0. In the scattering or decay amplitude, the vector bosons propagate, as in QED and QCD, between vertices. However, unlike photons and gluons, the weak vector bosons are massive, with masses of the order of 100 GeV. Consequently, at low energies the two vertices appear merged as one and the interaction is point-like. Such was the first proposal of a previously unknown fundamental interaction advanced by Fermi in 1933. From here we shall start, and present the general structure of the Lagrangian as the product of two currents.

We shall then see that parity and particle–antiparticle conjugation are not conserved; rather, they are maximally violated in CC weak interactions, as shown by the beautiful experiment of C. S. Wu and collaborators. The consequence is that the spinor fields in the CC weak interactions are eigenstates of the Dirac matrix γ_5 with negative eigenvalue, namely they are the negative chirality projections. We shall subsequently see how the helicity structure of the solutions of the Dirac equation (Section 2.8), joined with angular momentum conservation, explains why the charged pions decay preferentially in an energy disfavoured channel, provided the current is any mixture of vector and axial vector covariants. Then we shall present the measurement of the neutrino helicity, another masterpiece of experimental physics, by M. Goldhaber and collaborators.

We shall then see that the W boson coupling, equivalent to the Fermi coupling, to all leptons is universal, independent of the family. The same is not true for quarks. Indeed, the quark states in the CC weak interactions are not the states of definite mass, but linear superpositions, we say *mixing*, of them. The concept of mixing of the hadronic currents was introduced by N. Cabibbo to restore universality of the Fermi coupling. A consequence was, however, that decays of strange hadrons in non-strange ones of the same electric charge should be much faster than observed. The GIM mechanism solved the problem, introducing the hypothesis that a fourth quark would exist, the charm, completing a doublet with the strange one. Charm was soon discovered, as we saw in Chapter 4. We also saw that two more quarks exist. Consequently, the quark *mixing matrix* is a 3 × 3 unitary matrix. In it, all the phase factors can be eliminated, except one. This is the origin of CP violation in the Standard Model. In this chapter we shall deal with the absolute values of the mixing matrix elements and two examples of their measurement.

Finally, we shall see how the weak neutral currents were discovered with the Gargamelle bubble chamber at CERN in 1973. This showed a close similarity between weak and electromagnetic interactions and opened the way to their unification.

7.1 The Fermi Theory

Historically, the first process governed by the weak interaction has been the 'beta radioactivity', classified as different from the alpha-radioactivity, by E. Rutherford in 1899. The process, now called β^- for reasons that we shall soon see, is the decay of a nuclide (A, Z) into $(A, Z+1)$ with the emission of an electron. Today we know that, at nucleon level, this is the decay

$$n \to p + e^- + \bar{v}_e, \tag{7.1}$$

but neutrinos were not known at the time. Again, in hindsight, the underlying process at the quark level is

$$d \to u + e^- + \bar{v}_e. \tag{7.2}$$

As we already recalled in Chapter 2, in 1930 W. Pauli had assumed the existence of an 'invisible' neutral particle as a 'desperate hypothesis' to explain why the beta decay spectra are continuous, rather that mono-energetic, as they had to be as two-body decays. In 1933 the positron was discovered by Anderson and just a few weeks later, F. and I. Joliot-Curie discovered the positron emission, namely the β^+ of a nuclide $\left({}^{30}_{15}P, \text{ to be precise} \right)$. At nucleon and quark level, the process is

$$p \to n + e^+ + v_e; \qquad u \to d + e^+ + v_e. \tag{7.3}$$

The decay cannot happen for free protons, because its mass is smaller than that of the neutron, but can proceed inside a nucleus, borrowing energy from the environment. Similarly for the quarks.

In December 1933, E. Fermi (1933) formulated a theory to explain the nuclear beta decay. It was the proposal of a new fundamental interaction, the weak one. Fermi trusted the Pauli hypothesis and dubbed the 'invisible' particle (only one of them at the time) the 'neutrino', meaning 'the small neutral one' in Italian. He assumed that, in a beta decay, a couple of particles are emitted: an electron and a neutrino. He considered the process to be analogous to the emission of a photon in the decay of an atom. As with the photon, the electron–neutrino pair is created at the moment of the decay, rather than being present in the original system, as assumed at the time. Similarly, to that of the photon, quantized fermion fields should exist, one for the electron, $\psi_e(x)$, and one for the neutrino, $\psi_v(x)$. With these, he built a 'current', choosing amongst the Dirac bilinear covariants (2.50) the simplest option, the vector one, which is analogous to the electromagnetic current. Doubts existed at the time as to whether nucleons did obey the Dirac equation,

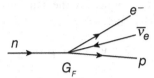

Fig. 7.1.   The neutron beta decay in the Fermi theory.

but soon after it became clear that the nucleons had to be treated similarly to the leptons, leading to the matrix element

$$\mathcal{M} = \frac{G_F}{\sqrt{2}} \left[\bar{\psi}_p(x) \gamma_\mu \psi_n(x) \; \bar{\psi}_e(x) \gamma^\mu \psi_\nu(x) \right]. \tag{7.4}$$

The constant G_F is called the Fermi constant and will be discussed soon. The interaction in Eq. (7.4) is the product of two vector currents, one hadronic and one leptonic. Both are 'charged' because they change the electric charge of the two fermions. We now speak of 'charged current (CC) weak interaction' to distinguish it from the 'neutral current (NC) weak interaction' that, as we shall see, was discovered later. In the Fermi theory, the four particles interact in a single point, like in Fig. 7.1 for the neutron beta decay. We shall see how this feature will be modified.

In the following two decades, many weak processes were found to be described by the Fermi current–current interaction, namely by a matrix element of the type

$$\mathcal{M} = \frac{G_F}{\sqrt{2}} J_{a\mu} J_b^\mu, \tag{7.5}$$

where J_a and J_b are two currents, different or equal, as in Eq. (7.4). We shall see in this chapter how discoveries in the second half of the twentieth century have shown that the structure of the weak currents is more complex, and rich, than what Fermi had assumed, and also than that of the currents of QED and QCD.

The CC processes are classified as follows.

(1) **Leptonic processes.** Only leptons are present, in both the initial and the final states. Examples are the μ^- decay

$$\mu^- \to e^- + \nu_\mu + \bar{\nu}_e, \tag{7.6}$$

and similarly for μ^+ and τ. Another example is the muon neutrino 'quasi-elastic' scattering of electron neutrino on electron

$$\nu_\mu + e^- \to \nu_e + \mu^-, \tag{7.7}$$

(2) **Semileptonic processes.** Both hadrons and leptons are present, as in the above-mentioned beta decay and in the charged pion decay

$$\pi^- \to \mu^- + \bar{\nu}_\mu, \qquad \pi^+ \to \mu^+ + \nu_\mu \tag{7.8}$$

Other important examples are the neutrino scatterings from nucleons or, correspondingly, from quarks, such as

$$\nu_\mu + n \to \mu^- + p, \qquad \nu_\mu + d \to \mu^- + u. \tag{7.9}$$

(3) **Non-leptonic processes.** Only hadrons are present both in the initial and in the final states. Still the process is weak. This class contains decays, for example

$$\Lambda^0 \to p + \pi^- \qquad s \to u + \bar{u} + d. \tag{7.10}$$

As we have already discussed, the weak nature of the process is easily recognized from the long decay times and also from the flavour violation.

We come back now to the Fermi constant, namely the coupling constant of the CC weak interactions at low energies. First we notice that, unlike α and α_s, G_F has physical dimensions. It is defined in such a way that the quantity $G_F / (\hbar c)^3$ has the dimensions of [energy^{-2}]. In NU, the dimensions are

$$[G_F] = \left[E^{-2}\right] = \left[L^2\right]. \tag{7.11}$$

As with all the fundamental constants, G_F must be measured with as much accuracy as possible. This is done using a purely leptonic process, the muon decay, Eq. (7.6). The calculation of the μ lifetime gives

$$\frac{\hbar}{\tau_\mu} = \Gamma\left(\mu^+ \to e^+ \bar{\nu}_\mu \nu_e\right) = \frac{1}{192\pi^3} \frac{G_F^2}{(\hbar c)^6} \left(m_\mu c^2\right)^5 (1 + \varepsilon), \tag{7.12}$$

or, in NU,

$$\Gamma\left(\mu^+ \to e^+ \bar{\nu}_\mu \nu_e\right) = \frac{1}{192\pi^3} G_F^2 m_\mu^5 (1 + \varepsilon), \tag{7.13}$$

where the small correction ε, which is zero if we neglect the electron mass, can be calculated exactly. Notice the dependence on the fifth power of the mass of the decaying particle (i.e. on the CM energy). The determination of the Fermi constant requires an accurate measurement of the muon lifetime and an extremely precise measurement of its mass. The present value is (Workman *et al.* 2022)

$$\frac{G_F}{(\hbar c)^3} = 1.1663788 \pm 0.0000006 \times 10^{-5} \; [510 \; \text{ppb}] \tag{7.14}$$

Consider now another purely leptonic process, the quasi-elastic scattering (7.7). The calculation of its cross-section gives

$$\sigma\left(\nu_\mu e^- \to \nu_e \mu^-\right) = \frac{G_F^2}{\pi} s = \frac{G_F^2}{\pi} 2 m_e E_\nu = 1.7 \times 10^{-45} E_\nu \, (\text{GeV}) \, \text{m}^2. \tag{7.15}$$

The cross-section grows linearly with the neutrino energy in the L frame, a behaviour that can be understood with the following dimensional argument. The physical dimensions of the cross-section are $[1/E]^2$. On the other hand, the cross-section is proportional to the Fermi constant squared, G_F^2, which has the dimension of $\left[1/E^2\right]^2$. Consequently G_F^2 must be multiplied by an energy squared. The only such quantity available is the CM energy square $s = \left(E_\nu + m_e\right)^2 + p_\nu^2 = m_e^2 + m_\nu^2 + 2m_e E_\nu$, which is $2m_e E_\nu$ in a very good approximation. However, no cross-section can increase indefinitely with energy, because the scattering probability cannot be larger than 100%. Actually, the weak interaction is not really point-like. It appears to be so at low energies, but is mediated by the W bosons, the propagator of which dumps the divergence at CM energies comparable with its mass, as we shall see in Section 7.6.

7.2 Parity Violation

We saw in Section 4.5 that the G-stack cosmic ray exposure in 1953 and the first experiments at accelerators showed the existence of two apparently identical particles, which were different only in their decay mode, namely the θ^+ decaying into $\pi^+\pi^0$, and the τ^+ (not to be confused with the much later discovered lepton) decaying into $\pi^+\pi^+\pi^-$. The spin parity of the former final state, a two-π system, belongs to the sequence $J^P = 0^+$, 1^-, 2^+,..., while, as we saw in Section 4.5, the Dalitz analysis of the three-pion final state of the τ decay gave $J^P = 0^-$. The problem became known as the $\theta\tau$ puzzle. The puzzle could be solved if parity were not conserved in the decay. This hypothesis was fairly acceptable to the experimentalists but sounded almost like blasphemy to theoreticians. Indeed, parity is a symmetry of space-time itself, just as the rotations are; 'it had to be absolutely conserved'. At the general conference on particle physics, the 'Rochester Conference', of 1956, R. Feynman asked C. N. Yang, after his speech, a question that he had been asked by M. Bloch, the co-discoverer of the η meson: 'is it possible to think that parity is not conserved?'. Yang answered that T. D. Lee and he had had a look at the issue, but did not reach any conclusions and would go back to it. A few months later, Lee and Yang showed that no experimental proof existed of parity conservation in weak interactions (Lee & Yang 1956).

Following their reasoning, let us consider the beta decay of a nucleus $N \to N' + e + \nu$ in the CM frame. The kinematic quantities are the three momenta $\mathbf{p}_N, \mathbf{p}_e, \mathbf{p}_\nu$. With them we can build:

- three scalar products such as $\mathbf{p}_N \cdot \mathbf{p}_e$; being scalar, the products do not violate P
- the mixed product $\mathbf{p}_N \cdot \mathbf{p}_e \times \mathbf{p}_\nu$; it is a pseudoscalar and, added to a scalar, would violate $\mathcal{P}$, but it is zero because the three vectors are coplanar.

Lee and Yang concluded that parity conservation could be tested only by using an axial vector. Such an axial vector is provided by polarization. One must polarize a sample of nuclei, inducing a non-zero expectation value of the intrinsic angular momentum $\langle \mathbf{J} \rangle$, and measure an observable proportional to the pseudoscalar $\langle \mathbf{J} \rangle \cdot \mathbf{p}$.

The experiment soon followed, done by C. S. Wu and collaborators at the National Bureau of Standards (Wu *et al.* 1957) with ^{60}Co. Polarization can be achieved by

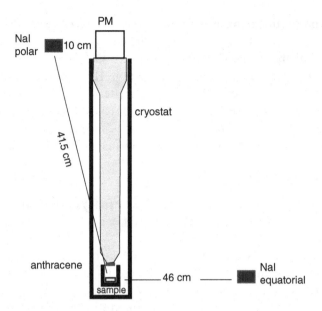

PM

NaI polar   10 cm

cryostat

41.5 cm

anthracene

46 cm

sample

NaI equatorial

Sketch of the experiment (simplified from Wu *et al.* 1957).

orienting the nuclear magnetic moments μ, which are parallel to the spins, in a magnetic field **B**, having energy $-\boldsymbol{\mu}\cdot\mathbf{B}$. The probability of a certain direction of the magnetic moment relative to the field is given by the Boltzmann factor

$$\exp\left(+\frac{\boldsymbol{\mu}\cdot\mathbf{B}}{kT}\right). \tag{7.16}$$

Now the magnetic moment of a particle is inversely proportional to its mass. Since the nuclear mass is large, compared with the electron, nuclei are difficult to polarize. We see from (7.16) that a very low temperature, in practice a few millikelvin, and a strong magnetic field are needed. The latter was obtained by embedding the cobalt in a paramagnetic crystal. If the crystal is in a magnetic field, even in a weak one, the electronic magnetic moments, which are large, become oriented in the field and generate inside the crystal local fields of dozens of tesla.

Fig. 7.2 shows a sketch of the experiment. The spin parity of the ^{60}Co nucleus is $J^P = 5^+$. The polarized nuclei beta decay into an excited state of ^{60}Ni with $J^P = 4^+$. The daughter nucleus keeps the polarization of the parent nucleus

$$^{60}\text{Co}\left(J^P = 5^+\right)\Uparrow \rightarrow\,^{60}\text{Ni**}\left(J^P = 4^+\right)\Uparrow +e^- +\bar{\nu}_e. \tag{7.17}$$

Two gamma decays of Ni follow in cascade to the fundamental level, maintaining polarization

$$^{60}\text{Ni**}\left(J^P = 4^+\right)\Uparrow\rightarrow\,^{60}\text{Ni*}(J^P = 2^+)\Uparrow +\gamma(1.173\,\text{MeV}) \tag{7.18}$$

$$^{60}\text{Ni*}\left(J^P = 2^+\right)\Uparrow\rightarrow\,^{60}\text{Ni}\left(J^P = 0^+\right)\Uparrow +\gamma\left(1.332\,\text{MeV}\right). \tag{7.19}$$

The two electromagnetic decays are not isotropic, namely the gamma emission probability is a function of the angle θ with the field. Therefore, we can monitor the polarization of the sample by measuring this anisotropy.

The polarizing magnetic field is oriented along the vertical axis of Fig. 7.2; its direction can be chosen to be upward or downward. Two counters are used to detect the photons, made from NaI crystals, which scintillate when absorbing a gamma. One counter (equatorial) is at 90° to the field, the other (polar) at about 0°. The electrons must be detected inside the cryostat. To this aim, a scintillating anthracene crystal is located at the tip of a plastic bar that guides the scintillation light to a photomultiplier (PM). In this way, the experiment counts the electrons emitted in the direction of the polarization or opposite to it, depending on the orientation of the polarizing field.

The operations start by switching the magnetic field on to polarize the nuclei. Once the polarization is obtained, in a few seconds, the field is switched off (at time zero) and the counting of the photons and of the electrons starts. The polarization slowly decays to disappear in a few minutes. The photon flux depends on the direction to the polarization axis, but does not change when it is reversed, because parity is conserved by electromagnetic interactions. As anticipated, the degree of polarization is measured by the gamma anisotropy, which, if W_γ are the counting rates, is defined as

$$\varepsilon_\gamma \equiv \frac{W_\gamma\left(90°\right) - W_\gamma\left(0°\right)}{W_\gamma\left(90°\right) + W_\gamma\left(0°\right)}. \tag{7.20}$$

Fig. 7.3(a) shows the measurements of $W_\gamma(0°)$ and $W_\gamma(90°)$, divided by their values at zero field, as functions of time. Both show the decay of the polarization, giving us the shape of the decay curve.

If the beta decay violates parity, the angular distribution of the emitted electrons is not symmetric under $\theta \Leftrightarrow \pi - \theta$. Therefore, the counting rate is expected to depend on the angle as

$$W_e\left(\theta\right) = \alpha\, P\, \beta_e\, \cos\left(\theta\right), \tag{7.21}$$

where the constant α is zero if parity is conserved and +1 or −1 if it is maximally violated, corresponding respectively to a $V + A$ or $V–A$ structure of the current in Eq. (7.4), where V and A are the vector and axial bilinear covariants, respectively. The initial polarization P of the Wu experiment was about 0.6. The factor β_e, which is the speed of the electrons divided by the speed of light, is explained by Eq. (2.66).

Fig. 7.3(b) shows the electron counting rates with the field direction upward and downward, divided by the counting rate without polarization. In the former configuration, the detector counts the electrons emitted at about 0°; in the latter, it counts those emitted at about 180°. Both ratios decay following the curve of the polarization. The fundamental observation is that the two rates are different: the electrons are emitted in directions (almost) opposite to the field much more frequently than (almost) along it. This was the experimental proof of parity violation. Moreover, the measurement of α gave

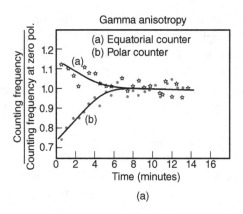 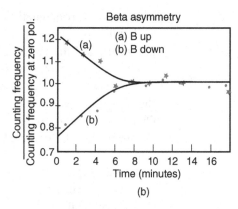

(a) (b)

Fig. 7.3. (a) The measurement of $W_\gamma(0°)$ and $W_\gamma(90°)$, divided by their values at zero field, as functions of time. (b) The electron counting rates with the field direction upward and downward, divided by the counting rate without polarization (Figure reprinted, with adaptation, from Wu, C. S. *et al.* Experimental test of parity conservation in beta decay. *Phys. Rev.* (1957) 105 1413, 1957 Fig. 2. © (1957) by American Physical Society), https://doi.org/10.1103/PhysRev.105.1413.

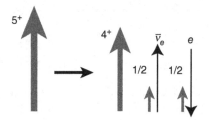

Fig. 7.4. Schematic of the spin 'directions' in the decay.

$$\alpha \approx -1, \tag{7.22}$$

which shows that the parity violation is, within the errors, maximal. If we assume the interaction to be $V + xA$, the result is compatible with $x = -1$. Taking the uncertainties of the measurement into account, the experiment gave

$$-1 < x < -0.7. \tag{7.23}$$

The Wu experiment showed that the space-time structure of the CC weak interaction is $V–A$. We try to illustrate the point in Fig. 7.4.

The thick arrows indicate the 'directions' of the spins; their lengths are such as to satisfy the conservation of the third component of the angular momentum. The thin arrows are the preferential directions of the motions. The result of the experiment is that the preferential motion of the electron is opposite to the field, and consequently of its spin. As the nuclei decay at rest, the preferential direction of the antineutrino is opposite to that of the electron. Therefore, the antineutrino spin is in the preferential direction of its velocity.

7.3 The Charged Current Lagrangian

In the same year of the discovery by Wu of parity violation, the V–A universality of the CC weak interaction was established by Sudarshan under the guidance of his mentor Marshak (Sudarshan & Marshak 1957), analysing all the experimental data. Feynman & Gell-Mann (1958) advanced the same hypothesis. The interaction Lagrangian has the general current–current structure

$$\frac{G_F}{\sqrt{2}} \, \bar{\psi}\gamma_\mu \left(1-\gamma_5\right)\psi \, \bar{\psi}\gamma^\mu \left(1-\gamma_5\right)\psi \tag{7.24}$$

where, to be general, we did not specify the particle, which can be any spin-1/2 fermion to which the four ψ operators belong.

The V–A structure implies the following. From the properties of the γ matrices, we have

$$\gamma_\mu \left(1-\gamma_5\right) = \frac{1}{2}\left(1+\gamma_5\right)\gamma_\mu \left(1-\gamma_5\right) \tag{7.25}$$

We see that only the left four-component spinor operators appear in the expressions (7.24).

In general, let us call ψ_f and ψ_i the fields of the final and the initial particles, respectively. Their left projections are

$$\bar{\psi}_{f,L} = \bar{\psi}_f \frac{1}{2}\left(1+\gamma_5\right); \qquad \psi_{i,L} = \frac{1}{2}\left(1-\gamma_5\right)\psi_i. \tag{7.26}$$

We can then write each of the currents in Eq. (7.24) as

$$\bar{\psi}_f\gamma_\mu \left(1-\gamma_5\right)\psi_i = 2\bar{\psi}_f \frac{1}{2}\left(1+\gamma_5\right)\gamma_\mu \frac{1}{2}\left(1-\gamma_5\right)\psi_i = 2\bar{\psi}_{f,L}\gamma_\mu \, \psi_{i,L} \tag{7.27}$$

In the latter form, the CC weak interaction is very similar to the electromagnetic one, but with the fundamental difference that the fields are now (only) the left chirality projections. Let us see the consequences.

7.4 Charged Pion Decay

Here we consider the charged pion decay as an example of a V–A phenomenon. Charged pions decay predominately (>99%) in the (semileptonic) channel:

$$\pi^+ \to \mu^+ + \nu_\mu \qquad \pi^- \to \mu^- + \bar{\nu}_\mu. \tag{7.28}$$

The second most probable channel is similar, with an electron in place of the muon:

$$\pi^+ \to e^+ + \nu_e \qquad \pi^- \to e^- + \bar{\nu}_e. \tag{7.29}$$

Since the muon mass is only a little smaller than that of the pion, the first channel is energetically disfavoured relative to the second; however, its decay width is the larger one

$$\frac{\Gamma(\pi \to e\nu)}{\Gamma(\pi \to \mu\nu)} = 1.2 \times 10^{-4}. \tag{7.30}$$

We saw in Section 1.7 (Eq. (1.74)) that the phase-space volume for a two-body system is proportional to the CM momentum. The ratio of the phase-space volumes for the two decays is then p_e^* / p_μ^*. Calling the charged lepton generically l, energy conservation is written as $\sqrt{p_l^{*2} + m_l^2} + p_l^* = m_\pi$, which gives $p_l^* = (m_\pi^2 - m_l^2)/2m_\pi$. The ratio of the momenta is then

$$\frac{p_e^*}{p_\mu^*} = \frac{m_\pi^2 - m_e^2}{m_\pi^2 - m_\mu^2} = \frac{140^2 - 0.5^2}{140^2 - 106^2} = 2.3. \tag{7.31}$$

As anticipated, the phase space favours decay into an electron. Given the experimental value (7.30), the ratio of the two matrix elements must be very small. This observation gives us very important information on the space-time structure of weak interactions.

We do not have the theoretical instruments for a rigorous discussion, but we can find the most general matrix element, $\mathcal{M}$, by using simple Lorentz invariance arguments. We do not assume for the moment the $V–A$ structure but leave the possibility of parity violation open. The matrix element may then be a scalar, a pseudoscalar or the sum of the two. For the lepton current we can use any of the bilinear covariants of Eq. (2.50) with the fields ψ_l of the charged lepton and $\psi_{\nu l}$ of its neutrino. For the hadronic current we have the field of the pion in the state, ϕ_π, which is a pseudoscalar and its four-momentum, p^μ. Finally, a scalar constant can be present, called the pion decay constant, which we denote by f_π.

We must now construct with the above-listed elements the possible matrix elements, namely scalar or pseudoscalar quantities. Using the pseudoscalar and the axial vector bilinears, we can build the two scalar quantities (the ellipses stand for uninteresting factors):

$$\mathcal{M} = \cdots f_\pi \phi_\pi \bar{\psi}_l \gamma_5 \psi_{\nu_l}; \qquad \mathcal{M} = \cdots f_\pi \phi_\pi \bar{\psi}_l p^\mu \gamma_\mu \gamma_5 \psi_{\nu_l}. \tag{7.32}$$

Using the scalar and vector bilinears, we can build two pseudoscalar terms:

$$\mathcal{M} = \cdots f_\pi \phi_\pi \bar{\psi}_l \psi_{\nu_l}; \qquad \mathcal{M} = \cdots f_\pi \phi_\pi \bar{\psi}_l p^\mu \gamma_\mu \psi_{\nu_l} \tag{7.33}$$

We have used four of the covariant quantities; there is no possibility of using the fifth one, the tensor.

Let us start with the vector current term. The pion four-momentum is equal to the sum of those of the charged lepton and of the neutrino $p^\mu = p_{\nu_l}^\mu + p_l^\mu$ hence

$$\begin{aligned}
\mathcal{M} &= \cdots f_\pi \phi_\pi \bar{\psi}_l \left(p_{\nu_l}^\mu + p_l^\mu \right) \gamma_\mu \psi_{\nu_l} \\
&= \cdots f_\pi \phi_\pi \bar{\psi}_l p_{\nu_l}^\mu \gamma_\mu \psi_{\nu_l} + \cdots f_\pi \phi_\pi \bar{\psi}_l p_l^\mu \gamma_\mu \psi_{\nu_l}
\end{aligned} \tag{7.34}$$

The fields obey the Dirac equation. As the neutrino mass is negligible, we have

$$\left(\gamma_\mu p^\mu_{\nu_l} - m_{\nu_l}\right)\psi_{\nu_l} = 0 \;\Rightarrow\; \gamma_\mu p^\mu_{\nu_l}\,\psi_{\nu_l} = 0$$
$$\bar{\psi}_l\left(\gamma_\mu p^\mu_l + m_l\right) = 0 \;\Rightarrow\; \bar{\psi}_l \gamma_\mu p^\mu_l = -\bar{\psi}_l\,m_l$$

$$(7.35)$$

In conclusion, we obtain

$$\mathcal{M} = \cdots m_l f_\pi \phi_\pi \bar{\psi}_l \psi_{\nu_l} \tag{7.36}$$

We see from the Dirac equation that the matrix element is proportional to the mass of the final lepton. Therefore, the ratio of the decay probabilities in the two channels is proportional to the ratio of their masses squared

$$m_e^2 / m_\mu^2 = 0.22 \times 10^{-4}. \tag{7.37}$$

This factor has the correct order of magnitude to explain the smallness of $\Gamma(\pi \to e\nu)/\Gamma(\pi \to \mu\nu)$. We shall complete the discussion at the end of this section.

Let us now examine the axial vector current term, namely

$$\mathcal{M} = \cdots f_\pi \phi_\pi \bar{\psi}_l p^\mu \gamma_\mu \gamma_5 \psi_{\nu_l} \tag{7.38}$$

Repeating the arguments of the vector case, we obtain

$$\mathcal{M} = \cdots m_l f_\pi \phi_\pi \bar{\psi}_l \gamma_5 \psi_{\nu_l} \tag{7.39}$$

and we again obtain the result (7.37).

Considering now the scalar and pseudoscalar covariants, we see immediately that they do not contain the factor m_l^2. Therefore, they cannot explain the smallness of (7.30).

In conclusion, the observed small value of the ratio between the probabilities of a charged pion decaying into an electron or into a muon proves that, at least in this case, the weak interaction currents are of type V, or of type A or of a mixture of the two.

The proportionality of $\mathcal{M}$ to m_l is a consequence of the helicity structure of the solutions of the Dirac equation and of angular momentum conservation; for example, for the π^-, the helicity of the practically massless antineutrino is positive. Because the total angular momentum is zero, the helicity of the electron, moving in opposite directions, must also be positive (see Fig. 7.5). But, as we saw in Section 2.8, this is the 'small' component, with amplitude m_l / E of a high-energy particle.

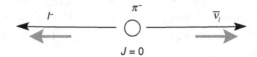

Fig. 7.5.   Angular momenta in the pion decay.

Assuming now that the structure of the CCs is *V–A*, the matrix element of the decays of the pion in lepton *l* is

$$\mathcal{M} = \cdots m_l f_\pi \phi_\pi \bar{\psi}_l \left(1 - \gamma_5\right) p^\mu \gamma_\mu \psi_{\nu_l} \tag{7.40}$$

To obtain the decay probabilities, we must integrate the absolute square of this quantity, for the electron and the muon, over phase space. We cannot do the calculation here, so we give the result directly as

$$\frac{\Gamma\left(\pi \to e\nu\right)}{\Gamma\left(\pi \to \mu\nu\right)} = \frac{p_e^* p_e^* m_e^2}{p_\mu^* p_\mu^* m_\mu^2} = \frac{m_e^2}{m_\mu^2} \left(\frac{m_\pi^2 - m_e^2}{m_\pi^2 - m_\mu^2}\right)^2 \tag{7.41}$$
$$= 0.22 \times 10^{-4} \times 2.3^2 \simeq 1.2 \times 10^{-4}$$

We conclude that the *V–A* structure is in agreement with the experiment.

Question 7.1 Knowing the experimental ratio for the K^+ meson

$$\Gamma\left(K \to e\nu\right) / \Gamma\left(K \to \mu\nu\right) = 1.6 \times 10^{-5} / 0.63 = 2.5 \times 10^{-5},$$

prove that the *V–A* hypothesis gives the correct prediction. □

7.5 Measurement of the Helicity of Leptons

We shall now describe the experiment of M. Goldhaber, L. Grodzins and A. Sunyar (Goldhaber *et al.* 1958) on the measurement of the helicity of the neutrino. The experiment was carried out at the Brookhaven National Laboratory in 1958. Let us see its logical steps.

The first element is the γ resonant emission and absorption by nuclei.

Consider a medium and let *N* be its nuclei. A nucleus can be excited to the *N* * level, and subsequently decays to the fundamental level by emitting a photon

$$N^* \to \gamma + N. \tag{7.42}$$

The 'resonance' process of interest is this emission followed by the absorption of the photon by another nucleus, which becomes excited in the *N* * level

$$\gamma + N \to N^*. \tag{7.43}$$

To be in resonance, the photon must have the right energy to give the transition energy *E* to the nucleus *N*. This is *E* augmented by the recoil energy of the final state. However, the energy of the γ from reaction (7.43) is *E* diminished by the recoil kinetic energy of the emitting nucleus. Therefore, the resonance process cannot take place if the excited nucleus *N* * is at rest. This condition is necessary for the experiment to succeed.

The resonant conditions can be satisfied if the initial *N* * moves relative to the medium when it decays. The energy of the γ in the reference frame of the medium depends on its direction relative to that of $N*$. As the γs have larger energies in the forward directions (Doppler effect), they can induce the resonance only in these directions. This is the second necessary condition for the experiment. So as not to interrupt the main discussion, we shall demonstrate at the end how the two conditions are satisfied. This is not yet enough: there is a third condition. Indeed, the recoiling nucleus should not change its direction (i.e. should not scatter hitting another nucleus) before decaying.

The second element of the experiment is the transfer of the neutrino helicity (h_ν) to a gamma and the measurement of the helicity of the latter (h_γ). To do this we first need a nuclide, which we call A, producing by K-capture the excited state $N*$ and the neutrino, the helicity of which we shall determine. Remember that K-capture is the capture by the nucleus of an atomic electron in S wave. The process is

$$A + e^- \rightarrow N* + \nu_e. \tag{7.44}$$

Obviously the condition that the energy of the $N*$ so produced is in resonance must be satisfied. This is not yet enough, because, as we shall immediately see, the angular momentum of A must be $J = 0$ and that of $N*$, $J = 1$. At this point, one might conclude that there are so many conditions that it is hopeless to seek two nuclides satisfying all of them. However, Goldhaber, Grodzins and Sunyar found that ^{152}Eu and ^{152}Sm have all the requested characteristics. Fortune favours the bold!

Let us see how helicity is transferred from the neutrino to the gamma. There are three steps.

(1) The ^{152}Eu decays by K-capture of an S wave electron

$$^{152}\text{Eu}\left(J = 0\right) + e^- \rightarrow {}^{152}\text{Sm}*\left(J = 1\right) + \nu_e \tag{7.45}$$

Let the neutrino direction be the quantization axis z. The Sm* direction is $-z$.

(2) Select the cases in which the Sm* decays emitting a gamma in the forward direction, namely $-z$, by use of the resonant emission–absorption process. The emission process is

$$^{152}\text{Sm}*\left(J = 1\right) \rightarrow {}^{152}\text{Sm}\left(J = 0\right) + \gamma, \tag{7.46}$$

which has a lifetime of 10 fs, short enough to satisfy the third condition.

The first three columns of Table 7.1 give all the combinations of the third components of the spins that satisfy the angular momentum conservation in the reaction (7.45). The fourth column gives the corresponding neutrino helicity. Taking into account that the projection of the gamma angular momentum on its velocity cannot be zero, only two cases remain.

We observe that both the gamma and the neutrino have the same helicity.

(3) Measurement of the gamma circular polarization, namely of its helicity. Fig. 7.6 shows a sketch of the experiment.

$s_z(e)$	$s_z(\text{Sm}^*)$	$s_z(\nu)$	h_ν	$s_z(\gamma)$	h_γ
+1/2	1	−1/2	−	1	−
+1/2	0	+1/2	+	0	x
−1/2	−1	+1/2	+	−1	+
−1/2	0	−1/2	−	0	x

Table 7.1.

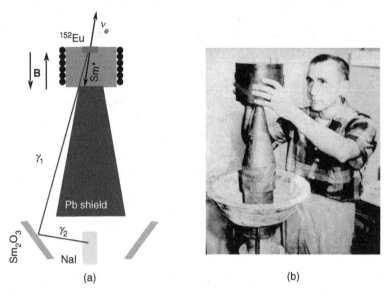

Fig. 7.6. (a) Sketch of the neutrino helicity experiment. (b) L. Grodzins holding the magnet of the neutrino-helicity apparatus (1958) (photo courtesy of L. Grodzins).

The europium source is located above an iron slab in a vertical magnetic field **B**, used to analyse the polarization state of the gamma. The flight direction of the samarium nuclei is approximately the downward vertical. Neglecting the small difference, we take the vertical as the z-axis. The direction of the magnetic field can be chosen as z or $-z$. Remember now that the spins of the electrons responsible for the ferromagnetism are oriented opposite to **B**. These electrons can easily absorb the photons, by flipping their spin, if the photon spin has the direction of **B**. On the contrary, they cannot do so for photons with spin in the direction opposite to **B**. Therefore, the iron slab absorbs substantially more of the former than of the latter.

The gamma detector is a NaI crystal. This cannot be reached directly by the gamma because it is shielded by a lead block. An adequately shaped samarium ring surrounds the detector.

If the resonance process takes place, a photon, call it γ_1, emitted by Sm* is absorbed by a Sm nucleus in the ring. The latter immediately de-excites, emitting a photon, γ_2, which reaches the detector (in a fraction of cases). The process can happen only if the Sm* was travelling in the right forward direction.

The measured quantity is the asymmetry R between the counting rate with the field oriented in one direction and the counting rate with the field oriented in the other direction, I_+ and I_- respectively, that is

$$R = \frac{I_+ - I_-}{I_+ + I_-}. \tag{7.47}$$

From this measurement the longitudinal polarization of the gamma is easily extracted. The final result is that the helicity of the neutrino is negative and compatible with -1. This proves the $V-A$ structure of the CC weak interaction.

We now prove that the resonance condition is not satisfied by Sm* decay at rest.

The energy difference between the two Sm levels is $E_{Sm} = 963$ keV. The recoil energy, $E_{K, Sm}$, is small; to find it we can use non-relativistic expressions. The recoil momentum p_{Sm} is equal and opposite to the photon momentum p_γ, which is also the photon energy. Since we are calculating a correction, we can approximate the photon energy with E_{Sm}. The Sm recoil kinetic energy is

$$E_{K, Sm} = \frac{p_{Sm}^2}{2M_{Sm}} = \frac{p_\gamma^2}{2M_{Sm}} \approx \frac{E_{Sm}^2}{2M_{Sm}} = \frac{0.963^2 \times 10^{12}\,\mathrm{eV}^2}{2 \times 1.52 \times 10^{11}\,\mathrm{eV}} = 3\,\mathrm{eV} \tag{7.48}$$

The recoil energy in the absorption process is substantially equal to this and, in conclusion, the photon energy is below the resonance energy by twice $E_{K, Sm}$, namely

$$\delta E = E_{Sm}^2 / M_{Sm} \Rightarrow \delta E = 6\,\mathrm{eV}. \tag{7.49}$$

Is this difference small or large? To answer this question we must compare δE with the resonance width. The natural width is very small, as for all nuclear electromagnetic transitions, about 20 meV, much less than δE. However, we must consider the Doppler broadening of the resonance due to thermal motion. At room temperature, $kT = 26$ meV, we have

$$\frac{\Delta E_{Sm}(\mathrm{thermic})}{E_{Sm}} = 2\sqrt{\frac{2\ln 2 \times kT}{M_{Sm}}} = 2\sqrt{\frac{1.4 \times 26 \times 10^{-3}}{1.52 \times 10^{11}}} \approx 10^{-6}$$

and

$$\Delta E_{Sm}(\mathrm{thermic}) = 10^{-6} \times 963\,\mathrm{keV} \approx 1\mathrm{eV}. \tag{7.50}$$

In conclusion, the energy of the photons emitted by Sm* at rest is smaller than the resonance energy by six times its width. The process does not take place. The first necessary condition is therefore satisfied.

We now prove that the resonance condition is satisfied for forward γ emission by Sm* in flight.

We compute the kinetic energy of the Sm* produced by the Eu K-capture. The reasoning is very similar to that above. The energy released in the transition is

$E_{Eu} = 911\,\text{keV}$. The recoil momentum and the neutrino momentum are equal and opposite. The neutrino momentum is equal to its energy, which we approximate by E_{Eu}. Consequently, the Sm * recoil kinetic energy is

$$E_{K,\,Sm^*} \sim \frac{E_{Eu}^2}{2M_{Sm}} = \frac{0.911^2 \times 10^{12} \times \text{eV}^2}{2 \times 1.52 \times 10^{11}\,\text{eV}} = 2.7\,\text{eV}. \tag{7.51}$$

The recoil speed in the laboratory reference frame is

$$\beta_{Sm^*} = \sqrt{\frac{2E_{K,\,Sm^*}}{M_{Sm}}} = \sqrt{\frac{2 \times 2.7}{1.52 \times 10^{11}}} = 5.8 \times 10^{-6}.$$

In this frame the photon energy E_L is maximum when the photon is emitted in the flight direction of the nucleus. Recalling Example 1.1 and setting $\gamma = 1$, we have

$$E_L - E_{Sm^*} = \beta_{Sm^*} \times E_{Sm^*} = 5.8 \times 10^{-6} \times 911\,\text{keV} = 5.3\,\text{eV}. \tag{7.52}$$

This is within about 1 eV from resonance; hence, taking into account the Doppler broadening, the resonance condition is satisfied. The second necessary condition is satisfied.

In the design stages of the experiment it was not at all guaranteed that these miraculous conditions would indeed be satisfied. To check this crucial point, L. Grodzins (Grodzins 1958) performed a preliminary experiment to measure the 'resonant' cross-section, using the same apparatus shown in Fig. 7.6 without the magnet. In this way he used the gammas coming from Sm * produced in reaction (7.46), exactly as in the final experiment. In conclusion, the success of the experiment is due to the exceptional kindness offered by Nature and to the equally exceptional boldness of the experimenters.

We recall that, in Chapter 2, we found in Eq. (2.66) that the expectation value of the helicity of a fermion of negative (or left) chirality is $-\beta$, where β is the ratio of its velocity to c. Now we learn that such are the electrons in the β^- decay of a nucleus. Let us see how the prediction can be experimentally verified. Measuring the helicity of the electrons is measuring their longitudinal polarization. In practice, it is much easier to measure the transverse polarization using a thin high-Z metal plate immersed in a magnetic field. The analysing power is due to the fact that, in the above conditions, the electron scattering cross-section in a metal depends on its transverse polarization. To change the electron polarization from longitudinal to transverse, we let the electrons go through a quarter of a circumference in a magnetic field. The direction of the momentum vector changes by 90° whilst the spin direction remains unaltered.

Fig. 7.7 shows the measurements of the helicity of electrons of different speeds coming from the beta decays of three nuclei. These have been chosen to cover three different speed intervals: tritium at small velocities, cobalt at intermediate ones and phosphorus at high velocities. We see that the agreement with the theoretical prediction is good.

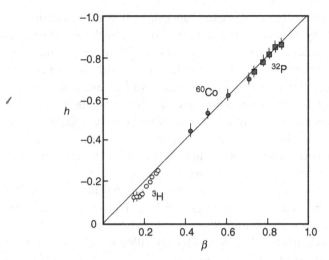

Fig. 7.7. The helicity of the electron as a function of its speed (adapted from F. Koks and J. van Klinken, *Nucl. Phys* A272 (1976) 61. Fig. 1. © 1976 with permission by Elsevier). https://doi.org/10.1016/0375-9474(76)90318-3.

7.6 Violation of the Particle–Antiparticle Conjugation

Weak interactions violate not only parity but also the particle–antiparticle conjugation.

In the discussion of parity violation we have used the fact that the Lagrangian contains the correlation term $\sigma \cdot \mathbf{p}$ between the spin and the momentum, which is not invariant under inversion of the axes. Observing that $\sigma \cdot \mathbf{p}$ does not vary under charge conjugation, the same reasoning would lead to the naïve conclusion that charge conjugation is conserved, but that conclusion is wrong.

Indeed, the operator $\sigma \cdot \mathbf{p}$ is, within a factor $1/p$, the helicity operator. The presence of this operator in the Lagrangian selects the left spinors and therefore violates C.

The Wu experiment gives only indirect evidence for C violation, assuming CPT invariance. We observe that under time-reversal $T : \mathbf{p} \to -\mathbf{p}, \sigma \to -\sigma$, and under $P : \mathbf{p} \to -\mathbf{p}, \sigma \to \sigma$, Therefore, under PT we have $\sigma \cdot \mathbf{p} \to -\sigma \cdot \mathbf{p}$. In conclusion, if CPT is conserved, $\sigma \cdot \mathbf{p}$ violates C.

To have direct evidence of C violation in a Wu-type experiment, one would need to study the beta decay of an antinucleus, which is impossible. However, we can use the decays of the mesons, in particular of the pions. The decay chains $\pi^- \to \mu^- + \bar{\nu}_\mu$ followed by $\mu^- \to e^- + \nu_\mu + \bar{\nu}_e$ and $\pi^+ \to \mu^+ + \nu_\mu$ followed by $\mu^+ \to e^+ + \bar{\nu}_\mu + \nu_e$ are reciprocally charge conjugated. The experiments show that the helicities of the electron in the former and of the positron in the latter have opposite expectation values. This is direct proof of C violation.

More generally, the C operator transforms a process with emission of a negative helicity neutrino into a process with emission of an antineutrino, which also has negative helicity. Indeed, C only changes particles in antiparticles, leaving the rest unvaried. However, experiments show that the antineutrinos have positive helicity. Namely, C is violated exactly as required by the $V - A$ structure of the interaction. In conclusion, the CC weak interactions violate both $\mathcal{P}$ and C maximally.

In order to recover the particle–antiparticle symmetry, L. Landau (Landau 1957) observed that all interactions were invariant under the 'combined parity' $C\mathcal{P}$, namely under the space inversion and the simultaneous particle–antiparticle substitution. Consider, for example, a fermion with positive helicity. Its $C\mathcal{P}$ counterpart is its antiparticle with negative helicity. For example, the β^- decay of a nucleon produces an antineutrino with positive helicity; in the $C\mathcal{P}$ mirror, the β^+ decay of the antinucleon would produce a neutrino with negative helicity.

The matter–antimatter symmetry that the discoveries of $\mathcal{P}$ and C non-conservation had broken was thus re-established. However, 7 years later, J. Christenson, J. Cronin, V. Fitch and R. Turlay (Christenson *et al.* 1964) observed that the 'long lifetime' neutral K meson, which is called K_L and is the $C\mathcal{P}$ eigenstate with $C\mathcal{P} = -1$, decays, even if rarely, into two pions, namely into a state with $C\mathcal{P} = +1$. Not even $C\mathcal{P}$ is perfect symmetry. Matter and antimatter are not exactly equal. We shall see this in Chapter 8.

7.7 Chirality and Helicity in *V–A* Interaction

In Section 7.3 we saw that the CC weak interactions are ruled by currents that have a universal V–A Lorentz structure. The fermion fields appearing in the currents are always left chirality solutions of the Dirac equation as in Eq. (7.27). The interaction violates, as we have just seen, the invariance under charge conjugation C, under parity $\mathcal{P}$ and even under $C\mathcal{P}$ (as we shall discuss amply in the next chapter), but does not violate $C\mathcal{P}\mathcal{T}$, the conservation of which, as we have seen in Section 3.4, is a completely general property of all the fundamental interactions. Let us see an important consequence, elaborating on a simple argument by P. B. Pal (Pal 2011).

$C\mathcal{P}\mathcal{T}$ requires that, if a process exists from a state A consisting of some particles, to a state B consisting of particles, which may be the same or different from those of A, then the process must exist, with the same amplitude, in which the $C\mathcal{P}$ conjugates of the particles in B (now initial state) go to the $C\mathcal{P}$ conjugates of the particles in A (now final state). Hence, a necessary condition for $C\mathcal{P}\mathcal{T}$ invariance is that, for every existing particle, a $C\mathcal{P}$ conjugate particle (its antiparticle we say) must exist. In the case of the CC weak interactions that we are discussing here, consider for example the process $e^- p \rightarrow \nu_e n$. As this is known to exist, then, also $\bar{\nu}_e \bar{n} \rightarrow e^+ \bar{p}$ must exist and proceed with the same amplitude of the former. The positron, for example, in the final state of the second process must be the $C\mathcal{P}$ conjugate of the electron in the initial state of the first one, and similarly for the other particles. The current–current interaction describing both processes is proportional to $\bar{\psi}_{e,L}\gamma_\mu\psi_{\nu_e,L}\,\bar{\psi}_{p,L}\gamma_\mu\psi_{n,L}$. Hence, the electron interacting with the proton is a negative chirality particle, the particle annihilated by the field $\bar{\psi}_{e,L}$. The same field, in the inverse reaction creates a positron, the CP conjugate of the electron, which is a positive chirality particle. To understand this, let us assume, to be concrete, that the electron and the positron are ultra-relativistic, and look at their helicity components. The helicity of a negative chirality electron is, as we have seen in Section 2.8, dominantly negative, with a small positive component. Under C the electron becomes a positron; under $\mathcal{P}$ the helicity changes sign. Hence the CP conjugate of the electron is a positron with dominant positive helicity. Such a particle, again as we saw in Section 2.8, is a positive chirality positron.

The conclusion is general: if the annihilation operator in a field operator annihilates a left chirality particle, the creation operator must create a positive chirality antiparticle, and the opposite if the field operator annihilates a right chirality particle. However, as far helicity is concerned, its content in a particle of defined chirality depends on its energy, as shown in Section 2.8 and clearly seen in Fig. 7.7.

In the above example, the initial and final nucleons are, say, non-relativistic negative chirality particles. As the energy is low, there is no dominant helicity component; rather, the $h = +1/2$ component is a bit smaller than the $h = -1/2$ component. Their CP conjugate antinucleons have positive chirality and the $h = +1/2$ helicity component is a bit larger than the $h = -1/2$ component. The opposite is true in the case of neutrinos. If they are massless, as assumed in the SM, helicity and chirality coincide. In the moment and in the point of the interaction, neutrinos have completely negative chirality and negative helicity; antineutrinos have positive chirality and positive helicity.

7.8 Lepton Universality

The Fermi point-like interaction is an approximate theory – a very good approximation at low energies – but actually it is mediated by a vector boson, the W, also called 'intermediate boson'. As we have anticipated in Section 7.1, this eliminates the divergence of the cross-sections at high energy. For example, the neutron beta decay is actually described by the Feynman diagram in Fig. 7.8(b), rather than by Fig. 7.8(a), considered in Section 7.1. Similarly, the quasi-elastic scattering of muon neutrino on electron is properly described by Fig. 7.9(b) rather than Fig. 7.9(a).

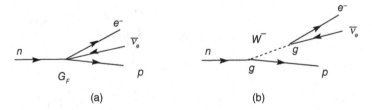

Fig. 7.8.   Neutron beta decay: (a) Fermi point-like interaction; (b) mediated by the W boson.

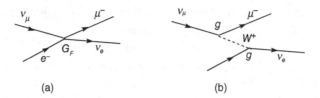

Fig. 7.9. Quasi-elastic scattering of ν_μ on electron: (a) Fermi point-like interaction; (b) mediated by the W boson.

The 'g' on the vertices of the diagrams is the 'weak charge', analogous to the electromagnetic and colour ones. As with the other ones, it is dimensionless. The matrix element contains, apart from numerical factors, the product of the couplings and of the propagator

$$\mathcal{M} \propto \frac{g\,g}{M_W^2 - t}. \tag{7.53}$$

If the momentum transfer is small, namely $-t \ll M_W{}^2$ (as is the case for a beta decay and may or may not be the case for quasi-elastic scattering) we can write, with a good approximation,

$$\mathcal{M} \propto \frac{g^2}{M_W^2}. \tag{7.54}$$

We see that the Fermi matrix element is small, and the interaction is feeble, because M_W is large. For example, with a value of g^2 of the same order of magnitude of the fine-structure constant, $\mathcal{M}$ is of the order of $10^{-6}\,\text{GeV}^{-2}$. In these conditions the momentum transfer is far too small to resolve the two vertices and the interaction behaves like a point-like four-fermion interaction. The experimental search for the intermediate boson started in the 1960s as part of the programmes of the neutrino beams produced in all the proton accelerator laboratories, but the intermediate boson was not found till 1983, with a different approach (Chapter 9), because its mass of about 80 GeV was too large. On the theory side, discovering how a gauge theory can have massive gauge bosons also took several years.

Question 7.2 Calculate the threshold energy needed to produce a W boson with a ν_μ beam on protons. □

The relationship between the Fermi constant and weak charge is, by definition

$$\frac{G_F}{(\hbar c)^3} = \frac{\sqrt{2}}{8} \frac{g^2}{\left(M_W c^2\right)^2} \qquad \text{(SI)} \tag{7.55a}$$

and in NU

$$G_F = \frac{\sqrt{2}}{8} \frac{g^2}{M_W^2} \qquad \text{(NU)}. \tag{7.55b}$$

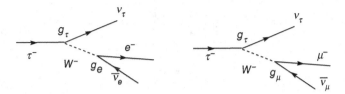

Fig. 7.10. Leptonic decays of the τ.

The CC weak interaction is universal, equal for all fermions. This property is immediately valid for leptons, but not for quarks. For these it holds after a 'rotation' of the states, as we shall see in the next section.

Let us see now a few examples of lepton universality.

The $e-\mu$ universality can be checked on the two leptonic decays of the τ

$$\tau^- \to e^- \overline{\nu}_e \nu_\tau \qquad \tau^- \to \mu^- \overline{\nu}_\mu \nu_\tau \tag{7.56}$$

Let us neglect, for simplicity, the electron and muon masses. As we are searching for possible differences, let us indicate the weak charges by different symbols, g_e, g_μ and g_τ (Fig. 7.10).

The two partial widths are (not mentioning constants that are the same for both)

$$\Gamma\left(\tau^- \to \mu^- \overline{\nu}_\mu \nu_\tau\right) \propto \frac{g_\tau^2}{M_W^2} \frac{g_\mu^2}{M_W^2} m_\tau^5$$

$$\Gamma\left(\tau^- \to e^- \overline{\nu}_e \nu_\tau\right) \propto \frac{g_\tau^2}{M_W^2} \frac{g_e^2}{M_W^2} m_\tau^5. \tag{7.57}$$

We measure their ratio by measuring the ratio between the corresponding branching ratios BR

$$\frac{\Gamma\left(\tau^- \to \mu^- \overline{\nu}_\mu \nu_\tau\right)}{\Gamma\left(\tau^- \to e^- \overline{\nu}_e \nu_\tau\right)} = \frac{BR\left(\tau^- \to \mu^- \overline{\nu}_\mu \nu_\tau\right)}{BR\left(\tau^- \to e^- \overline{\nu}_e \nu_\tau\right)} = \frac{g_\mu^2 \rho_\mu}{g_e^2 \rho_e}, \tag{7.58}$$

where the last factor is the ratio of the phase-space volumes, which can be precisely calculated. Using the measured quantities (Yao *et al.* 2006) we have

$$\frac{BR\left(\tau^- \to \mu^- \overline{\nu}_\mu \nu_\tau\right)}{BR\left(\tau^- \to e^- \overline{\nu}_e \nu_\tau\right)} = \frac{(17.36 \pm 0.05)\%}{(17.84 \pm 0.05)\%} = 0.974 \pm 0.004, \tag{7.59}$$

which gives

$$g_\mu / g_e = 1.001 \pm 0.002. \tag{7.60}$$

The $\mu - \tau$ universality can be checked from the muon and tau beta decay rates (Fig. 7.11). Taking into account that the μ decays 100% of the time in this channel, we have

$$\frac{\Gamma\left(\mu^- \to e^- \overline{\nu}_e \nu_\mu\right)}{\Gamma\left(\tau^- \to e^- \overline{\nu}_e \nu_\tau\right)} = \frac{1}{\tau_\mu} \frac{\tau_\tau}{BR\left(\tau^- \to e^- \overline{\nu}_e \nu_\tau\right)}. \tag{7.61}$$

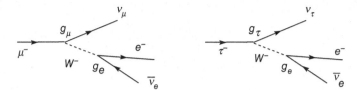

Fig. 7.11.   Beta decay of the μ and of the τ.

On the other hand, the theoretical ratio is

$$\frac{\Gamma\left(\mu^- \to e^-\overline{\nu}_e\nu_\mu\right)}{\Gamma\left(\tau^- \to e^-\overline{\nu}_e\nu_\tau\right)} = \frac{g_e^2 g_\mu^2 m_\mu^2 \rho_\mu}{g_e^2 g_\tau^2 m_\tau^5 \rho_\tau} = \frac{g_\mu^2 m_\mu^5 \rho_\mu}{g_\tau^2 m_\tau^5 \rho_\tau} \qquad (7.62)$$

and we have

$$\frac{g_\mu^2}{g_\tau^2} = \frac{1}{\tau_\mu} \frac{\tau_\tau}{BR\left(\tau^- \to e^-\overline{\nu}_e\nu_\tau\right)} \frac{m_\tau^5 \rho_\tau}{m_\mu^5 \rho_\mu}. \qquad (7.63)$$

In conclusion, we need to measure the two lifetimes, the two masses and the branching ratio $BR\left(\tau^- \to e^-\overline{\nu}_e\nu_\tau\right)$ The measurements give

$$g_\mu/g_\tau = 1.001 \pm 0.003. \qquad (7.64)$$

7.9 Quark Mixing

While the coupling of the W boson to leptons is universal, as we have just seen, the same is not true for the quarks, if their states of definite flavour are considered, namely those we have met so far. In 1963, N. Cabibbo (Cabibbo 1963) found out that universality can be recovered if the quarks in the CC weak interactions are not in definite flavour states, but quantum superpositions of the latter. This property is now known as 'quark mixing'. For historical precision, at that time the quark hypothesis had not yet been advanced and Cabibbo introduced the mixing for the hadronic currents, which is completely equivalent to quark mixing. In addition, only hadrons made up of the 'light flavours' (u, d and s) were known. Similar mixing phenomena exist amongst neutrinos, in contradiction with the SM assumptions. This will be discussed in Chapter 9.

The problem considered by Cabibbo was the following. There are two types of beta decays of the strange hadrons: those that conserve strangeness and those that violate it by $|\Delta S| = 1$. Although universality requires the corresponding matrix elements to be equal, those of the latter are substantially smaller than those of the former. For example, the $\Delta S = 0$ decay

$$n \to p e^-\overline{\nu}_e \qquad (7.65)$$

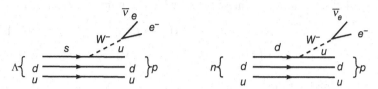

Fig. 7.12. Strangeness-changing and strangeness-non-changing beta decays.

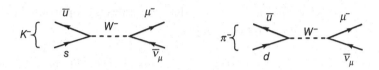

Fig. 7.13. Strangeness-changing and strangeness-non-changing decays.

has a probability that is much larger than that of the similar $|\Delta S| = 1$ decay

$$\Lambda \to p e^- \bar{\nu}_e. \tag{7.66}$$

Using the knowledge we have today, Fig. 7.12 shows the diagrams at the quark level of the two decays. The two quarks present in both cases before and after the decay are, in a rough approximation, simple 'spectators'. The final quarks are the same in the two cases; the only difference is that in one case an s quark decays, in the other a d quark.

From now on we simplify the notation for the spinor fields. For example, for the electron left chirality field, in place of $\psi_{e,L}$, we shall simply write e_L; for its conjugate, in place of $\bar{\psi}_{e,L}$, we shall write $\bar{e}_L$.

Universality would require their matrix elements to be

$$\mathcal{M} \propto G_F \cdot \bar{e}_L \gamma_\alpha \nu_{e\,L} \cdot \bar{u}_L \gamma^\alpha d_L \qquad \mathcal{M} \propto G_F \cdot \bar{e}_L \gamma_\alpha \nu_{e\,L} \cdot \bar{u}_L \gamma^\alpha s_L \tag{7.67}$$

with the same coupling constant.

As a second example, consider the $\Delta S = 0$ decay of the pion

$$\pi^- \to \mu^- \bar{\nu}_\mu \tag{7.68}$$

and the similar $|\Delta S| = 1$ decay of the kaon

$$K^- \to \mu^- \bar{\nu}_\mu. \tag{7.69}$$

Fig. 7.13 shows their quark diagrams. Again, the only difference is the decaying quark: s or d.

The 'universal' matrix elements would be

$$\mathcal{M} \propto G_F \cdot \bar{\mu}_L \gamma_\alpha \nu_{\mu\,L} \cdot \bar{u}_L \gamma^\alpha s_L \qquad \mathcal{M} \propto G_F \cdot \bar{\mu}_L \gamma_\alpha \nu_{\mu\,L} \cdot \bar{u}_L \gamma^\alpha d_L \tag{7.70}$$

Let us focus on the meson case, which is simpler. The measured partial decay rates are

$$\Gamma(\pi \to \mu\nu) = BR(\pi \to \mu\nu)/\tau_{\pi^+} = 1/(2.6 \times 10^{-8}) \text{s}^{-1}$$
$$\Gamma(K \to \mu\nu) = BR(K \to \mu\nu)/\tau_{K^+} = 0.64/(1.24 \times 10^{-8}) \text{s}^{-1}$$

(7.71)

giving the ratio

$$\Gamma(K \to \mu\nu)/\Gamma(\pi \to \mu\nu) = 1.34.$$

(7.72)

However, if the coupling constants of the $\bar{u}s$ pair and of the $\bar{u}d$ pair to the W are the same as in (7.70), the ratio of the decay rates is proportional to the ratio of the phase-space volumes and is given by

$$\frac{\Gamma(K \to \mu\nu)}{\Gamma(\pi \to \mu\nu)} = \frac{m_K \left[1 - (m_\mu/m_K)^2 \right]^2}{m_\pi \left[1 - (m_\mu/m_\pi)^2 \right]^2} = 8.06.$$

(7.73)

Actually, the situation is not so simple, because the quarks decay inside the hadrons. We discussed in Section 7.3, for the pion decay, how the effects of the strong interaction can be factorized into the pion decay constant f_π. The same can be done for the kaon decay with another decay constant f_K. These factors cannot be measured directly and are difficult to calculate exactly, but we can say something about their ratio, which is what we need. Actually if the $SU(3)_f$ symmetry were exact, we would have $f_K/f_\pi = 1$. It can be shown that the observed symmetry breaking implies $f_K/f_\pi > 1$. Therefore, the effect of the strong interactions is to worsen the disagreement between the experiment and the universality. The ratio between the semileptonic decay rates of the K and of the pion is an order of magnitude smaller than expected.

The analysis of the semileptonic decays of the nucleons and of the hyperons, with and without change of strangeness, must take into account the hadronic structure as well and its approximate $SU(3)_f$ symmetry. We say here only that the conclusion is that, again, the $|\Delta S| = 1$ decays are suppressed by about an order of magnitude, compared with the $\Delta S = 0$ decays. Notice that it is the change in strangeness that matters, not the strangeness itself. For example, the decay $\Sigma^\pm \to \Lambda e^\pm \nu$ is not suppressed.

Another issue is that the value of the coupling constant in the beta decay of the neutron is somewhat smaller than that in the muon decay.

All of this is explained if we assume, like Cabibbo, that the down-type quarks entering the CC weak interactions are not d and s, but, say, d' and s'. Each (d,s) and (d',s') pair is an orthonormal base. The latter is obtained from the former by the rotation of a certain angle, called the 'Cabibbo angle', θ_C. This is shown schematically in Fig. 7.14. In a formula, the down-type quark that couples to the W is a quantum superposition of d and s, namely the state

$$d' = d\cos\theta_C + s\sin\theta_C.$$

(7.74)

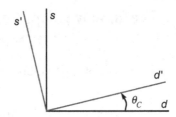

Fig. 7.14. The Cabibbo rotation.

Indeed, the coefficients of d and s must satisfy the normalization condition, namely the sum of their square must be one. Therefore, they can be thought of as the sine and the cosine of an angle.

In the Cabibbo theory there is only one matrix element for (7.65) and (7.66) in which d' appears, namely

$$\mathcal{M} \propto G_F \cdot \overline{e}_L \gamma_\alpha v_{e\,L} \cdot \overline{d'}_L \gamma^\alpha u_L. \tag{7.75}$$

Using the (7.74) we obtain, for the two decays,

$$\begin{aligned} \mathcal{M} &\propto G_F \cos\theta_C \cdot \overline{e}_L \gamma_\alpha v_{eL} \cdot \overline{d}_L \gamma^\alpha u_L \quad \text{for } \Delta S = 0 \\ \mathcal{M} &\propto G_F \sin\theta_C \cdot \overline{e}_L \gamma_\alpha v_{eL} \cdot \overline{s}_L \gamma^\alpha u_L \quad \text{for } \Delta S = 1 \end{aligned} \tag{7.76}$$

Since the angle θ_C is small, the $|\Delta S| = 1$ transition probabilities, which are proportional to $\sin^2 \theta_C$, are smaller than the $\Delta S = 0$ ones that have the factor $\cos^2 \theta_C$ by about an order of magnitude. Moreover, the constant of the neutron decay is $G_F^2 \cos^2 \theta_C$, which is somewhat smaller than the pure G_F^2 of the muon decay.

If the theory is correct, a single value of the Cabibbo angle must agree with the rates of all the semileptonic decays, of the nuclei, of the neutron, of the hyperons and of the strange and non-strange mesons. Both experimental and theoretical work is needed for this verification. Experiments must measure decay rates and other relevant kinematic quantities with high accuracy. Theoretical calculations must consider the fact that the elementary processes at the quark level, such as those shown in Figs. 7.12 and 7.13, take place inside hadrons. Consequently, the transition probabilities are not given simply by the matrix elements in (7.76). The evaluation of the interfering strong interactions effects is not easy because the QCD coupling constant α_s is large in the relevant momentum transfer region. LQCD calculations help here.

We shall discuss the measurement of $\sin\theta_C$ and $\cos\theta_C$ in Section 7.12 in two examples. We mention here that all the measurements give consistent results. The values are

$$\theta_C = 13.16° \quad \cos\theta_C = 0.97373 \pm 0.00031 \quad \sin\theta_C = 0.02627 \tag{7.77}$$

In conclusion, the CC weak interactions are also universal in the quark sector, provided that the 'quark mixing' phenomenon is taken into account.

7.10 The Glashow, Iliopoulos and Maiani Mechanism

An immediate consequence of the Cabibbo theory is the presence, in the Lagrangian, of the term

$$
\begin{aligned}
\bar{d}'_L \gamma_\alpha d'_L = {}& \cos^2\theta_C \bar{d}_L \gamma_\alpha d_L + \sin^2\theta_C \bar{s}_L \gamma_\alpha s_L \\
& + \cos\theta_C \sin\theta_C \left[\bar{d}_L \gamma_\alpha s_L + \bar{s}_L \gamma_\alpha d_L \right],
\end{aligned}
\tag{7.78}
$$

which describes neutral current transitions. In particular, the last term implies neutral currents that change strangeness (SCNC = strangeness-changing neutral currents) because they connect s and d quarks. However, the corresponding physical processes are strongly suppressed. For example, the two NC and CC decays

$$
K^+ \to \pi^+ + \nu_e + \bar{\nu}_e \quad K^+ \to \pi^0 + \nu_e + e^+
\tag{7.79}
$$

should proceed with similar probabilities, as understood from the diagrams shown in Fig. 7.15. On the contrary, the former decay is strongly suppressed, the measured values of the branching ratios (Workman $et\ al.$ 2022) being

$$
\mathrm{BR}\left(K^+ \to \pi^+ \nu \bar{\nu}\right) = 1.14^{+0.40}_{-0.33} \times 10^{-10}; \quad \mathrm{BR}\left(K^+ \to \pi^0 e^+ \nu_e\right) < 3 \times 10^{-3}
\tag{7.80}
$$

S. Glashow, I. Iliopoulos and L. Maiani observed in 1970 (Glashow $et\ al.$ 1970) that the d' and u states can be thought of as the members of the doublet $\begin{pmatrix} u \\ d' \end{pmatrix}$. Now, they argued, a fourth quark might exist, the 'charm' c, as the missing partner of s', to form a second similar doublet $\begin{pmatrix} c \\ s' \end{pmatrix}$.

Since s' is orthogonal to d', we have

$$
s' = -d\sin\theta_C + s\cos\theta_C.
\tag{7.81}
$$

We anticipated this situation in Fig. 7.14. Clearly, the relationship between the two bases is the rotation

$$
\begin{pmatrix} d' \\ s' \end{pmatrix} = \begin{pmatrix} \cos\theta_C & \sin\theta_C \\ -\sin\theta_C & \cos\theta_C \end{pmatrix} \begin{pmatrix} d \\ s \end{pmatrix}.
\tag{7.82}
$$

This was the prediction of a new flavour. We saw in Section 4.9 how it was discovered.

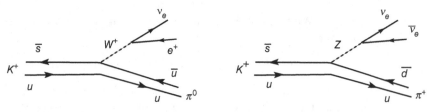

Fig. 7.15. Strangeness-changing CC and NC decays.

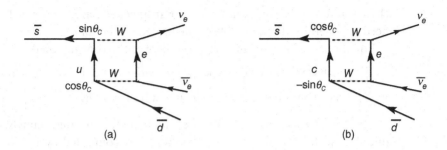

Fig. 7.16. Second-order quark-level diagrams for the decay $K^+ \rightarrow \pi^+ + \nu_e + \overline{\nu}_e$. Line of the spectator quark u not shown.

Let us now see how the 'GIM' mechanism succeeds in suppressing the strangeness-changing neutral currents. In addition to the terms (7.78), we now have

$$\overline{s}'_L \gamma_\alpha s'_L = \sin^2\theta_C \overline{d}_L \gamma_\alpha d_L + \cos^2\theta_C \overline{s}_L \gamma_\alpha s_L$$
$$- \cos\theta_C \sin\theta_C \left[\overline{d}_L \gamma_\alpha s_L + \overline{s}_L \gamma_\alpha d_L \right]. \tag{7.83}$$

Summing the two, we obtain

$$\overline{s}'_L \gamma_\alpha s'_L + \overline{d}'_L \gamma_\alpha d'_L = \overline{s}_L \gamma_\alpha s_L + \overline{d}_L \gamma_\alpha s d_L. \tag{7.84}$$

The SCNC cancel out. However, the issue is not so simple, because we must consider the higher-order contributions. The relevant second-order diagram is shown in Fig. 7.16(a), where the line of the 'spectator' u quark has been omitted for clarity. The corresponding calculated rate is much larger than the experimental value in (7.80).

It is here that the beauty of the GIM solution shows up. Indeed, if a fourth quark exists, the diagram of Fig. 7.16(b) must be included too. The factors at the quark vertices are $\cos\theta_C \sin\theta_C$ and $-\cos\theta_C \sin\theta_C$ respectively. The rest is the same and the two contributions cancel each other out. To be precise, the cancellation would be perfect if the masses of the u and c quarks were equal. They are not, and the sum of the two diagrams is not zero, but small enough to be perfectly compatible with observations.

GIM showed that the cancellation takes place at all orders.

We observe, finally, that a neutral current term remains in the Lagrangian, namely the NC between equal quarks or, in other words, the strangeness-conserving neutral current. As we shall see in Section 7.13, the corresponding physical processes were indeed discovered, in 1973. We observe here that the Cabibbo rotation is irrelevant for the NC term. In other words this term is the same in the two bases.

7.11 The Quark Mixing Matrix

The GIM mechanism explains the suppression of the SCNC in the presence of two families. Later, the third family, with its two additional quark flavours, was discovered, as we have seen. It was also found that the flavour-changing neutral currents

(FCNC), for all flavours, not only for strangeness, are suppressed. Therefore, we need to generalize the concepts of the preceding sections.

Equation (7.82) is a transformation between two orthogonal bases. The doublet $\begin{pmatrix} d \\ s \end{pmatrix}$ is the base of the down-type quarks with definite mass. These are the states, let us say, that would be stationary, if they were free. The doublet $\begin{pmatrix} d' \\ s' \end{pmatrix}$ is the base of down-type quarks that are the weak interaction eigenstates, namely the states produced by such an interaction. The two bases are connected by a unitary transformation that we now call V, to develop a formalism suitable for generalization to three families. The elements of V are real in the two-family case, as we shall soon show. We rewrite (7.82) as

$$\begin{pmatrix} d' \\ s' \end{pmatrix} = \begin{pmatrix} V_{ud} & V_{us} \\ V_{cd} & V_{cs} \end{pmatrix} \begin{pmatrix} d \\ s \end{pmatrix} = \begin{pmatrix} \cos\theta_C & \sin\theta_C \\ -\sin\theta_C & \cos\theta_C \end{pmatrix} \begin{pmatrix} d \\ s \end{pmatrix}. \tag{7.85}$$

The generalization to three families was done by M. Kobayashi and K. Maskawa in 1973 (Kobayashi & Maskawa 1973). The quark mixing transformation is

$$\begin{pmatrix} d' \\ s' \\ b' \end{pmatrix} = \begin{pmatrix} V_{ud} & V_{us} & V_{ub} \\ V_{cd} & V_{cs} & V_{cb} \\ V_{td} & V_{ts} & V_{tb} \end{pmatrix} \begin{pmatrix} d \\ s \\ b \end{pmatrix}. \tag{7.86}$$

The matrix is called the Cabibbo–Kobayashi–Maskawa (CKM) matrix. It is unitary, namely

$$VV^+ = 1. \tag{7.87}$$

The three-family expression of the CC interaction is

$$\sum_{i=1}^{3} \bar{u}^i \gamma_\mu \left(1 - \gamma_5\right) V_{ik} d^k = \sum_{i=1}^{3} \bar{u}_L^i \gamma_\mu V_{ik} d_L^k, \tag{7.88}$$

where we have set $u^1 = u, u^2 = c, u^3 = t, d^1 = d, d^2 = s, d^3 = b$. Focussing on the flavour indices, the structure is

$$\begin{pmatrix} \bar{u} & \bar{c} & \bar{t} \end{pmatrix} \begin{pmatrix} V_{ud} & V_{us} & V_{ub} \\ V_{cd} & V_{cs} & V_{cb} \\ V_{td} & V_{ts} & V_{tb} \end{pmatrix} \begin{pmatrix} d \\ s \\ b \end{pmatrix}. \tag{7.89}$$

This justifies the names of the indices of the matrix elements.

We shall now determine the number of independent elements of the matrix. A complex 3×3 matrix has, in general, 18 real independent elements, 9 if it is unitary. If it were real, it would have been orthogonal, with three independent elements, corresponding to the three rotations, namely the Euler angles. The six remaining elements of the complex matrix are therefore phase factors of the $\exp(i\delta)$ type. Not all of them are physically meaningful.

Indeed, the particle fields, the quarks in this case, are defined modulo an arbitrary phase factor. Moreover, (7.89) is invariant for the substitutions

$$d^k \rightarrow e^{i\theta_k} d^k \qquad V_{ik} \rightarrow e^{-i\theta_k} V_{ik}. \tag{7.90}$$

With three such substitutions we can absorb a global phase for each row in the d-type quarks, eliminating three phases. Similarly, we can absorb a global phase factor for each column in a u-type quark. Doing so, it seems, at first, that other three phase factors can be eliminated. However, only two of them are independent. Indeed V does not change when all the down-type and all the up-type change by the same phase. Consequently, the six phases we used to redefine the fields must satisfy a constraint. Only five of them are independent. In conclusion, the number of physically meaningful phases is $6 - 5 = 1$. Summing up, the three-family mixing matrix has four free parameters, which can be taken to be three rotation angles and one phase factor $\exp(i\delta)$. The latter, a complex quantity, explains CP violation as we shall see soon.

Going back to two families, a 2×2 unitary matrix, has four real independent elements. Three of them are phase factors, of which two can be absorbed in the down-type quarks and two in the up-type ones. Similarly, to above, the four phases used to redefine the fields must satisfy a constraint. Only three of them are independent, enough to absorb all the phase factors. As anticipated, the matrix is real, with only one independent parameter, the Cabibbo angle, and no CP violation is provided.

Coming back to three families, we define the rotations as follows. We take three orthogonal axes (x, y, z) and we let each of them correspond to a down-type quark (d, s, b), as in Fig. 7.17. We rotate in the following order: the first rotation is by θ_{12} around z, the second by θ_{13} around the new y, the third by θ_{23} around the latest x. The product of three rotation matrices, which are orthogonal, describes the sequence. Writing, to be brief, $c_{ij} = \cos\theta_{ij}$ and $s_{ij} = \sin\theta_{ij}$, we have

$$V = \begin{pmatrix} 1 & 0 & 0 \\ 0 & c_{23} & s_{23} \\ 0 & -s_{23} & c_{23} \end{pmatrix} \begin{pmatrix} c_{13} & 0 & s_{13} \\ 0 & 1 & 0 \\ -s_{13} & 0 & c_{13} \end{pmatrix} \begin{pmatrix} c_{12} & -s_{12} & 0 \\ s_{12} & c_{12} & 0 \\ 0 & 0 & 1 \end{pmatrix}.$$

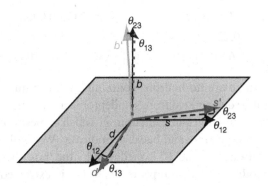

Fig. 7.17. The quark rotations.

We must still introduce the phase. This cannot be simply a factor, which would be absorbed by a field, becoming non-observable. Actually, there are several equivalent procedures. We shall use the following expression:

$$V = \begin{pmatrix} 1 & 0 & 0 \\ 0 & c_{23} & s_{23} \\ 0 & -s_{23} & c_{23} \end{pmatrix} \begin{pmatrix} c_{13} & 0 & s_{13}e^{-i\delta_{13}} \\ 0 & 1 & 0 \\ -s_{13}e^{+i\delta_{13}} & 0 & c_{13} \end{pmatrix} \begin{pmatrix} c_{12} & -s_{12} & 0 \\ s_{12} & c_{12} & 0 \\ 0 & 0 & 1 \end{pmatrix}. \tag{7.91}$$

and, performing the products

$$V = \begin{pmatrix} V_{ud} & V_{us} & V_{ub} \\ V_{cd} & V_{cs} & V_{cb} \\ V_{td} & V_{ts} & V_{tb} \end{pmatrix}$$

$$= \begin{pmatrix} c_{12}c_{13} & s_{12}c_{13} & s_{13}e^{-i\delta_{13}} \\ -s_{12}c_{23} - c_{12}c_{23}s_{13}e^{i\delta_{13}} & c_{12}c_{23} - s_{12}s_{23}s_{13}e^{i\delta_{13}} & s_{23}c_{13} \\ s_{12}s_{23} - c_{12}c_{23}s_{13}e^{i\delta_{13}} & -c_{12}s_{23} - s_{12}c_{23}s_{13}e^{i\delta_{13}} & c_{23}c_{13} \end{pmatrix}. \tag{7.92}$$

The expressions on the right-hand sides of (7.91) and (7.92) are valid only if the mixing matrix is unitary. This must be experimentally verified. 'New physics', not included in the SM, may induce violations of unitarity and, consequently, invalidate the expression of the mixing matrix in terms of three rotation angles and a phase factor. Therefore, the theory must be tested by measuring all the elements of the mixing matrix, nine amplitudes and a phase, and by checking whether or not the unitary conditions between them are satisfied.

Let us start by considering the absolute values of the matrix elements. The most precise determination of $|V_{ud}|$ comes from the study of super-allowed $0^+ \to 0^+$ nuclear beta decays, which are pure vector transitions; more information comes from the neutron lifetime and the pion semileptonic decay, as we shall see below. Five more absolute values, $|V_{us}|, |V_{cd}|, |V_{cs}|, |V_{ub}|$ and $|V_{cb}|$, have been determined by measuring the semileptonic decay rates of the hadrons of different flavours, strangeness, charm and beauty, as we shall see in an example. The values are (Workman *et al.* 2022):

$$\begin{array}{lll} |V_{ud}| = 0.97373 \pm 0.00031, & |V_{us}| = 0.2243 \pm 0.0008, & |V_{ub}| = (3.82 \pm 0.20) \times 10^{-3} \\ |V_{cd}| = 0.221 \pm 0.004, & |V_{cs}| = 0.975 \pm 0.006, & |V_{ub}| = (40.8 \pm 1.4) \times 10^{-3} \\ |V_{td}| = (8.6 \pm 0.2) \times 10^{-3}, & |V_{ts}| = (41.5 \pm 0.9) \times 10^{-3}, & |V_{tb}| = 1.014 \pm 0.029 \end{array} \tag{7.93}$$

Since there are no hadrons containing the top quark, the elements of the third row cannot be determined from semileptonic decays. In the next chapter we shall study the meson oscillation phenomenon, in which a meson initial state changes with time in an antimeson state and back, periodically. We shall see, in particular, that the oscillation between B^0 and $\bar{B}^0$ is mediated by the 'box' diagram shown in Fig. 8.6, in which the product $|V_{tb}||V_{td}|$ appears. Its value is extracted from the measured oscillation frequency, with uncertainties dominated by the theoretical uncertainties in hadronic effects. Similarly, $|V_{tb}||V_{ts}|$ is extracted from the $B_s^0 \bar{B}_s^0$ oscillation frequency.

The ninth element, $|V_{tb}|$, is very close to 1, namely the top quark decays in 100% of the cases, within the present experimental accuracy, in the channel $t \to b + W$. In the hadron colliders, the Tevatron and the LHC, top quarks are most probably produced in $t\bar{t}$ pairs by strong interactions. However, the 'single top' production has also been observed. In this process a W boson is produced, which then decays as $W \to b + t$. From the measurement of the single top production cross-section, the value of $|V_{tb}|$ independent of the unitarity assumption is extracted.

Using the independently measured absolute values of the elements, four unitarity checks can be done with small uncertainties

$$
\begin{aligned}
|V_{ud}|^2 + |V_{us}|^2 + |V_{ub}|^2 - 1 &= -0.0015 \pm 0.0007 & \text{1st row} \\
|V_{cd}|^2 + |V_{cs}|^2 + |V_{cb}|^2 - 1 &= 0.001 \pm 0.012 & \text{2nd row} \\
|V_{ud}|^2 + |V_{cd}|^2 + |V_{td}|^2 - 1 &= -0.0028 \pm 0.0020 & \text{1st column} \\
|V_{us}|^2 + |V_{cs}|^2 + |V_{ts}|^2 - 1 &= 0.004 \pm 0.0012 & \text{2nd column.}
\end{aligned}
\tag{7.94}
$$

We see that the conditions are satisfied within the statistics.

It is also useful to consider the angles and their sines. Their values, obtained with a global fit now assuming unitarity of the mixing matrix, are the following:

$$
\begin{aligned}
\sin\theta_{12} &= 0.22500 \pm 0.00067 & \theta_{12} &= 12.87° \\
\sin\theta_{13} &= 0.00369 \pm 0.00011 & \theta_{13} &= 0.21° \\
\sin\theta_{23} &= 0.04182^{+0.00085}_{-0.00074} & \theta_{23} &= 2.40°
\end{aligned}
\tag{7.95}
$$

We see that the angles are small or very small (the rotations in Fig. 7.17 have been exaggerated to make them visible). Moreover, there is a hierarchy in the angles, namely $s_{12} \gg s_{23} \gg s_{13}$. We do not know why.

Therefore, the diagonal elements of the CKM matrix are very close to one; the mixing between the second and third families is smaller than that between the first two; the mixing between the first and third families is even smaller. In practice, we might say, the hadrons prefer to decay semileptonically within the same family. This implies that the 2×2 submatrix of the first two families is very close to being unitary and therefore the Cabibbo angle is almost equal to θ_{12} and, finally, that $|V_{ud}| \approx |V_{cs}| \approx \cos\theta_C$ and $|V_{us}| \approx |V_{cd}| \approx \sin\theta_C$.

Question 7.3 Check if the unitarity conditions are satisfied, within the errors, considering the first two families only, that is, not adding the last addendum. □

We anticipate here that, as we shall see in Chapter 10 the pattern of the neutrino mixing angles is completely different.

M. Kobayashi and K. Maskawa observed in 1972 (Kobayashi & Maskawa 1973) that the phase factor present in the mixing matrix for three (but not for two) families implies CP violation. This is because the phase factor $\exp(i\delta)$ appears in the wave function that becomes $\exp[i(\omega t + \delta)]$. The latter expression is obviously not invariant under time reversal if $\delta \neq 0$ and $\delta \neq \pi$. Since CPT is conserved, CP must be violated.

We shall see in the next chapter how the phase is measured. We report here that the CP-violating phase in (7.91) is large. It value is

$$\delta_{13} = (1.144 \pm 0.027) \text{ rad} = (65.6 \pm 1.5)° \tag{7.96}$$

Further tests, which include the matrix elements in their absolute values and phases, can be done by looking at the products of different lines and columns, which by unitarity should be zero. In practice, the most sensitive case, of the six ones, is when the three terms in the sum have the same order of magnitude. Otherwise the triangle is quite narrow. We are left with the product of the first and third lines

$$V_{ud}V_{ub}^* + V_{cd}V_{cb}^* + V_{td}V_{tb}^* = 0. \tag{7.97}$$

We can consider each term in this expression as a vector in the complex plane and read it as stating that the sum of the three vectors should be zero. Geometrically this means that the vectors make up a triangle, called the 'unitary triangle'. Each term is a complex number with an absolute value and a phase. The latter are functions of the unique δ_{13} phase, if we assume unitarity. This assumption cannot be done if we want to check it. The three angles of the unitary triangle have been called α, β and γ at the SLAC laboratory in the USA and ϕ_2, ϕ_1 and ϕ_3 respectively at the KEK laboratory in Japan, so that each has two names. They are defined as:

$$\alpha = \phi_2 \equiv \arg\left(-\frac{V_{td}V_{tb}^*}{V_{ud}V_{ub}^*} \right);$$

$$\beta = \phi_1 \equiv \arg\left(-\frac{V_{cd}V_{cb}^*}{V_{td}V_{tb}^*} \right); \tag{7.98}$$

$$\gamma = \phi_3 \equiv \arg\left(-\frac{V_{ud}V_{ub}^*}{V_{cd}V_{cb}^*} \right).$$

Let us now draw the unitary triangle in Fig. 7.18. Traditionally, the side lengths are normalized dividing by $|V_{cd}V_{cb}^*|$, so that the absolute value of the second term in (7.97) is equal to one and, being the position on the plane arbitrary, we can take it as the x axis. In this way we fix the positions of the two vertices in $(0,0)$ and $(1,0)$. Much effort has been dedicated to measuring, with increasing accuracy, physical quantities that overdetermine the constraints in the triangle plane, checking that they are coherently consistent with a single triangle. We shall come back to this in Section 8.11.

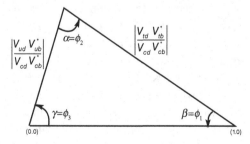

Fig. 7.18.   Sketch of the unitary triangle.

7.12 Measuring the Cabibbo Angle

We now give two examples of measurement of the absolute values of the mixing matrix elements, namely of $|V_{ud}|$ and $|V_{us}|$, which are very close to the cosine and the sine of the Cabibbo angle.

$|V_{ud}| \approx \cos\theta_C$ is measured in three different types of processes:

- in the super-allowed beta transitions of several nuclei (i.e. $J^P = 0^+ \to J^P = 0^+$ transitions between two members of the same isospin multiplet, with $\Delta I_z = \pm 1$); this is currently the most accurate method
- in the beta decay of the neutron
- in the so-called $\pi e3$ decay of the pion.

We now give some hints on $\pi e3$, which is

$$\pi^+ \to \pi^0 + e^+ + \nu_e. \tag{7.99}$$

This channel is theoretically very clean, being free from nuclear physics effects and having, as a purely vector transition, a simple matrix element. However, it is experimentally challenging because it is extremely rare, with a branching ratio of 10^{-8}. Consequently, in order to have a statistical uncertainty of, say, 10^{-3} one needs to collect a total of 10^{14} pion decays and to be able to discriminate with the necessary accuracy the $\pi e3s$. Clearly, the reason for the rareness of the decay is the smallness of the Q value, considering that (Workman *et al.* 2022)

$$\Delta \equiv m_{\pi^+} - m_{\pi^0} = 4.5936 \pm 0.0005 \text{ MeV}. \tag{7.100}$$

Notice that the uncertainty is only 10^{-4}.

Fig. 7.19 shows the two Feynman weak interaction diagrams at the quark level. However, the non-decaying quark does not behave simply as a 'spectator' as Fig. 7.19 suggests. On the contrary, it strongly interacts with the companions, before and after the decay, as exemplified in Fig. 7.20.

We set $p_\mu = p_\mu^{\pi^+} + p_\mu^{\pi^0}$ and $q_\mu = p_\mu^{\pi^+} - p_\mu^{\pi^0}$ where $p_\mu^{\pi^+}$ and $p_\mu^{\pi^0}$ are the four-momenta of the initial and final pions respectively. It can be shown that the matrix element is

$$\begin{aligned}
\mathcal{M} &= f_+\left(q^2\right)G_F V_{ud}\, p^\mu \bar{\nu}_e \gamma_\mu \left(1-\gamma_5\right)e \\
&= f_+\left(0\right)G_F V_{ud}\, p^\mu \bar{\nu}_e \gamma_\mu \left(1-\gamma_5\right)e,
\end{aligned} \tag{7.101}$$

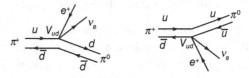

Fig. 7.19. Quark-level diagrams for $\pi e3$ decay.

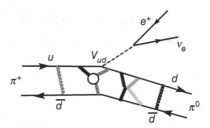

Fig. 7.20.   Hadronic complications in $\pi e3$.

where $f_+(q^2)$ is a function of the square modulus of the four-momentum transfer, which takes into account the effects of strong interactions. Since in the $\pi e3$ the Q value is so small, we have approximated the form factor with its value at $q^2 = 0$ in the last expression. Moreover, it turns out that $f_+(0)$ is determined by the $SU(2)$ symmetry, because π^0 and π^+ belong to the same isospin multiplet. The QCD processes, such as those in Fig. 7.20, that determine the form factors cannot be calculated perturbatively, but LQCD calculations have reached sub-per-cent precisions, which are matched to experimental uncertainties.

The partial width is given by

$$\Gamma\left(\pi^+ \to \pi^0 e^+ \nu_e\right) \equiv \frac{\mathrm{BR}\left(\pi_{e3}^+\right)}{\tau_\pi} = \frac{G_F^2 \Delta^5}{30\pi^3} |V_{ud}|^2 \left(1 - \frac{\Delta}{2m_{\pi^+}}\right)^3 f(\varepsilon)(1 + \delta_{EM}), \qquad (7.102)$$

where Δ is the Q value of the decay, δ_{EM} is a 'radiative correction' at the loop level of a few per cent and $\varepsilon \equiv (m_e / \Delta)^2 \approx 10^{-2}$.

The function f is the 'Fermi function', which is known. The overall theoretical uncertainty is small $\simeq 10^{-3}$. As Δ appears at the fifth power, its uncertainty is relevant, but is also small as in Eq. (7.100). The pion lifetime, in the denominator, is known with precision too. As for $BR\left(\pi_{e3}^+\right)$, the most precise measurement of this branching ratio comes from the PIBETA experiment, performed at the PSI Laboratory in Zurich (Pocanic *et al.* 2004). The positive pions are stopped in the middle of a sphere of CsI crystals, used to detect the two γs from the π^0 decay, measure their energies and normalize to the $\pi^+ \to e^+ \nu_e$ rate. The measured value is

$$\mathrm{BR}\left(\pi_{e3}^+\right) = (1.036 \pm 0.006) \times 10^{-8}. \qquad (7.103)$$

With this value and the above-mentioned inputs from theory, we obtain

$$|V_{ud}| = 0.9739 \pm 0.0029 \qquad (7.104)$$

We now consider $|V_{us}| \approx \sin\theta_C$. It can be determined in different processes, as follows.

- Semileptonic decays of hyperons with change of strangeness.
- Semileptonic decays of the K-mesons, which currently give the most precise values. The initial particle can be a charged or a neutral kaon, the daughters can be $\pi e \nu_e$

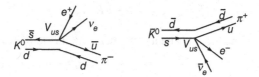

Fig. 7.21. Quark-level diagrams for $Ke3$.

or $\pi\mu\nu_\mu$ in their different charge states. We anticipate here that the neutral kaons' states of definite lifetime are not the states of definite strangeness, K^0 and $\bar{K}^0$, but two linear combinations of them, called K_S and K_L, having shorter and longer lifetimes, respectively.

The KLOE experiment (Ambrosino *et al.* 2006) extracted a value of V_{us} from the measurements of the K_S lifetime, τ_s, and of the branching ratio of the decay

$$K_S \to \pi^- e^+ \nu_e \quad \text{and} \quad \pi^+ e^- \bar{\nu}_e \tag{7.105}$$

The two diagrams shown in Fig. 7.21 contribute at the quark level. The quantity that is measured is the product $|V_{us}| f_+(0)$, where $f_+(0)$ is the form factor of the decay, a function of the square of the momentum transfer $q_\mu = p_\mu^{K^0} - p_\mu^{\pi^\pm}$, calculated at $q_\mu = 0$.

The factor includes the hadronic contributions, similar to what we just mentioned for the pion decay. The required calculations are in the non-perturbative regime of QCD. They are performed with LQCD, taking into account the isospin symmetry breaking, with accuracy matching the experimental one, resulting in (Carrasco *et al.* 2016)

$$f_+(0) = 0.9698 \pm 0.0017 \ \left[1.8 \times 10^{-3}\right]. \tag{7.106}$$

The KLOE experiment collected a pure sample of about 13 000 semileptonic K_S decay events working at the DAΦNE ϕ-factory at Frascati. A ϕ-factory is a high-luminosity $e^+ e^-$ collider that operates at the centre of mass energy $\sqrt{s} = m_\phi = 1.020$ GeV. The pure ϕ-meson initial state decays $\approx 34\%$ of the time into two neutral kaons. A single wave function describes the time evolution of both particles. It is antisymmetric under their exchange because the orbital angular momentum is $L = 1$. Consequently, the two bosons cannot be equal; if one of them decays as K_S, the other one must decay as K_L. Moreover, being in the centre of mass frame, the two decays are back to back. Consequently, if we detect a K_L, we know that a K_S is present in the opposite direction. The mean decay paths of K_S and K_L are $\lambda_S \approx 0.6$ cm and $\lambda_L \approx 350$ cm. The latter makes the decay vertex of the K_L well visible in the detector.

The KLOE detector consists mainly of a large cylindrical TPC in a magnetic field surrounded by a lead scintillating-fibre sampling calorimeter. A sample of about 400 million $K_S K_L$ pairs was collected. The two rates $K_S \to \pi^+ e^- \bar{\nu}_e$ and $K_S \to \pi^- e^+ \nu_e$ were separately measured and normalized to the branching ratio into $K_S^0 \to \pi^+ \pi^-$, which is known with $\approx 0.1\%$ accuracy. The result is

$$BR(K_S \to \pi^- e^+ v_e) + BR(K_S \to \pi^+ e^- \bar{v}_e) = (7.046 \pm 0.091) \times 10^{-4}. \quad (7.107)$$

With this value and the above-mentioned calculations, one obtains

$$|V_{us}| = 0.2240 \pm 0.0024. \quad (7.108)$$

Example 7.1 Estimate the following ratios: $\Gamma(D^0 \to K^+ K^-)/\Gamma(D^0 \to \pi^+ K^-)$, $\Gamma(D^0 \to \pi^+ \pi^-)/\Gamma(D^0 \to \pi^+ K^-)$ and $\Gamma(D^0 \to K^+ \pi^-)/\Gamma(D^0 \to \pi^+ K^-)$.

We start by recalling the valence quark composition of the hadrons of the problem: $D^0 = c\bar{u}, K^+ = u\bar{s}, K^- = s\bar{u}, \pi^+ = u\bar{d}, \pi^- = d\bar{u}$. We draw the tree-level diagrams for the decaying quark for each process.

The first diagram is favoured because the coefficients at the two vertices $|V_{cs}|$ and $|V_{ud}|$ are both large, being $\approx \cos\theta_C$ (the diagrams are said to be 'Cabibbo favoured' (CF)). In the second and third diagrams, the coefficient at one vertex is large while the coefficient at the other is small: respectively $|V_{us}|$ and $|V_{cd}| \approx \sin\theta_C$ ('singly Cabibbo suppressed' (SCS)). In the fourth diagram the coefficients at both vertices are small, its amplitude is proportional to $|V_{us}||V_{cd}| \approx \sin^2\theta_C$ ('doubly Cabibbo suppressed' (DCS)). See Figs. 7.22 and 7.23.

Summing up, we have

$$\frac{\Gamma(D^0 \to K^+ K^-)}{\Gamma(D^0 \to \pi^+ K^-)} \propto \frac{|V_{cs}|^2 |V_{us}|^2}{|V_{cs}|^2 |V_{ud}|^2} \approx \tan^2\theta_C \approx 0.05;$$

$$\frac{\Gamma(D^0 \to \pi^+ \pi^-)}{\Gamma(D^0 \to \pi^+ K^-)} \propto \frac{|V_{cd}|^2 |V_{ud}|^2}{|V_{cs}|^2 |V_{ud}|^2} \approx \tan^2\theta_C \approx 0.05;$$

$$\frac{\Gamma(D^0 \to K^+ \pi^-)}{\Gamma(D^0 \to \pi^+ K^-)} \propto \frac{|V_{cd}|^2 |V_{us}|^2}{|V_{cs}|^2 |V_{ud}|^2} \approx \tan^4\theta_C \approx 0.0025.$$

In a proper calculation of the decay rates, one must take into account the phase space (easy) and the colour field effects (difficult). With this caveat, anyhow, the experimental values confirm the hierarchy

$$\frac{\Gamma(D^0 \to K^+ K^-)}{\Gamma(D^0 \to \pi^+ K^-)} \approx 0.10;$$

$$\frac{\Gamma(D^0 \to \pi^+ \pi^-)}{\Gamma(D^0 \to \pi^+ K^-)} \approx 0.04; \quad \square$$

$$\frac{\Gamma(D^0 \to K^+ \pi^-)}{\Gamma(D^0 \to \pi^+ K^-)} < 0.02.$$

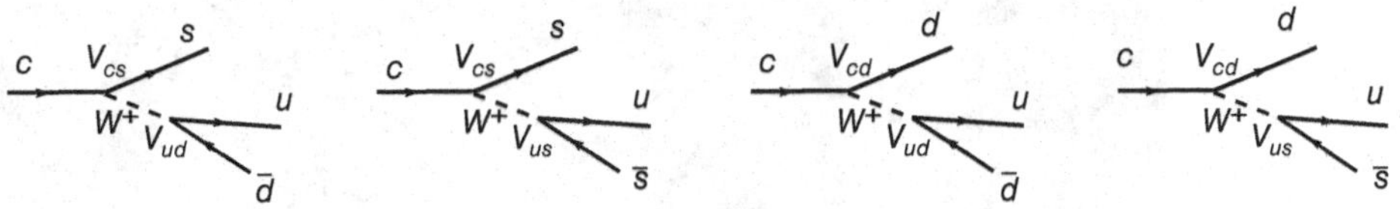

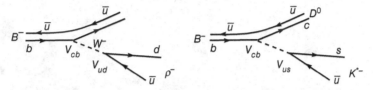

Fig. 7.22. Quark diagrams for c decays.

Fig. 7.23. Quark diagrams for two b decays.

Example 7.2 Estimate the ratio: $\Gamma\left(B^- \to D^0 K^{*-}\right)/\Gamma\left(B^- \to D^0 \rho^-\right)$.

The valence quark compositions are: $B^- = b\bar{u}$, $D^0 = c\bar{u}$, $\rho^- = d\bar{u}$, $K^{*-} = s\bar{u}$. We draw the diagrams.

Looking at the vertex coefficients we have

$$\frac{\Gamma\left(B^- \to D^0 K^{*-}\right)}{\Gamma\left(B^- \to D^0 \rho^-\right)} \propto \frac{\left|V_{us}\right|^2 \left|V_{cb}\right|^2}{\left|V_{ud}\right|^2 \left|V_{cb}\right|^2} = \frac{\left|V_{us}\right|^2}{\left|V_{ud}\right|^2} \approx \tan^2 \theta_C \approx 0.05.$$

The experimental value is ≈ 0.05. $\square$

7.13 Weak Neutral Currents

We have seen that flavour-changing neutral current processes are strongly suppressed. However, flavour-conserving neutral current processes exist in nature. The experimental search for such processes went on for many years. Starting in 1965, A. Lagarrigue led the construction of a 'giant' bubble chamber, with a broad research programme mainly in neutrino physics. It was called Gargamelle, after the name of the mother of the giant Gargantua, to pay homage to Rabelais. Gargamelle was filled with 15 t of CF_3Br, which is a freon, a heavy liquid that provides both the mass necessary for an appreciable neutrino interaction rate and a good γ detection probability, with a short radiation length, $X_0 = 11$ cm. The neutrino beam was built at CERN from the proton synchrotron (PS in Fig. 1.17). The experiments with this beam made many contributions to neutrino physics in the 1970s, in particular the discovery of neutral currents in 1973 (Hasert *et al.* 1973). Let us see how.

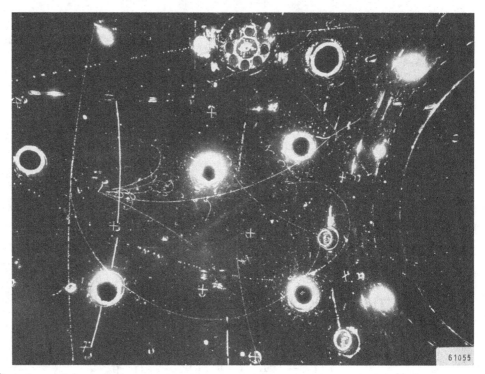

61055

Fig. 7.24.   A neutral current event in Gargamelle (© CERN).

The incident beam contains mainly v_μ (with a small v_e contamination). All the CC events have a μ^- in the final state, which is identified by its straight non-interacting minimum ionizing track.

If neutral currents exist, the following process can happen on a generic nucleus N

$$v_\mu + N \to v_\mu + \text{hadrons}. \tag{7.109}$$

This type of event is identified by the absence of the muon in the final state, which contains only hadrons (the neutrino cannot be seen). Fig. 7.24 is an example discussed by Perkins (2004).

Analysing the image, we identify all the tracks as hadrons and none as a muon. Neutrinos enter from the left of the picture and one of them interacts. Around the vertex we see the following (clockwise): a short dark track directed upward, which is recognized as a stopping proton; two e^+e^- pairs that are the materialization of the two photons from a decay $\pi^0 \to \gamma\gamma$; and two charged tracks of opposite signs. The track moving upwards is negative (as inferred by the known direction of the magnetic field) and interacts (it passes below two eye-shaped images; the interaction is near the second one); therefore, it is a hadron. The positive track is a π^+ that ends with a charge-exchange reaction producing a π^0, as recognized from the electron, originated by the Compton scattering of one of the γs from its decay. The electron is the small vertical track under the 'eye' pointing to the end point of the π^+ track.

The neutral currents differ from the CCs in two very important aspects

(1) they are 'diagonal', namely they transform a particle in itself
(2) they contain both the negative (left) and positive (right) components of the Dirac
    spinors.

Processes that would exist if non-diagonal terms in the neutral currents were present
have been, and are, experimentally searched for, but none has been found. Stringent
limits exist (Workman *et al.* 2022), for example, in the muon decays

$$\frac{\Gamma\left(\mu^- \to e^-\gamma\right)}{\Gamma_{tot}} < 4.2\times10^{-13} \tag{7.110}$$

and

$$\frac{\Gamma\left(\mu^- \to e^-e^-e^+\right)}{\Gamma_{tot}} < 1.0\times10^{-12}, \tag{7.111}$$

and in the cross-sections of the so-called muon conversion processes on nuclei,
compared with the total capture cross-section, for example

$$\frac{\sigma\left(\mu^-Au \to e^-Au\right)}{\sigma\left(\mu^-Au \to capture\right)} < 7\times10^{-13} \tag{7.112}$$

The second difference is that the NC interactions contain both left, $g_L^i\bar{\psi}_{i,L}\gamma_\mu\psi_{i,L}$,
and right, $g_R^i\bar{\psi}_{i,R}\gamma_\mu\psi_{i,R}$, currents, where the index i runs on all the fermions, the six
quarks and the six leptons for the left ones, on the charged fermions only for the right
currents, and g_L^i and g_R^i are numerical constants. In Chapter 9 we shall see how the
electroweak theory connects these constants to the electric charges of the fermions.

Problems

7.1 Draw the Feynman quark diagrams of the following decays:
$K^{*+} \to K^0 + \pi^+; n \to p + e^- + \bar{\nu}_\mu; \pi^+ \to \mu^+ + \nu_\mu$.

7.2 Draw the Feynman quark diagrams of the following decays:
$\pi^+ \to \pi^0 + e^+ + \nu_e; \rho^+ \to \pi^0 + \pi^+; K^0 \to \pi^- + \pi^+; \Lambda \to p + e^- + \bar{\nu}_e$.

7.3 Find the value of the Fermi constant G_F in SI units, knowing that
$G_F / \left(\hbar c\right)^3 = 1.17 \times 10^{-5}\,\text{GeV}^{-2}$.

7.4 The PEP was a collider in which two beams of e^+ and e^- collided in the CM reference
frame. Consider the beam energy $E_{cm} = 29$ GeV and the reaction $e^+ + e^- \to \tau^+ + \tau^-$.
Find the average distance the τ will fly before decaying.

7.5 Consider the decays $\mu^+ \to e^+ + \nu_e + \bar{\nu}_\mu$ and $\tau^+ \to e^+ + \nu_e + \bar{\nu}_\tau$. The branching ratios
are 100% for the first and 16% for the second. The μ lifetime is $\tau_\mu = 2.2\ \mu s$. Calculate
the τ_τ lifetime.

7.6 Neglecting the masses, calculate the cross-section of the process: $e^+e^- \to \tau^+\tau^-$ at $\sqrt{s} = 10$ GeV and at $\sqrt{s} = 100$ GeV.

7.7 What are the differences between a neutrino and an antineutrino? What are the conserved quantities in neutrino scattering? Complete the missing particle in $\nu_\mu + e^- \to \mu^- + ?$. If neutrinos are massless, what is the direction of their spin? And for antineutrinos? The Universe is full of neutrinos at a temperature of about 2 K. What is the neutrino average speed if their mass is 50 meV?

7.8 Write the reaction (or the reactions, if they are more than one) by which a ν_μ can produce a single pion hitting: (a) a proton and (b) a neutron. Does the decay $\mu^+ \to e^+ + \gamma$ exist? Does $\mu^+ \to e^+ + e^+ + e^-$ exist? Give reasons for your answers.

7.9 We send a π^- beam onto a target and we observe the inclusive production of Λ. We measure the momentum p_Λ and the polarization σ_Λ of the hyperon. How can we check if parity is conserved in these reactions? What do you expect to happen?

7.10 How can you observe parity violation in the decay $\pi \to \mu\nu$?

7.11 The muons have the same interactions, electromagnetic and weak, as the electrons. Why does a μ with energy of a few GeV pass through an iron slab, while an electron of the same energy does not?

7.12 What is the minimum momentum of the electron from the decay of a μ at rest? What is the maximum momentum?

7.13 Cosmic rays are mainly protons. Their energy spectrum decreases with increasing energy. Their interactions with the atmospheric nuclei produce mesons, which give rise, by decaying, to ν_μ and ν_e. In a sample of $N_\nu = 10^6 \nu_\mu s$ with 1 GeV of energy, how many interact in crossing the Earth along its diameter? (Note: $\sigma \approx 7$ fb, $\rho \approx 5 \times 10^3$ kg m^{-3}, $R \approx 6000$ km.)

7.14 Consider the neutrino cross-section on an electron $\sigma\left(\nu_\mu e^- \to \nu_\mu e^-\right) \approx \dfrac{G_F^2}{\pi} s$ and on an 'average nucleon' (namely the average cross-section on protons and neutrons) $\sigma\left(\nu_\mu N \to \mu^- h\right) \approx 0.2 \times \dfrac{G_F^2}{\pi} s$ at energies $\sqrt{s} \gg m$, where m is the target mass and h is any hadronic state (the factor 0.2 is due to the quark distribution inside the nucleon). Calculate their ratio at $E_\nu = 50$ GeV. How does this ratio depend on energy? Calculate σ / E_ν for the two reactions.

7.15 Draw the Feynman diagrams at tree level for the elastic scattering $\bar{\nu}_e e^-$. What is the difference in $\nu_e e^-$?

7.16 The GALLEX experiment at the Gran Sasso laboratory measured the ν_e flux from the Sun by counting the electrons produced in the reaction $\nu_e + {}^{71}\text{Ga} \to {}^{71}\text{Ge} + e^-$. Its energy threshold is $E_{\text{th}} = 233$ keV. From the solar luminosity one finds the expected neutrino flux $\Phi = 6 \times 10^{14}$ m^{-2}s^{-1}. For a rough calculation, assume the whole flux to be above threshold and the average cross-section to be $\sigma = 10^{-48}$ m^2. Assuming the detection efficiency $\varepsilon = 40\%$, how many ^{71}Ga nuclei are necessary to have one neutrino interaction per day? What is the corresponding ^{71}Ga mass? What is the natural gallium mass if the abundance of the ^{71}Ga isotope is $a = 40\%$? (The measured flux turned out to be about one half of the expected one. This was a fundamental observation in the process of discovering of neutrino oscillations.)

7.17 How many metres of Fe must a v_μ of 1 GeV penetrate to interact, on average, once? How long does this take? Compare that distance with the diameter of the Earth's orbit. (Note: $\sigma = 0.017$ fb, $\rho = 7.7 \times 10^3$ kg m^{-3}, $Z = 26$, $A = 56$.)

7.18 Write down a Cabibbo favoured (CF) and a singly Cabibbo suppressed (SCS) semileptonic decay of the c quark. Write three CF and three SCS decays of D^+.

7.19 Draw the Feynman diagram for anti-bottom quark decay, favoured by the mixing. Write three favoured decay modes of the B^+.

7.20 Draw the principal Feynman diagrams for the top quark decay.

7.21 Draw the Feynman diagrams for bottom and charm decays. Estimate the ratio $\Gamma(b \to c + e + v_e) / \Gamma(b \to c)$.

7.22 Consider the measured decay rates $\Gamma(D^+ \to \bar{K}^0 e^+ v_e) = (7 \pm 1) \times 10^{10}$ s^{-1} and $\Gamma(\mu^+ \to e^+ v_e \bar{v}_\mu) = 1/(2.2 \mu s)$. Justify the ratio of the two quantities.

7.23 Consider the decays: (1) $D^+ \to \bar{K}^0 + \pi^+$; (2) $D^+ \to K^+ + \bar{K}^0$; (3) $D^+ \to K^+ + \pi^0$. Find the valence quark composition and establish whether it is favoured, suppressed or doubly suppressed for each of them.

7.24 Consider the measured values of the ratio $\Gamma(\Sigma^- \to ne^- \bar{v}_e)/\Gamma_{tot} \approx 10^{-3}$ and of the upper limit $\Gamma(\Sigma^+ \to ne^+ v_e)/\Gamma_{tot} < 5 \times 10^{-6}$. Give the reason for such a difference.

7.25 Consider the decays: (1) $B^0 \to D^- + \pi^+$; (2) $B^0 \to D^- + K^+$; (3) $B^0 \to \pi^- + K^+$; (4) $B^0 \to \pi^- + \pi^+$. Find the valence quark composition of each of them, establish the dependence of the partial decay rates on the mixing matrix element and sort them in decreasing order of these rates.

7.26 A pion with momentum $p_\pi = 500$ MeV decays in the channel $\pi^+ \to \mu^+ + v$. Find the minimum and maximum values of the μ momentum. What are the flavour and the chirality of the neutrino?

7.27 Consider a large water Cherenkov detector for solar neutrinos. The electron neutrinos are detected by the reaction $v_e + e^- \to v_e + e^-$. Assume the cross-section (at about 10 MeV) $\sigma = 10^{-47}$ m^2 and the incident flux in the energy range above threshold $\Phi = 10^{10}$ m^{-2}s^{-1}. What is the water mass in which the interaction rate is 10 events a day if the detection efficiency is $\varepsilon = 50\%$?

7.28 An iron-core star ends its life in a supernova explosion, if its mass is large enough. The atomic electrons are absorbed by nuclei by the process $e + Z \to (Z-1) + v_e$. The star core implodes and its density grows enormously. Assume an iron core with density $\rho = 100\,000$ t mm^{-3}. Consider the neutrino energy $E_v = 10$ MeV and its cross-section on iron $\sigma \approx 3 \times 10^{-46}$ m^2. Find the neutrino mean free path. ($A_{Fe} = 56$.)

7.29 In 1959, B. Pontecorvo proposed an experimental idea to establish whether or not $\bar{v}_e$ and $\bar{v}_\mu$ are different particles. To produce $\bar{v}_\mu$, a low-energy π^+ beam is brought to rest in a target. The μ^+s from their decays come to rest too and then decay.

(1) What is the lowest energy-threshold reaction that would be permitted if $\bar{v}_e = \bar{v}_\mu$ but forbidden if $\bar{v}_e \neq \bar{v}_\mu$?

(2) What is its energy threshold?

(3) Does the considered process provide any $\bar{v}_\mu$ above threshold?

7.30 Give a cascade of 'Cabibbo favoured' decays through flavoured hyperons of the following charmed hyperons: $\Sigma_c^{++}(uuc)$, $\Xi_c^+(usc)$ and $\Omega_c^0(ssc)$.

7.31 Give a cascade of 'Cabibbo favoured' decays through flavoured hyperons of the following beauty hyperons: $\Sigma_b^+ (uub)$, $\Xi_b^- (dsb)$ and $\Lambda_b^0 (udb)$

7.32 Consider the two beta decays of the Σ^- hyperon $\Sigma^- \to n + e^- + \nu_e$ of branching ratio $1.017 \pm 0.034 \times 10^{-3}$ and $\Sigma^- \to \Lambda + e^- + \nu_e$ of branching ratio $5.73 \pm 0.27 \times 10^{-5}$. State qualitatively the reason for the difference.

Summary

This is the first chapter on weak interactions. We started with the CC interactions and have seen the following:

• the point-like Fermi interaction approximation at low energies and the Fermi constant
• the current–current structure of the CC Lagrangian
• the parity and particle–antiparticle conjugation violation
• the measurement of the helicity of the leptons in the weak decays
• the lepton universality
• the Cabibbo mixing of the quarks and the quark mixing matrix.

We then discussed the discovery of the neutral current weak interactions.

Further Reading

Haidt, D. & Pullia, A. (2013) The weak neutral current discovery and impact. *Riv. del Nuov. Cim.* 36 335–395

Lee, T. D. (1957) Nobel Lecture; *Weak Interactions and Nonconservation of Parity*

Okun, L. B. (1981) *Leptony i kwarki* Nauka Moscow [English translation: *Leptons and Quarks*], North-Holland, Amsterdam (1982)]

Pullia, A. (1984) Structure of charged and neutral weak interactions at high energy. *Riv. del Nuov. Cim.* 7 *Series* 3

Yang, C. N. (1957) Nobel Lecture; *The Law of Parity Conservation and other Symmetry Laws of Physics*

8

Oscillations and $C\mathcal{P}$ Violation in Quarks

In this chapter we shall discuss two important aspects, different but correlated, of the weak interactions. One is the oscillations between members of flavoured, electrically neutral meson pairs. The other is the $C\mathcal{P}$ violation phenomena that are strictly connected with the mixing, but are not exclusive of this, having been observed also in the decay of charged mesons (Section 8.10).

We begin with an elementary discussion of the neutral K system that will elucidate the states of definite strangeness, those of definite $C\mathcal{P}$ and those with definite mass and lifetime. We shall describe the oscillation between the states of definite strangeness, giving the relevant mathematical expressions and discussing the experimental evidence, including the observation of the regeneration of the initial flavour.

We then define the different modes of $C\mathcal{P}$ violation: in the wave function (known also as 'in the mixing'), in the interference between decay with and without mixing and in the decays (known also as 'direct violation'). All modes happen for neutral mesons; the last one also happens for charged mesons.

In Section 8.6 we describe oscillations and $C\mathcal{P}$ violation in the B^0 system, which needs a somewhat more advanced formalism. Beautiful experimental results have been obtained at high-luminosity electron–positron colliders, KEKB in Japan and PEP2 in California, built on purpose and called *beauty factories*. PEP2 concluded its life in 2008. KEKB went through an upgrade programme to SuperKEKB to increase its luminosity by about two orders of magnitude up to the design figure of $8 \times 10^{35} \, \text{cm}^{-2}\text{s}^{-1}$. In June 2022, after 2 years of commissioning, SuperKEKB achieved the world-record luminosity of $4.7 \times 10^{34} \, \text{cm}^{-2}\text{s}^{-1}$.

Beauty physics is also the target of the dedicated experiment LHCb at LHC, and part of the programme of the general purpose experiments ATLAS and CMS, in particular, but not only, for the B_s^0, which is not accessible to beauty factories. We shall see in Section 8.7 examples of $C\mathcal{P}$ violation in this system. $C\mathcal{P}$ violation in the decay of neutral mesons in general is discussed in Section 8.8, for the charm sector in Section 8.9 and for charged mesons in Section 8.10. Finally, in Section 8.11, we shall look at how the many different measurements can be put together to test the SM with the unitary triangle.

Oscillations, mixing and $C\mathcal{P}$ violation have also been discovered, contrarily to the prediction of the SM, in the neutrino sector. This will be studied in Chapter 10.

8.1 Flavour Oscillations, Mixing and CP Violation

The physics of the flavoured, electrically neutral meson–antimeson pairs is an important chapter of particle physics, providing beautiful examples of quantum two-state systems.

Since the top quark does not bind inside hadrons, there are four such meson doublets: three made of down-type quarks, the K^0s (d and s), the B^0s (d and b) and the B_s^0s (s and b), and one of up-type quarks, the D^0s (u and c). In each case, the states with definite flavour differ from the states with definite mass and lifetime, that is, the stationary states.

Consider the neutral kaons. They are produced by a strong interaction as K^0, with positive strangeness, or $\overline{K}^0$, with negative strangeness, antiparticles of one another. Both of them are quantum superpositions of two states of definite masses and lifetimes, called K_S and K_L (for shorter and longer lifetime). This is called 'mixing'. The two states have slightly different masses, and this originates a quantum oscillation phenomenon, specifically the strangeness oscillation. Consider a beam of mesons originally containing only K^0s. We observe that, as time goes by, the probability of observing in it a $\overline{K}^0$, which was initially zero, gradually increases, reaches a maximum and decreases again, and so on. The opposite happens for the probability of observing a K^0. It is initially 100%, then decreases, reaches a minimum and increases again. Flavour oscillations have been observed, in historical order, in the K^0, in the B^0 and in the B_s^0. We speak of 'survival probability' of the originally present particle and 'appearance probability'. For the D^0, very small mixing and oscillations have been observed.

The neutral kaon oscillations were theoretically predicted in 1955 and experimentally established in 1960, as we shall see in Sections 8.3 and 8.4. Only in 1987, 27 years later, was the B^0 oscillation observed by the UA1 experiment at CERN (see Section 9.6) and ARGUS at the positron–electron collider DORIS at DESY. Almost another 20 years were needed to observe B_s^0 oscillations, discovered in 2006, by the CDF and D0 experiments at the Tevatron collider at Fermilab, the D^0 mixing was discovered by BaBar and Belle in 2007.

A second very important phenomenon present in the neutral flavoured meson system is the violation of CP symmetry. It had already been discovered in 1964 in the neutral kaon system (Section 8.5). This remained the only example of CP violation for 37 years, until 2001. That year, two high-luminosity e^+e^- colliders were in operation, designed to produce neutral B pairs with high statistics, called 'beauty factories', with the associated experiments: PEP2 with BaBar at SLAC, in the USA, and KEKB with Belle at KEK, in Japan. In 2021, CP violation was observed by the LHCb experiment in the B_s^0 and in 2019 in the D^0 system. Notice, however, that oscillations and CP violation are independent phenomena. As a matter of fact, CP violation had been established in 2008 in decays of the charged B.

The quantum mechanical description of the oscillation phenomenon is identical for all the neutral meson pairs. However, the phenomenology and the corresponding experimental techniques vary from case to case due to the largely different values of two characteristic times, the oscillation period and the shorter lifetime. The angular frequency of the oscillation is equal, in natural units, to the mass difference Δm between the two mass eigenstates. Here we shall give the orders of magnitude of the relevant quantities, and in Table 8.1 in Section 8.11 we shall give the precise values. The oscillation period $T = 2\pi / \Delta m$ differs by orders of magnitude between the four

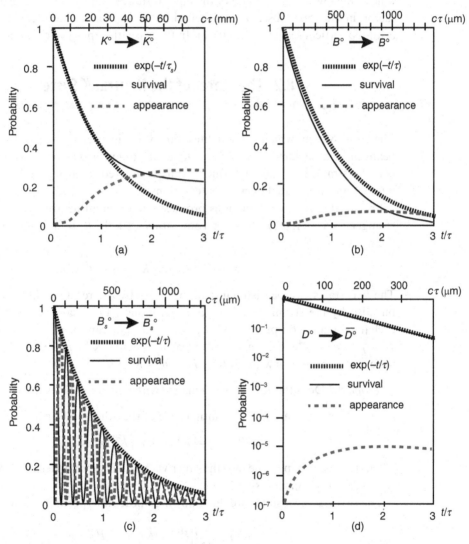

Fig. 8.1. Appearance, survival, oscillation and decay probabilities for neutral mesons as functions of the number of lifetimes (lower axis) and $c\tau$ (upper axis). (a) K^0 system, (b) B^0 system, (c) B_s^0 system, (d) D^0 system. Note the logarithmic vertical scale.

mesons, being of the order of 1000 ps for the K^0, 400 ps for the D^0, 12 ps for the B^0 and 0.3 ps for the B_s^0. Moreover, the phenomenon can be observed only when both eigenstates are present, namely a few times the shorter lifetime (after which only one eigenstate is present). The shorter lifetime is very different from the much longer lifetime in the case of the K^0, but almost equal to it in the other cases. The ratio of the oscillation period to the shorter lifetime is of the order of 10 for the K^0 and the B^0 (about 12 and 8 respectively), two orders smaller for the B_s^0 (about 0.2) and very large, of the order of 1000, for the D^0. The appearance and survival probabilities are shown in Fig. 8.1, in historical order of discovery, as functions of time normalized to the lifetime in the lower abscissa and of the flight distance, to be multiplied by the Lorentz γ factor, in the upper abscissa, together with the exponential decay curve. Note the logarithmic vertical scale for the charm. In this case, the appearance signal is very small.

8.2 The States of the Neutral K System

The kaon system, which is the lightest one, was historically the first to be studied, at beam energies of a few GeV. K^0 and $\bar{K}^0$ are distinguished by only one quantum number, the strangeness flavour, which, while conserved in strong and electromagnetic interactions, is violated by weak interactions.

Specifically, strong interactions produce two different neutral K mesons, one with strangeness $S = +1$, the $K^0 = d\bar{s}$, and one with $S = -1$, the $\bar{K}^0 = s\bar{d}$, through, for example, the reactions

$$K^+ + n \rightarrow K^0 + p \text{ and } K^- + p \rightarrow \bar{K}^0 + n. \tag{8.1}$$

The two states can be distinguished, not only by the reaction that produced them, but also by the strong reactions they can induce. For example, a K^0 may 'charge exchange' scatter on protons $K^0 + p \rightarrow K^+ + n$, but cannot produce a hyperon as in $K^0 + p \rightarrow \pi^0 + \Sigma^+$, while a $\bar{K}^0$ produces a hyperon $\bar{K}^0 + p \rightarrow \pi^0 + \Sigma^+$ but does not have charge exchange with protons $\bar{K}^0 + p \rightarrow K^+ + n$.

Question 8.1 Does the K^0 charge exchange with neutrons? Does the $\bar{K}^0$ do that? □

The two mesons are each the antiparticle of the other, and we have

$$CP \,|\, K^0 \rangle = |\, \bar{K}^0 \rangle, \quad CP \,|\, \bar{K}^0 \rangle = |\, K^0 \rangle. \tag{8.2}$$

K^0 and $\bar{K}^0$ can change one into the other via virtual common decay modes, mainly as $K^0 \leftrightarrow 2\pi \leftrightarrow \bar{K}^0$ and $K^0 \leftrightarrow 3\pi \leftrightarrow \bar{K}^0$.

The two CP eigenstates are the following linear superpositions of K^0 and $\bar{K}^0$

$$\left| K_1^0 \right\rangle = \frac{1}{\sqrt{2}} \left(\left| K^0 \right\rangle + \left| \bar{K}^0 \right\rangle \right) \quad CP = +1$$

$$\left| K_2^0 \right\rangle = \frac{1}{\sqrt{2}} \left(\left| K^0 \right\rangle - \left| \bar{K}^0 \right\rangle \right) \quad CP = -1. \tag{8.3}$$

Let us now consider the 2π and 3π neutral systems. As we know, the CP eigenvalue of a neutral two-π system is positive. Actually, we recall that

$$CP(\pi^0\pi^0) = \left[CP(\pi^0)\right]^2 = (-1)^2 = +1$$
$$CP(\pi^+\pi^-) = C(\pi^+\pi^-)P(\pi^+\pi^-) = (-1)^l(-1)^l = +1. \tag{8.4}$$

As a consequence, if CP is conserved, only the K_1^0, the CP eigenstate with the eigenvalue $CP = +1$, can decay into 2π.

Let us now consider the neutral three-π systems. The case of three π^0 is easy. We have

$$CP(\pi^0\pi^0\pi^0) = \left[CP(\pi^0)\right]^3 = (-1)^3 = -1. \tag{8.5}$$

The state $\pi^+\pi^-\pi^0$ requires more work. Let us call **l** the angular momentum of the two-pion $\pi^+\pi^-$ system in their centre of mass reference, and **L** the π^0 angular momentum relative to the two-pion system in the overall centre of mass frame. The total angular momentum of the 3π system is the sum of the two and must be zero, namely $\mathbf{J} = \mathbf{l} \oplus \mathbf{L} = \mathbf{0}$, implying that $\mathbf{l} = \mathbf{L}$. Therefore, the parity is $P = P^3(\pi)(-1)^l(-1)^L = -1$. As for the charge conjugation we have $C(\pi^0) = +1$ and $C(\pi^+\pi^-) = (-1)^l$. In total we have

$$CP(\pi^+\pi^-\pi^0) = (-1)^{l+1}.$$

We now take into account the fact that the difference between the K mass and the mass of three pions is small, $m(K) - 3m(\pi) = 80$ MeV. Therefore, the phase-space volume in the decay is very small. This strongly favours the S wave, namely $l = 0$ and then $CP = -1$. In principle, the $CP = +1$ decays might occur, but with minimum angular momenta $l = L = 1$; in practice their kinematic suppression is so large, that they do not exist and we have

$$CP(\pi^+\pi^-\pi^0) = (-1)^{l+1} = -1. \tag{8.6}$$

In conclusion, if CP is conserved, only the CP eigenstate with the eigenvalue $CP = -1$, the K_2^0, can decay into 3π. Summing up, if CP is conserved, we have

$$K_1^0 \to 2\pi, \, K_2^0 \nrightarrow 2\pi, \, K_1^0 \nrightarrow 3\pi, \, K_2^0 \to 3\pi. \tag{8.7}$$

If CP were absolutely conserved, K_1^0 and K_2^0 would be the states of definite mass and lifetime. As we shall see, CP is very slightly violated; therefore, the states of definite mass and lifetimes, called K_S and K_L ('K short' and 'K long' respectively) are not exactly K_1^0 and K_2^0 However, the difference is very small, and we shall neglect it for the time being.

Experimentally, the lifetime of the (short) state decaying into 2π, τ_S, is about 570 times shorter than the lifetime of the (long) state decaying into 3π, τ_L. The values are

$$\tau_S = 89.54 \pm 0.04 \, \text{ps} \quad \tau_L = 51.16 \pm 0.21 \, \text{ns}. \tag{8.8}$$

The long life of K_L is due to the fact that its decay into 2π is forbidden by CP while its CP-conserving decay into 3π is hindered by the small Q value of the decay. This very fact shows that the CP violation by weak interactions is small, if it occurs at all.

Let us also look at the widths and at the difference between them. From (8.8) we have

$$\Gamma_S = \frac{1}{\tau_S} = 7.4 \,\mu\text{eV}; \quad \Gamma_L = \frac{1}{\tau_L} = 0.013 \,\mu\text{eV}$$
$$\Delta\Gamma = \Gamma_S - \Gamma_L \simeq \Gamma_S = 7.4 \,\mu\text{eV} = 11.2 \,\text{ns}^{-1} \tag{8.9}$$

Other related quantities are

$$c\tau_S = 2.67 \,\text{cm}; \quad c\tau_L = 15.5 \,\text{m}. \tag{8.10}$$

Suppose we produce a neutral K beam sending a proton beam extracted from an accelerator onto a target. Initially it contains both K_S and K_L. However, the beam composition varies with the distance from the target. Take, for example, a K beam momentum of 5 GeV, corresponding to the Lorentz factor $\gamma \approx 10$. In a lifetime the K_S travel $\gamma c\tau_S = 27$ cm. Therefore, at a distance of a few metres (in a vacuum) we have a pure K_L beam, no matter the initial composition.

Let us now consider the masses, which, we recall, are defined for the states K_L and K_S. It happens that the difference between them is extremely small, so small that it cannot be measured directly. The measured mean value of the neutral kaon masses (Workman *et al.* 2022) is

$$m(K^0) = 497.611 \pm 0.013 \,\text{MeV} \tag{8.11}$$

The mass difference is indirectly measured from the strangeness oscillation period, which we shall see in the next section. Its value is

$$\Delta m = m_L - m_S = 3.481 \pm 0.006 \,\mu\text{eV}$$
$$= 5.293 \pm 0.009 \,\text{ns}^{-1}, \tag{8.12}$$

Which, in relative terms, is only 7×10^{-15} of the K^0 mass. Notice that $\Delta m > 0$, which is not a consequence of the definitions, but means that the larger mass K^0 lives longer.

8.3 Strangeness Oscillations

In 1955, Gell-Mann and Pais, A. (Gell-Man & Pais 1955) pointed out that a peculiar phenomenon, strangeness oscillations, should happen in an initially pure K^0 beam prepared, for example, by using the reaction $\pi^- p \to K^0 \Lambda$. Let us look at the probability of finding a K^0 and that of finding a $\bar{K}^0$ as functions of the proper time t. From an experimental point of view, the time corresponds to the distance from the target. The states of definite mass m_i and definite lifetime, or equivalently definite width Γ_i, have the time dependence $\exp\left[-i\left(m_i - i\Gamma_i/2\right)\right]t$. These are neither the K^0 nor the $\bar{K}^0$ but, provided CP is conserved, the CP eigenstates

$$\left|K_1^0\right\rangle = \frac{1}{\sqrt{2}}\left(\left|K^0\right\rangle + \left|\bar{K}^0\right\rangle\right), \quad \left|K_2^0\right\rangle = \frac{1}{\sqrt{2}}\left(\left|K^0\right\rangle - \left|\bar{K}^0\right\rangle\right). \tag{8.13}$$

The K^0 is a superposition of these, namely

$$\left|K^0\right\rangle = \frac{1}{\sqrt{2}}\left(\left|K_1^0\right\rangle + \left|K_2^0\right\rangle\right), \quad \left|\bar{K}^0\right\rangle = \frac{1}{\sqrt{2}}\left(\left|K_1^0\right\rangle - \left|K_2^0\right\rangle\right). \tag{8.14}$$

Therefore, the temporal evolution of the wave function (the suffix 0 is to remind us that, at $t = 0$, the state is K^0, as opposed to $\bar{K}^0$) is

$$\Psi_0(t) = \frac{1}{2}\left[\left(\left|K^0\right\rangle + \left|\bar{K}^0\right\rangle\right)e^{-im_S t - \frac{\Gamma_S}{2}t} + \left(\left|K^0\right\rangle - \left|\bar{K}^0\right\rangle\right)e^{-im_L t - \frac{\Gamma_L}{2}t}\right]. \tag{8.15}$$

To understand the phenomenon better, assume for the time being that the mesons are stable, $\Gamma_S = \Gamma_L = 0$. Expression (8.15) becomes

$$\Psi_0(t) = \frac{1}{2}\left[e^{-im_S t} + e^{-im_L t}\right]|K^0\rangle + \frac{1}{2}\left[e^{-im_S t} - e^{-im_L t}\right]|\bar{K}^0\rangle. \tag{8.16}$$

The probability of finding a K^0 in the beam at time t is

$$\begin{aligned}
\left|\left\langle K^0 \middle| \Psi_0(t)\right\rangle\right|^2 &= \frac{1}{4}\left|e^{-im_S t} + e^{-im_L t}\right|^2 \\
&= \frac{1}{2}[1 + \cos(\Delta m \cdot t)] = \cos^2\left(\frac{\Delta m}{2}t\right).
\end{aligned} \tag{8.17}$$

A correlated feature is the appearance in time of $\bar{K}^0$s in the initially pure K^0 beam. The probability of finding a $\bar{K}^0$ is

$$\begin{aligned}
\left|\left\langle \bar{K}^0 \middle| \Psi_0(t)\right\rangle\right|^2 &= \frac{1}{4}\left|e^{-im_S t} - e^{-im_L t}\right|^2 \\
&= \frac{1}{2}[1 - \cos(\Delta m \cdot t)] = \sin^2\left(\frac{\Delta m}{2}t\right).
\end{aligned} \tag{8.18}$$

Summing up, the probabilities of finding a K^0 and a $\bar{K}^0$ are initially one and zero, respectively. As time passes, the former decreases, the latter increases, so much so that at time $T/2$, the probability of finding a K^0 becomes zero and that of finding a $\bar{K}^0$ is one. Then the process continues with inverted roles. The two-state quantum system 'oscillates' between the two opposite flavour states. It is a 'beat' phenomenon between the monochromatic waves corresponding to the two eigenstates. In NU the two angular frequencies are equal to the masses, as seen in (8.18). Therefore, the oscillation period is $T = 2\pi/|\Delta m| \approx 1.2$ ns. As anticipated, the measurement of the period gives the mass difference but, notice, only in absolute value.

To appreciate the order of magnitude, consider a beam energy of 10 GeV. The first oscillation maximum is at the distance $\gamma cT/2 = 3.6$ m.

As for the sign of Δm, we give only the following hint. If the K^0 beam travels in a medium, its refraction index is different than in vacuum, as happens for photons. Since the index depends on Δm in magnitude and sign, the latter can be determined. The result is that $\Delta m > 0$.

We talked above of the probability of observing a K^0 or a $\bar{K}^0$, but how can we distinguish them? We cannot do this by observing the 2π or 3π decay, because these channels select the states with definite $\mathcal{CP}$, not those of definite strangeness.

To select definite strangeness states, we must observe their semileptonic decays. These decays obey the '$\Delta S = \Delta Q$ rule' which is as follows: the difference between the strangeness of the hadrons in the final and initial states is equal to the difference of their electric charges. The rule, which was established experimentally, is a consequence of the quark contents of the states

$$
\begin{aligned}
K^0 = \bar{s}d \quad \bar{s} \to \bar{u}l^+\nu_l &\Rightarrow K^0 \to \pi^-l^+\nu_l; \quad K^0 \text{ not} \to \pi^+l^-\bar{\nu}_l \\
\bar{K}^0 = s\bar{d} \quad s \to ul^-\bar{\nu}_l &\Rightarrow \bar{K}^0 \to \pi^+l^-\bar{\nu}_l; \quad \bar{K}^0 \text{ not} \to \pi^-l^+\nu_l
\end{aligned}
\tag{8.19}
$$

We see that the sign of the charged lepton flags the strangeness of the K. The semileptonic decays are called K^0_{e3} and $K^0_{\mu3}$ depending on the final charged lepton. It is easy to observe them due to their large branching ratios, namely

$$
BR\left(K^0_{e3}\right) \simeq 39\%; \quad BR\left(K^0_{\mu3}\right) \simeq 27\%.
\tag{8.20}
$$

Let us now call $P^\pm(t)$ the probabilities of observing a $+$ and a $-$ lepton at time t. These are the survival probability of the initial flavour and the appearance probability of the other flavour (see Fig. 8.1(a)). Considering unstable kaons now, the probabilities are

$$
P^+(t) = \left|\langle K^0 | \Psi_0(t)\rangle\right|^2 = \frac{1}{4}\left[e^{-\Gamma_S t} + e^{-\Gamma_L t} + 2e^{-\frac{\Gamma_S + \Gamma_L}{2}t}\cos(\Delta m\, t)\right],
\tag{8.21a}
$$

$$
P^-(t) = \left|\langle \bar{K}^0 | \Psi_0(t)\rangle\right|^2 = \frac{1}{4}\left[e^{-\Gamma_S t} + e^{-\Gamma_L t} - 2e^{-\frac{\Gamma_S + \Gamma_L}{2}t}\cos(\Delta m\, t)\right].
\tag{8.21b}
$$

Both expressions are the sums of two decreasing exponentials and a damped oscillating term. The damping is dominated by the smaller lifetime $\tau_S = 90$ ps. Therefore, the phenomenon is observable only within a few τ_S. Over such short times we can consider the term $e^{-\Gamma_L t}$ as a constant (remember that $\tau_L = 51$ ns). Observe finally that τ_S is much smaller than the oscillation period $T \approx 1.2$ ns. Therefore the damping is strong. Fig. 8.1(a) shows the two probabilities.

Experimentally one measures the charge asymmetry, namely the difference between the numbers of observed $K^0 \to \pi^-l^+\nu_l$ events and $\bar{K}^0 \to \pi^+l^-\bar{\nu}_l$ events. We see from (8.21) that this is a damped oscillation

$$
\delta(t) \equiv P^+(t) - P^-(t) = e^{-\frac{\Gamma_S}{2}t}\cos(\Delta m \cdot t).
\tag{8.22}
$$

The experimental results are shown in Fig. 8.2.

The fit of the experimental points gives us Γ_S and $|\Delta m|$. We have already given their values.

Let us look more carefully at the data. We see that at very late times $(t \gg \tau_S)$ when only K_L survive, the asymmetry does not go to zero as it should, according to (8.22). This implies that the two components K^0 and $\bar{K}^0$ did not become equal and,

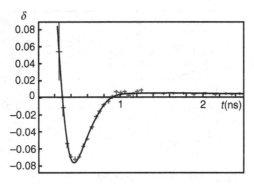

Fig. 8.2. Charge asymmetry (Reprinted from S. Gjesdal *et al. Phys. Lett.* B52 (1974) 113, Fig. 1 © 1974 with permission by Elsevier. https://doi.org/10.1016/0370-2693(74)90734-5).

consequently, that the longer lifetime state is not a CP eigenstate. The wave function of the eigenstate contains a small 'impurity' with the 'wrong' CP. We shall come back to CP violation in Section 8.5.

8.4 Regeneration

The crucial test of the Gell-Mann and Pais theory discussed in the previous section was proposed by Pais and Piccioni in 1955 (Pais & Piccioni 1955) and performed by Piccioni and collaborators in 1960 (Muller *et al.* 1960).

Fig. 8.3 shows a scheme of the experiment. A π^- beam bombards the thin target A producing K^0 by the reaction $\pi^- p \rightarrow K^0 \Lambda$. The K^0 state is the mixture (8.14) of K_1^0 and K_2^0. The former component decays mainly into 2π and does so at short distances, the latter survives for longer times and does not decay into 2π, provided CP is conserved. We observe the 2π decays immediately after the target, with a decreasing rate as we move farther away. When the short component has disappeared, for all practical purposes we are left with a pure K_2^0 beam with half of the original intensity. If we insert a second target B here, an absorber, the surviving neutral kaons interact with the nuclei in this target by strong interactions.

Strong interactions distinguish between the states of different strangeness, namely between K^0 and $\bar{K}^0$. Indeed, if the energy is as low as we suppose, the only inelastic reaction of the K^0 is the charge exchange, whilst the $\bar{K}^0$ can also undergo reactions with hyperon production, such as

$$\bar{K}^0 + p \rightarrow \Lambda + \pi^+ \qquad \bar{K}^0 + n \rightarrow \Lambda + \pi^0, \tag{8.23}$$

And similarly with a Σ hyperon in place of the Λ. Therefore, the total inelastic cross-section is much larger for $\bar{K}^0$ than for K^0 and the absorber B preferentially absorbs the former, provided its thickness is large enough. To simplify the discussion,

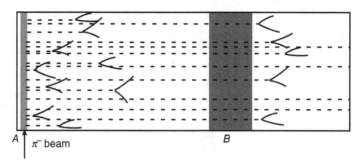

A π^- beam B

Fig. 8.3.   Logical scheme of the Pais–Piccioni experiment.

let us consider an idealized absorber that completely absorbs the $\bar{K}^0$ while transmitting the K^0 component without attenuation. After B we then again have a pure K^0 beam with intensity exactly 1/4 that of the original one. After the absorber we observe the reappearance of 2π decays. The absorber has regenerated the short lifetime component. The phenomenon is very similar to that exhibited by polarized light. Its observation established the nature of the short and long lifetime neutral kaons as coherent superpositions of states of opposite strangeness.

Question 8.2 Consider two pairs of mutually perpendicular linear polarization states of light with polarizations at 45° to each other. Consider the following analogy: let the K^0, $\bar{K}^0$ system be analogous to the first pair of axes, and let the K_1^0, K_2^0 be analogous to the second pair. Is this analogy correct? Design an experiment analogous to the experiment of Pais and Piccioni, using linear polarizers. □

8.5 CP Violation

Violation of the CP symmetry has been observed in weak interaction processes only; it was discovered in neutral K decays in 1964, and later established in B meson (2001), D meson (2019) and B_s^0 meson (2021) decays. There are three kinds of CP violation.

(1) **Violation in the wave function**, also called **violation in mixing**. It happens when the wave functions, or the fields, of the free Hamiltonian are not CP eigenstates. It is a small, but important, effect. In the neutral kaon system, for example, the shorter lifetime state is not exactly K_1^0 (the CP eigenstate with eigenvalue +1), but contains a small K_2^0 component (the CP eigenstate with eigenvalue −1); symmetrically, the long lifetime state is not exactly K_2^0 but contains a bit of K_1^0 We shall discuss this phenomenon in this section.

(2) **Violation in decays** or direct violation. Let M be a meson and f the final state of one of its decays. Let $\bar{M}$ be its antimeson and $\bar{f}$ the conjugate state of f. If CP is conserved, the two decay amplitudes are equal, namely $A(M \rightarrow f) = A(\bar{M} \rightarrow \bar{f})$.

The equality holds both for the square moduli, namely for the decay probabilities, and for the phases. The phase is detectable by the interference between different amplitudes contributing to the matrix element, provided that the conditions we shall discuss on the phases both of the weak decays and of the strong final-state interactions are satisfied. This type of CP violation occurs in the decays of both the neutral and the charged mesons. We shall discuss this type of violation in Sections 8.8 and 8.9.

(3) **Violation in the interference between decays with and without mixing, with oscillations.** This may happen for a neutral meson decay into a final state f that is a CP eigenstate and that can be reached by both flavours. It occurs even if CP is conserved both in the mixing and in the decay, provided there is a phase difference between the mixing and decay amplitudes. The phenomenon has been observed in the K^0, B^0, B_s^0 and D^0 mesons, as we shall discuss in an example, in Section 8.6.

J. Christenson, J. Cronin, V. Fitch and R. Turlay first discovered CP violation in 1964 (see Christenson *et al.* 1964). Specifically, they observed that the long lifetime neutral K-mesons decay, in a few cases per thousand, into 2π.

The first element of the experiment is the neutral beam, containing the kaons, obtained by steering the proton beam extracted from a proton synchrotron (the AGS of the Brookhaven National Laboratory) onto a target. A dipole magnet deflects the charged particles produced in the target, while the neutral ones travel undeflected. A collimator located beyond the magnet selects the neutral component. After a few metres, this contains the K_L long-life mesons and no K_S. To this must be added an unavoidable contamination of neutrons and gammas.

The experiment aims to establish whether the CP-violating decay

$$K_L \to \pi^+ + \pi^- \tag{8.24}$$

exists. Experimentally the topology of the event consists of two opposite sign tracks. This decay, if it does exist, is expected to be much rarer than other decays with two charged tracks in the final state, K_{e3}^0, $K_{\mu3}^0$ and

$$K_L \to \pi^+ + \pi^- + \pi^0. \tag{8.25}$$

However, the last is a three-body decay containing a non-observed neutral particle, and one takes advantage of this kinematic property to select the events (8.24) as we shall see.

Fig. 8.4 is a drawing of the experiment. The volume in which the decays are expected, a few metres long, should ideally be empty, to avoid K_S^0 regeneration and interactions of beam particles simulating the decay. In practice, it is filled with helium gas that, with its light atoms, acts as a 'cheap vacuum'. The measuring apparatus is a two-arm spectrometer, adjusted to accept the kinematic of the decay (8.24). Each arm is made of two spark chamber sets before and after a bending magnet. In this way the momentum and charge of each particle are measured. The spark chambers are photographed like the bubble chambers but, unlike these, can be triggered by an electronic signal. The trigger signal was originated in two Cherenkov counters at the ends of the arms.

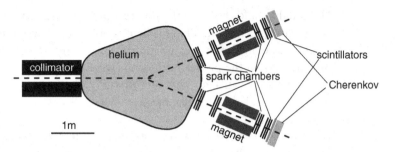

Fig. 8.4. Schematic view of the Christenson *et al.* experiment.

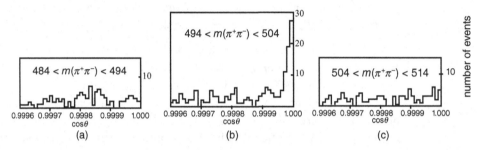

Fig. 8.5. Distribution of cos θ (defined in the text) in three different ranges of $m(\pi^+\pi^-)$: (a) below the K mass; (b) around the K mass; (c) above the K mass (M. Fitch, Nobel Lecture 1980, Fig. 3: M. Fitch, © The Nobel Foundation 1980 Figure 3 in this link www.nobelprize.org/uploads/2018/06/fitch-lecture.pdf).

In the data analysis, the three-body events are suppressed, imposing two conditions: (1) the angle θ between the direction of the sum of the momenta of the two tracks and the beam direction should be compatible with zero, and (2) the mass $m(\pi^+\pi^-)$ of the two-particle system should be compatible with the K mass.

Fig. 8.5 shows three cos θ distributions. Part (b) is for the events with $m(\pi^+\pi^-)$ near to the K mass. Panels (a) and (c) are for two control zones with $m(\pi^+\pi^-)$ immediately below and above the K mass. In the central panel, and only in this, a clear peak is visible at $\theta = 0$ above the background. This is the evidence that the long-lifetime neutral kaon also decays into $\pi^+\pi^-$, a state with $CP = +1$.

The measured value of the branching ratio in the CP-violating channel is

$$BR\left(K_L \to \pi^+\pi^-\right) = 2 \times 10^{-3}. \tag{8.26}$$

Summarizing, the experiment shows that the two CP eigenstates, K_1^0 and K_2^0, are not the states with definite mass and lifetime. The states with definite mass and lifetime can be written as

$$\left|K_S\right\rangle = \frac{1}{\sqrt{1+|\varepsilon|^2}}\left(\left|K_1^0\right\rangle - \varepsilon\left|K_2^0\right\rangle\right)$$

$$\left|K_L\right\rangle = \frac{1}{\sqrt{1+|\varepsilon|^2}}\left(\varepsilon\left|K_1^0\right\rangle + \left|K_2^0\right\rangle\right). \tag{8.27}$$

The ε parameter measures the small impurity of the wrong *CP*. The experiment of Fitch and Cronin (see Christenson *et al.* 1964) is sensitive to the absolute value of this complex parameter. Let us now define the ratio of the transition amplitudes of K_L and of K_S into $\pi^+\pi^-$ as

$$\eta_{+-} \equiv |\eta_{+-}| e^{i\phi^{+-}} \equiv \frac{A\left(K_L \to \pi^+\pi^-\right)}{A\left(K_S \to \pi^+\pi^-\right)}. \tag{8.28}$$

Its square modulus is the ratio of the decay rates. If *CP* violation is only due to the wave function impurity, one finds that

$$|\varepsilon|^2 \equiv |\eta_{+-}|^2 = \frac{\Gamma\left(K_L \to \pi^+\pi^-\right)}{\Gamma\left(K_S \to \pi^+\pi^-\right)}. \tag{8.29}$$

We have just seen how the numerator was measured. The denominator is easily determined, being the main decay of the K_S. The present value of $|\varepsilon|$ is (Workman *et al.* 2022)

$$|\varepsilon| = |\eta_{+-}| = (2.232 \pm 0.011) \times 10^{-3}. \tag{8.30}$$

We now go back to the observation we made at the end of the previous section. Fig. 8.2 shows that, at late times, when only K_L survive, they decay through $K_L \to \pi^- l^+ \nu_l$ a little more frequently than through the *CP* conjugate channel $K_L \to \pi^+ l^- \bar{\nu}_l$. Quantitatively it is

$$\delta_L = \frac{N\left(K_L \to \pi^- l^+ \nu_l\right) - N\left(K_L \to \pi^+ l^- \bar{\nu}_l\right)}{N\left(K_L \to \pi^- l^+ \nu_l\right) + N\left(K_L \to \pi^+ l^- \bar{\nu}_l\right)} = (3.32 \pm 0.06) \times 10^{-3} \tag{8.31}$$

This shows, again and independently, that matter and antimatter are, somewhat, different. Let us suppose that we wish to tell an extraterrestrial being what we mean by matter and by antimatter. We do not know whether their world is made of the former or of the latter. We can tell them, 'prepare a neutral *K*-meson beam and go far enough from the production point to be sure to have been left only with the long-lifetime component.' At this point they are left with K_L mesons, independently of the matter or antimatter constitution of their world. We continue, 'count the decays with a lepton of one or the other charge and call positive the charge of the sample that is about three per thousand larger. Humans call matter the one that has positive nuclei.' If, after a while, our correspondent answers that their nuclei have the opposite charge, and comes to meet you, be careful: apologize, but do not shake their hand.

The measurement of the charge asymmetry (8.31) determines the real part of ε. Actually we can write the K_L wave function as

$$|K_L\rangle = \frac{1}{\sqrt{1+|\varepsilon|^2}} \left(\varepsilon |K_1^0\rangle + |K_2^0\rangle\right) \simeq \varepsilon |K_1^0\rangle + |K_2^0\rangle$$
$$= \frac{1}{\sqrt{2}} (1+\varepsilon) |K^0\rangle - \frac{1}{\sqrt{2}} (1-\varepsilon) |\bar{K}^0\rangle. \tag{8.32}$$

Consequently,

$$\delta_L = \frac{|1+\varepsilon|^2 - |1-\varepsilon|^2}{|1+\varepsilon|^2 + |1-\varepsilon|^2} = 2\frac{\text{Re}\,\varepsilon}{1+|\varepsilon|^2} \simeq 2\,\text{Re}\,\varepsilon. \tag{8.33}$$

The present value is (Workman *et al.* 2022)

$$\text{Re}(\varepsilon) = (1.596 \pm 0.013) \times 10^{-3}. \tag{8.34}$$

Comparing with (8.30), we see that the ε phase is about $\pi/4$. Its value is (Workman *et al.* 2022)

$$\phi^{+-} = (43.52 \pm 0.05)^{\circ}. \tag{8.35}$$

8.6 Oscillation and CP Violation in Interference with Mixing in B^0

In this section we shall discuss two phenomena in the B^0 system, the beauty oscillations and the CP violation in the interference between decays without and with oscillation. Both phenomena have been discovered at the beauty factories by the Belle and BaBar experiments. We shall see how CP violation is originated by the presence of a phase factor in the quark mixing matrix and of a difference between this phase and the phase of the decay amplitude.

The neutral-B system behaves very similarly to the neutral-K system. However, we shall describe its evolution in time with a slightly different formalism. In the kaon case, it would have been

$$\begin{aligned} |K_S\rangle &= p|K^0\rangle + q|\bar{K}^0\rangle \\ |K_L\rangle &= p|K^0\rangle - q|\bar{K}^0\rangle, \end{aligned} \tag{8.36}$$

where p and q are two complex numbers satisfying the normalization condition

$$|p|^2 + |q|^2 = 1 \tag{8.37}$$

Note that $\arg(p/q*)$ is a phase common to K_S and K_L and does not have a physical meaning.

The formalism of the previous section is recovered with

$$p = (1+\varepsilon)/\sqrt{2} \quad q = (1-\varepsilon)/\sqrt{2}. \tag{8.38}$$

There are two important differences between the B and the K systems. The first is that the lifetimes of the two B are equal within the errors (see Table 8.1 at the end of the chapter). The reason is that the Q values of the decays of both particles are large, rather than one large and one small as in the case of K. We label the two eigenstates according to their larger and smaller masses as B_H and B_L (heavy and light). Their mass difference is

$$\Delta m_B \equiv m_H - m_L > 0, \tag{8.39}$$

which is positive by definition. We shall call Γ_B the common value of the widths

$$\Gamma_B = \Gamma_{B_H} = \Gamma_{B_L} \simeq 0.43 \,\text{meV}. \tag{8.40}$$

The second difference is the suppression of the common decay channels of B^0 and $\bar{B}^0$ due to the smallness of the corresponding mixing elements. The important consequence

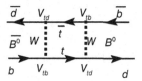 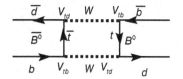

Fig. 8.6.   The dominant 'box' diagrams of the neutral B system.

is that $|p/q| \approx 1$, namely the CP violation in the mixing is small. Transitions between B^0 and $\bar{B}^0$ can happen, at the lowest order through the 'box' diagrams shown in Fig. 8.6.

The box diagrams with u or c quarks replacing one or two t quarks should also be considered. However, the contribution of a quark internal line is proportional to the square of its mass. Consequently, the diagrams with quarks different from top are negligible. The Standard Model gives the rules to compute the mass difference from a box diagram. In particular, the product $|V_{td}|^2 |V_{tb}|^2$ is proportional to Δm_B and can be determined by measuring the oscillation period as anticipated in Section 7.10.

Question 8.3 Evaluate the space distance between vertices in both diagrams of Fig. 8.6. □

We now describe the neutral B system starting from the expressions

$$|B_L\rangle = p|B^0\rangle + q|\bar{B}^0\rangle$$
$$|B_H\rangle = p|B^0\rangle - q|\bar{B}^0\rangle. \tag{8.41}$$

Setting

$$m \equiv (m_H + m_L)/2 \tag{8.42}$$

the evolution in time of the L and H eigenstates is given by the factors $e^{-\frac{\Gamma_B}{2}t}e^{-imt}e^{+i\frac{\Delta m_B}{2}t}$ and $e^{-\frac{\Gamma_B}{2}t}e^{-imt}e^{-i\frac{\Delta m_B}{2}t}$ respectively, where t is the proper time.

Let us call $\Psi_0(t)$ and $\Psi_{\bar{0}}(t)$ the wave functions of the states that are purely B^0 and purely $\bar{B}^0$ at $t=0$ respectively. A calculation similar to that we made in Section 8.3 leads to the following expressions

$$\Psi_0(t) = h_+(t)B^0 + \frac{q}{p}h_-(t)\bar{B}^0 \tag{8.43}$$

and

$$\Psi_{\bar{0}}(t) = \frac{p}{q}h_-(t)B^0 + h_+(t)\bar{B}^0, \tag{8.44}$$

where (notice the imaginary unit in the second expression)

$$h_+(t) = e^{-\frac{\Gamma_B}{2}t}e^{-imt}\cos\left(\frac{\Delta m_B}{2}t\right); h_-(t) = ie^{-\frac{\Gamma_B}{2}t}e^{-imt}\sin\left(\frac{\Delta m_B}{2}t\right). \tag{8.45}$$

If we have a pure B^0 state at $t=0$, the probability to find a B^0 at a generic t is

$$\left|\left\langle B^0 \mid \Psi_0(t)\right\rangle\right|^2 = \left|h_+(t)\right|^2 = e^{-\Gamma_B t} \cos^2\left(\frac{\Delta m_B}{2}t\right)$$
$$= \frac{1}{2}e^{-\Gamma_B t}\left(1+\cos\Delta m_B t\right) \tag{8.46}$$

and the probability to find a $\bar{B}^0$ is

$$\left|\left\langle \bar{B}^0 \mid \Psi_0(t)\right\rangle\right|^2 = \left|h_-(t)\right|^2 = \left|\frac{q}{p}\right|^2 e^{-\Gamma_B t}\sin^2\left(\frac{\Delta m_B}{2}t\right)$$
$$= e^{-\Gamma_B t}\sin^2\left(\frac{\Delta m_B}{2}t\right) = \frac{1}{2}e^{-\Gamma_B t}\left(1-\cos\Delta m_B t\right) \tag{8.47}$$

in the approximation $|p/q| = 1$ (see Fig. 8.1(b)). Similar expressions are valid starting from a pure $\bar{B}^0$ state, that is, with the wave function $\Psi_{\bar{0}}(t)$. The difference between the probabilities of observing opposite-flavour and same-flavour decays, normalized to their sum, called flavour asymmetry,

$$\frac{P_{OF} - P_{SF}}{P_{OF} + P_{SF}} = \cos\left(\Delta m_B t\right), \tag{8.48}$$

is measurable as a function of time, as we shall see, determining Δm_B.

To measure the phase of p/q we need a second phase to use as a reference. Only phase differences have a physical meaning. Consider for this purpose a CP eigenstate f of eigenvalue η_f into which both B^0 and $\bar{B}^0$ can decay. Let A_f be the amplitude for $B^0 \to f$ and $\bar{A}_f$ the amplitude for $\bar{B}^0 \to f$. If $A_f \neq \bar{A}_f$, CP is violated. If $|A_f| \neq |\bar{A}_f|$, we observe the violation as a difference between the two decay rates. However, the absolute values of the two amplitudes are equal in the important case that we shall discuss. The CP violation is due to the phase difference between two ratios, p/q (i.e. the mixing) and $A_f/\bar{A}_f$, the ratio of the decay amplitudes. In conclusion we must measure the phase of

$$\lambda_f \equiv \eta_f \frac{p}{q}\frac{A_f}{\bar{A}_f}. \tag{8.49}$$

Notice that $|\lambda_f| = 1$.

The amplitudes for the decay into the final state f of $\Psi_0(t)$ and of $\Psi_{\bar{0}}(t)$ are, respectively,

$$\langle f \mid \Psi_0(t)\rangle = A_f h_+(t) + \frac{q}{p}\bar{A}_f h_-(t)$$
$$= \frac{A_f}{\lambda_f}e^{-imt}e^{-\frac{\Gamma_B}{2}t}\left[\lambda_f\cos\left(\frac{\Delta m_B}{2}t\right) + i\sin\left(\frac{\Delta m_B}{2}t\right)\right] \tag{8.50}$$

and

$$\langle f \mid \Psi_{\bar{0}}(t)\rangle = \frac{p}{q}A_f h_-(t) + \bar{A}_f h_+(t)$$
$$= \bar{A}_f e^{-imt}e^{-\frac{\Gamma_B}{2}t}\left[i\lambda_f\sin\left(\frac{\Delta m_B}{2}t\right) + \cos\left(\frac{\Delta m_B}{2}t\right)\right]. \tag{8.51}$$

The CP-violating observable is the ratio between the difference and the sum of the two probabilities. After a few iterations, taking into account that $|A_f| = |\bar{A}_f|$ and that $|\lambda_f| = 1$, we obtain

$$\left| \langle f | \Psi_0(t) \rangle \right|^2 + \left| \langle f | \Psi_{\bar{0}}(t) \rangle \right|^2 = 2|A_f|^2 e^{-\Gamma_B t}$$

and

$$\left| \langle f | \Psi_0(t) \rangle \right|^2 - \left| \langle f | \Psi_{\bar{0}}(t) \rangle \right|^2 = |A_f|^2 e^{-\Gamma_B t} 2\eta_f \, \mathrm{Im}(\lambda_f) \sin(\Delta m_B \cdot t) \qquad (8.52)$$

and finally

$$a_{fCP} = \frac{\left| \langle f | \Psi_0(t) \rangle \right|^2 - \left| \langle f | \Psi_{\bar{0}}(t) \rangle \right|^2}{\left| \langle f | \Psi_0(t) \rangle \right|^2 + \left| \langle f | \Psi_{\bar{0}}(t) \rangle \right|^2} = \eta_f \, \mathrm{Im}\, \lambda_f \sin(\Delta m_B \cdot t). \qquad (8.53)$$

We see that CP is violated if $\mathrm{Im}\, \lambda_f \neq 0$.

We now consider the measurements of the mass difference Δm_{B^0} and of the CP asymmetry a_{fCP} at the beauty factories. We recall here that these high-luminosity $e^+ e^-$ colliders provide hundreds of millions of $B^0 \bar{B}^0$ pairs in a pure $J^{PC} = 1^{--}$ state. The factories operate at the $\Upsilon\left(4^1 S_3\right)$ resonance that is only 20 MeV above $m_{B^0} + m_{\bar{B}^0}$. Consequently, the B move slowly in the centre of mass frame and their decay vertices are difficult to be resolve in this frame. The beauty factories are consequently built 'asymmetric', meaning that the energies of the two beams are not equal, in order to have the centre of mass moving in the laboratory. The electron and positron momenta are $p(e^-) = 9$ GeV and $p(e^+) = 3.1$ GeV in PEP2 and $p(e^-) = 8$ GeV and $p(e^+) = 3.5$ GeV in KEKB, corresponding to an average Lorentz factor of the B of $\langle \beta\gamma \rangle = 0.56$ and $\langle \beta\gamma \rangle = 0.425$, respectively. The average distance between the production and the decay vertices is about $\Delta z \approx 200$ μm. It is measured by surrounding the collision point with a 'vertex detector'. The vertex detector is made up of several layers of silicon micro-strip tracking devices assembled with high mechanical accuracy. The accuracy in the vertex reconstruction is typically $80-120$ μm, corresponding to about one half of the flight length in a lifetime. The proper time, the variable that appears in the expressions above, is the distance measured in the laboratory divided by $c\langle \beta\gamma \rangle$.

We know that the two neutral B produced in any $e^+ e^-$ annihilation are one B^0 and one $\bar{B}^0$, but we do not know which is which. The time evolution of the two-state system is given by a single wave function that describes both particles. In other words, the phase difference between the B^0 and the $\bar{B}^0$ does not vary in time. However, one of the B can identify itself as a particle or an antiparticle when and if it decays semileptonically. Just as we have discussed for the kaons, we have

$$\begin{array}{llll} B^0 = \bar{b}d & \bar{b} \to \bar{c}l^+\nu_l & \Rightarrow B^0 \to D^- l^+\nu_l & \\ \bar{B}^0 = b\bar{d} & b \to cl^-\bar{\nu}_l & \Rightarrow \bar{D}^0 \to D^+ l^-\bar{\nu}_l. & \end{array} \qquad (8.54)$$

Consequently, by observing the sign of the lepton or by reconstructing the D, we 'tag' the neutral B as a B^0 or $\bar{B}^0$. We measure the time of this decay relative to the time of production by measuring the distance between production and decay vertices, and we measure the velocity of the particle by measuring the momenta of its daughters. We

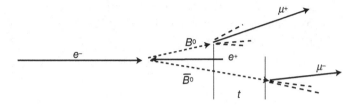

Fig. 8.7.   In cases of semileptonic decay, the sign of the lepton tags the flavour of the meson.

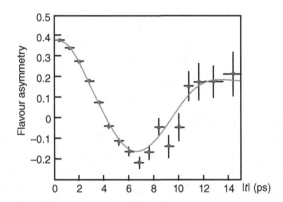

Fig. 8.8.   Flavour asymmetry as a function of time (absolute value) from the Belle experiment.

then take as $t = 0$ the time of the tagging decay. If the tagging B is a $\bar{B}^0$, the companion is a B^0 at that time, and its wave function evolves as $\Psi_0(t)$, and vice versa. Strange as it may appear, its evolution is given by $\Psi_0(t)$ even before the tag decay, namely for $t < 0$. Indeed, the evolution of the wave functions is completely deterministic in quantum mechanics. Once known at an instant, the wave function is known at any time.

Let us now consider firstly the beauty oscillation, which is very similar to that of the kaons. We consider the cases in which the flavours of both B can be identified by the final states of their decays, for example by the signs of the leptons, as sketched in Fig. 8.7. The times between the production of the $B\bar{B}$ pair and each of the two decays are measured, obtaining the time interval t between the decays. Since the decay time of one or of the other B can be taken as $t = 0$ indifferently, t is known in absolute value only. Fig. 8.8 shows the flavour asymmetry measured by the Belle experiment (Abe *et al.* 2005) (BaBar has similar results) as a function of $|t|$. Notice that the lifetime is about one order of magnitude smaller than the oscillation period. Consequently, the number of events per unit time decreases at longer times and the error bars increase accordingly.

By fitting to the data expression (8.48), corrected to take into account the experimental effects that are partially different for the two flavours, the decay width and the mass difference are obtained. The value of the latter averaged on all the experiments by the Heavy Flavor Averaging Group Collaboration (HFAG) (Amhis *et al.* 2023)

$$\Delta m_B = 0.5065 \pm 0.0019 \text{ps}^{-1} = 0.3340 \pm 0.0013 \qquad (8.55)$$

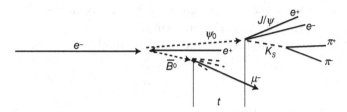

Fig. 8.9. Kinematics for $e^+ e^- \rightarrow B^0 \overline{B}^0$. One of the neutral B is tagged as a $\overline{B}^0$ by its decay $\overline{B}^0 \rightarrow \mu^- + \dots$. This instant is defined as the origin of the time. The other B decays into $J / \psi + K_S$ at time t (which may be positive or negative).

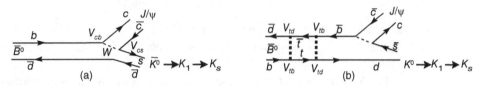

Fig. 8.10. (a) Feynman diagram at quark level for the decay $\overline{B}^0 \rightarrow J / \psi + \overline{K}^0$ without oscillations. (b) Box diagram with oscillation for the same process.

From this measurement we can extract (Beringer *et al.* 2012)

$$|V_{td}||V_{tb}| = (8.4 \pm 0.6) \times 10^{-3}. \tag{8.56}$$

The CP asymmetry a_{fCP} has been observed in different channels. We shall consider only the final CP eigenstates $f = J / \psi + K_S$ and $f = J / \psi + K_L$. The final orbital momentum is $L = 1$ for angular momentum and parity conservation. As a consequence, the CP eigenvalues are $\eta_{J/\psi + K_L} = +1$ and $\eta_{J/\psi + K_S} = -1$. Notice that the branching ratios are small, about 0.9×10^{-3}. The peak luminosity of the beauty factories is larger than $10^{34}\,\mathrm{cm^{-2}s^{-1}}$, corresponding to the production of $10^6\, B\overline{B}$ pairs a day. Belle and BaBar have collected about 5×10^8 events.

The experiments tag the events as we described and select the cases in which the companion B decays into one of the CP eigenstates. An example is sketched in Fig. 8.9.

The Standard Model gives a very clean prediction for Im λ_f in both cases. We shall give a plausibility argument but not a proof. To be concrete, consider the decay of the tagged state $\Psi_{\overline{0}}(t)$. It may decay directly as $\overline{B}^0$, with the diagram of Fig. 8.10(a), or oscillate into a B^0 and then decay, with the diagram of Fig. 8.10(b). The two amplitudes do not interfere at the level of the diagrams shown in the figure, because in one case there is a K^0 and in the other a $\overline{K}^0$, which can be distinguished. However, if the kaon decays as a CP eigenstate, namely, in good approximation, as a K_1^0 (or a K_2^0), the final states are identical and the two amplitudes do interfere. In the present discussion the difference between K_1^0 and $K_S(K_2^0$ and $K_L)$ can be safely neglected.

We have shown in the figure the relevant elements of the mixing matrix. All of them, and those relative to the transition to K_1^0 that are not shown, are real except V_{td}, to a very good approximation. The decay amplitudes, which do not depend on V_{td}, as can be seen in Fig. 8.10, are also real. Hence the phase of $A_{J/\psi + K} / \overline{A}_{J/\psi + K}$ is zero. On the other hand, the mixing is governed by the box diagram in Fig. 8.10(b), which contains V_{td} twice. The other element in the box is $V_{tb} \approx 1$ with very good approximation. Taking all this into account and recalling Eq. (8.39), we find

$$\lambda_{J/\psi+K_S} = \frac{p}{q} \frac{A_{J/\psi+K}}{\bar{A}_{J/\psi+K}} = e^{2i\beta}. \tag{8.57}$$

The final result is

$$\mathrm{Im}\, \lambda_{J/\psi+K_S} = 2\,\mathrm{Im}\, V_{td} = \sin(2\beta) \tag{8.58}$$

and

$$\mathrm{Im}\, \lambda_{J/\psi+K_L} = -\sin(2\beta). \tag{8.59}$$

In conclusion we expect that the observables $a_{\mathcal{CP},J/\psi+K_S}(t)$ and $a_{\mathcal{P},J/\psi+K_L}(t)$ to be two sinusoidal functions of time with the same period, the same amplitude and opposite phases.

Fig. 8.11 shows the $\mathcal{CP}$ asymmetry $a_{f\mathcal{CP}}$, as defined by Eq. (8.53), measured by Belle (upper panels) (Adachi *et al.* 2012) and BaBar (lower panels) (Aubert *et al.* 2009), corresponding to the full data sets of the two experiments, 4.65×10^8 and 7.72×10^8 $B\bar{B}$

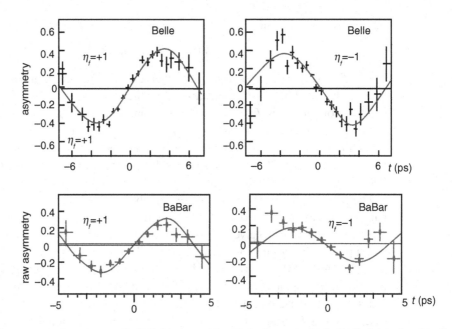

Fig. 8.11. The $a_{f\mathcal{CP}}$ asymmetry as measured by Belle (upper panels) and BaBar (lower panels), for the final states $f = J/\psi K_S$ (left panels) and $f = J/\psi K_L$ (right panels). The asymmetry is 'raw' for BaBar; the background is subtracted for Belle. Here η_f is the $\mathcal{CP}$ eigenvalue of the final state. (Figures reprinted from Adachi, I. *et al.* Precise measurement of the $\mathcal{CP}$ violation parameter $\sin 2\phi_1$ in $B^0 \to (c\bar{c})K^0$ decays. *Phys. Rev. Lett.* 108 171802 2012 Fig. 2, © (2012) by American Physical Society, https://doi.org/10.1103/PhysRevLett.108.171802, and Aubert, B. *et al.* Measurement of time-dependent $\mathcal{CP}$ asymmetry in $B^0 \to (c\bar{c})K^{(*)0}$ decays. *Phys. Rev.* D79 072009 2009 Fig. 2 © (2009) by American Physical Society, https://doi.org/10.1103/PhysRevD.79.072009).

pairs, respectively; η_f is the CP eigenvalue of the final state. The measured asymmetry is not the ideal one for three principal experimental reasons: (1) the presence of backgrounds, (2) the experimental resolution in the measurement of time (1–1.5 ps) and (3) the presence of miss-tags, meaning B^0 wrongly tagged as $\bar{B}^0$ and vice versa. Notice that the background is larger in the case of the K_L because this particle does not decay in the detector because of its long lifetime. The K_L is detected when it interacts in the calorimeter as a hadronic shower. All these effects reduce the amplitude of the sinusoidal time dependence. The asymmetry in Fig. 8.11 is 'raw' for BaBar, but the background is subtracted for Belle. The values of the two experiments are (first error is statistical, and the second systematic):

$$\begin{aligned} \text{BaBar} \quad & \sin(2\beta) = 0.687 \pm 0.028 \pm 0.012 \\ \text{Belle} \quad & \sin(2\beta) = 0.667 \pm 0.023 \pm 0.012. \end{aligned} \tag{8.60}$$

8.7 Oscillation and CP Violation in B_s^0

The B_s^0 is the heaviest of the neutral ground state mesons, being made of the $s\bar{b}$ quarks and having a mass $m_{B_s^0} = 5366.92 \pm 0.10$ MeV. As such, detailed studies became possible only at LHC, where a dedicated experiment, LHCb, was built. In addition, as we have seen for the B^0, because the mass is so large, many decay channels are open and large samples of B_s^0 must be collected to study each of them; the full luminosity of LHC is necessary.

In this section we shall start discussing the measurement of the mass difference between the two mass eigenstates, which are called B_{sH}^0, the heavier, and B_{sL}^0, the lighter. We also recall, from the introduction to this chapter, that, compared with the other neutral mesons, B_s^0 has the shortest oscillation period $T = 2\pi / \Delta m_s$, about 0.3 ps, where $\Delta m_s = m_H - m_L$ is the mass difference. The system also has the smallest ratio of oscillation period to shorter lifetime, about 20%, allowing us to observe the oscillation over many oscillation periods (see Fig. 8.1(c)).

Question 8.4 Compute the ratio between oscillation period and lifetime in the four cases. □

The B_s^0 mesons produced at the LHC energies travel an average of about 1 cm distance before decaying. In addition, because they are considerably heavier than most of the particles produced in the collision, their decay products have larger displacements relative to the collision point and a larger momentum transverse to the beam axis than the other particles. These features are exploited by LHCb (Aaij *et al.* 2022) to select a clean sample of B_s^0 and $\bar{B}_s^0$.

To study the $B_s^0 \leftrightarrow \bar{B}_s^0$ oscillation, the flavour, namely B_s^0 or $\bar{B}_s^0$, must be known both at production and at decay, as must the proper time between the two events. The proper time is obtained by measuring the distance between production and decay points and reconstructing the momentum from the measured momenta of the decay products. The flavour at decay is determined by the sign of the pion in $B_s^0 \to D_s^- \pi^+$ or $\bar{B}_s^0 \to D_s^+ \pi^-$. In both cases, the $D_s^\pm$ is observed in both $K^- K^+ \pi^\pm$ and $\pi^- \pi^+ \pi^\pm$ final states, obtaining a sample of almost 400 000 events. At LHC the identification at production is not as simple as at an $e^+ e^-$ collider. Two separate flavour tagging algorithms are used. The first exploits the fact that B_s^0 or $\bar{B}_s^0$ are almost always produced in pairs. Thus the flavour of the decaying one can be determined by looking at the decay products of the other. For example, if the latter decays semileptonically, we look at the charge of the final lepton; if the decay chain at the quark level is $b \to c \to s$, we look at the charge of the K meson or of the charm hadron. A second algorithm is based on the identification of the particles produced in the hadronization of the beauty quarks.

The oscillations are seen in the decay time distributions of mesons that decay as produced, say $B_s^0 \to D_s^- \pi^+$, and of those that decay after changing their flavour, say $\bar{B}_s^0 \to B_s^0 \to D_s^- \pi^+$, and similarly for those born as $\bar{B}_s^0$. The equations describing the phenomenon are the same we studied in the $K^0 \bar{K}^0$ case, Eq. (8.21). Consider, for example, the case of initial B_s^0, and call $P^-(t)$ and $P^+(t)$ the probabilities of observing a $-$ and $+$ pion respectively, at time t. These are the survival probability of the initial flavour and the appearance probability of the other one, respectively. If Γ_H and Γ_L are the widths of B_{sH}^0 and B_{sL}^0, the probabilities are

$$P^-(t) = \frac{1}{4} \left[e^{-\Gamma_H t} + e^{-\Gamma_L t} + 2 e^{-\frac{\Gamma_H + \Gamma_L}{2} t} \cos(\Delta m_s \cdot t) \right] \tag{8.61a}$$

$$P^-(t) = \frac{1}{4} \left[e^{-\Gamma_H t} + e^{-\Gamma_L t} - 2 e^{-\frac{\Gamma_H + \Gamma_L}{2} t} \cos(\Delta m_s \cdot t) \right], \tag{8.61b}$$

and similar if the initial state is $\bar{B}_s^0$. One recognizes an almost exponentially decreasing factor and an oscillating one with opposite sign for disappearance and appearance respectively.

Fig. 8.12 shows the decay time distributions of the surviving and oscillated events (for the entire sample of both initial B_{sH}^0 and B_{sL}^0). Clearly, $t = 0$ cannot be reached experimentally because at this time production and decay points are coincident. The detection efficiency gradually increases with increasing time, namely distance between the two points, corresponding to the increasing part of Fig. 8.12, reaching a plateau at 2 ps. The resolution in the decay time measurement is, on average, about 44 fs. The figure beautifully shows the initial flavour reaching a maximum, decreasing, reaching a minimum, and increasing again, and so on, while the initially absent flavour gradually increases, reaches a maximum and decreases again, and so on, in opposite phase with the former.

By fitting Eq. (8.61) (modified to take into account efficiency and other experimental features), the value of the mass difference is

$$\Delta m_s = 17.7683 \pm 0.0051 \,(\text{stat}) \pm 0.0032 \,(\text{syst}) \, \text{ps}^{-1} \tag{8.62}$$

We come now to the CP violation, a phenomenon for which the largest experimental contributions come from the LHC experiments, in particular from LHCb.

The SM expectations for the B_s^0 system are of a negligible CP violation in mixing, implying that $|p/q| \cong 1$. This had been confirmed by previous experiments. The violation

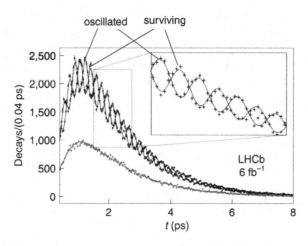

is observed in the decay and in the interference between mixing and decay. We shall limit the discussion here to a single example of the latter. As previously discussed, we consider the decay to the same final state f, which is a $\mathcal{CP}$ eigenstate, that can be reached from the initial flavour or directly (namely $B_s^0 \to f$, or $\overline{B}_s^0 \to f$) or after oscillation (namely $B_s^0 \to \overline{B}_s^0 \to f$, or $\overline{B}_s^0 \to B_s^0 \to f$). The transition probabilities are the absolute squares of the sum of the two amplitudes that interfere. $\mathcal{CP}$ violation shows up as a time-dependent difference between the two, the $\mathcal{CP}$ asymmetry a_{fCP} in Eq. (8.53).

In the experiment we shall discuss (Aaji *et al.* 2021a), the final state is $f = K^+ K^-$. A sample of 70 310 ± 320 events is obtained selecting the events in the peak in the $K^+ K^-$ mass distribution at the B_s^0 mass. Considering that the branching ratio of the channel is only $(2.66 \pm 0.22) \times 10^{-5}$, one understands, as already remarked, that a very large data set must be collected. The flavour tagging at production and the measurement of the decay time are done as discussed above for the oscillation.

Fig. 8.13(a) shows the decay time distributions in $K^+ K^-$ for beauty mesons tagged as B_s^0 and $\overline{B}_s^0$ at production. The initial rise is a consequence of the increase of efficiencies at short decay paths, as in the case of the oscillation; the following decrease is a consequence of the exponential decay. The oscillations, with opposite phases, are clearly visible, together with the fit results. These include values of the period $2\pi / \Delta m_s$ and of the time origin $t = 0$. Fig. 8.13(b) shows the asymmetry folded in a single period using these values.

The fit provides, as the best value of the asymmetry,

$$a_{CP,B_s^0} = 0.236 \pm 0.013, \tag{8.63}$$

which clearly shows $\mathcal{CP}$ violation in the system

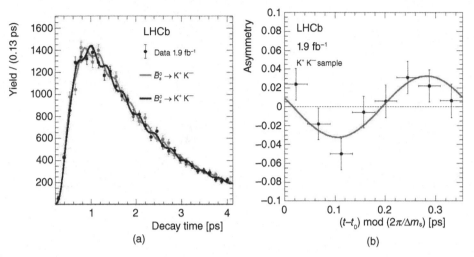

Fig. 8.13. (a) $B_s^0 \to K^+K^-$ decay time distribution for flavour-tagged mesons. (b) Asymmetry folded into one mixing period. The LHCb collaboration., Aaij, R., Abellán Beteta, C. *et al. J. High Energ. Phys.* 2021, 75 (2021) Fig. 8 left panels. Under CC BY 4.0.

8.8 $\mathcal{CP}$ **Violation in Neutral Meson Decays**

$\mathcal{CP}$ violation in the decays has been observed both in the neutral and in the charged mesons, namely $B^\pm$. The phenomenon is the same for all neutral mesons, but the measured observables are different because the violation is very small, of the order of 10^{-6}, in the K^0 system, larger in the B and intermediate for the D^0. For the B it is directly observed as a difference between the rates of two charge conjugated decays; for the K^0 it is observed by comparing four decay rates: the decays of the two neutral kaons into $\pi^+\pi^-$ and into $\pi^0\pi^0$, which are $\mathcal{CP}$ eigenstates. The charm case will be discussed in the next section.

Since the two pions from a kaon decay are in a spatially symmetric state, their isospin wave function must be symmetric, that is, the total isospin can be $I = 0$ or $I = 2$. Let us consider the amplitudes for the decay of the K^0 into states of definite isospin, A_0 and A_2. Taking into account the Clebsch–Gordan coefficients, the amplitudes for the (weak) decays into the two charge states are

$$A_W\left(K^0 \to \pi^+\pi^-\right) = \frac{1}{\sqrt{3}}\left(A_2 + \sqrt{2}A_0\right)$$
$$A_W\left(K^0 \to \pi^0\pi^0\right) = \frac{1}{\sqrt{3}}\left(\sqrt{2}A_2 - A_0\right). \tag{8.64}$$

We state here without proof that the $\mathcal{CPT}$ invariance requires that the corresponding amplitudes for the $\bar{K}^0$ are

$$A_W\left(\bar{K}^0 \to \pi^+\pi^-\right) = -\frac{1}{\sqrt{3}}\left(A_2^* + \sqrt{2}A_0^*\right)$$

$$A_W\left(K^0 \to \pi^0\pi^0\right) = \frac{1}{\sqrt{3}}\left(-\sqrt{2}A_2^* + A_0^*\right). \tag{8.65}$$

We write explicitly the two weak decay amplitudes as

$$A_0 \equiv |A_0|e^{i\phi_0} \quad A_2 \equiv |A_2|e^{i\phi_2}. \tag{8.66}$$

An overall phase is arbitrary but the phase difference is not and has physical meaning. As a consequence, at least two amplitudes, as in the case under discussion, must be present for *CP* violation. The condition is necessary but not sufficient, however. The violation occurs because more phase factors, due to strong interactions, are present.

We can consider the strong interaction between the two pions in the final state as an elastic scattering. Since strong interactions conserve the isospin, the scattering amplitudes are unit in absolute value and are pure phase factors, say $e^{i\delta_0}$ and $e^{i\delta_2}$. Since strong interactions conserve *CP*, the 'strong phases' have the same sign in a decay and in its charge conjugated.

Consequently, the complete transition amplitudes are

$$A\left(K^0 \to \pi^+\pi^-\right) = \frac{1}{\sqrt{3}}\left(e^{i\delta_2}A_2 + \sqrt{2}e^{i\delta_0}A_0\right)$$

$$= \frac{1}{\sqrt{3}}\left(|A_2|e^{i(\delta_2+\phi_2)} + \sqrt{2}|A_0|e^{i(\delta_0+\phi_0)}\right)$$

$$A\left(\bar{K}^0 \to \pi^+\pi^-\right) = -\frac{1}{\sqrt{3}}\left(e^{i\delta_2}A_2^* + \sqrt{2}e^{i\delta_0}A_0^*\right) \tag{8.67a}$$

$$= -\frac{1}{\sqrt{3}}\left(|A_2|e^{i(\delta_2-\phi_2)} + \sqrt{2}|A_0|e^{i(\delta_0-\phi_0)}\right)$$

and

$$A\left(K^0 \to \pi^0\pi^0\right) = \frac{1}{\sqrt{3}}\left(\sqrt{2}e^{i\delta_2}A_2 - e^{i\delta_0}A_0\right)$$

$$= \frac{1}{\sqrt{3}}\left(\sqrt{2}|A_2|e^{i(\delta_2+\phi_2)} - |A_0|e^{i(\delta_0+\phi_0)}\right)$$

$$A\left(\bar{K}^0 \to \pi^0\pi^0\right) = \frac{1}{\sqrt{3}}\left(-\sqrt{2}e^{i\delta_2}A_2^* - e^{i\delta_0}A_0^*\right) \tag{8.67b}$$

$$= \frac{1}{\sqrt{3}}\left(-\sqrt{2}|A_2|e^{i(\delta_2-\phi_2)} - |A_0|e^{i(\delta_0-\phi_0)}\right).$$

We see that all the amplitudes have similar structure. If *CP* is conserved, the weak phases are zero and $\left|A\left(K^0 \to \pi^+\pi^-\right)\right| / \left|A\left(\bar{K}^0 \to \pi^+\pi^-\right)\right| = 1$ and $\left|A\left(K^0 \to \pi^0\pi^0\right)\right| / \left|A\left(\bar{K}^0 \to \pi^0\pi^0\right)\right| = 1$. A very small *CP* violation in the neutral kaon system is observed as a small difference from 1 in one of these ratios. Let us see how.

It has been found experimentally that the A_0 amplitude dominates, namely that $|A_2| / |A_0| \approx 1/22$. Indeed the latest PDG average (Workman *et al.* 2023) is

$$\Gamma\left(K_S \to \pi^+\pi^-\right) / \Gamma\left(K_S \to \pi^0\pi^0\right) = 2.55 \pm 0.005, \tag{8.68}$$

which is close to the value predicted for a pure $I = 0$ final state (taking into account the small difference between pion masses). We now fix the arbitrary phase of A_0 and A_2, following the suggestion of Wu and Yang (Wu & Yang 1964), in such a way that the larger amplitude A_0 is a real positive number.

The next step is to express the observables in terms of the real numbers A_0, δ_0 and δ_2 and the complex number A_2. The observables are the two complex amplitudes ratios

$$\eta_{+-} \equiv |\eta_{+-}| e^{i\phi^{+-}} \equiv A(K_L \to \pi^+\pi^-) / A(K_S \to \pi^+\pi^-), \tag{8.69}$$

which we have already met in (8.28), and

$$\eta_{00} \equiv |\eta_{00}| e^{i\phi^{00}} \equiv A(K_L \to \pi^0\pi^0) / A(K_S \to \pi^0\pi^0). \tag{8.70}$$

The calculation is not difficult, but is long. The result, obtained by neglecting terms of order higher than the first in Re ε, Im ε and $|A_2 / A_0|$, where ε is the CP-violating parameter discussed in Section 8.5, is

$$\eta_{+-} = \varepsilon + \varepsilon' \quad \eta_{00} = \varepsilon - 2\varepsilon' \tag{8.71}$$

and

$$\varepsilon' = \frac{i}{\sqrt{2}} \frac{\text{Im}\, A_2}{A_0} e^{i(\delta_2 - \delta_0)}. \tag{8.72}$$

We see that CP is violated only if A_2 is not zero and not real. Moreover, since, as we shall see, the experimentally accessible observable is Re ε', a non-zero strong phase difference is needed. Actually, from scattering experiments, we know that $\delta_2 - \delta_0 = 48.3 \pm 1.5°$.

In the absence of CP violation in the decays, we have $\eta_{+-} = \eta_{00}$. Consequently, we may look for a violation by searching for a difference in the absolute values or in the phases of these parameters. We already mentioned the measurements of η_{+-} in Section 8.4. Considering similar measurements for η_{00}, we obtain (Workman *et al.* 2022)

$$|\eta_{00} / \eta_{+-}| = 0.9950 \pm 0.0007; \quad \phi^{00} - \phi^{+-} = (0.01 \pm 0.07)°. \tag{8.73}$$

The measurements are very precise, but not enough to show a difference. Let us then try another way, namely to seek, for example, a possible difference between $\Gamma(K^0 \to \pi^+\pi^-)$ and $\Gamma(\bar{K}^0 \to \pi^+\pi^-)$. Neglecting powers higher than the first of the real and imaginary parts of ε and ε', their ratio is given by (see Problem 8.9)

$$\left| A(\bar{K}^0 \to \pi^+\pi^-) / A(K^0 \to \pi^+\pi^-) \right| \approx 1 - 2\,\text{Re}\,\varepsilon'. \tag{8.74}$$

Consider next the corresponding ratio for the $\pi^0\pi^0$ final state, that is,

$$\left| A(\bar{K}^0 \to \pi^0\pi^0) / A(K^0 \to \pi^0\pi^0) \right| \approx 1 + 4\,\text{Re}\,\varepsilon', \tag{8.75}$$

and the difference between the two ratios,

$$\left| A(\bar{K}^0 \to \pi^0\pi^0) / A(K^0 \to \pi^0\pi^0) \right| - \left| A(\bar{K}^0 \to \pi^+\pi^-) / A(K^0 \to \pi^+\pi^-) \right| = 6\,\text{Re}\,\varepsilon'. \tag{8.76}$$

However, when measuring decay rates, we deal with the free Hamiltonian eigenstates, namely with K_S and K_L. The observable that is directly related to the difference (8.76) is the 'double ratio', that is, the ratio between the ratios of the decay rates into $\pi^+\pi^-$ and $\pi^0\pi^0$ of K_S and K_L. It easy to show that

$$\mathrm{Re}\left(\frac{\varepsilon'}{\varepsilon}\right) = \frac{1}{6}\left(1 - \frac{|\eta_{00}|^2}{|\eta_{+-}|^2}\right) = \frac{1}{6}\left[1 - \frac{\Gamma\left(K_L \to \pi^0\pi^0\right)\Gamma\left(K_S \to \pi^+\pi^-\right)}{\Gamma\left(K_L \to \pi^+\pi^-\right)\Gamma\left(K_S \to \pi^0\pi^0\right)}\right]. \qquad (8.77)$$

The experimental search for a possible non-zero value of $\mathrm{Re}(\varepsilon'/\varepsilon)$ started in the 1970s with a sensitivity of the order of 10^{-2}, too small to detect the effect. The struggle to reduce the systematic and statistic uncertainties continued both at CERN and at FNAL until the sensitivity reached a few parts in tens of thousands and the effect was discovered. We shall mention here only the principal experimental difficulties and the results.

A first difficulty is the rareness of the *CP* -violating decays of the K_L, $K_L \to \pi^+\pi^-$ (with a branching ratio of 2×10^{-3}) and $K_L \to \pi^0\pi^0\left(1 \times 10^{-3}\right)$. Moreover, the last decay suffers from possible contamination from the $K_L \to \pi^0\pi^0\pi^0$ decay, which is 200 times more frequent. Another problem is the large difference in the average decay paths of the two kaons, which, for example, at 110 GeV are $\beta\gamma\tau_L \approx 3.4$ km and are $\beta\gamma\tau_S \approx 6$ m, while the two decay distributions along the detector should be as similar as possible to avoid instrumental asymmetries.

In practice the experiments have used the following procedures.

- Simultaneous detection of $\pi^+\pi^-$ and $\pi^0\pi^0$ in order to cancel the uncertainty on the incident fluxes in the ratio.
- Having two beams of K_L and of K_S simultaneously in the detector. The beams must have energy spectra as equal as possible, have the same direction and produce spatial distributions of the decays as similar as possible.
- Good spatial and outstanding energy resolutions to reduce the contamination from other decay channels.

The *CP* violation in the decay was discovered by the NA31 experiment at CERN (Barr *et al.* 1993) on a sample of 428 000 $K_L \to \pi^0\pi^0$ decays. In 1993, NA31 published the result $\mathrm{Re}(\varepsilon'/\varepsilon) = (2.30 \pm 0.65) \times 10^{-3}$. This value is at 3.5 standard deviations from zero; however, the contemporary experiment at FNAL, E731, with similar statistics, obtained (Gibbons *et al.* 1993) $\mathrm{Re}(\varepsilon'/\varepsilon) = (0.74 \pm 0.56) \times 10^{-3}$, which is compatible with zero. The issue was solved by the next generation of experiments with a number of $K_L \to \pi^0\pi^0$ decays of the order of 10^7 and with improved systematic accuracy. KTeV at FNAL (Alavi-Harati *et al.* 2003) obtained $\mathrm{Re}(\varepsilon'/\varepsilon) = (2.07 \pm 0.28) \times 10^{-3}$ and NA48 at CERN (Batlay *et al.* 2002) obtained $\mathrm{Re}(\varepsilon'/\varepsilon) = (1.47 \pm 0.22) \times 10^{-3}$. The two values agree. The weighted average of the four measurements gives (Workman *et al.* 2022)

$$\mathrm{Re}(\varepsilon'/\varepsilon) = (1.66 \pm 0.23) \times 10^{-3}. \qquad (8.78)$$

As already mentioned, larger *CP* violation in the decay has been observed, by BaBar and Belle, also for the *B* mesons, both charged and neutral. This observation does not require measuring a double ratio. Actually, *B* mesons can decay in a huge number of different channels, due to their large masses. Consequently, each branching ratio is small, typically of the order of 10^{-5}, and even with the samples of several 10^8 $B\bar{B}$ pairs provided by the beauty factories, a few thousand events per channel are available. This gives statistical sensitivities to asymmetries of the order of several 10^{-2}.

We already mentioned that necessary conditions for CP violation in the decay are the presence of more than one contribution in the amplitude and of strong phases. We look closer to the issue now. Consider, for simplicity, the case of two contributions to the weak amplitude. The two CP-conjugated weak decay amplitudes are

$$A_f \equiv A\left(B \to f\right) = \left|A_1\right|e^{i\delta_{W1}} + \left|A_2\right|e^{i\delta_{W2}}$$
$$\bar{A}_f \equiv A\left(\bar{B} \to \bar{f}\right) = \left|A_1\right|e^{-i\delta_{W1}} + \left|A_2\right|e^{-i\delta_{W2}}. \tag{8.79}$$

Calling again δ_{S1} and δ_{S2} the strong phases present in the final states, the total amplitudes are

$$A_f = \left|A_1\right|e^{i(\delta_{S1}+\delta_{W1})} + \left|A_2\right|e^{i(\delta_{S2}+\delta_{W2})};$$
$$\bar{A}_{\bar{f}} = \left|A_1\right|e^{i(\delta_{S1}-\delta_{W1})} + \left|A_2\right|e^{i(\delta_{S2}-\delta_{W2})} \tag{8.80}$$

The difference between the two decay rates is proportional to the difference between their absolute values. It is easy to calculate that

$$\left|A_f\right|^2 - \left|\bar{A}_{\bar{f}}\right|^2 = -2\left|A_1\right|\left|A_2\right|\sin\left(\delta_{S1}-\delta_{S2}\right)\sin\left(\delta_{W1}-\delta_{W2}\right) \tag{8.81}$$

This shows, once more, that a phase difference between weak and strong phases must be present for a CP-violating asymmetry.

8.9 CP Violation in Charm

CP violation is expected to be small in the neutral charm system. As a consequence, the mass eigenstates, D_1 and D_2, are defined as the almost pure positive and negative CP states, respectively, independently of which is the heavier. Charm oscillations are expected to exist as for the other neutral mesons, but have not been directly observed yet. The reasons for the difficulty are seen in Fig. 8.1(d). The value of the mass difference can however be extracted from a measurement by LHCb (Aaij *et al.* 2021b), which we do not discuss here, of $\Delta m_c / \Gamma$, where Γ is the total width

$$\Delta m_c \equiv m_1 - m_2 = \left(9.67 \pm 1.4\right) \times 10^{-3} \text{ ps}^{-1}$$
$$= \left(6.37 \pm 0.92\right) \text{ meV} \tag{8.82}$$

The charm sector is the only one that allows us to probe CP violation in the up-type quarks sector. Historically, it was the last to be observed, in 2019, by the LHCb experiment (Aaij *et al.* 2019). The violation was observed in singly Cabibbo suppressed (SCS) decay modes, namely in which the charm quark decay is $c \to u\bar{d}d$ or $c \to u\bar{s}s$.

The measured quantity is the time-integrated asymmetry in the neutral charm meson decay in final CP eigenstates f, namely

$$A_{CP}(f) = \frac{\Gamma(D^0 \to f) - \Gamma(\bar{D}^0 \to f)}{\Gamma(D^0 \to f) + \Gamma(\bar{D}^0 \to f)} \qquad (8.83)$$

Two such SCS final states are considered, $f = K^+K^-$ and $f = \pi^+\pi^-$. In these channels, the SM expectation is of a pure 'direct' *CP* violation, because the other types, in the mixing and in the interference, if present, cancel out in the integration over time.

Two samples of flavour-tagged D mesons were collected. In the first sample the D mesons come from the D^* mesons (the vector mesons analogous to the K^* mentioned in Section 4.5), which decay, via strong interaction, practically at the collision point, as $D^{*+} \to D^0\pi^+$ and the charge conjugate $\bar{D}^{*-} \to \bar{D}^0\pi^-$. The sign of the pion, which is 'soft' (meaning low energy), the Q value of the decay being only 146 MeV, flags the flavour of the D. In the second sample the D mesons come from a semileptonic B decay, namely $B^+ \to D^0\mu^+v_\mu X$ or $B^- \to \bar{D}^0\mu^-\bar{v}_\mu X$, where X stands for possible other particles, and the sign of the muon tags the flavour of the D.

An interesting channel is the SCS $f = K^+K^-$. The effect is expected to be small, of the order of 10^{-3}; instrumental asymmetries must be eliminated. These asymmetries are of two types: first, the production cross-sections for the two conjugate channels are not equal; second, the efficiencies in the detection of the opposite sign tags (π^+ vs. π^- in one case, μ^+ vs. μ^- in the other) are not exactly equal. The trick is to consider the difference between the asymmetries in two similar decay channels, the other one, SCS also, being $f = \pi^+\pi^-$, namely

$$\begin{aligned}\Delta A_{CP} = &\frac{\Gamma(D^0 \to K^+K^-) - \Gamma(\bar{D}^0 \to K^+K^-)}{\Gamma(D^0 \to K^+K^-) + \Gamma(\bar{D}^0 \to K^+K^-)} \\ &- \frac{\Gamma(D^0 \to \pi^+\pi^-) - \Gamma(\bar{D}^0 \to \pi^+\pi^-)}{\Gamma(D^0 \to \pi^+\pi^-) + \Gamma(\bar{D}^0 \to \pi^+\pi^-)}.\end{aligned} \qquad (8.84)$$

Doing so, the production asymmetries of the D^* in one sample of B and in the other are the same for the two final states and cancel out. Similarly, the asymmetries in the detection of different charge muons are also the same for the two final states and cancel out. Fig. 8.14 shows the tree-level Feynman diagram for the four decays.

However, one might doubt that, taking the difference, any *CP* asymmetry might cancel out too. Fortunately, however, in the limit of U-spin symmetry, the decay amplitudes in K^+K^- and $\pi^+\pi^-$ are equal and opposite, so that the effect, if present, is not suppressed, but increased. Th U-spin is a $SU(2)$ subgroup of flavour $SU(3)$, similar to the isospin. In Fig. 4.18 we saw the octet of pseudoscalar mesons. Fig. 8.15 shows the same, without the $U = I = 0$ mesons for simplicity, rotated by 60°. The axes are now the third component of U-spin and the electric charge. We have one triplet and two doublets, each containing particles of the same charge. While the I-spin symmetry is broken by the electromagnetic interaction and conserved by the strong one, the U-spin is already broken, as $SU(3)$ in general, by the strong interaction. The violation is consequently sizeable, and not completely known, but it cannot be so large to make the two signs equal.

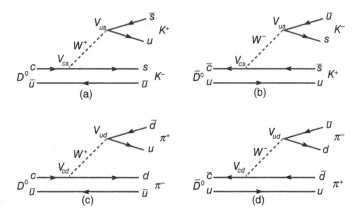

Fig. 8.14. Tree-level Feynman diagram for the SCS decays of the D^0 system considered in the experiment.

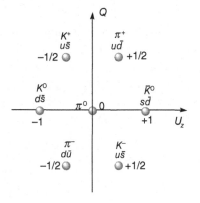

Fig. 8.15. The three U-spin multiplets of the pseudoscalar mesons.

The decay amplitudes in K^+K^- and $\pi^+\pi^-$, two doublets of $U = 1/2$, are equal and opposite for the following reason. In both cases, the total U-spin state $|U,U_z\rangle = |0,0\rangle$. Hence we have

$$|U,U_z\rangle = |0,0\rangle = \frac{1}{\sqrt{2}}\left(\left|\frac{1}{2},+\frac{1}{2}\right\rangle - \left|\frac{1}{2},-\frac{1}{2}\right\rangle\right). \tag{8.85}$$

That, in the two cases, gives

$$|0,0\rangle = \frac{1}{\sqrt{2}}\left(|\pi^+\rangle - |\pi^-\rangle\right) \text{ and } |0,0\rangle = \frac{1}{\sqrt{2}}\left(|K^-\rangle - |K^+\rangle\right). \tag{8.86}$$

The positive particles, π^+ in one case, K^+ in the other, appear with opposite signs, because they have opposite U_z values, and similarly for the negative ones. Hence the opposite sign in the amplitudes.

The final result is

$$\Delta A_{CP} = (-15.7 \pm 2.9) \times 10^{-4}, \tag{8.87}$$

in agreement with the SM expectations.

8.10 CP Violation in Charged Meson Decay

CP violation and oscillations between opposite-flavour neutral mesons discussed so far are two strongly connected, but different, phenomena. As a matter of fact, CP violation has also been observed in charged meson decays, specifically in the B mesons, where the phenomenon is obviously independent of mixing. In this section we shall discuss an important example, which leads to a value of the unitary triangle angle $\gamma = \arg\left(-V_{ud}V_{ub}^* / V_{cd}V_{cb}^*\right)$. This is the only angle that can be measured in decays that are described by tree-level diagrams; this makes the SM prediction free of theoretical uncertainties, differently from the cases in which box diagrams are present, as in Section 8.6.

The angle γ is experimentally accessible through the interference of the $\bar{b} \to \bar{c}u\bar{s}$ and $\bar{b} \to \bar{u}c\bar{s}$ decay amplitudes (and their CP conjugate). This interference is observed in the decays $B^\pm \to DK^\pm$ where D represents a quantum superposition of D^0 and $\bar{D}^0$ states, when the decay channel of the D is common to both D^0 and $\bar{D}^0$ We consider here the decays into the CP eigenstates $D \to K_s^0 K^+ K^-$ (Aaij *et al.* 2021c). Fig. 8.16 shows the tree-level diagrams for the decays $B^+ \to DK^+$. The reader can draw the corresponding ones for $B^- \to DK^-$.

The absolute square of the matrix element of the decay, that contains the interference term, depends on only two kinematical variables, the coordinates of the Dalitz plot of the three-body final state, say $m^2\left(K_s^0 K^+\right)$ vs. $m^2\left(K_s^0 K^-\right)$. No dependence on angles is physically possible, because, the spin of the D being equal to zero, no polarization of the meson can exist. As we know, because the phase-space factor is proportional to the area element of the Dalitz plot, the absolute square of the matrix element is proportional to the density of events across the plot.

As we noticed in similar cases, γ is a phase and only phase differences have physical meaning; therefore, the presence of another phase is necessary. In this case, this is the phase difference due to strong interaction between the amplitudes of $B^- \to \bar{D}^0 K^-$ and $B^- \to D^0 K^-$, that we call $\delta_B^{DK}\left(m^2_{K_s^0 K^+}, m^2_{K_s^0 K^-}\right)$. This is a function of the Dalitz plot coordinates, while γ is not. The equivalent expressions for the CP conjugate decays $B^+ \to \bar{D}^0 K^+$ and $B^+ \to D^0 K^+$ is obtained by making the substitution $\gamma \to -\gamma$ and exchanging the coordinates of the Dalitz plot with one another. The strong phase δ_B^{DK}

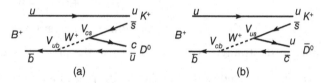

Fig. 8.16. Tree-level Feynman diagram for (a) $B^+ \to D^0 K^+$ and (b) $B^+ \to \bar{D}^0 K^+$. If D^0 and $\bar{D}^0$ decay in the same channel, the two processes have the same initial and the same final states and, consequently, the probability is the square modulus of the sum of the two amplitudes.

does not change, with the consequence that the observable phase difference changes from $\delta_B^{DK}\left(m_{K_s^0 K^+}^2, m_{K_s^0 K^-}^2\right) - \gamma$ to $\delta_B^{DK}\left(m_{K_s^0 K^-}^2, m_{K_s^0 K^+}^2\right) + \gamma$; notice the flipping of the arguments.

In summary, we have two Dalitz plots for the $K_s^0 K^+ K^-$ final state of the decay of the D meson, one originating from the B^+ decay and one from the B^- decay, namely from $B^+ \rightarrow \bar{D}^0 K^+$ and from $B^- \rightarrow D^0 K^-$. The structures of both Daliz plots are complex and contain a lot of information. This can be extracted provided a large enough data set is available to show, in all the necessary detail, the strong phase variation across the Dalitz plot. The luminosity of LHC, once more, was essential to allow LHCb to collect about 1900 events of $D \rightarrow K_s^0 K^+ K^-$. With these statistics, it has been possible to extract, in a model-independent way, the strong δ_B^{DK} phase variation across the Dalitz plots, and the unitary triangle phase

$$\gamma = \left(68.7_{-5.1}^{+5.2}\right)^\circ. \tag{8.88}$$

The value is consistent with previous indirect determinations. Combining all the measurements (Workman *et al.* 2022), one obtains

$$\gamma = \left(65.9_{-3.5}^{+3.3}\right)^\circ. \tag{8.89}$$

8.11 The Unitary Triangle

In Section 7.11 we summarized the measured absolute values of the CKM matrix elements and of the δ_{13} phase. The latter is the sole source of all the $\mathcal{CP}$ violation phenomena in the quark sector. A fundamental assumption of the SM is that the CKM matrix, with three quark families, is unitary. Unknown physics, beyond the SM, might induce non-unitarity of the matrix. For example, a fourth family of quarks would have such an effect. Consequently, much experimental and theoretical effort has been dedicated to test the mixing matrix unitarity with ever increasing accuracy. In this chapter, we have discussed, even if not exhaustively, a number of relevant examples.

In Section 7.11 we have already shown that the sums of the absolute squares of the elements of the first two rows and the first two columns are equal to one, within the uncertainties. We have also seen that the unitary condition on two different lines, or columns, in practice consists of a single testable constraint, namely

$$V_{ud}V_{ub}^* + V_{cd}V_{cb}^* + V_{td}V_{tb}^* = 0. \tag{8.90}$$

This leads to the checks on the unitary triangle (sketched in Fig. 7.18). By convention the triangle is usually drawn in the plane of the complex quantity $\dfrac{V_{ud}V_{ub}^*}{V_{cd}V_{cb}^*}$ (because this is independent of where the phase factor is located in the mixing matrix). We recall that its sides are normalized dividing each term of Eq. (8.90) by $\left|V_{cd}V_{cb}^*\right|$, so that the absolute value of the second term is equal to one and we can fix the positions of its two extremes on the real axis in (0,0) and (1,0), as in Fig. 8.17.

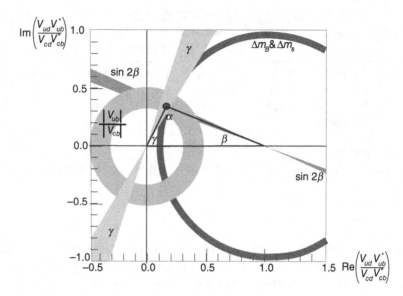

Fig. 8.17. The 'unitary triangle'.

Several measurements exist of the internal angles. We discussed one of these for γ in Section 8.10 and one of $\sin 2\beta$ in Section 8.6. The latter determines β within the fourfold ambiguity $\beta, \beta + \pi, \pi/2 - \beta, -\pi/2 - \beta$. However, further measurements by the BaBar and Belle experiments have resolved all of them. Using also the measured value of α (not discussed here), we can check the value of the sum, which is (Workman *et al.* 2022) $\alpha + \beta + \gamma = (173 \pm 6)°$, compatible with 180°.

The most stringent constraints are shown in Fig. 8.17 (data from Workman *et al.* 2022) the regions (shown at 95% confidence level) overlap consistently in a single 'point', the small dark region in the diagram. In the SM the length of the left side is $R_b \equiv |V_{ud}V_{ub}^*| / |V_{cd}V_{cb}^*|$ and the length of the right side is $R_t \equiv |V_{td}V_{tb}^*| / |V_{cd}V_{cb}^*|$.

In R_b, the ratio $|V_{ud}|/|V_{cd}|$ is known with better accuracy (1.8%, see Eq. (7.93)) than the other factors; as $|V_{ud}|/|V_{cd}| = 4.406 \pm 0.079$, we write

$$R_b = (4.406 \pm 0.079)\left|\frac{V_{ub}}{V_{cb}}\right|. \tag{8.91}$$

The remaining factor, the ratio $\left|\dfrac{V_{ub}}{V_{cb}}\right|$, is obtained by measuring the semileptonic decays, at the quark level, of the b quark:

$$\frac{|V_{ub}|^2}{|V_{cb}|^2} \propto \frac{\Gamma(b \to u l \nu_l)}{\Gamma(b \to c l \nu_l)}. \tag{8.92}$$

As for R_t, it can be shown to be proportional to the square root of the ratio of the B^0 and B_s^0 mass differences, namely $\sqrt{\Delta m_B / \Delta m_s}$. We have seen how these quantities are measured. Notice also that in the ratio the uncertainties in the hadronic factors largely cancel out.

Table 8.1. Lifetimes, total widths and mass differences of the pseudoscalar neutral flavoured mesons.

	$\tau(\text{ps})$	$c\tau(\mu\text{m})$	$\Gamma\left(\text{ps}^{-1}\right)$	$\Gamma(\text{meV})$	$\Delta m\left(\text{ps}^{-1}\right)$	$\Delta m(\text{meV})$
K_L	$51.16 \pm 0.21 \times 10^3$	15.3×10^6	2.0×10^{-5}	1.3×10^{-5}	$5.292 \pm 0.009 \times 10^{-3}$	$3.483 \pm 0.006 \times 10^{-3}$
K_S	89.54 ± 0.004	2.67×10^4	0.011	7.4×10^{-3}		
D_1	0.4103 ± 0.0010	123	2.4	1.61	$9.67 \pm 0.14 \times 10^{-3}$	$6.13 \pm 0.86 \times 10^{-3}$
D_2	0.4103 ± 0.0010					
B_H	1.520 ± 0.005	459	0.65	0.43	0.5065 ± 0.0019	0.3337 ± 0.0033
B_L	1.520 ± 0.005					
B_{sH}^0	1.497 ± 0.015	439	0.86	0.57	17.765 ± 0.06	11.5 ± 0.5
B_{sL}^0	1.497 ± 0.015					

With this experimental information, we draw two circles: one of centre $(0,0)$ and radius R_b, and one of centre $(1,0)$ and radius R_t. The two circles cross one another, showing that, indeed, we have a triangle. Then we draw the constraints from β and γ, which cross each other just in one of the circles' intersections, so choosing one solution. For simplicity we do not show the constraints of α and others, that are less stringent, and in agreement. The coherent convergence of all the measurements in a single point provides a stringent test of the SM.

In concluding this section, we summarize in Table 8.1 the values of the lifetimes, widths and mass differences for the four flavoured pseudoscalar meson pairs, which we have partly discussed in this chapter.

Problems

8.1 The DAΦNE Φ-factory at Frascati is an e^+e^- collider at the centre of mass energy equal to the ϕ mass. Calculate the ratio between the annihilation rates into K^+K^- and $K^0\bar{K}^0$ neglecting the mass difference between charged and neutral kaons. Is this a good approximation? Considering the case $K^0\bar{K}^0$, calculate the relative frequency of $K_1^0 K_1^0$, $K_1^0 K_2^0$ and $K_2^0 K_2^0$.

8.2 From which of the $\bar{p}p$ initial states $^1S_0, {}^3S_1, {}^1P_1, {}^3P_0, {}^3P_1$ and $^3P_2$ can each of the following reactions, $\bar{p}p \to K^+K^-$, $\bar{p}p \to K_1^0 K_1^0$ and $\bar{p}p \to K_1^0 K_2^0$ proceed?

8.3 A π^- is sent onto a target producing neutral K mesons and Λ hyperons. Consider the component of the resulting K beam with momentum $p = 10$ GeV. What is the ratio between K_S and K_L at the production point? What is it at $l = 10$ m from the production point? Determine the fraction of decays into 2π that would be observed in the absence of CP violation.

8.4 An experiment needs an almost monochromatic K^+ beam with momentum $p = 2$ GeV. We obtain it by building a magnetic spectrometer and a system of slits. However, the total length of the beam is limited by the lifetime of the K^+. At what distance is the K^+ intensity reduced to 10% of the initial value?

8.5 Consider the reactions $\pi^- p \to K^0 + X$ and $\pi^- p \to \bar{K}^0 + Y$ and establish the minimum mass of the states X and Y that are compatible with the conservation laws and the two corresponding energy thresholds.

8.6 Consider a neutral K-meson beam with momentum $p_K = 400$ MeV impinging on a liquid hydrogen target and determine the reaction channels open for each component K^0 and $\bar{K}^0$. Estimate which has the larger cross-section.

8.7 An asymmetric beauty factory operates at the $\Upsilon(4^1S_3)$, namely at $\sqrt{s} = 10\,580$ MeV, to study the process $e^+ e^- \to B^0 \bar{B}^0$. 'Asymmetric' means that the centre of mass moves in the reference frame of the collider. Consider an event in which both mesons are produced with the Lorentz factor $\beta\gamma = 0.56$. The decay vertex of one of them is at 120 μm from the principal vertex; amongst the B decay products there is a μ^+. How many lifetimes did the particle live? What can we say about the two flavours? The second B decays at 0.5 mm, again with a μ in its final state. How many lifetimes did it live? Can the μ be positive? Why?

8.8 Consider a sample of events collected an integrated luminosity of 100 ev/fb of the BaBar experiment. In the laboratory frame the centre of mass moves with the average Lorentz factor $\beta\gamma = 0.56$. How many seconds at a luminosity $\mathcal{L} = 10^{34}$ cm^{-2} s^{-1} would be needed to collect such a sample? Assuming the value $\Delta R = 3$ at the $\Upsilon(4S)$ resonance, how many $B^0 \bar{B}^0$ pairs have been collected? What is the average separation between production and decay vertices?

8.9 Prove expression (8.71) neglecting terms of orders above the first in ε and ε'.

8.10 Prove expression (8.72) neglecting terms of orders above the first in ε and ε'.

8.11 Prove expression (8.74) neglecting terms of orders above the first in ε and ε'.

8.12 Consider the decays $K_1^0 \to \pi^+ \pi^- \pi^0$ and $K_2^0 \to \pi^+ \pi^- \pi^0$. Call l the angular momentum of the $\pi^+ \pi^-$ system, $I_{\pi\pi}$ its isospin and L the angular momentum of the π^0 relative to it. Suppose that the mass of the K were much larger that its actual value and that higher values of l and L were not suppressed. Establish the possible values of $I_{\pi\pi}$ and I assuming CP is conserved.

Summary

In this chapter we have studied CP-violation phenomena in meson physics and in neutral mesons' oscillations. We have seen in particular:

- the states of definite flavour, CP and mass and lifetime
- the $K^0 \bar{K}^0$ (strangeness) oscillations and the regeneration
- that CP violation can happen in the wave function, in the interference with oscillations and in the decay
- the CP violation in the $K^0 \bar{K}^0$ system
- the oscillations and CP violation in the $B^0 \bar{B}^0$ system
- the oscillations and CP violation in the $B_s^0 \bar{B}_s^0$ system
- the CP violation in meson decays, neutral K, neutral and charged B

- CP violation in the charm in the $D^0\bar{D}^0$ system
- the unitary triangle.

Further Reading

Cavoto, G. L. & Pelliccioni M. (2011) Quark mixing, last came charm. *Riv. Nuov. Cim.* 34 643

Cronin, J. W. (1980) Nobel Lecture; *CP Symmetry Violation. The Search for its Origin*

Fitch, V. L. (1980) Nobel Lecture; *The Discovery of Charge-Conjugation Parity Asymmetry*

Giorgi, M. A., Neri, N. & Rama, M. (2013) B physics at e+e− flavor factories. *Riv. Nuov. Cim.* 36 6

Kleinknecht, K. (2003) *Uncovering CP Violation.* Springer Tracts in Modern Physics 195

9

The Standard Model

The electroweak theory describes with a gauge theory, with the symmetry group $SU(2) \otimes U(1)$, the electromagnetic and weak interactions, both NC and CC. In particular, the electromagnetic and weak coupling constants are not independent but correlated by the theory. However, electroweak theory and QCD, both being gauge theories, are unified by the theoretical framework while their coupling constants are independent. Electroweak theory and QCD together form the Standard Model of fundamental interactions.

In Sections 9.1–9.3 we shall introduce the electroweak theory, as usual without any theoretical rigour. The unification characteristics appear mainly in the neutral current processes. The transition probabilities of all these processes are predicted by the theory in terms of the weak mixing angle, or simply 'weak angle', also called 'Weinberg angle'. We shall mention the several processes in which it has been determined and discuss one of them in detail as an example.

A crucial prediction of the theory is the existence of three vector bosons, W^+, W^- and Z^0. Even if the theory does not predict their masses, it precisely states the relations of the two masses with two measured quantities: the Fermi constant and the weak mixing angle. We shall describe the UA1 experiment and the discovery of the vector bosons in Sections 9.6 and 9.7. As with all the coupling constants, the weak mixing angle, which is a ratio of coupling constants, depends on the energy scale. We shall see its evolution in Section 9.8.

Once the masses of the vector bosons are known, the theory predicts their branching ratios in all the decay channels. All these predictions have been experimentally verified with a high degree of accuracy. We shall see the experimental proof of the fact that the vector bosons have weak charges themselves and that, consequently, they interact directly with one another. We shall discuss the precision tests of the electroweak theory, performed in the last decennium of past century at the LEP electron–positron collider and at the Tevatron proton–antiproton collider, which operated from 1983 to 2011.

The electroweak Lagrangian is initially written without mass terms, implying that both the gauge bosons and the fermions are massless. The proof of the renormalizability of the electroweak theory, namely that the divergence appearing in the calculations can be mathematically controlled, was given by 't Hooft (1971). The masses of the vector bosons are theoretically generated, without destroying the renormalizability, by the spontaneous breaking of the gauge symmetry, through a mechanism discovered in 1964 by Englert and Brout (1964), Higgs (1964a, b) and, somewhat later, by Guralnik, Hagen and Kibble (1964). This is usually called the 'BEH mechanism', from the initials of the discoverers, or also the 'Higgs mechanism'. We shall discuss it, without any rigour, in Section 9.12. The fundamental prediction of the mechanism is the existence of a scalar boson, which became known as the Higgs boson, H, or simply 'the Higgs'. All the couplings of the

Higgs boson to the gauge bosons and to the fermions are expressed by the theory as functions of the Higgs mass, which, however, the theory does not fix. In Section 9.14 we shall describe the LHC *pp* collider and its two general-purpose experiments, ATLAS and CMS. After having discussed the search for the Higgs boson at LEP and at the Tevatron, we shall come to its discovery in 2012 by ATLAS and CMS. The measurements of its properties, its mass and width, its spin and parity, its couplings to the bosons and to the fermions will be discussed in Sections 9.16–9.19. All of them have been found to agree with the predictions of the Standard Model, so completing the experimental verification of its basic building blocks in the spinor, vector and scalar sectors. Finally, in Section 9.20 we shall summarize the results of the global fit of the SM parameters, as they are known today, with a hint to the perspective of the next phases of the LHC experiments.

9.1 The Electroweak Interaction

We discuss now the properties of the gauge bosons, $W^\pm, Z^0$ and the photon. The fundamental representation of $SU(2) \otimes U(1)$ hosts three and one gauge fields, respectively. A quantity called *weak isospin*, which we shall indicate by I_W, corresponds to $SU(2)$. From now on we shall call it simply isospin. The quantity corresponding to $U(1)$ is called *weak hypercharge* or simply hypercharge, Y_W. All the members of the same isospin multiplet have the same hypercharge. Notice that weak isospin and hypercharge have nothing to do with those of the hadrons.

Hypercharge can be defined in two equivalent ways: as twice the average electric charge of the multiplet or as

$$Y_W \equiv 2(Q - I_{Wz}). \tag{9.1}$$

Let us call $W = (W_1, W_2, W_3)$ the triplet of fields corresponding to $SU(2)$. Clearly, W has $I_W = 1$ and $Y_W = 0$. It interacts with the isospin of the particles.

Let us call B the field corresponding to $U(1)$. Its isospin, its electric charge and its hypercharge are zero. It interacts with the hypercharge of the particles.

These four fields are not the physical fields that mediate the interactions. The weak CC interactions are mediated, as we shall see immediately, by W^+ and W^-, which are linear combinations of W_1 and W_2, while the mediators of the electromagnetic and weak NC interactions, the photon and the Z, are linear combinations of W_3 and B.

The experiments we have discussed in Chapter 7 (and many others) showed that the charged W, the mediator of the CC weak interactions, couples to the negative chirality bispinors. We must take this into account in assigning isospin and hypercharge to the particles.

Let us start with the leptons. There are two left chirality leptons in every family; we lodge them in the same isospin doublet ($I_W = 1/2$) to make them able, due to their non-zero isospin, to couple with $W^\pm$ and write the equations

$$\begin{pmatrix} I_{Wz} = +1/2 \\ I_{Wz} = -1/2 \end{pmatrix} = \begin{pmatrix} \nu_{eL} \\ e_L^- \end{pmatrix}, \quad = \begin{pmatrix} \nu_{\mu L} \\ \mu_L^- \end{pmatrix}, \quad = \begin{pmatrix} \nu_{\tau L} \\ \tau_L^- \end{pmatrix}. \tag{9.2}$$

As anticipated in Section 7.13, unlike the charged current, the neutral current also interacts with right charged fermions, with different couplings, but not with right neutrinos. The charged right lepton of each family is an isospin singlet ($I_W = 0$)

$$e_R^-, \quad \mu_R^-, \quad \tau_R^-. \tag{9.3}$$

There is no corresponding neutral current for hypothetical right neutrinos.

The situation of the quarks is similar, provided we take CKM mixing into account, that is, the W couples universally to the CKM rotated quark states d', s' and b'. For every colour, there are three isospin doublets, one for each family (nine in total)

$$\begin{pmatrix} I_{Wz} = +1/2 \\ I_{Wz} = -1/2 \end{pmatrix} = \begin{pmatrix} u_L \\ d'_L \end{pmatrix}, \quad = \begin{pmatrix} c_L \\ s'_L \end{pmatrix}, \quad = \begin{pmatrix} t_L \\ b'_L \end{pmatrix} \tag{9.4}$$

and the singlets (18 in total)

$$d_R, \quad u_R, \quad s_R, \quad c_R, \quad b_R, \quad t_R. \tag{9.5}$$

The quark mixing is irrelevant for the neutral current weak interactions. So, we can write it in terms of the 'rotated' quarks or of the quarks of definite flavours, with the same result. Indeed, what we have observed for two families at the end of Section 7.10 is also true for three families.

Notice that the weak isospin of the left quarks of the first family is equal, by chance, to its flavour isospin.

All the quantum numbers of the antiparticles are equal and opposite to those of the corresponding particles. In the charged current sector the operator that creates a negative chirality (left) particle creates its positive chirality (right) antiparticle. They belong to three doublets

$$\begin{pmatrix} I_{Wz} = +1/2 \\ I_{Wz} = -1/2 \end{pmatrix} = \begin{pmatrix} e_R^+ \\ \overline{\nu}_{eR} \end{pmatrix}, \quad = \begin{pmatrix} \mu_R^+ \\ \overline{\nu}_{\mu R} \end{pmatrix}, \quad = \begin{pmatrix} \tau_R^+ \\ \overline{\nu}_{\tau R} \end{pmatrix}. \tag{9.6}$$

The antiparticles of (9.3), which appear in the NC, are isospin singlets

$$e_L^+, \quad \mu_L^+, \quad \tau_L^+. \tag{9.7}$$

Again, there is no term for left antineutrinos.

The antiquark doublets are

$$\begin{pmatrix} I_{Wz} = +1/2 \\ I_{Wz} = -1/2 \end{pmatrix} = \begin{pmatrix} \overline{d}'_R \\ \overline{u}_R \end{pmatrix}, \quad = \begin{pmatrix} \overline{s}'_R \\ \overline{c}_R \end{pmatrix}, \quad = \begin{pmatrix} \overline{b}'_R \\ \overline{t}_R \end{pmatrix}. \tag{9.8}$$

Their singlets are

$$\overline{d}_L, \quad \overline{u}_L, \quad \overline{s}_L, \quad \overline{c}_L, \quad \overline{b}_L, \quad \overline{t}_L. \tag{9.9}$$

Table 9.1 summarizes the values of isospin, hypercharge and electric charge of the fundamental fermions. The values are identical for every colour. In the table each particle and antiparticle is labelled with a subscript that specifies its chirality, L for negative, R for positive. Remember, however, that chirality, which is very important in the coupling, namely in the processes of creation and of destruction of the particle, is not a conserved quantum number. In fact γ_5 does not commute even with the free Hamiltonian.

Table 9.1 Fermions isospin, hypercharge, electric charge and 'Z-charge factors' $c_Z = I_{Wz} - s^2 Q$ (see Section 9.3).

	I_W	I_{Wz}	Q	Y_W	c_Z		I_W	I_{Wz}	Q	Y_W	c_Z
ν_{lL}	1/2	+1/2	0	−1	1/2	$\bar{\nu}_{lR}$	1/2	−1/2	0	1	−1/2
l_L^-	1/2	−1/2	−1	−1	$-1/2 + s^2$	l_R^+	1/2	+1/2	+1	1	$1/2 - s^2$
l_R^-	0	0	−1	−2	s^2	l_L^+	0	0	+1	2	$-s^2$
u_L	1/2	+1/2	2/3	1/3	$1/2 - (2/3)s^2$	$\bar{u}_R$	1/2	−1/2	−2/3	−1/3	$-1/2 + (2/3)s^2$
d_L'	1/2	−1/2	−1/3	1/3	$-1/2 + (1/3)s^2$	$\bar{d}_R'$	1/2	+1/2	1/3	−1/3	$1/2 - (1/3)s^2$
u_R	0	0	2/3	4/3	$-(2/3)s^2$	$\bar{u}_L$	0	0	−2/3	−4/3	$(2/3)s^2$
d_R'	0	0	−1/3	−2/3	$(1/3)s^2$	$\bar{d}_L'$	0	0	1/3	2/3	$-(1/3)s^2$

Example 9.1 Establish whether any of the following processes exists: $W^- \to e_L^- + \bar{\nu}_{eR}$, $W^- \to d_L' + \bar{u}_R$, $Z^0 \to \bar{u}_R + u_R$, $W^+ \to \bar{d}_R' + u_L$, $Z^0 \to \bar{u}_R + u_L$, $Z^0 \to \bar{u}_L + u_L$.

Electric charge and hypercharge are absolutely conserved quantities. The former conservation is satisfied by all the above processes. Let us check hypercharge conservation. Consider $W^- \to e_L^- + \bar{\nu}_{eR}$. The initial W^- has $Y = 0$ as all the gauge bosons, the left electron has $Y = -1$, the right antineutrino has $Y = +1$; synthetically $0 \to -1 + 1$. The process exists. For $W^- \to d_L' + \bar{u}_R$ we have $0 \to 1/3 - 1/3$, OK. For $Z^0 \to \bar{u}_R + u_R$ we have $0 \to -1/3 + 4/3$; hypercharge is not conserved, the process does not exist. For $W^+ \to \bar{d}_R' + u_L$ we have $0 \to -1/3 + 1/3$, OK. For $Z^0 \to \bar{u}_R + u_L$ we have $0 \to -1/3 + 1/3$, OK. For $Z^0 \to \bar{u}_L + u_L$ we have $0 \to -4/3 + 1/3$; this does not exist. □

9.2 Structure of the Weak Neutral Currents

The neutral current weak interactions are mediated by the Z boson. They have two important characteristics, already mentioned in Section 7.2.

(1) Neutral current couples each fermion with itself only, for example ee and not $e\mu$. If they are quarks, they must have the same colour: $^B u^B u$, not $^R u^B u$, because the Z, like the W, does not carry any colour. Fig. 9.1 shows four non-existent vertices.
(2) Neutral currents do not have the space-time $V-A$ structure.

Example 9.2 Consider the couplings $u_R \to Z^0 + u_R$ and $u_R \to Z^0 + u_L$. Are they possible? Electric charge is conserved in both cases. The former process is possible because the hypercharge balance is $4/3 \to 0 + 4/3$, the latter is not because $4/3 \to 0 + 1/3$. □

For every family there are seven currents coupling Z to every fermion: six for the left and right charged fermions, and one for neutrinos, which are only left. We write down the seven currents of the first family in the equations in (9.10) and draw the vertices in Fig. 9.2.

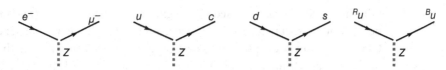

Fig. 9.1. Four couplings that do not exist.

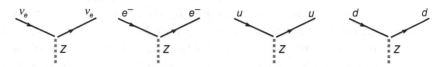

Fig. 9.2. The existing couplings for the first family.

$$\frac{1}{2}g_L^{\nu_e}\bar{\nu}_e\gamma_\mu\left(1-\gamma_5\right)\nu_e = g_L^{\nu_e}\bar{\nu}_{eL}\gamma_\mu\nu_{eL}$$

$$\frac{1}{2}g_L^e\bar{e}\gamma_\mu\left(1-\gamma_5\right)e+\frac{1}{2}g_R^e\bar{e}\gamma_\mu\left(1+\gamma_5\right)e = g_L^e\bar{e}_L\gamma_\mu e_L + g_R^e\bar{e}_R\gamma_\mu e_R$$

$$\frac{1}{2}g_L^u\bar{u}\gamma_\mu\left(1-\gamma_5\right)u+\frac{1}{2}g_R^u\bar{u}\gamma_\mu\left(1+\gamma_5\right)u = g_L^u\bar{u}_L\gamma_\mu u_L + g_R^u\bar{u}_R\gamma_\mu u_R$$

$$\frac{1}{2}g_L^d\bar{d}\gamma_\mu\left(1-\gamma_5\right)d+\frac{1}{2}g_R^u\bar{d}\gamma_\mu\left(1+\gamma_5\right)d = g_L^d\bar{d}_L\gamma_\mu d_L + g_R^u\bar{d}_R\gamma_\mu d_R.$$

(9.10)

Notice that the first term in every row has the structure of the CC, namely $V-A$ coupling with left fermions. The second terms couple with right fermions. Every term corresponds to a different physical process and the corresponding coupling constant might be, a priori, different. Therefore, we have used different symbols. With three families, we have, in total, 21 NC 'weak charges'. The power of the electroweak theory is to give all these charges in terms of two constants, the elementary electric charge and the weak mixing angle θ_W, which we shall meet soon.

The coupling of the Z is a universal function of the charge Q and of the third isospin component I_z of the particle, as we shall see in the next section. We immediately state that the Z

- couples to both left and right fermions
- couples to the $W^\pm$
- also couples to electrically neutral particles, provided they have $I_z \neq 0$, such as left neutrinos
- does not couple to states with both $Q = 0$ and $I_z = 0$, such as the γ and itself.

9.3 Electroweak Unification

Theoretical attempts to unify electromagnetic and weak interactions started in the late 1950s, but the first substantial step forward was achieved by S. Glashow in 1961

(Glashow 1961), who assumed as the gauge group $SU(2) \otimes U(1)$. Having two gauge bosons, one associated with $U(1)$, one associated with the neutral generator of $SU(2)$, Glashow assumed the photon to be a superposition of the two. As the square moduli of the two amplitudes in the superposition must sum to 1, he wrote them as the sine and the cosine of a 'mixing angle', called θ, now θ_W. A similar model was proposed by A. Salam and J. Ward in 1964 (Salam & Ward 1964). These were the first correct steps in electroweak unification, but a fundamental problem remained unsolved: the range of the electromagnetic interaction is infinite, hence the photon is massless, while the range weak interaction is very short, hence their gauge bosons must be massive. In 1964 the 'BEH mechanism' leading to the spontaneous breaking of a gauge theory was proposed, as we shall discuss in Section 9.12. By that year the two pillars of the future theory, the $SU(2) \otimes U(1)$ symmetry and the mechanism of spontaneous gauge symmetry breaking, were known, but it took 3 years to put them together. The synthesis was independently done by S. Weinberg (Weinberg 1967) and A. Salam. Salam presented the synthesis in lectures at the Imperial College London in the autumn of 1967, published the following year (Salam 1968), in which he introduced the name 'electroweak theory'. Both showed how the spontaneous symmetry breaking of a gauge symmetry can lead to massive bosons. The proof that the theory can be renormalized was found in 1971 by 't Hooft and Veltman ('t Hooft 1971). Let us now see the relationships between the fields W and B and the physical fields $W^\pm$, Z and γ together with their coupling to fermions.

The field $W^\mu \equiv \left(W_1^\mu, W_2^\mu, W_3^\mu \right)$ is a four-vector in space-time (index μ) and a vector in isotopic space. The fields of the physical charged bosons are

$$W^\pm = \frac{1}{\sqrt{2}} \left(W_1 \pm i W_2 \right). \tag{9.11}$$

For every fermion doublet, there is a space-time four-vector, isospin vector, called 'weak current' $j_\mu \equiv \left(j_{1\mu}, j_{2\mu}, j_{3\mu} \right)$. The field W^μ couples to j_μ as $g W^\mu j_\mu$ with the dimensionless coupling constant g. The charged currents are linear combinations of two components of the current

$$j^\pm = j_1 \pm i j_2. \tag{9.12}$$

Considering, for example, the doublet $\begin{pmatrix} v_{eL} \\ e_L^- \end{pmatrix}$, the corresponding charged currents are

$$j_{e\mu}^+ = \overline{v}_{eL} \gamma_\mu e_L^-; \quad j_{e\mu}^- = \overline{e}_L^- \gamma_\mu v_{eL}. \tag{9.13}$$

The field B^μ is a space-time four-vector, isoscalar. It couples with the hypercharge current j_μ^Y, which is also four-vector and isoscalar. The coupling constant is g'. The hypercharge current is twice the difference between the electromagnetic current j_μ^{EM} and the neutral component of the weak NC, in agreement with Eq. (9.1)

$$j_\mu^Y = 2 j_\mu^{EM} - 2 j_{3\mu}. \tag{9.14}$$

The first term is the electromagnetic current that we know, which, for the charged fermion f is

$$j_{f\mu}^{EM} = \bar{f}\gamma_\mu f. \tag{9.15}$$

Chirality is not specified because the electromagnetic interaction does not depend on it.

Let us call A and Z the physical fields that mediate the electromagnetic and the weak NC interactions respectively. They are two mutually orthogonal linear superpositions of W_3 and B. We shall determine them by imposing that the photon does not couple to neutral particles, while the Z^0 does. The transformation is expressed in terms of the two coupling constants g and g' or, equivalently, as a rotation through an angle θ_W, the weak mixing angle

$$\begin{pmatrix} Z^0 \\ A \end{pmatrix} = \frac{1}{\sqrt{g^2 + g'^2}} \begin{pmatrix} g & -g' \\ g' & g \end{pmatrix} \begin{pmatrix} W_3 \\ B \end{pmatrix} = \begin{pmatrix} \cos\theta_W & -\sin\theta_W \\ \sin\theta_W & \cos\theta_W \end{pmatrix} \begin{pmatrix} W_B \\ B \end{pmatrix}. \tag{9.16}$$

The weak mixing angle is defined by the relationship

$$\theta_W \equiv \tan^{-1}\frac{g'}{g}. \tag{9.17}$$

The rotation is not small, being $\theta_W \approx 29°$, as we shall see. The interaction Lagrangian, being symmetrical under the gauge group, is an isoscalar, namely

$$\mathcal{L} = g\left(j_\mu^1 W_1^\mu + j_\mu^2 W_2^\mu + j_\mu^3 W_3^\mu\right) + \frac{g'}{2} j_\mu^Y B^\mu. \tag{9.18}$$

We can write this expression as

$$\mathcal{L} = \frac{g}{\sqrt{2}}\left(j_\mu^- W_+^\mu + j_\mu^+ W_-^\mu\right) + j_\mu^3\left(g W_3^\mu - g' B^\mu\right) + g' j_\mu^{EM} B^\mu.$$

Also introducing the neutral physical fields and grouping terms, we obtain

$$\mathcal{L} = \frac{g}{\sqrt{2}}\left(j_\mu^- W_+^\mu + j_\mu^+ W_-^\mu\right) + \frac{g}{\cos\theta_W}\left(j_\mu^3 - \sin^2\theta_W j_\mu^{EM}\right)Z^\mu + g\sin\theta_W A^\mu. \tag{9.19}$$

Let us examine this fundamental expression. Its terms are, in order, the CC weak interaction, the NC weak interaction and the electromagnetic interaction.

The constant in front of the last term must be proportional to the electric charge, assuring that the photon does not couple to neutral particles. Actually, the relationship with the elementary electric charge is

$$g\sin\theta_W = \frac{q_e}{\sqrt{\epsilon_0 \hbar c}} = \sqrt{4\pi\alpha}. \tag{9.20}$$

This expression 'unifies' the weak charge and the electric charge. As anticipated, all the interactions mediated by the four vector bosons are expressed in terms of two constants, the electric charge q_e and the weak mixing angle θ_W. However, the model does not predict the values of the two fundamental parameters. They must be determined experimentally.

From (9.17) and (9.20) we immediately have the relationship between the coupling constant of $U(1)$ and the electric charge

$$g' \cos \theta_W = \sqrt{4\pi\alpha}. \tag{9.21}$$

From (9.20) and (9.21) we also have

$$\frac{1}{\alpha} = \frac{4\pi}{g'^2} + \frac{4\pi}{g^2}, \tag{9.22}$$

which shows how the couplings of both gauge groups contribute to $1/\alpha$. At low energies where $1/\alpha \approx 137$, with $\sin^2 \theta_W \approx 0.232$, we have

$$4\pi / g'^2 = 105.2 \quad \text{and} \quad 4\pi / g^2 = 31.8. \tag{9.23}$$

The second term of (9.19) gives the coupling of Z with fermions. We see that it is universal in the sense that it is a universal function of the charge and of the third isospin component

$$\begin{aligned}
g_Z &\equiv \frac{g}{\cos \theta_W} \left(I_{Wz} - Q \sin^2 \theta_W \right) \\
&= \frac{\sqrt{4\pi\alpha}}{\sin \theta_W \cos \theta_W} \left(I_{Wz} - Q \sin^2 \theta_W \right) = \frac{g}{\cos \theta_W} c_Z.
\end{aligned} \tag{9.24}$$

On the right-hand side in the last equation, we introduced the 'Z-charge factor' c_Z

$$c_Z \equiv I_{Wz} - Q \sin^2 \theta_W. \tag{9.25}$$

The structure of c_Z is determined by the gauge group $SU(2) \otimes U(1)$, as the colour factors $\kappa_\lambda^{c_i c_j}$ are determined by $SU(3)$. We gave the fermion Z-charge factors in Table 9.1.

The first term of Eq. (9.19) describes the charged current weak processes we discussed in Chapter 7. As we know, the coupling constant g is given in terms of the Fermi constant and of the W mass by (7.55b), which we repeat here

$$G_F = \frac{\sqrt{2} g^2}{8 M_W^2}. \tag{9.26}$$

Using Eq. (9.20) we have the prediction of the W mass in terms of the fine-structure constant, the Fermi constant and the weak mixing angle

$$M_W = \left(\frac{g^2 \sqrt{2}}{8 G_F} \right)^{1/2} = \sqrt{\frac{\pi\alpha}{\sqrt{2} G_F}} \frac{1}{\sin \theta_W} = \frac{37.3}{\sin \theta_W} \text{ GeV}. \tag{9.27}$$

In this model, a measurement of the weak angle gives the W mass and vice versa.

The standard model gives a precise prediction of the relation between vector boson masses, the Fermi constant and the weak angle. Beyond (9.27), we have, at the lowest perturbative order, the ratio of the masses

$$M_W / M_Z = \cos \theta_W. \tag{9.28}$$

We shall discuss the (indirect) measurement of the weak angle in Section 9.5. Its value is $\sin^2 \theta_W \approx 0.232$ and we have the prediction

$$M_W \approx 80 \text{ GeV} \qquad M_Z \approx 91 \text{ GeV}. \tag{9.29}$$

We shall describe the discovery of the vector bosons in Section 9.7.

In complete generality, we can say that the Standard Model provides a unified description of all the known elementary processes of nature (but see Chapter 10). It is the most comprehensive theoretical structure ever built and the most accurately tested one. Its electroweak section, in particular, contains the following.

- The CC weak processes. We have studied a few examples at low energies, where the standard model coincides with the four-fermion interaction.
- The NC weak processes. These are the processes in which unification appears directly, especially at energies comparable to the mediator masses. We shall see a few experimental tests in this chapter.
- The direct interaction between mediators. It was tested precisely at LEP, as we shall see in Section 9.10.
- The 'generation' of the masses of the gauge bosons, leptons and quarks by the BEH mechanism. The Higgs boson was discovered in 2012 at LHC, as we shall see at the end of the chapter. The details of the mechanism are being further tested by the LHC experiments.

9.4 The Weak Mixing Angle

Electroweak unification appears in the neutral current weak processes, where the 'weak charges' are predicted in terms of $\sin^2 \theta_W$. If the theory is correct, the values of this parameter extracted from measurements of cross-sections or decay rates of different processes must agree. The extraction itself is made according to the prescriptions of the theory, which always imply the calculation of Feynman diagrams. For precise measurements, calculations must go beyond the tree level and include the radiative corrections at the level needed to reach the required precision.

A long series of high-accuracy experiments has tested the universality of the interaction in a wide energy range, from keV to hundreds of GeV, and for many different couplings. We shall mention only the main experiments here, without entering into details, and come back to the most precise determinations in Section 9.8.

- The gauge boson masses have been measured with a high degree of accuracy, as we shall see. Their ratio gives the most precise value of the weak mixing angle.
- Parity violation in atoms. The atomic electrons are bound to the nucleus not only by the electromagnetic interaction, exchanging a photon with its quarks, but also by the weak NC one, exchanging a Z with them. The effect of the latter is extremely small and not observable as a shift of the levels. However, the interference between the amplitudes is observable (Zel'dovich 1959) as a parity-violating effect (of the order of a part per million, ppm), providing a test of the theory at the MeV energy scale (Noecker *et al.* 1988; Grossman *et al.* 2005).
- Polarized electron elastic scattering on deuterium. Both photon exchange and Z exchange contribute to the process. The latter contribution is too small to be observable in the cross-section. However, the interference between the two amplitudes can be measured as a small difference (asymmetry) between the differential

cross-sections of the two electron polarization states (of the order of a few ppm). Indeed, the elementary process to be measured would be the electron–quark scattering. However, quarks are inside nucleons that, in turn, are inside the deuterium nucleus. Consequently, the accuracy in the determination of the weak angle is limited by the theoretical uncertainties in the QCD calculations. The first experiment was done at SLAC at $|Q^2| = 1.6$ GeV^2 (Prescott *et al.* 1978).

- Electron–electron (Møller) elastic scattering with polarized beam on non-polarized target. The process is similar to the previous one, but it is purely leptonic. Consequently, the extraction of the weak angle from the measured asymmetry A_{PV} is free of hadronic uncertainties. The experiment is more difficult because the asymmetry is very small, only a fraction of a ppm, and because the Møller events must be separated from the more frequent electron–proton scatterings. The first experiment was done at SLAC in 2004 on a 50 GeV high-intensity polarized electron beam at $|Q^2| = 0.026$ GeV^2, measuring A_{PV} with 20% precision (Anthony *et al.* 2004).

- Forward–backward asymmetry in the differential cross-section of the electron–positron annihilation into a fermion pair. The effect is due, again, to the interference between photon and Z exchanges, which violates parity resulting in an asymmetry around 90°. The asymmetry has been measured in a wide energy range between 10 GeV and 200 GeV. It is large at energies comparable to M_Z.

- Deep inelastic scattering of ν_μ and $\bar{\nu}_\mu$ on nuclei. In this case, since the probe is a neutrino, which has only the weak interaction, we can determine the weak angle directly from the measurement of the cross-sections. However, the use of a complex hadronic target limits the accuracy that can be reached in the weak angle determination. Typical momentum transfer values are of several GeV^2.

- Scattering of ν_μ and $\bar{\nu}_\mu$ on electrons. This is a purely leptonic process, free from the problem just mentioned. The measurement of its cross-section consequently provides a clean means of determining the weak angle without theoretical uncertainties. There is, of course, a price to be paid: the neutrino–electron cross-sections are four orders of magnitude smaller than the neutrino–nucleus ones. The most precise experiment of this type has been CHARM2 experiment performed at CERN in the 1980s–1990s, which we shall now describe.

Its aim was to measure the ratio of the total cross-sections of the two elastic scattering processes

$$\nu_\mu e^- \to \nu_\mu e^-; \quad \bar{\nu}_\mu e^- \to \bar{\nu}_\mu e^-. \tag{9.30}$$

These are similar to the quasi-elastic scattering (7.7), with the Z as a mediator instead of the W, as shown in Fig. 9.3 for the first of them. Even at neutrino beam energies of 100 GeV, the momentum transfers are small compared to M_Z. Therefore, as discussed in relation to Eq. (7.15) the cross-sections are proportional to s, namely to the product of neutrino energy and target mass, hence to $G_F^2 m_e E_\nu$. As the electron mass is several thousand times smaller than the mass of a nucleus, neutrino–electron cross-sections are, as anticipated, four orders of magnitude smaller than those on light nuclei.

Let us now find how the ratio of the cross-sections of the processes in (9.30) depends on the weak angle. Even if we do not have the theoretical instruments for the calculation, we

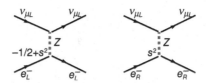

Fig. 9.3.   Diagrams for the $\nu_\mu e^-$ scattering. The Z-charge factors are shown at the lower vertex.

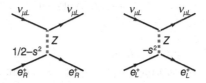

Fig. 9.4.   Diagrams for the $\nu_\mu e^+$ scattering. The Z-charge factors are shown at the lower vertex.

can obtain the result by physical arguments. Since we are interested in the ratio, we can ignore common factors. Since the energies are very high we shall consider the electrons as massless. Under this hypothesis, both negative chirality (L) electrons and neutrinos have negative helicity and positive chirality (R) electrons and neutrinos have positive helicity.

Let us start with the first reaction, to which the two diagrams in Fig. 9.3 contribute, namely the scattering of the left neutrino (the only one that exists) on left electrons and on right electrons. The two contributions can in principle be distinguished, by measuring the helicity, and therefore do not interfere. We take the sum of the squares of the two amplitudes.

Since the upper vertex is the same, the two contributions are proportional to the square of the Z-charge factors, c_Z, of the lower vertices. These are written in Fig. 9.3, with the notation $s^2 = \sin^2 \theta_W$

Moving on to the second reaction, we observe that its cross-section is equal, for the $C\mathcal{PT}$ invariance, to the cross-section of $\nu_\mu e^+ \to \nu_\mu e^+$. The corresponding diagrams, shown in Fig. 9.4, have the same upper vertex as those in Fig. 9.3. Therefore, in the ratio we only have to consider the lower vertices, summing the squares of the Z-charge factors.

We now observe that, for both processes, the two contributions are a scattering of a left fermion on a left fermion ($L + L$) and a left one on a right one ($L + R$). The two cases are different. Let us see why, with reference to the neutrino scattering (the antineutrino case is similar).

We analyse the two contributions with reference to Fig. 9.5, in which the kinematical variables are shown in the CM frame. In the ($L + L$) term, shown in Fig. 9.5(a), the angular momenta both of the incoming and outgoing pairs are zero. Therefore, since the interaction is point-like, all the scattering angles have the same probability, namely the angular cross-section is constant. In the ($L + R$) case of Fig. 9.5(b), the total angular momentum is $\mathbf{J} = 1$ with third component $J_z = -1$ (the quantization axis is the neutrino flight line). Therefore, only one out of the $2J + 1 = 3$ a priori possible spin states is allowed. In conclusion, there is a 1/3 factor in the ($L + R$) cross-section.

Before going on we observe that the conclusion just reached is general and that it can be written in a form valid for all the neutral currents: the total NC cross-sections

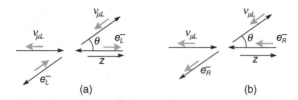

(a) The three-momenta and the spins in the left neutrinos on left electrons scattering in the centre of mass frame. (b) The same for left neutrinos on right electrons scattering. All particles are treated as massless.

between fermions $L + R$ and $R + L$ are, when considered at the same energy, three times smaller than the corresponding $L + L$ and $R + R$.

Returning to our calculation, and summing up the results, we have, aside from a common constant (which, for the curious, is $2 / \pi$),

$$
\sigma_{v_\mu e} \propto G_F^2 m_e E_v \left[\left(-\frac{1}{2} + \sin^2 \theta_W \right)^2 + \frac{1}{3} \sin^4 \theta_W \right]
$$

$$
\sigma_{\bar{v}_\mu e} \propto G_F^2 m_e E_{\bar{v}} \left[\frac{1}{3} \left(-\frac{1}{2} + \sin^2 \theta_W \right)^2 + \sin^4 \theta_W \right].
$$

Taking the ratio, at the same energy for neutrinos and antineutrinos, we have

$$
R = \frac{\sigma_{v_\mu e}}{\sigma_{\bar{v}_\mu e}} = 3 \frac{1 - 4 \sin^2 \theta_W + \frac{16}{3} \sin^4 \theta_W}{1 - 4 \sin^2 \theta_W + 16 \sin^4 \theta_W}. \tag{9.31}
$$

To measure the ratio, we expose the detector to a neutrino beam and an antineutrino beam. Let us call $N(v_\mu e)$ and $N(\bar{v}_\mu e)$ the numbers of neutrino–electron scattering events obtained in the two exposures. The incident neutrino energy is not well defined; rather, the beams have a wide energy spread. Since the cross-sections are proportional to the energy, the observed numbers of events are normalized to the ratio F of the energy-weighted fluxes

$$
F \equiv \int \Phi_{\bar{v}_\mu} \left(E_{\bar{v}} \right) E_{\bar{v}} dE_{\bar{v}} / \int \Phi_{v_\mu} \left(E_v \right) E_v dE_v \tag{9.32}
$$

obtaining the empirical ratio

$$
R_{\exp} = \frac{N(v_\mu e)}{N(\bar{v}_\mu e)} F. \tag{9.33}
$$

Let us now see the main characteristics the detector needs to have. As an example, the CHARM2 set-up is shown in Fig. 9.6. First, its sensitive mass must be large, given the smallness of the cross-sections; in practice, with the available neutrino beam intensities, it is of the order of hundreds of tons. Second, the detector must visualize the tracks of the events and measure their energy. At the same time, it must provide the target for neutrino interactions. In practice a 'fine-grain calorimeter' must be built.

The third problem is the backgrounds. Neutrinos interact both with the nuclei and with the electrons of the detector, but the latter process, which we are interested in,

The CHARM2 set-up (© Photo CERN).

happens only once in ten thousand. Moreover, the 'signal' is a single electron track, a topology that can be easily simulated by background events. There are two principal types of background.

(1) The muon neutrino beam contains an unavoidable contamination of electron neutrinos. It is small, about 1%, but the probability of the process

$$v_e + N \rightarrow e + X \tag{9.34}$$

is 10 000 times that of the elastic scattering, which is our signal. Consequently, the background-to-signal ratio is of the order of 100.

(2) The CC neutrino interactions with the nuclei are recognized by the μ track, which is straight and deeply penetrating. However, this is not the case for the NC interactions. Consider in particular

$$v_\mu + N \rightarrow v_\mu + \pi^0 + X \qquad \pi^0 \rightarrow \gamma\gamma. \tag{9.35}$$

Sometimes the hadronic part, called X, does not have enough energy and escapes detection; sometimes one of the photons materializes in a positron–electron pair that is confused with a single electron, simulating the signal.

Both backgrounds can be discriminated on kinematic grounds. Indeed, the electron hit by a neutrino maintains its direction within a very small angle, because its mass is small. Let us see.

The collision kinematic is depicted in Fig. 9.7.

Let us write the energy and momentum conservation

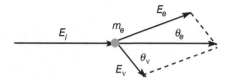

Fig. 9.7. Kinematic of the neutrino electron scattering.

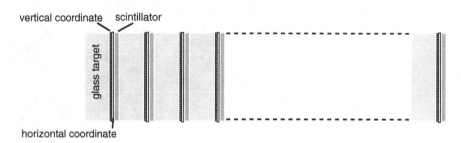

Fig. 9.8. Schematic of the CHARM2 experiment.

$$E_i + m_e = E_e + E_v;$$
$$0 = E_v \sin\theta_v + E_e \sin\theta_e;$$
$$E_i = E_v \cos\theta_v + E_e \cos\theta_e. \tag{9.36}$$

The last equation can be written in the form

$$E_i = E_v + E_e - E_v(1 - \cos\theta_v) - E_e(1 - \cos\theta_e).$$

Using the first equation in (9.36) we have

$$E_i = E_i + m_e - E_v(1 - \cos\theta_v) - E_e(1 - \cos\theta_e)$$

and in conclusion

$$E_e(1 - \cos\theta_e) = m_e - E_v(1 - \cos\theta_v) \le m_e$$

$$\Rightarrow 1 - \cos\theta_e \le \frac{m_e}{E_v}.$$

Finally, with a very good approximation,

$$E_e\theta_e^2 \le 2m_e. \tag{9.37}$$

We see that, in the elastic scattering events, the product of the square of the angle of the electron with its energy is very small. We conclude that our detector must have a good energy resolution and, to measure the electron direction with high accuracy, a very good spatial resolution. The latter condition implies a low atomic mass medium in order to minimize multiple scattering. A further condition is a good granularity to distinguish electrons from π^0s.

The detector has a modular structure, as shown schematically in Fig. 9.8. Its mass is 792 t, with a 4 m × 4 m section and 33 m length. Every module is made of a glass (a low-Z material) slab 48 cm thick, followed by a pair of tracking chambers, each measuring

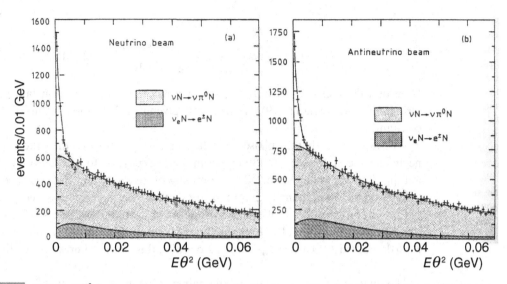

Fig. 9.9.   CHARM2: $E_e\theta_e^2$ distribution for neutrinos and antineutrinos (Reprinted from D. Geiregat *et al. Phys. Lett.* B259 (1991) 499, Fig. 1 © 1991 with permission by Elsevier).

one coordinate, and an array of scintillator counters. These have two functions, to trigger the read-out electronics if an event takes place and to measure the energy.

Fig. 9.9 shows the $E_e\theta_e^2$ distributions for both neutrino and antineutrino exposures. The peak close to 0 is the signal of neutrino–electron scattering, above the background. The contributions of the above-mentioned two principal background sources are evaluated and their sum compared to the experimental data at large angles, where the signal is absent. Having found agreement, the background function is extrapolated in the peak region and its contribution subtracted, to obtain the size of the signal.

The result is (Vilain *et al.* 1994)

$$\sin^2\theta_W = 0.2324 \pm 0.0083. \tag{9.38}$$

9.5 The Intermediate Vector Bosons

A crucial prediction of the electroweak theory is the existence of the vector bosons W and Z and of their characteristics: masses and total and partial widths. Let us see them in detail. As we saw in Section 9.3, the predicted values of the masses are

$$M_W \approx 80\,\text{GeV} \quad M_Z \approx 91\,\text{GeV}. \tag{9.39}$$

As we know, the couplings of W and Z to all the leptons and quarks are predicted by the theory. The calculation of the partial widths requires the knowledge of quantum field theory and cannot be done here. However, once we know one of them, we can calculate the others with elementary arguments. We start from the partial width of the decay

$$W^- \rightarrow l^- + \nu_l, \tag{9.40}$$

that is

$$\Gamma_{ev} = \Gamma_{\mu v} = \Gamma_{\tau v} = \left(\frac{g}{\sqrt{2}}\right)^2 \frac{M_W}{24\pi} = \frac{1}{2}\frac{G_F M_W^3}{3\sqrt{2\pi}} \approx 225 \text{ MeV}. \tag{9.41}$$

Note that the factor $g/\sqrt{2}$ in the above expression is simply the constant in the CC term in the Lagrangian (9.19). We have also given its expression in terms of the Fermi constant, by using Eq. (7.55).

We now assume that the phase-space factors are equal because the lepton masses are negligible in comparison to the energies in the W decay. As for the decays into quark–antiquark pairs, we must consider the mixing on one side and the existence of three colours on the other, namely that there are three possibilities for each decay channel.

Charge conservation implies that the quarks of the pair must be one of up-type and one of down-type. Not all the channels are open, namely the decays into $\bar{t} + d$, $\bar{t} + s$, $\bar{t} + b$ do not exist because $M_W < m_t$ (historically, this was not known at the time of the W discovery).

The quark and the antiquark of the pair may be in the same or different families. Given the smallness of the mixing angles, the partial widths in the quark–antiquark pairs are small. Neglecting the quark masses and recalling the colour factor 3, we have

$$\begin{aligned}\Gamma_{us} &\equiv \Gamma(W \to \bar{u}s) = 3 \times |V_{us}|^2 \Gamma_{ev} \\ &= 3 \times 0.224^2 \times \Gamma_{ev} \approx 35 \text{ MeV}\end{aligned} \tag{9.42}$$

and

$$\begin{aligned}\Gamma_{cd} &\equiv \Gamma(W \to \bar{c}d) = 3 \times |V_{cd}|^2 \Gamma_{ev} \\ &= 3 \times 0.22^2 \times \Gamma_{ev} \approx 33 \text{ MeV},\end{aligned} \tag{9.43}$$

where Γ_{ub} and Γ_{cb} are very small.

The widths in a quark–antiquark pair of the same family are

$$\begin{aligned}\Gamma_{ud} &\equiv \Gamma(W \to \bar{u}d) = 3 \times |V_{ud}|^2 \Gamma_{ev} \\ &= 3 \times 0.974^2 \times \Gamma_{ev} = 2.84 \times \Gamma_{ev} \approx 640 \text{ MeV}\end{aligned} \tag{9.44}$$

and

$$\begin{aligned}\Gamma_{cs} &\equiv \Gamma(W \to \bar{c}s) = 3 \times |V_{cs}|^2 \Gamma_{ev} \\ &= 3 \times 0.99^2 \times \Gamma_{ev} \approx 660 \text{ MeV}.\end{aligned} \tag{9.45}$$

We obtain the total width by summing the partial ones

$$\Gamma_W \approx 2.04 \text{ GeV}. \tag{9.46}$$

The coupling of the Z is proportional to the Z-charge factors c_Z (Table 9.1), as we see by rewriting Eq. (9.19):

$$g_Z \equiv \frac{g}{\cos\theta_W}\left(I_3^W - Q\sin^2\theta_W\right) = \frac{g}{\cos\theta_W}c_Z. \tag{9.47}$$

Let us start with the neutrino–antineutrino channels. We can obtain the partial width expression from (9.41), with the constant $g = \cos\theta_W$ in place of $g/\sqrt{2}$, because this is the factor of the NC term in the Lagrangian (9.19), and with M_Z in place of M_W. We have

$$\Gamma_\nu \equiv \Gamma\left(Z \to \nu_l \bar{\nu}_l\right) = \left(\frac{g}{\cos\theta_W}\right)^2 \frac{M_Z}{24\pi}\left(\frac{1}{2}\right)^2$$
$$= \frac{G_F M_W^2 M_Z}{\cos^2\theta_W \, 3\sqrt{2}\pi}\left(\frac{1}{2}\right)^2.$$

We now use Eq. (9.28) to eliminate M_W, obtaining

$$\Gamma_\nu = \frac{G_F M_Z^3}{3\sqrt{2}\pi}\left(\frac{1}{2}\right)^2 \approx 660 \times \frac{1}{4} \text{ MeV} = 165 \text{ MeV}. \tag{9.48}$$

The two-neutrino final states are not observable. Considering that other invisible particles might exist, one defines as 'invisible width' the total width in the invisible channels. If the only contribution to this is given by three neutrinos, the width is

$$\Gamma_{\text{inv}} = 3\Gamma_\nu \approx 495 \text{ MeV}. \tag{9.49}$$

The measurement of Γ_{inv} provides a way of testing whether there are more 'light' neutrinos, that is, with masses smaller than $M_Z/2$, and if there are other invisible particles.

Going now to the charged leptons and setting $s^2 = \sin^2\theta_W$, we have

$$\Gamma_l = \Gamma_e = \Gamma_\mu = \Gamma_\tau = \frac{G_F M_Z^3}{3\sqrt{2}\pi}\left[\left(-\frac{1}{2}+s^2\right)^2 + s^4\right] \tag{9.50}$$
$$\approx 660 \times 0.125 \text{ MeV} = 83 \text{ MeV}.$$

For the quark–antiquark decays we do not need to worry about mixing but we must remember the three colours. The $t\bar{t}$ channel is closed. For the other two up-type pairs, neglecting the quark masses, we have

$$\Gamma_u = \Gamma_c = 3\frac{G_F M_Z^3}{3\sqrt{2}\pi}\left[\left(\frac{1}{2}-\frac{2}{3}s^2\right)^2 + \left(-\frac{2}{3}s^2\right)^2\right] \tag{9.51}$$
$$\approx 660 \times 0.42 \text{ MeV} = 280 \text{ MeV}.$$

Finally, for the three down-type pairs, we obtain

$$\Gamma_d = \Gamma_s = \Gamma_b = 3\frac{G_F M_Z^3}{3\sqrt{2}\pi}\left[\left(-\frac{1}{2}+\frac{1}{3}s^2\right)^2 + \left(\frac{1}{3}s^2\right)^2\right] \tag{9.52}$$
$$\approx 660 \times 0.555 \text{ MeV} = 370 \text{ MeV}.$$

From the experimental point of view, it is not generally possible to distinguish the different quark–antiquark channels. Indeed, this is only possible, in some instances, for the $c\bar{c}$ and $b\bar{b}$ channels. Therefore, the total hadronic cross-section is measured. The predicted value is

$$\Gamma_h = 2\Gamma_u + 3\Gamma_d \approx 1.67 \text{ GeV}. \tag{9.53}$$

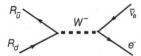

Fig. 9.10.   Diagram for process (9.56). The upper index labels the colour.

Summing up, the total width is

$$\Gamma_Z = \Gamma_{\text{inv}} + 3\Gamma_l + \Gamma_h \approx 2.42 \text{ GeV}. \tag{9.54}$$

We still need other predictions, namely those of the vector boson production cross-sections. To be precise, let us consider the W and Z formation experiments. Both vector bosons can be obtained by a quark–antiquark annihilation and they were discovered this way at the 'quark–antiquark collider' at CERN, which was, of course, the proton–antiproton collider. A second possibility for Z formation is given by the e^+e^- colliders. Two such colliders were built for precision studies following the Z discovery: the storage ring LEP at CERN and the linear collider SLC at SLAC.

In a proton–antiproton collision, it may happen that a quark and an antiquark come very close to each other and annihilate into a Z or a W. The probability of this process, which is weak, is very small compared to the much more frequent strong reactions, even at resonance. Let us call x_q the momentum fraction carried by the quark and $\bar{x}_q$ the momentum fraction carried by the antiquark. The collision is observed in the proton–antiproton centre of mass system, which is not, in general, the quark–antiquark centre of mass system. If $\sqrt{s}$ is the centre of mass energy of the proton–antiproton collision, the quark–antiquark centre of mass energy is

$$\sqrt{\hat{s}} = x_q x_{\bar{q}} \sqrt{s}. \tag{9.55}$$

Let us start by considering the process

$$\bar{u} + d \rightarrow e^- + \bar{\nu}_e \tag{9.56}$$

in the neighbourhood of the resonance, namely for $\sqrt{\hat{s}} \approx M_W$. The dominant diagram is given in Fig. 9.10.

The situation is analogous to that of $e^+ + e^- \rightarrow e^+ + e^-$ near a resonance. We can use, as a first approximation, the Breit–Wigner expression of the cross-section with two spin 1/2 particles in the initial and final states, through an intermediate vector state. The expression is Eq. (4.67), considering that the two quarks must have the same colour. For a given colour, this happens one time out of nine. On the other hand, we have already taken into account that there are three colours when evaluating the partial width. Therefore, the cross-section, summed over the colours, is

$$\sigma\left(\bar{u}d \rightarrow e^-\bar{\nu}_e\right) = \frac{1}{9} \frac{3\pi}{\hat{s}} \frac{\Gamma_{ud}\Gamma_{ev}}{\left(\sqrt{\hat{s}} - M_W\right)^2 + \left(\Gamma_W / 2\right)^2}. \tag{9.57}$$

At the resonance peak $\sqrt{\hat{s}} = M_W$, using the values we have just computed for the widths, we have

$$\sigma_{\max}\left(\bar{u}d \to e^-\bar{v}_e\right) = \frac{4\pi}{3} \frac{1}{M_W^2} \frac{\Gamma_{ud}\Gamma_{ev}}{\Gamma_W^2} \tag{9.58}$$

$$= \frac{4\pi}{3} \frac{1}{81^2} \frac{0.640 \times 0.225}{2.04^2} \times 388 \left[\mu b / \text{GeV}^{-2}\right] \approx 8.8 \text{ nb}.$$

Obviously the charge conjugate process $u + \bar{d} \to e^+ + v_e$ contributes to the W^+ with an equal cross-section.

We now consider the Z production followed by its decay into e^+e^-. Two processes contribute:

$$\bar{u} + u \to e^- + e^+; \quad \bar{d} + d \to e^- + e^+. \tag{9.59}$$

Their cross-sections in resonance are

$$\sigma_{\max}\left(\bar{u}u \to e^-e^+\right) = \frac{4\pi}{3} \frac{1}{M_Z^2} \frac{\Gamma_u \Gamma_e}{\Gamma_Z^2}$$

$$= \frac{4\pi}{3} \frac{1}{91^2} \frac{0.280 \times 0.083}{2.42^2} \times 388 \,\mu b \approx 0.8 \text{ nb} \tag{9.60}$$

and

$$\sigma_{\max}\left(\bar{d}d \to e^-e^+\right) = \frac{4\pi}{3} \frac{1}{M_Z^2} \frac{\Gamma_d \Gamma_e}{\Gamma_Z^2} \approx 1 \text{ nb}. \tag{9.61}$$

Notice that the cross-sections for the Z are almost an order of magnitude smaller than those for the W. This is because of the fact that the Z partial widths are smaller and the mass is larger.

9.6 The UA1 Experiment

In 1976, C. Rubbia, D. Cline and P. McIntyre (Rubbia *et al.* 1976) proposed transforming the CERN Super Proton Synchrotron (SPS) into a storage ring in which protons and antiprotons would counter-rotate and collide head on, as we have already discussed in Section 1.10. In this way, with 270 GeV per beam, the energy needed to create the W and the Z could be reached. To this aim a large number of antiprotons had to be produced, concentrated in a dense beam and collided with an intense proton beam. Let us evaluate the necessary luminosity.

We can think of the proton and the antiprotons as two groups of partons, quarks, antiquarks and gluons, travelling in parallel directions, as shown in Fig. 9.11, neglecting, in a first approximation, the transverse momentum of the partons. Let us consider the valence quarks and antiquarks, respectively. They carry the largest fraction of the total momentum, about 1/6 on average, with a rather broad distribution (see Fig. 6.12). It is important to note that the width of the $\sqrt{s}$ distribution is much larger than the widths of the W and the Z resonances. Therefore, the W and Z production cross-sections grow with collision energy because the larger $\sqrt{s}$ is, the greater the probability of finding a quark–antiquark pair with $\sqrt{\hat{s}}$ close to resonance. In conclusion,

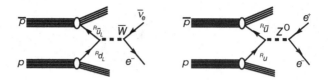

Fig. 9.11. W and Z production in a $p\bar{p}$ collider. Upper left indices label the colour.

the higher the energy, the better. The initial design centre of mass energy at CERN was $\sqrt{s} = 540$ GeV; it later reached 630 GeV.

The calculation of the proton–antiproton cross-sections starts from those at the quark level and takes into account the quark distribution functions and the effects of the colour field. The evaluation made in the design phase gave the values

$$\sigma\left(\bar{p}p \to W \to ev_e\right) \approx 530 \text{ pb}; \quad \sigma(\bar{p}p \to Z \to ee) \approx 35 \text{ pb}. \quad (9.62)$$

To be precise, both the valence and the sea quarks contribute to the process; however, at $\sqrt{s} = 540$ GeV, the average momentum fraction at the W and Z resonances is $\langle x \rangle \approx M_W / \sqrt{s} \approx 0.15$. Therefore, the process is dominated by the valence quarks, while the sea quarks have momentum fractions that are too small. We thus know that the annihilating quark is in the proton, and the antiquark is in the antiproton. This information is lost at higher collision energies.

As we mentioned in Section 1.12, the stochastic cooling technique was developed at CERN to increase the density of particles within bunches at the collision point. Starting from this experience, in 1983 an advanced accelerator physics programme was launched under the guidance of S. van der Meer that made it possible to reach the luminosity $\mathcal{L} = 10^{28}$ cm^{-2} s^{-1} large enough to search for W and Z.

Example 9.3   How many $W \to ev$ events and how many $Z \to e^+e^-$ events are observed in 1 year with the luminosity $\mathcal{L} = 10^{28}$ cm^{-2} s^{-1} and 50% detection efficiency? We apply the mnemonic rule that 1 year $= \pi \times 10^7$ s. Taking into account the time needed to fill the machine, for maintenance, etc., we take 10^7s. With the above-mentioned cross-sections we have $\approx 25\, W \to ev$ and $\approx 2\, Z \to ee$. Actually the W was discovered several months before the Z. $\square$

The production of weak vector bosons is a rare event. Indeed the cross-sections in (9.62) are eight and nine orders of magnitude smaller than the total proton–antiproton cross-section, which is 60 mb at the energies we are considering. Weak interactions are weak indeed! Consequently, the detector must be able to detect the interesting events with a discriminating power of at least 10^{10}. This is the reason why we considered only the leptonic channels above, which can be discriminated. The hadronic channels $W \to \bar{q}q'$, $Z \to \bar{q}q$ are more frequent but are submerged in a huge background due to strong interaction processes, such as

$$gg \to gg; \quad gq \to gq; \quad g\bar{q} \to g\bar{q}; \quad q\bar{q} \to q\bar{q}. \quad (9.63)$$

The leptonic channels are

$$\bar{p}p \to W \to e\nu_e; \quad \bar{p}p \to W \to \mu\nu_\mu; \quad \bar{p}p \to W \to \tau\nu_\tau \tag{9.64a}$$

and

$$\bar{p}p \to Z \to ee; \quad \bar{p}p \to Z \to \mu\mu; \quad \bar{p}p \to Z \to \tau\tau. \tag{9.64b}$$

Leptons can be present in the strong interaction processes too, being produced indirectly by hadron decays, but they can be discriminated. The crucial variable is the 'transverse momentum', p_T, namely the momentum component perpendicular to the colliding beams. In most cases, the proton–antiproton collision is soft, namely it gives origin to low transverse momentum hadrons. Consider one of them, for example a charm, which decays into a charged lepton. This lepton might simulate one of the (9.64) processes. However, in the rest frame of the decaying particle, the lepton momentum is a fraction of the charm mass, less than 1 GeV. The Lorentz transformation to the laboratory frame does not alter the lepton component normal to the charm velocity, which is about that of the beams. The transverse momenta of the kaons are even smaller, while those of the beauties are somewhat larger.

However, there are cases in which two partons come very close to each other and collide violently, namely with a large momentum transfer, by one of the processes (9.63). The hit parton appears as a jet at high transverse momentum. A possible semileptonic decay of a hadron produces a high p_T lepton. However, these leptons are inside a jet, while those from the W and Z decays are not. In conclusion, we search for leptons that have a high p_T and are 'isolated', namely without other particles in a 'cone' around its direction, which is defined as follows. The direction of each track is defined using the azimuth angle ϕ and, in place of the polar angle θ, the 'pseudorapidity', which we have already defined in Eq. (6.71), but which we recall here

$$\eta = -\ln\left[\tan\left(\frac{\theta}{2}\right)\right]. \tag{9.65}$$

A distance in the ϕ, η plane is then defined as

$$\Delta R \equiv \sqrt{\Delta\phi^2 + \Delta\eta^2}. \tag{9.66}$$

Finally, a track is defined as isolated if no other track is present within a properly defined value of ΔR. Typical values are between 0.3 and 0.7.

The same criteria also apply to the neutrino, in the case of the W. Even if neutrinos cannot be detected, we can infer their presence indirectly. To this aim we must build a hermetic detector, which completely surrounds the interaction point with homogeneous calorimeters, in order to intercept all the hadrons and charged leptons and to measure their energies. Moreover, the calorimeters are divided into cells. Each cell is seen from the collision point in a known direction. A 'vector energy' element with absolute value equal to the measured energy and the direction of the cell is then defined. The sum of all these vectors is called, in experimental jargon, 'vector energy'. Considering that the particles are ultra-relativistic, the vector energy is, in practice, a momentum, but it is measured with the calorimeters that measure energies. However, the energy of the high-energy muons cannot be measured with calorimetric means because these particles cannot be absorbed in a reasonable length. We solve the problem by determining their momenta by measuring their trajectories in a magnetic field. The muon momenta are then added to the vector energy and we check if this vector sum is compatible with zero or not. In the presence of one (or more) neutrinos we find an imbalance and we say that the 'missing momentum' is the momentum of the neutrino(s). The same vector quantity is also called 'missing energy'.

In practice the detector cannot be closed at small angles with the beams, where the physical elements needed to drive the beam itself are located. As a consequence, the procedure just described can be done only for the transverse components, obtaining the transverse missing energy flow, E_T^{miss}, and the transverse missing momentum, p_T^{miss}, to which the undetected particles at small angles make a negligible contribution.

Summarizing, the principal channels for the W and Z search and the corresponding topologies are

$$W \rightarrow e^{\pm}\nu_e \text{ isolated electron at high } p_T \text{ and high } p_T^{\text{miss}} \tag{9.67a}$$

$$W \rightarrow \mu^{\pm}\nu_{\mu} \text{ isolated muon at high } p_T \text{ and high } p_T^{\text{miss}} \tag{9.67b}$$

$$Z \rightarrow e^+e^- \text{ two isolated electrons; opposite sign; at high } p_T \tag{9.67c}$$

$$Z \rightarrow \mu^+\mu^- \text{ two isolated muons; opposite sign; at high } p_T. \tag{9.67d}$$

This discussion determines the main specifications of the experimental apparatus.

Two experiments were built at the CERN proton–antiproton collider, called UA1 and UA2. The W and the Z were observed by UA1 (Arnison *et al.* 1983a, b) first and immediately afterwards by UA2 (Banner *et al.* 1983; Bagnaia *et al.* 1983). The results of the two experiments are in perfect agreement and of the same quality. We shall describe here those of UA1.

Fig. 9.12 shows an artist's view of UA1 when open, and Fig. 9.13 shows the UA1 logic structure. The two beams travelling in the vacuum pipe enter the detector from the left and the right, respectively, colliding at the centre of the detector. A particle produced in the collision meets, in series, the following elements.

(1) The central detector, which is a large cylindrical time projection chamber providing electronic images of the charged tracks, is immersed in a horizontal magnetic field in the plane of the drawing, perpendicular to the beams.

(2) The electromagnetic calorimeters, made up of a sandwich of lead plates alternated with plastic scintillator plates; in the calorimeter, electrons and photons lose all their energy, which is measured.

(3) The other particles penetrate the hadron calorimeter, which is a sandwich of iron and plastic scintillator plates. The iron plates on the left and right sides of the beams also act as the yoke of the magnet driving the magnetic return flux. In the calorimeter, the hadrons lose all (or almost all) of their energy, which is measured.

(4) In practice the highest-energy hadronic showers, especially in the forward directions, are not completely contained in the calorimeters, as ideally they should be. They are absorbed in iron absorbers.

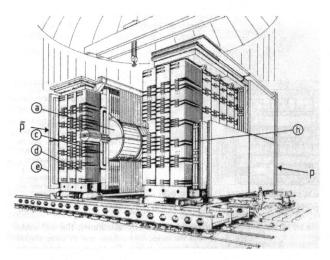

Fig. 9.12. Artist's view of the UA1 experiment, shown in its open configuration. The labels indicate the components: (a) tracking central detector, (c) magnetic field coil, (d) hadronic calorimeters, (e) drift chambers for μ detection, (h) Fe absorber. (C. Rubbia, Nobel lecture 1984. Fig. 8a: C. Rubbia, © The Nobel Foundation 1984 Figure 8a in this link www.nobelprize.org/uploads/2018/06/rubbia-lecture.pdf).

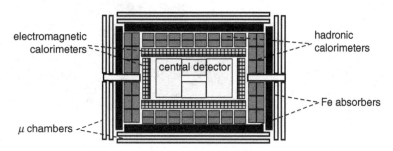

Fig. 9.13. Simplified horizontal cross-section of UA1.

(5) The particles that survive after the iron absorbers are neutrinos and muons. Large tracking drift and streamer chambers detect the muons.

The detector is hermetic, but at small angles with the beams; the response of the calorimeters is made as homogeneous as possible.

9.7 The Discovery of *W* and *Z*

Fig. 9.14 shows the reconstruction of one of the first $W \to e\nu$ events observed by UA1. We observe many tracks that make the picture somewhat confusing. These are particles pertaining to the 'rest of the event', that is, coming from the interaction of partons

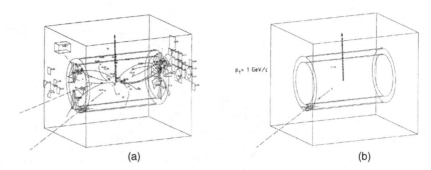

A $W \rightarrow e\nu$ event. (a) The tracks, the hit calorimeter cells and the missing transverse momentum are shown. (b) Only tracks with $p_T > 1$ GeV. (C. Rubbia, Nobel lecture 1984. Fig. 16(a, b): C. Rubbia, © The Nobel Foundation 1984 Figure 16(a, b) in this link www.nobelprize.org/uploads/2018/06/rubbia-lecture.pdf).

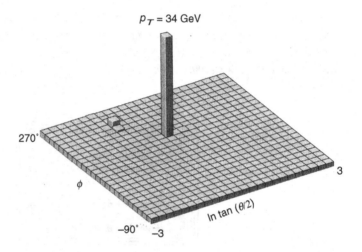

Lego plot of a $W \rightarrow e\nu$ event in an electromagnetic calorimeter: ϕ azimuth, θ anomaly to the beam direction. © John Wiley & Sons Inc. 1994. Reprinted with permission.

different from those that produced the W. They are soft and can be easily eliminated simply by neglecting all tracks with p_T smaller than a few times Λ_{QCD}, in practice with $p_T < 1$ GeV, as shown in Fig. 9.14(b).

With this simple 'cut', we are left with a clean picture of a single charged track with the characteristics of an electron. Its momentum, measured from its curvature, and its energy, measured in the calorimeter, are equal within the errors. We also find that the transverse momentum is not balanced. The transverse missing momentum is shown in Fig. 9.14. The calorimeters give a complementary view of the events, namely they show the energy flow from the collision point as a function of the angles.

Fig. 9.15 shows such a view for a $W \rightarrow e\nu$ event. The cells of the diagram, called a 'lego plot', correspond to the physical electromagnetic calorimeter cells. The two coordinates are the azimuth ϕ and a function of the anomaly θ with the beam direction

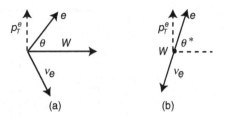

The momenta (a) in the laboratory and (b) in the centre of mass of the *W*.

as polar axis. Since the density of tracks in the forward directions, namely for $\theta = 0$ and $\theta = \pi$, is very high, the function $\ln \tan \theta / 2$ is used to obtain a smooth distribution. In this event there is practically a single, large, localized energy deposit. This is how the calorimeter sees the electron.

Summarizing, we see that simple kinematic selection criteria allow unambiguous identification of the very rare cases in which a *W* is produced. It subsequently decays into $e\nu$.

The situation is similar for the decays into $\mu\nu$ and into $\tau\nu$, which we shall not discuss. We mention, however, that the comparison of the three cross-sections gives a test of lepton universality, namely

$$g_\mu / g_e = 1.00 \pm 0.07 \,(\text{stat}) \pm 0.04 \,(\text{syst})$$
$$g_\tau / g_e = 1.01 \pm 0.10 \,(\text{stat}) \pm 0.06 \,(\text{syst}). \tag{9.68}$$

Let us now consider the measurement of the *W* mass. As the calorimetric measurement of the electron energy is more precise than the muon momentum measurement, we choose the $e\nu$ channel. We cannot reconstruct the electron–neutrino mass because only the transverse component of the neutrino momentum is known. However, we can measure M_W with the 'Jacobian peak' method.

Fig. 9.16(a) gives a scheme of the *W* decay kinematic in the laboratory frame. The *W* momentum component transverse to the beam is very small in general. Neglecting it in the first approximation, the flight direction of the *W* is the direction of the beams. Consider the electron momentum, which is measured. Its component normal to the *W* motion, p_T, is equal in the laboratory and in the centre of mass frame (Fig. 9.16(b)):

$$p_T = \frac{M_W}{2} \sin\theta^*. \tag{9.69}$$

Let $\dfrac{dn}{d\theta^*}$ be the decay angular distribution in the rest frame of the *W*. The transverse momentum distribution is then given by

$$\frac{dn}{dp_T} = \frac{dn}{d\theta^*} \frac{d\theta^*}{dp_T}. \tag{9.70}$$

The quantity $\dfrac{d\theta^*}{dp_T}$ is called the Jacobian of the variable transformation. Its expression is such that

$$\frac{dn}{dp_T} = \frac{1}{\sqrt{(M_W/2)^2 - p_T^2}} \frac{dn}{d\theta^*}. \tag{9.71}$$

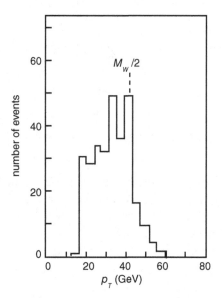

Fig. 9.17. Electron p_T distribution for $W \to e\nu$ events (adapted from Albajar *et al.* *Z. Phys.* C44 (1989) 15, Fig. 16a. © 1989 with permission by Elsevier. https://doi.org/10.1016/0370-2693(91)91760-S).

The essential point is that the first factor on the right-hand side diverges for

$$p_T = M_W/2. \tag{9.72}$$

Consequently, the p_T distribution has a sharp maximum at $M_W/2$. Notice that the conclusion does not depend on the longitudinal momentum of the W, which may be large. The position of the maximum does, on the contrary, depend on the transverse momentum of the W, which, as we have said, is small but not completely negligible. Its effect is a certain broadening of the peak.

The electron transverse momentum distribution for the W events is shown in Fig. 9.17, where the Jacobian peak is clearly seen. From this distribution, UA1 measured $M_W = 83 \pm 3$ GeV, where the uncertainty is substantially determined by the systematic uncertainty on the energy calibration. UA2 measured $M_W = 80 \pm 1.5$ GeV.

A further test of the electroweak theory is the measurement of the electron helicity in the decay $W \to e\nu$. Consider the process in the W rest frame as in Fig. 9.18.

As we know, the $V-A$ structure of the CC weak interaction implies that if the energy of the electrons (which we are considering now) is much larger than their masses, as in the present case, then the electrons have dominant helicity -1 and the positrons have dominant helicity $+1$. We take z, the direction of the beams, as the quantization axis for the angular momenta in the initial state, as in Fig. 9.18. The total angular momentum is $J = 1$. As already seen, since the W production is due to valence quarks, we know that the initial quark has the direction of the proton, the antiquark that of the antiproton. Therefore, the third component of the angular momentum is $J_z = -1$.

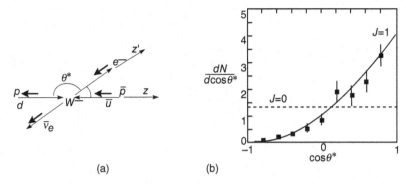

Fig. 9.18.   (a) Kinematics of the W production and decay. (b) Angular distribution measured by UA1 (adapted from Albajar *et al.*
Z. Phys. C44 (1989) 15, Fig. 48. © 1989 with permission by Elsevier. https://doi.org/10.1016/0370-2693(91)91760-S).

We take the electron direction z' as the quantization axis in the final state. By the same token the third component is $J_z = -1$. Therefore, the angular dependence of the differential cross-section is given by

$$\frac{d\sigma}{d\Omega} \propto \left[d^1_{1,-1}\right]^2 = \left[\frac{1}{2}(1+\cos\theta\,^*)\right]^2 . \tag{9.73}$$

The distribution measured by UA1 is shown in Fig. 9.18(b); the curve is Eq. (9.73), which is in perfect agreement with the data. The dashed horizontal line is the prediction for W spin $J = 0$. In this way we measure the W spin.

Notice that the observed asymmetry shows that parity is violated but does not prove that the CC structure is $V - A$. The $V + A$ structure predicts the same angular distribution. Only polarization measurements can distinguish the two cases.

Question 9.1 Prove the last statement. □

We now consider the discovery of the Z. Fig. 9.19 shows the UA1 tracking view of a typical $Z \to e^- e^+$ event. Again, the confused view becomes clear with the selection $p_T > 1$ GeV. Only two tracks remain. One of them is positive, the other negative; for both, the energy as measured in the electromagnetic calorimeter is equal to the momentum measured from curvature.

Fig. 9.20 shows the calorimetric view of a $Z \to e^- e^+$ event: two localized, isolated energy deposits appear in the electromagnetic calorimeter.

The mass of Z is obtained by measuring the energies of both electrons in the electromagnetic calorimeters and the angle between their tracks in the central detector. Fig. 9.21 is the M_Z distribution of the first 24 UA1 events. The average is $M_Z = 93$ GeV with a systematic uncertainty of ±3 GeV; the UA2 measurement gave $M_Z = 91.5$ GeV with a systematic uncertainty of ±1.7 GeV.

In conclusion, by 1983 the UA1 and UA2 experiments confirmed that the vector bosons predicted by the electroweak theory exist and have exactly the predicted characteristics.

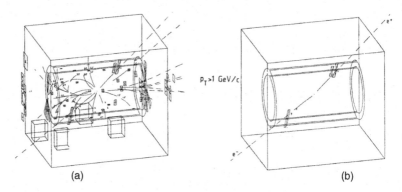

Fig. 9.19. (a) A $Z \to e^- e^+$ event. (b) Only tracks with $p_T > 1$ GeV (C. Rubbia, Nobel lecture 1984. Fig. 25 and 26: C. Rubbia, © The Nobel Foundation 1984 Figure 25 and 26 in this link www.nobelprize.org/uploads/2018/06/rubbia-lecture .pdf).

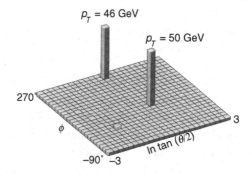

Fig. 9.20. Lego plot of a $Z \to e^- e^+$ event in the electromagnetic calorimeter. © John Wiley & Sons Inc. 1994. Reprinted with permission.

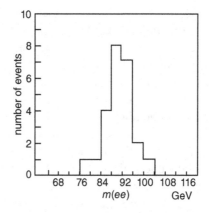

Fig. 9.21. Distribution of $m\left(e^+ e^-\right)$ for the first 24 UA1 events (adapted from C. Albajar *et al. Z. Phys.* C36 (1987) Fig. 3a © 1987 with permission by Elsevier. https://doi.org/10.1016/0370-2693(87)91509-7).

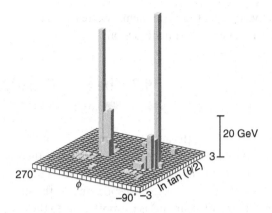

Fig. 9.22.   Two quark jets in the UA1 hadronic calorimeter (adapted from Albajar *et al.* 1987).

Particularly important is the ratio of the two masses, experimentally because it is not affected by the energy scale calibration, and theoretically because it provides directly the weak angle. Indeed, Eq. (9.27), valid at the tree level, gives

$$\cos^2 \theta_W = 1 - \left(M_W / M_Z \right)^2 . \tag{9.74}$$

The ratio of the masses measured by UA1 and UA2 gives

$$\begin{aligned} \text{UA1}: \quad & \sin^2 \theta_W = 0.211 \pm 0.025 \\ \text{UA2}: \quad & \sin^2 \theta_W = 0.232 \pm 0.027. \end{aligned} \tag{9.75}$$

These values are in agreement with the low energy measurements we mentioned in Section 9.4. We shall come back to this point in Section 9.8 for a more accurate discussion.

Question 9.2 Can the Z decay into two equal pseudoscalar mesons? And into two scalar mesons? □

Question 9.3 In their first data-taking period, UA1 and UA2 collected $\approx 300\,W$ and $\approx 30\,Z$ each. What is the principal source of uncertainty on the W mass? On the Z mass? On their ratio? □

Before closing this section, let us see how the quarks appear in a hadron collider. As we know, to 'observe' a quark we should not try to break a nucleon in order to extract one of them; rather, we must observe the hadronic energy flow in a high-energy collision at high momentum transfer. One of the first observations of UA2 (Banner *et al.* 1982) and UA1 was that of events with two hadronic jets in back-to-back directions. They are violent collisions between two quarks, which in the final state hadronize into jets. More rarely a third jet was observed, due to the radiation of a gluon.

The lego plot of a two-jet event as seen in the UA1 calorimeter is shown in Fig. 9.22. Comparing it with Fig. 9.20, we see that the two quarks, as seen in the calorimeter, are very similar to electrons, with some differences: the peaks are wider and more activity is present outside them, two features that are well understood thinking of the anti-screening QCD phenomenon.

9.8 The Evolution of $\sin^2\theta_W$

As already stated, the accurate comparison of the weak mixing angle values extracted from different physical processes requires a theoretical calculation beyond the tree level, one that includes higher-order terms. The most important radiative corrections are those to the W mass, corresponding to the diagrams in Fig. 9.23. The correction to M_W due to diagram (a) is proportional to the difference between the squares of the masses of the two quarks in the loop, namely to

$$G_F\left(m_t^2 - m_b^2\right) \approx G_F m_t^2. \tag{9.76}$$

Therefore, an accurate measurement of M_W allows us to predict the top mass.

The correction due to diagrams (b) and (c) depends on the H boson mass M_H. The dependence is, however, only logarithmic and consequently the prediction of the H boson mass is less precise.

We shall discuss the precision measurement of the W and top masses in Section 9.11.

In Section 5.8 we studied the evolution of the QED 'constant' α, which is proportional to the square of the electric charge, and in Section 6.5 we studied the evolution of the QCD 'constant' α_s, which is proportional to the colour charges squared. The corresponding gauge groups are $U(1)$ and $SU(3)$. The gauge group of the electroweak theory is $SU(2) \otimes U(1)$. The 'electroweak charges' are g for $SU(2)$ and g' for $U(1)$. The tangent of the weak angle is the ratio of the $U(1)$ and $SU(2)$ charges g'/g. These two charges need to be renormalized in the theory in a manner similar to the other ones and, consequently, are functions of the momentum transfer. Like the other charges, the electroweak charges cannot be measured directly. The quantity that can be extracted from the observable in the most direct way is their ratio, that is, the weak mixing angle.

The evolution of $\sin^2\theta_W$ as a function of the momentum transfer $|Q|$ is more complicated than that of α or α_s, because both numerator and denominator vary. It is shown in Fig. 9.24.

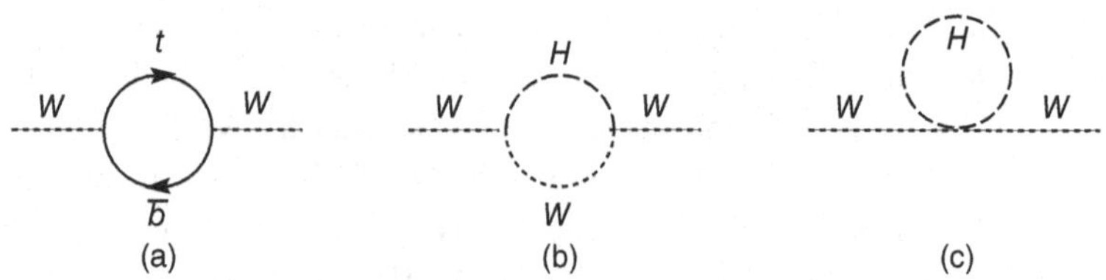

Fig. 9.23. Principal radiative correction diagrams to the W mass.

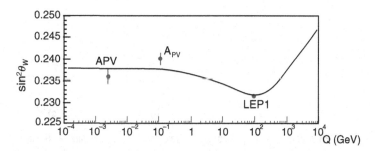

Fig. 9.24. $\sin^2\theta_W$ as a function of $|Q|$. APV: atomic parity violation; A_{PV}: asymmetry in polarized Møller scattering from Porsev *et al.* (2009); LEP1: Z-pole measurements; Prediction of the SM from F. Jegerlehner (2017).

First of all notice that the variation of $\sin^2\theta_W$ is very small, only a few per cent, even in the huge range of Q^2 we are considering. Consequently, only the most precise determinations of those we have mentioned in Section 9.4 are reported in Fig. 9.24.

The curve is the Standard Model prediction (Jegerlehner 2017). Its change at M_W is due to the following reason. Since the gauge bosons carry weak charges, both fermion and boson loops contribute to the renormalization of the weak charge, with opposite effects on its slope. This situation is similar to that we have met in QCD. However, unlike in QCD, the weak gauge bosons are massive. Therefore, only fermionic loops are important at energies below M_W. Here the W loops set in. However, they contribute to the evolution of the $SU(2)$ constant g, not to the evolution of the $U(1)$ constant g'. This inverts the slope of the weak mixing angle evolution.

9.9 Precision Tests at LEP

As we have just discussed, at the beginning of the 1990s, all the crucial predictions of the electroweak theory had been experimentally verified, with the very important exception of the H boson, which we shall deal with at the end of the chapter.

The following steps were the high-precision tests. For these, the ideal instrument is the e^+e^- collider. For this purpose the LEP machine was designed and built at CERN with a 27 km circumference and with energy and luminosity adequate for studying not only all the features of the Z resonance but also the crucial processes $e^+e^- \rightarrow W^+W^-$.

In the same period, B. Richter and collaborators designed a novel type of collider at Stanford, called the Stanford Linear Collider (SLC). A big problem of the circular e^+e^- colliders is the large amount of power radiated by the orbiting particles due to their centripetal acceleration. The electrical power that must be spent just to maintain them circulating in the rings at a constant energy grows with the fourth power of the energy, at fixed orbit radius. Clearly, the construction costs of a machine increase with the length of its tunnel and, above certain energy, it may become convenient to accelerate electrons and positrons in a linear structure and collide them head on only once.

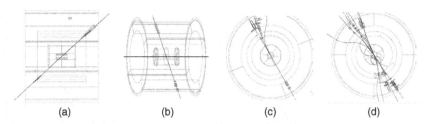

DELPHI events: (a) e^+e^- pair, (b) $\mu^+\mu^-$ pair, (c) $\tau^+\tau$ pair, (d) quark pair (images © CERN).

In such a way one spends more electrical energy on accelerating the particles, which are 'used' only once, but does not spend energy keeping them in orbit. In practice the trade-off is reached around 200 – 300 GeV.

An added advantage was that a linear accelerator already existed at SLAC. However, several technological developments were necessary in order to produce extremely dense and thin bunches, a few micrometres across at the collision point. The SLC with the Mark II and later with the SLD experiments, started producing physical results at the same time as LEP in 1989. The SLC luminosity was much smaller than that of LEP but its beams could be polarized allowing different tests of the theory, which, however, we shall not discuss.

The e^+e^- colliders are precision instruments, providing collisions that are always, and not only rarely as in a hadron collider, between elementary, point-like objects. This has the following consequences: all the events are interesting, not one in a billion or so. Moreover, the events are very clean, and no 'rest of the event' is present, as in a hadron collider. Finally, the e^+e^- annihilation leads to a pure quantum state, of definite quantum numbers, $J^{PC} = 1^{--}$.

Four experiments, called ALEPH, DELPHI, L3 and OPAL, worked at LEP from 1989 on the Z peak (collecting 4 million events each) and from 1996 to 2000 at increasing energies up to 209 GeV. Even if rather different in important details, the basic features of all the set-ups are similar. Each of them has a central tracking chamber in a magnetic field oriented in the direction of the beams, electromagnetic and hadronic calorimeters and large muon chambers. Micro-strip silicon detectors are located between the central detector and the beam pipe to provide a close-up image of the vertex region, with 10 μm resolution, necessary to look for secondary vertices of charm and beauty decays.

Fig. 9.25 shows four events of different types at DELPHI.

A fundamental measurement is the shape of the resonance line of the Z. In practice, it is taken by measuring the hadronic cross-section, since it is the largest one, as a function of the machine energy. An outstanding experimental effort by the scientists working on the experiments and on the machine led to the astonishing precision of $\Delta\sqrt{s} = \pm 2$ MeV, which is 40 ppm.

Consider the generic f^+f^- final state, different from e^+e^-, to which t channel exchange also contributes. Two s channel diagrams are present in general at tree level, with γ and Z exchange, as shown in Fig. 9.26. Near the resonance, where the Z exchange dominates, we can express the cross-section in the Breit–Wigner approximation as

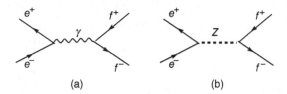

Fig. 9.26. Tree-level Feynman diagrams for the process $e^+e^- \to f^+f^-$ with exchange in the s channel: (a) a photon; (b) a Z boson.

$$\sigma\left(e^+e^- \to f^+f^-\right) = \frac{3\pi}{s} \frac{\Gamma_e\Gamma_f}{\left(\sqrt{s}-M_Z\right)^2 + \left(\Gamma_Z/2\right)^2}. \tag{9.77}$$

At the peak, namely for $\sqrt{s} = M_Z$ we have

$$\sigma\left(e^+e^- \to f^+f^-\right) = \frac{12\pi}{M_Z^2} \frac{\Gamma_e\Gamma_f}{\Gamma_Z^2}. \tag{9.78}$$

Example 9.4   Calculate the cross-section for $e^+e^- \to \mu^+\mu^-$ at the peak. How many Z are produced in this channel at the typical LEP luminosity, $\mathcal{L} = 10^{31}\,\mathrm{cm}^{-2}\mathrm{s}^{-1}$?

$$\sigma\left(e^+e^- \to \mu^+\mu^-\right) = \frac{12\pi}{M_Z^2} \frac{\Gamma_e\Gamma_\mu}{\Gamma^2} = \frac{12\pi}{91^2} \frac{84^2}{2450^2}$$
$$= 5.3\times10^{-6}\,\mathrm{GeV}^{-2} \times 339\,\mu\mathrm{b}\,\mathrm{GeV}^{-2} = 2.1\,\mathrm{nb}.$$

About one Z per minute. □

Example 9.5   Repeat the calculation for the hadronic cross-section.
$$\sigma\left(e^+ + e^- \to \mathrm{hadrons}\right) = \frac{12\pi}{m_Z^2} \frac{\Gamma_e\Gamma_h}{\Gamma^2}$$
$$= \frac{12\pi}{91^2} \frac{84\times1690}{2450^2} = 40.2\,\mathrm{nb}.$$

About 1000 Z per hour. □

However, as we discussed in Section 4.9, the Breit–Wigner approximation is a rather bad one in this case, as we can see in Fig. 9.27, which shows Eq. (9.77) as a dashed line together with the experimental data. The disagreement is due mainly to the bremsstrahlung of a photon from one of the initial or final particles, as in Fig. 9.28.

Other, smaller but theoretically more interesting, corrections are due to the diagrams in Fig. 9.29.

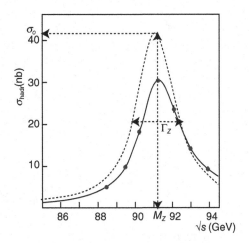

The Z resonance at LEP. Error bars have been enlarged by 10 (adapted from LEP & SLD 2006).

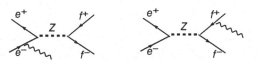

Initial and final bremsstrahlung diagrams.

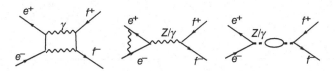

Higher-order photonic and non-photonic corrections.

The analysis starts from the continuous curve in Fig. 9.27 obtained by interpolating the experimental points. The curve is then corrected by considering the calculated contribution of the electromagnetic corrections of Fig. 9.28. The result is the dashed curve.

From this we extract the mass M_Z, the total width Γ_Z and the peak value σ^0. The most precise values of these observables and of those that we shall now discuss are extracted from an over-constrained fit to all the available data, including the Standard Model radiative corrections. At the time of LEP decommissioning, the best values were the following. The Z-mass (LEP & SLD 2006), with its 20 ppm accuracy, is

$$M_Z = 91.1875 \pm 0.0021 \text{ GeV}. \tag{9.79}$$

The total width is

$$\Gamma_Z = 2.4952 \pm 0.0023 \text{ GeV}. \tag{9.80}$$

The peak hadronic cross-section is

$$\sigma^0 = 41.540 \pm 0.037 \text{ nb.} \tag{9.81}$$

We already see that these values are in agreement with our approximate evaluations in Section 9.5.

We now consider the partial widths, obtained by measuring the partial cross-sections in the corresponding channels.

The values of the three leptonic widths measured by LEP are equal within 2‰ accuracy, a fact that verifies the universality of the coupling of the leptons to the neutral current, while the tests of Section 7.8 were for the charged current. Their average is

$$\Gamma_l = 83.984 \pm 0.086 \text{ MeV.} \tag{9.82}$$

The total hadronic width is

$$\Gamma_h = 1744.4 \pm 2.0 \text{ MeV.} \tag{9.83}$$

It is in general impossible to establish the nature of the quark–antiquark pair in the final state. Exceptions are the charm and beauty cases, in which the presence of a short lifetime particle can be established by observing secondary vertices inside the hadronic jets with the vertex detector. The kinematic reconstruction of the event can then distinguish between the two on statistical grounds. The two partial widths, given as fractions of the hadronic width, are again in agreement with the Standard Model predictions

$$R_c \equiv \Gamma_c / \Gamma_h = 0.1721 \pm 0.0030 \tag{9.84}$$

and

$$R_b \equiv \Gamma_b / \Gamma_h = 0.21629 \pm 0.00036. \tag{9.85}$$

Example 9.6   Calculate the distances travelled by a D^0 and by a B^0 with 50 GeV energy. ☐

A fundamental contribution given by the LEP experiments is the precision measurement of the number of 'light neutrino' types, N_ν. The more neutrino species there are, the larger the Z partial width for decay into neutrinos is. As neutrinos are not detected, this is called 'invisible' width, Γ_{inv}. It is determined by subtracting from the total width the measured partial widths, namely those in the hadronic and charged leptonic channels, $\Gamma_{\text{inv}} = \Gamma_Z - 3\Gamma_l - \Gamma_h$. We assume the invisible width to be due to N_ν neutrino types each contributing with the partial width as given by the Standard Model $(\Gamma_\nu)_{\text{SM}}$. We notice here that the height of the peak of line shapes of the resonance curves are more sensitive to the total width than the width at half maximum of the curve themselves. This is since, as the above equations show (e.g. Eq. (9.78)), the height is inversely proportional to the *square* of the total width, so, for example, a 1% change in the width results in a 2% change in the height. This is clearly visible in Fig. 9.30. In practice the line shape of the resonance curve into charged leptons' final states is used, with the result

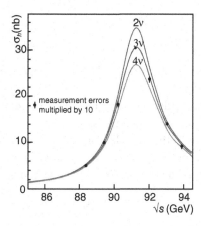

Fig. 9.30. The Z line shape and expectations for different numbers of neutrinos. By permission of Particle Data Group, Lawrence Berkeley National Lab.

$$\left(\Gamma_v / \Gamma_l\right)_{\mathrm{SM}} = 1.991 \pm 0.001, \tag{9.86}$$

obtaining

$$N_v = \frac{\Gamma_{\mathrm{inv}}}{\Gamma_l}\left(\frac{\Gamma_l}{\Gamma_v}\right)_{\mathrm{SM}}. \tag{9.87}$$

The combined result from the four LEP experiments and SLD is

$$N_v = 2.984 \pm 0.008. \tag{9.88}$$

Fig. 9.30 shows the resonance curve measured by LEP, compared with the curves calculated for two, three and four neutrino flavours.

In conclusion there are no other neutrinos beyond the three we know, at least if their mass is smaller than $M_Z / 2$. In the hypothesis that the structure of the families is universal, there are no more families beyond those we know.

9.10 Intermediate Bosons' Self-coupling

As we have already recalled, LEP was designed to test the Standard Model not only at the Z pole but also at higher energies, above the threshold of the W (pair) production through the reaction processes $e^+e^- \rightarrow W^+W^-$. This study is important because it tests a fundamental aspect of the electroweak theory, namely the fact that the vector bosons self-interact.

Let us start by considering the weak process

$$\nu_\mu + e^- \rightarrow \mu^- + \nu_e. \tag{9.89}$$

As we saw in Section 7.1, its cross-section would grow indefinitely with increasing energy if the interaction was the Fermi point-like interaction represented in Fig. 9.31(a).

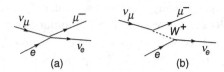

Fig. 9.31. Feynman diagram for the scattering $v_\mu + e^- \to \mu^- + v_e$: (a) in the low-energy point-like Fermi approximation; (b) as mediated by the W boson.

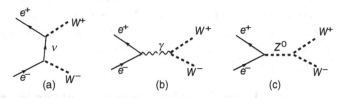

Fig. 9.32. The three tree-level diagrams of $e^+ + e^- \to W^+ + W^-$.

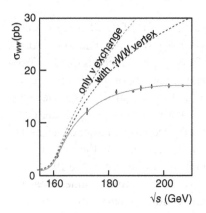

Fig. 9.33. Cross-section for $e^+ + e^- \to W^+ + W^-$. By permission of Particle Data Group, Lawrence Berkeley National Lab.

The W boson mediating the interaction, as in Fig. 9.31(b), solves the problem. However, it induces a new problem. Indeed, if the W exists, then the process

$$e^+ + e^- \to W^+ + W^- \tag{9.90}$$

exists. It is mediated by a neutrino, as shown in Fig. 9.32(a). Now, computing this diagram we find a diverging cross-section, the intermediate curve in Fig. 9.33. We have forgotten that the W is charged. Does the inclusion in our calculation of the photon exchange shown in the diagram in Fig. 9.32(b) solve the problem? The answer is no; the sum of the two diagrams again gives a diverging cross-section, the intermediate curve in Fig. 9.33.

In the electroweak theory there is another neutral vector boson beyond the photon, the Z, and it couples directly to the Ws, because the weak vector bosons carry weak charges. Finally, the cross-section, calculated also including the diagram of Fig. 9.32(c), does not diverge. It is the continuous curve in Fig. 9.33.

Fig. 9.33 shows the cross-section of reaction (9.90) as measured by the LEP experiments up to $\sqrt{s} = 209$ GeV. The perfect agreement with the predictions tests another crucial aspect of the theory, namely the weak charge of the weak interaction mediators.

Another important result obtained by measuring the energy dependence of the W production cross-section is the accurate determination of the W mass and width. The determination of the energy at which the cross-section first becomes different from zero, namely the energy threshold, gives the W mass. The rapidity of the initial growth determines the W width, because if Γ_W is larger the growth is slower. However, the mass and the width of the W were later measured with higher precision in hadronic colliders, as we shall see in the next sections.

9.11 Precision Measurements of the W and Top Masses at the Tevatron

Precise measurements of the W and top masses are important because of the radiative corrections to the W mass. The lowest-order diagrams are shown in Fig. 9.23. The diagram of Fig. 9.23(a) depends quadratically on the top mass; those of Fig. 9.23(b, c) depend logarithmically on the H mass. Consequently, an accurate knowledge of the W mass, joined to that of the other electroweak parameters, allowed a prediction of the top mass before its discovery. Later on, precise knowledge of both the W and the top masses allowed us to predict constraints on the H mass.

As we saw in Section 4.10, the top quark was discovered by the CDF experiment in 1995. Previously, at the end of the 1980s, Amaldi *et al.* (1987) and Costa *et al.* (1988) had analysed the weak angle values measured at different Q^2 scales and the relevant radiative corrections. The top mass was left as a free parameter. The result was that the data were in perfect agreement with the Standard Model, provided the top mass was not too large, namely $m_t < 200$ GeV. To explore all the range up to this very high value of the mass, a collider of higher energy and higher luminosity than the Tevatron in its initial configuration was needed. While the energy of an existing machine cannot be increased much, the luminosity can. A high luminosity was important for the whole physics programme, not only for the top.

The peak luminosity of the Tevatron (the initial luminosity after having injected the beams) increased during the first operation periods, called Run Ia and Run Ib, from typically $\mathcal{L} = 5.4 \times 10^{30}\,\mathrm{cm}^{-2}\mathrm{s}^{-1}$ to $\mathcal{L} = 1.6 \times 10^{31}\,\mathrm{cm}^{-2}\mathrm{s}^{-1}$. A major upgrade programme of the accelerator complex took place between 1996 and 2001. The most important addition was a new ring, the 'main injector', designed to substantially increase the antiproton beam intensity. In the subsequent Run II, the centre of mass energy had somewhat increased from $\sqrt{s} = 1.8$ TeV to $\sqrt{s} = 1.96$ TeV. The typical peak luminosity increased much more, to $\mathcal{L} = 3.4 \times 10^{32}\,\mathrm{cm}^{-2}\mathrm{s}^{-1}$, almost two orders of magnitude larger than in the first run. The Tevatron was finally shut down in 2011, when LHC had

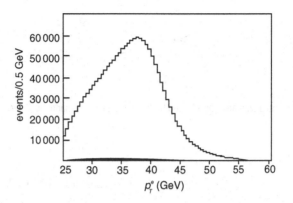

Distribution of the electron transverse momentum p_T^e from the D0 experiment. Background is the black shallow region. Measured points and Monte Carlo simulation are indistinguishable.

become fully operational. The CDF, designed and built in parallel with the collider, took data from 1985 to 2011, corresponding to a total integrated luminosity of about 10 events/fb. A second experiment, called D0, from the name of the intersection at which it was located, started taking data in 1992. Both experiments went through major upgrades in preparation for the Run II.

The precision measurements of the W and top mass imposes stringent requirements to the detectors. Both the experiments, CDF and D0, were upgraded to cope with that. A high spatial resolution tracking is needed to identify the decay vertices of the beauty hadrons, which are very close, of the order of a fraction of a millimetre, to the primary vertex. This is achieved with several layers of silicon micro-strip detectors surrounding the beam pipe. Outside the silicon detectors further tracking elements provide precise measurements of the track points over a length of the order of a metre (somewhat different in the two experiments) in a magnetic field. The electromagnetic and hadronic calorimeters follow at larger distances. Finally, after a thick Fe filter provided by the yoke of the magnet, tracking chambers are used to identify the penetrating muons. As we saw discussing the UA1 experiment, hermeticity and granularity of the calorimeter system are essential for the measurement of the energy imbalance and its direction, called the missing energy vector, or missing momentum, p_T^{mis}.

Measurements of the W boson mass were made at Tevatron by both experiments with different methods, mainly with the Jacobian peak method, which we saw when discussing UA1. The value of M_Z of LEP was used as a calibration of the energy scales of the calorimeters. Fig. 9.34 shows the electron transverse momentum p_T^e distribution from data of a 4.3 events/fb exposure of the D0 experiment (Abazov *et al.* 2012). Compare this with Fig. 9.17 at the time of the discovery. Data points and Monte Carlo simulation are indistinguishable. CDF has similar results (Aaltonen *et al.* 2012). Precision measurements of the W mass were performed by the ATALAS, CMS and LHCb experiments at LHC. We shall not discuss them here, but directly quote the average value from PDG (Workman *et al.* 2022)

$$M_W = 80.377 \pm 0.012 \text{ GeV} \tag{9.91}$$

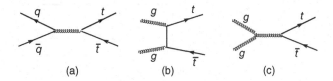

Fig. 9.35. First-order diagrams contributing to the $t\bar{t}$ production at hadron colliders: (a) quark annihilation; (b) and (c) gluon fusion.

and

$$\Gamma_W = 2.085 \pm 0.042 \text{ GeV}. \tag{9.92}$$

Coming now to the top, we mention that the precision measurements of the vector boson masses at LEP gave a prediction of the top mass, which by 1993, assuming the H boson mass in the range $60 < M_H < 700$ GeV, was

$$m_t = 166 \pm 27 \text{ GeV}. \tag{9.93}$$

As we saw in Chapter 4, two years later, CDF at FNAL discovered the top in the predicted mass range. The top production cross-section was measured with increasing precision, with a final value of

$$\sigma_{t\bar{t}} \left(\sqrt{s} = 1.96 \text{ TeV} \right) = 7.50 \pm 0.48 \text{ pb}, \tag{9.94}$$

in good agreement with the QCD predictions. At the Tevatron energies, α_s is already small enough to allow reliable and perturbative calculations. Notice that top production is a very rare phenomenon, the total $p\bar{p}$ cross-section being $\sigma_{p\bar{p},tot} \left(\sqrt{s} = 1.96 \text{ TeV} \right) = 81.9 \pm 2.3$ mb. Top quarks are produced once every 10^{10} interactions.

A single top quark, rather than a pair, can be produced via weak interactions. The corresponding cross-section is about a factor of two smaller than (9.94). The process gives information, in particular, on the mixing matrix element $|V_{tb}|$, as we mentioned in Section 7.10, and which will not be further discussed in this book.

Fig. 9.35 illustrates the most important diagrams that contribute to $t\bar{t}$ in hadron colliders. In a $\bar{p}p$ collider, the process is dominated (85% at Tevatron) by the quark annihilation diagram (Fig. 9.35(a)). On the contrary, in a pp collider, such as LHC, the gluon fusion diagrams (Fig. 9.35(b, c)) dominate.

Top has practically a unique decay channel

$$t \rightarrow W + b, \tag{9.95}$$

so that the intermediate state of a $t\bar{t}$ event is $W^+ b W^- \bar{b}$. Each of the two Ws will then decay in a lepton–neutrino pair, or in a quark–antiquark pair. The quarks produced in the W decay, in the latter case, and the b-quarks, in any case, hadronize and are observed as jets in the detectors. We then define the following topologies.

(1) If both Ws decay in a charged lepton neutrino pair, the topology is called 'dilepton'; it consists of two charged leptons of opposite charge, which can be the same

or different ones, two neutrinos and two b jets, namely jets containing a beauty hadron (this happens in 10.6% of the cases):

$$t \to W^+ + b \to W^+ + \text{jet}(b); \quad \bar{t} \to W^- + \bar{b} \to W^- + \text{jet}(\bar{b})$$
$$W^- \to e^- \bar{\nu}_e \text{ or } \to \mu^- \bar{\nu}_\mu \quad \text{and} \quad W^+ \to e^+ \nu_e \text{ or } \to \mu^+ \nu_\mu. \tag{9.96}$$

(2) If one W decays into a lepton pair and the other into a quark–antiquark pair, the topology is called 'lepton + jets'; it consists of a charged lepton, a neutrino and four jets, two of which are b jets (43.9%):

$$t \to W^+ + b \to W^+ + \text{jet}(b); \quad \bar{t} \to W^- + \bar{b} \to W^- + \text{jet}(\bar{b})$$
$$W \to e\nu_e \text{ or } \to \mu\nu_\mu \quad \text{and} \quad W \to q\bar{q}' \to \text{jet} + \text{jet}. \tag{9.97}$$

(3) If both Ws decay into quark–antiquark pairs, the topology is called 'all-hadronic' and consists of six jets, of which two are b jets (45.5%):

$$t \to W^+ + b \to W^+ + \text{jet}(b); \quad \bar{t} \to W^- + \bar{b} \to W^- + \text{jet}(\bar{b})$$
$$W^+ \to q\bar{q}' \to \text{jet} + \text{jet} \quad \text{and} \quad W^- \to q\bar{q}' \to \text{jet} + \text{jet}. \tag{9.98}$$

In practice, the experiments are well suited to detect and identify electrons and muons. They are not so well suited for the τ leptons, because they quickly decay into final states containing neutrinos, which cannot be detected.

Both CDF and D0 have used all the above-mentioned topologies for their measurements of the top mass. However, we shall limit our discussion to the one that is best suited, namely lepton + jets, where the lepton is an electron or a muon. In comparison, the dilepton has less background, but is not kinematically constrained due to the presence of two neutrinos, as we shall immediately see, while for all-hadrons all the momenta are measured but the QCD background is substantially larger.

The following selection criteria are used to suppress backgrounds in the lepton + jets channel. An electron or muon should be present, with a large transverse momentum ($p_T > 20$ GeV) and should be isolated. A b jet should be identified by the silicon detectors. Finally, a large enough missing momentum must be present $\left(p_T^{\text{miss}} > 20 \,\text{GeV} \right)$.

Consider now the final particles or jets in the lepton + jets topology (9.97). The top mass, m_t, is equal to the invariant mass of its decay products: one μ or e, its neutrino and a b jet for one of the tops, one b jet and two normal jets for the other. The energy–momenta are measured for all of these, with the exception of the neutrino. This makes three unknown kinematic quantities. In the initial state two partons have collided. Their transverse momenta are small and can be neglected, in first approximation, while the longitudinal momenta are unknown. In total, five kinematical quantities are not known. They are eliminated using the available kinematical constraints. Four constraints are given by energy–momentum conservation, two more by imposing that the mass of the decay products of the two candidate Ws be equal to M_W. Finally, another constraint is obtained by imposing that the masses of the decay products of the two tops have the same, unknown, value m_t. In total there are seven constraints and five unknowns. The problem is overdetermined and can be solved by searching

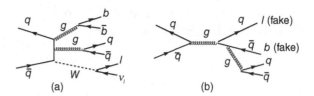

Fig. 9.36. First-order diagrams for the dominating background processes to the lepton + jets topology.

for the χ^2 minimum in a fitting procedure. Notice that, in the fitting process, the missing momentum p_T^{miss} cannot be used, because is not a measured quantity; rather it is calculated starting from measured quantities already considered in the above budget. As a matter of fact, a precise value of p_T^{miss} is the result of the fitting process.

Notice that the dilepton topology, which has less background, with its three more unknowns, is underdetermined by one.

The relative uncertainty of m_t at the Tevatron is impressively small, about 0.5%. We cannot discuss here in any detail how that has been achieved. However, we mention two important ingredients.

The knowledge of the energy scale, and its evolution in time, at better than the per cent level is mandatory. The energy scales of the electrons and the muons are separately calibrated as follows. When a Z is present and decays in a pair of those leptons, its mass is reconstructed from the measured energy–momenta of the pair. These are then scaled to obtain the M_z value measured at LEP. The jet energy scale has larger uncertainties. An in situ calibration technique was developed, taking advantage of the fact that, in the lepton + jets topology, two of the jets are the decay products of a W, which has a precisely known mass. The jet energy scale is obtained by imposing to the reconstructed M_W to be equal to the known mass of the W.

A very accurate understanding of the background is the second important issue. The parton-level diagrams of the dominating backgrounds are shown in Fig. 9.36. In (a) we see the (weak) production of a W associated with jets. In (b) there is no W; hence neither leptons nor b quarks, but two hadronic jets, are misinterpreted as such. These, and other contributions, are calculated with perturbative QCD. Monte Carlo simulation programs are then used to pass from parton to hadron level and to generate the distributions of the experimentally observed quantities, such as the m_t reconstructed, as explained above.

Both experiments have measured m_t in the lepton + jets and in the other two topologies and then averaged the results, taking into account the correlations between them. The result is

$$m_t = 173.5 \pm 1.0 \ \mathrm{GeV}. \tag{9.99a}$$

We now observe that, since the top quark is coloured and only colour singlets are observable, its mass is not a physical observable, as that of all the quarks. The top mass measured as just described, sometimes called 'jet mass', is related to the mass parameter in the Lagrangian, the 'pole mass' by known theoretical arguments. However, the latter has associated uncertainties, which are somewhat larger than the uncertainties associated with the jet mass quoted in Eq. (9.99).

Increasingly accurate measurements of the top properties, in particular of its mass, have been subsequently done by the ATLAS and CMS experiments at LHC. The present world average (Workman *et al.* 2022) is

$$m_t = 172.69 \pm 0.30 \, \text{GeV}. \tag{9.99b}$$

If the ambiguities in the top mass definition are taken into account and the colour structure of the fragmentation process is taken into account, then additional theoretical uncertainty of about 0.5 GeV must be added.

9.12 The Spontaneous Breaking of the Gauge Symmetry

In the preceding chapters we have seen how the electromagnetic, the strong and the weak interactions are ruled by Lagrangians, each invariant under a gauge transformation, $U(1)$, $SU(3)$ and $SU(2)$ respectively. While the gauge fields of the first two are massless, those of the weak interaction, $W^\pm$ and Z are not. How can massive mediators be implemented in a gauge theory?

A second issue is the non-zero mass of the fermions. Indeed, for massless particles, the Dirac equation separates into two independent equations, one for the left spinor and one for the right spinor. This property allowed us to consider them as different particles and to classify them as being one in an isotopic doublet and one in a singlet. That is, the two stationary states are invariant under two different gauge transformations. This is impossible if the fermion is massive, because in this case the chirality does not commute with the Hamiltonian.

Both problems are solved by assuming a spontaneous breakdown of the gauge symmetry. Notice, however, that different terms in the Lagrangian are responsible for the masses of the gauge bosons on one side and of the fermions, leptons and quarks on the other. Both sides should be tested experimentally, as we shall see.

The electroweak Lagrangian is perfectly gauge invariant, but the physical vacuum (i.e. the lowest energy state), is a member of a set of physically equivalent states. These states respect the symmetry, transforming according to a representation of the group. This degeneracy of the vacuum induces scalar fields having non-zero vacuum expectation values.

We now quote very briefly the most relevant historical steps, also to justify why we call BEH the mechanism, Higgs the boson. In 1961 Nambu (Nambu & Jona-Lasinio 1961) and Goldstone published independent articles on models of symmetry breaking in particle physics based on analogy with superconductivity. Goldstone (1961) had proven (Goldstone theorem) that the spontaneous breakdown of an unitary symmetry implies the existence of one, or more, depending on the symmetry group, scalar bosons, which are massless, if the symmetry is not explicitly broken, as is the case here (different from the case of the chiral symmetry discussed in Section 6.9). But none of these bosons was known to exist. Did any possibility exist to evade the Goldstone theorem?

The next step was again in condensed matter theory; in 1963, Phil Anderson (Anderson 1963) pointed out that in a superconductor the longitudinal electromagnetic modes become the Goldstone mode, which is massive, the partner of the transverse modes, which are also massive. In superconductivity, in a non-relativistic regime, the Goldstone theorem could then be evaded. But was that also possible in Lorentz-invariant theory, when there is no matter in the vacuum? On 26 June 1964, the USA journal *Physical Review Letters* received a manuscript in which F. Englert and R. Brout (Englert & Brout 1964), also condensed matter theorists, showed how the Anderson mechanism, suitably modified, could be applied in a relativistic gauge theory. Less than 1 month later, on 24 July, P. Higgs sent to the European journal *Physics Letters* a paper in which he also independently showed, in an example, how the Goldstone massless mode becomes the longitudinal mode of a massive vector boson, just as in the Anderson scheme. The paper was accepted (Higgs 1964a). On 31 July, still unaware of Englert and Brout, Higgs sent to the same journal a second paper, with further developments of the model. The journal, controlled by CERN, rejected the paper, considered irrelevant for particle physics. Higgs decided to send the paper to *Physical Review Letters*, with some further remarks. At the time, the drafts were physically written on paper and took time to cross the Atlantic. The revised paper reached the journal on 31 August, exactly the date in which the one of Englert and Brout was published and accepted (Higgs 1964b). One of the remarks added after the *Physics Letters* rejection was, 'It is worth noting that an essential feature of this type of theory is the prediction of incomplete multiplets of scalar and vector bosons'. While the theoretical mechanism of Brout and Englert and of Higgs is the same, the prediction of the scalar boson is by Higgs. Joking about these events later, Higgs said that the rejection by *Physics Letters*, which had made him furious, had been, after all, a stroke of luck.

Coming back to physics, we start with an analogy known to the reader, considering the vector potential in classical electromagnetism. Its divergence can be arbitrarily fixed. Only two components of the four-potential are gauge invariant, the other two are not, but are not observable. The same happens in QED. Consider a photon propagating along the z-axis. The z and the time components of its field, A_z and A_t, are not physical and are gauge dependent. The 'transverse' components, A_x and A_y, are observable and do not change under a gauge transformation. Similarly, the longitudinal states of the vector gauge fields are not observable. In the BEH mechanism, the Goldstone bosons become the non-observable longitudinal states of the massive vector bosons. The transverse states of the latter become the massive gauge bosons we observe.

The Goldstone bosons of the spontaneously broken symmetry that we considered in Section 6.9, the chiral symmetry of QCD, are the pions. They have non-zero masses because that symmetry is also explicitly broken by the non-zero quark masses. That symmetry is global. However, the gauge symmetry we are considering here is exact. The masses of the states corresponding to the Goldstone phenomenon may be non-zero because the symmetry is local.

We shall see now how these general ideas are implemented in the electroweak theory. We can do that only without any theoretical rigor.

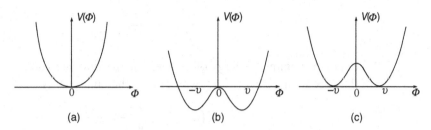

Fig. 9.37. The potential of Eq. (9.102); (a) $\mu^2 > 0$; (b) $\mu^2 < 0$; (c) as (b) with an additional constant to have the energy in the vacua equal to 0.

In quantum field theory, a system is described by its Lagrangian density or by its Hamiltonian energy density. The former is the difference between kinetic and potential energy densities; the latter is their sum. In $SU(2) \otimes U(1)$, the spontaneous symmetry breaking is obtained by including a potential density $V(\Phi)$ that contains a scalar complex field Φ, which is an isospin doublet.

The mechanism is not very simple. Before tackling it, let us start with the simpler case of a real scalar field, with the Lagrangian

$$\mathcal{L} = \frac{1}{2}\partial_\mu \Phi \partial^\mu \Phi - V(\Phi). \tag{9.100}$$

On the right-hand side, the first term is the kinetic energy, the second is the potential

$$V(\Phi) = \frac{1}{2}\mu^2 \Phi^2 + \frac{1}{4}\lambda \Phi^4. \tag{9.101}$$

Higher powers of Φ are not included because they would introduce infinity quantities that cannot be eliminated by renormalization. The potential is invariant under the substitution $\Phi \leftrightarrow -\Phi$. The term proportional to Φ^4 makes the field quanta interact with each other. V is an energy per unit volume, with physical dimension $\left[E^4\right]$, Φ has the dimensions $[E]$ and the constant λ is dimensionless. We take $\lambda > 0$ to guarantee vacuum stability, namely to have it a lower bound for $\Phi \to \infty$. If only the term proportional to Φ^2 were present, the potential would have corresponded to scalar particles of mass μ. However, for $\Phi = 0$ the potential does not have a minimum, but a maximum, hence there cannot be any small oscillations around it, namely those oscillations that are interpreted as the particles of the field. The system is unstable and moves towards a minimum, one of the two that are present. We find them by solving the equation

$$\left.\frac{\partial V}{\partial \Phi}\right|_{\Phi_{min}} = \Phi\left(\mu^2 + \lambda \Phi^2\right) = 0. \tag{9.102}$$

In the usual case in which μ is real, hence $\mu^2 > 0$, we have $\Phi_{min} = 0$ and $V(\Phi_{min}) = 0$. As expected, in the minimum energy state both the field and the potential energy are zero.

The curve in Fig. 9.37(a) illustrates the situation. The important step is to consider μ imaginary, corresponding to $\mu^2 < 0$. As we shall see, there will be no physical object with imaginary mass.

If $\mu^2 < 0$, Eq. (9.102) has two solutions

$$\Phi_{\min} = \pm\sqrt{\frac{-\mu^2}{\lambda}} \equiv \pm\upsilon. \tag{9.103}$$

The state of minimum energy of a system in quantum mechanics is its vacuum state. In this case we have two vacua, with exactly the same energy

$$V(\Phi_{\min}) = -\frac{1}{4}\frac{\mu^4}{\lambda} = -\frac{\lambda}{4}\upsilon^4, \tag{9.104}$$

the vacuum of the system is 'degenerate'. This quantity υ is called vacuum expectation value (VEV). The potential is shown in Fig. 9.37(b).

We now take advantage of the fact that a constant can always be added to a potential by adding $\frac{\mu^4}{4\lambda}$ to have an expression in which the minima of the potential are zero, $V(\Phi_{\min}) = 0$. Equation (9.101) becomes

$$V(\Phi) = \frac{\lambda}{4}\left(\Phi^2 - \frac{\mu^2}{\lambda}\right)^2 = \frac{\lambda}{4}\left(\Phi^2 - \upsilon^2\right)^2. \tag{9.105}$$

The new potential is shown in Fig. 9.12(c).

After this exercise we can go to the BEH mechanism, in which several of the features we have just seen are also present. Here Φ is the Higgs field that is a complex scalar in the space-time and a doublet in $SU(2)$

$$\Phi = \begin{pmatrix} \varphi^+ \\ \varphi^0 \end{pmatrix}, \tag{9.106}$$

where the apexes are the electric charges, and we define

$$\varphi^+ = \frac{\varphi_1 + i\varphi_2}{\sqrt{2}}; \quad \varphi^0 = \frac{\varphi_3 + i\varphi_4}{\sqrt{2}}. \tag{9.107}$$

The Lagrangian is now

$$\mathcal{L} = \frac{1}{2}\left(\partial^\mu\Phi\right)^\dagger\left(\partial_\mu\Phi\right) - \left(\frac{1}{2}\mu^2\Phi^\dagger\Phi + \frac{1}{4}\lambda^2\left(\Phi^\dagger\Phi\right)^2\right), \tag{9.108}$$

and we define

$$\begin{aligned}\Phi^2 \equiv \Phi^\dagger\Phi &= \begin{pmatrix} \varphi^{+*} & \varphi^{0*} \end{pmatrix}\begin{pmatrix} \varphi^+ \\ \varphi^0 \end{pmatrix} \\ &= \varphi^{+*}\varphi^+ + \varphi^{0*}\varphi^0 = \frac{1}{2}\left(\varphi_1^2 + \varphi_2^2 + \varphi_3^2 + \varphi_4^2\right).\end{aligned} \tag{9.109}$$

Similarly to the simple scalar case – the equations are substantially the same – for $\mu^2 < 0$, the minima are at

$$\Phi^2_{\min} = -\frac{\mu^2}{\lambda} = \upsilon^2 \tag{9.110}$$

The field in Eq. (9.106) has two complex degrees of freedom, corresponding to four real ones. However, imposing to the VEV that it is electrically neutral, as logic requires, the upper components of Φ should be zero $\left(\varphi_1 = \varphi_2 = 0\right)$, reducing the degrees of freedom

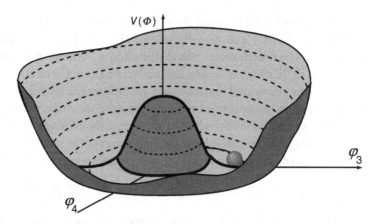

The potential and the symmetry breakdown.

to two, φ_3 and φ_4 and we can represent it in a plane in Fig. 9.38. It is immediately seen that V is symmetrical under the rotations in this plane. The rotations leave Φ^2 (i.e. the distance from the V axis) invariant. We now break the symmetry under rotations and choose one of the degenerate minima, to have it on the real axis, namely $\varphi_3 \neq 0$, $\varphi_4 = 0$, and write

$$\Phi = \begin{pmatrix} \varphi^+ = 0 \\ \varphi^0 = (\upsilon + H)/\sqrt{2} \end{pmatrix}, \tag{9.111}$$

where H is the Higgs field. For its shape, the potential is called 'Mexican hat' or, as Peter Higgs prefers, 'wine bottle'.

The H mass is obtained by considering the perturbations of the potential for small oscillations of the field around the minimum, say from $\Phi = \upsilon$ to $\Phi = \upsilon + \sigma$ in the direction of the φ_3 axis

$$\Phi^2 - \upsilon^2 = 2\upsilon\sigma + \sigma^2$$

and substitute in the second term of (9.105), which is valid now too, obtaining

$$\delta V = \lambda \upsilon^2 \sigma^2 + \lambda \left(\upsilon \sigma^3 + \frac{1}{4}\sigma^4 \right).$$

Considering that the mass term of a particle of the 'field' σ has the form $\frac{1}{2}m^2\sigma^2$, we recognize it in the first addendum. The second is the self-interaction energy density. In conclusion the H mass is

$$M_H = \sqrt{2\lambda}\,\upsilon. \tag{9.112}$$

Notice that the potential is constant for small movements in the direction of the φ_4 axis; consequently, the corresponding oscillations are massless. The gauge transformation will absorb them in the gauge bosons.

In writing (9.111) and choosing a direction, we have broken three times the original $O(4)$ symmetry $\varphi_1^2 + \varphi_2^2 + \varphi_3^2 + \varphi_4^2 =$ invariant to $\varphi_1 = 0$, $\varphi_2 = 0$, $\varphi_3 \neq 0, \varphi_4 = 0$. This spontaneous breakdown of the symmetry transforms three components of Φ into the physical W^+, W^- and Z^0, after the rotation by the weak mixing angle, giving them masses. The fourth vector boson, the photon, remains massless. The remaining component of Φ, which is neutral and scalar, becomes the physical H. Notice that, in such a way, the vector and scalar bosons are different components of the same system. The H boson can be considered as an excited state above the vacuum chosen by the system.

A drawback of the Standard Model is that the constant λ is not defined by the theory, remaining arbitrary. The BEH mechanism gives the relations between the vector boson masses on one side, and the coupling constants g and g' and the VEV on the other:

$$M_W = \frac{1}{2} g \upsilon \qquad\qquad M_Z = \frac{1}{2}\sqrt{g^2 + g'^2}\,\upsilon. \tag{9.113}$$

Consequently, the VEV is extracted from the known values of the vector boson masses or, better, from the equivalent relation to the Fermi constant, which is known with a very high accuracy:

$$\upsilon = \sqrt{\frac{1}{\sqrt{2}G_F}} = 246 \text{ GeV}. \tag{9.114}$$

Notice that this value corresponds to an enormous energy density of Eq. (9.104). The latter is proportional to the fourth power of the VEV, being an energy divided by a volume, which in natural units has the dimensions of an inverse energy cube.

The theory does not give any relation between M_H and the vector boson masses.

Consider now the fundamental fermions. The BEH mechanism gives mass to them too. As we have already stated, before symmetry breaking they are massless. Consider, for example, the electron. Clearly the mass term in the Lagrangian must be built with the electron fields e_L and e_R. It must be a scalar and isoscalar. It is immediately seen that the terms $m_e \bar{e}_L e_L$ and $m_e \bar{e}_R e_R$ are identically zero for the properties of the left and right projectors. Moreover, neither $m_e \bar{e}_L e_R$ or $m_e \bar{e}_R e_L$ can be used because they are isodoublets, not isoscalars. The spontaneous symmetry breakdown surviving (neutral) components of the isodoublet field Φ solve the problem, as Weinberg showed (Weinberg 1967). The terms $\bar{e}_L \varphi^0 e_R$ and $\bar{e}_R \bar{\varphi}^0 e_L$ are isoscalar and Lorentz-invariant. Their sum $\bar{e}_L \varphi^0 e_R + \bar{e}_R \bar{\varphi}^0 e_L$ is used in the theory to write the electron mass term in the Lagrangian:

$$\mathcal{L}_e = \frac{1}{\sqrt{2}} f_e \left(\bar{e}_L \varphi^0 e_R + \bar{e}_R \bar{\varphi}^0 e_L \right). \tag{9.115}$$

This is proportional to the VEV and to the dimensionless constant f_e, called Yukawa coupling. Again, without giving any demonstration, we state that this term leads to the value of the electron mass

$$m_e = \frac{1}{\sqrt{2}} f_e \upsilon. \tag{9.116}$$

Similar expressions are found for all the charged fermions (f), both leptons and quarks, each with its Yukawa coupling (λ_f). The Yukawa coupling measures the strength of the interaction of each fermion with the Higgs field, but is not determined by the theory. Being proportional to the masses, the couplings are very different, as are the masses. The inability of the theory to give any prediction on purpose points to a serious limitation of the Standard Model.

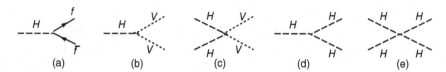

Fig. 9.39.   The vertices of the H boson.

Question 9.4 Find the values of the Yukawa coupling of the electron and of the top. □

Question 9.5 Evaluate what the proton mass would be if the Yukawa couplings of the u and d quarks were 0, rather than having the masses of Table 4.5 [$m_p \approx 928$ MeV]. □

Question 9.6 How large would the range of the nuclear forces, λ in Eq. (2.1), be in the same hypothesis as in Question 9.5? How large would the deuterium radius be, assuming an effective coupling constant (corresponding to α in the Coulomb force) of 10? Compare with the hydrogen atom (the deuterium radius would be about 1.5 times the Bohr radius). Evaluate the binding energy in such conditions. □

As for neutrinos, we see immediately that the same recipe does not work, because right neutrinos have never been observed. Consequently, neutrinos remain massless after the symmetry breakdown.

It is worth noticing that, contrary to what one often finds written, the Higgs boson is not the origin of all the mass, but simply of that of the gauge bosons and of the fundamental fermions (except neutrinos). Only a small fraction of the mass of the nucleons, and consequently of the nuclei (and of the matter), is due to the mass of the quarks (and electrons). The mass of the nucleons is rather the energy of the gluon field corresponding to the quark confinement inside, as we saw in Section 6.7.

The BEH mechanism predicts all the couplings of the H to the fermions, to the vector bosons and to itself as functions of M_H. Actually, the coupling amplitude to fermions is proportional to the fermion mass, the couplings to the vector bosons and to itself are proportional to the squares of the masses. In any case, the decays into larger-mass particles are favoured. Calling generically V the vector bosons, the coupling of the H are

$$g_{Hf\bar{f}} = \frac{m_f}{\upsilon}, g_{HVV} = \frac{2M_V^2}{\upsilon}, g_{HHVV} = \frac{2M_V^2}{\upsilon}, g_{HHH} = \frac{3M_H^2}{\upsilon}, g_{HHHH} = \frac{3M_H^2}{\upsilon^2}. \quad (9.117)$$

The corresponding vertices are shown in Fig. 9.39(a–e).

The partial width in the charged lepton–antilepton pairs corresponding to the vertex in Fig. 9.39(a), when the channel is open $\left(\text{i.e.} M_H > 2m_l \right)$ is

$$\Gamma\left(H \to l^+l^- \right) = \frac{G_F m_l^2}{4\pi\sqrt{2}} M_H \beta_l^3, \quad (9.118)$$

where $\beta_l = \sqrt{1 - 4m_l^2 / M_H^2}$ is the lepton velocity. There is no coupling of H to neutrinos.

The partial width in a quark–antiquark pair, when the channel is open, is similar, but with three important differences: (1) it contains a factor three, due to the

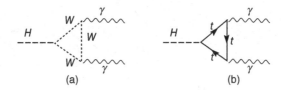

Fig. 9.40. Principal contributions to the $H \rightarrow \gamma\gamma$ decay.

colour; (2) the running of the quark mass must be taken into account; and (3) there are higher-order QCD corrections of the order of $\alpha_s(M_H)$ (i.e. of about 10%) which we will not discuss, but which must be taken into account for precise calculations. The partial width is, at the first order

$$\Gamma(H \rightarrow q\bar{q}) = \frac{3G_F m_q^2(M_H)}{4\pi\sqrt{2}} M_H \beta_q^3. \tag{9.119}$$

Question 9.7 Calculate the ratio between the values of $\Gamma(H \rightarrow b\bar{b})$ without and with assuming running of m_b, for $M_H = 100$ GeV. $\square$

When the decay channel in W^+W^- is open, the decay width is

$$\Gamma(H \rightarrow W^+W^-) = \frac{G_F M_H^3}{8\pi\sqrt{2}} \left(1 - \frac{4M_W^2}{M_H^2}\right)^{1/2} \left[1 - 4\left(\frac{M_W^2}{M_H^2}\right) + 12\left(\frac{M_W^2}{M_H^2}\right)^4\right] \tag{9.120}$$

and similarly for ZZ when open

$$\Gamma(H \rightarrow ZZ) = \frac{G_F M_H^3}{8\pi\sqrt{2}} \left(1 - \frac{4M_Z^2}{M_H^2}\right)^{1/2} \left[1 - 4\left(\frac{M_Z^2}{M_H^2}\right) + 12\left(\frac{M_Z^2}{M_H^2}\right)^4\right]. \tag{9.121}$$

Below threshold, namely for $M_H < 2M_W$ or $M_H < 2M_Z$, respectively, the H can still decay in a pair of vector bosons, one real and one virtual. They are named W^* and Z^*. We do not give the full expressions here, but the corresponding widths $\Gamma(H \rightarrow WW^*)$ and $\Gamma(H \rightarrow ZZ^*)$ decrease rapidly as the process goes farther and farther below threshold.

The H is indirectly coupled to photon pairs, via fermions and vector bosons loops. The most important contributions come from the heaviest loops, shown in Fig. 9.40. Of the two, the dominant one is the W loop (Fig. 9.40(a)), which is about five times larger than the top loop. The signs of both amplitudes are opposite, as is always the case between bosonic and fermionic loops.

Once all the partial widths have been calculated within the Standard Model with the required accuracy, the total H width is obtained, as their sum, as a function of M_H. Below the WW threshold, namely if $M_H < 2M_W$, the width is small. For example, for $M_H \approx 130$ GeV, Γ_H is a few MeV, corresponding to a lifetime of 10^{-22} s. The width increases rapidly for $M_H > 2M_W$ to about 1 GeV for $M_H \approx 200$ GeV and a few hundred GeV for $M_H \approx 1000$ GeV. Obviously, the H has a definite mass and a definite width, but before knowing that, experiments had to cope with such a large spread of possibilities.

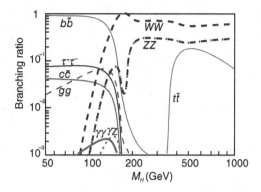

Fig. 9.41. The H branching ratios as functions of M_H. Notice that both scales are logarithmic.

Finally, Fig. 9.41 gives the branching ratios of the H as functions of its mass. Again, these are important to design the search strategies in the full range of possible H masses.

9.13 The Search for the Higgs at LEP and at the Tevatron

As already discussed, indirect information on M_H is obtained, assuming the SM, from fits to precision measurements of electroweak observables. The diagrams in Fig. 9.23(b, c) give the first-order contributions of the H boson to the ratio of the W and Z masses, which is logarithmically sensitive to the H mass. A global fit to the precision electroweak data gave the prediction (LEP EW Working Group 2012)

$$M_H = 94^{+29}_{-24}\,\text{GeV},$$

corresponding to an upper limit $M_H < 152$ GeV at the 95% confidence level. Obviously, the experimental search for the H cannot be limited to this mass range. Rather it must be extended up to the value at which the theory becomes logically inconsistent. The cross-section for the process $WW \to H \to ZZ$ exceeds the unitary limit for M_H larger than about 700 GeV. Taking some margin, LHC was designed to be able to reach $M_H = 1000$ GeV.

Sensitive searches for the H boson were carried out in the LEP experiments. At the LEP energies, and corresponding range of reachable M_H values, the dominant decay channel is $H \to b + \bar{b}$, as can be seen in Fig. 9.41. The main search channel is shown in Fig. 9.42. The diagram corresponds to two different situations, depending on the energy. At the Z peak, that is, for energy $\sqrt{s} \approx M_Z$, the first Z is a real particle and the second is a virtual one. At higher energies, $\sqrt{s} > M_Z$, the first Z is virtual and the second is real. Notice that the change from 'real' to 'virtual', and vice versa, is a gradual process.

At the Z peak, the cross-section is appreciable if the virtual Z is not very far from resonance, namely if M_H is too small compared to M_Z. The search at the Z resonance did not find the H, providing a limit on its mass.

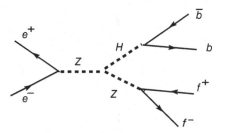

H production at an e^+e^- collider followed by $b\bar{b}$ decay.

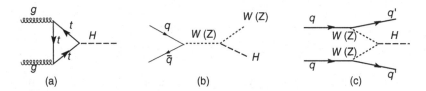

Principal H production diagram at the Tevatron $\bar{p}p$ collider: (a) gluon fusion; (b) H bremsstrahlung; (c) vector boson fusion.

For similar reasons the limit that can be reached on M_H at higher energies is about $\sqrt{s} - M_Z$. The LEP energy was increased as much as possible by installing as many superconductive radio-frequency cavities as could be fitted in the ring to provide the power necessary to compensate for the increasing synchrotron radiation, up to $\sqrt{s} = 209$ GeV. A total integrated luminosity of about 0.5 events/fb was delivered, but the H boson was not found. The final limit on the mass at 95% confidence level is (LEP 2006)

$$M_H > 114.4 \text{ GeV}. \tag{9.122}$$

Question 9.8 Knowing that the cross-section $\sigma\left(e^+ + e^- \rightarrow H + Z\right) \simeq 1$ pb at $\sqrt{s} = 209$ GeV for $M_H = 70$ GeV, how many events are collected, assuming 100% efficiency, in a run of 100 pb^{-1}? □

The CDF and D0 experiments collected in the Run II about 10 fb^{-1} integrated luminosity. The search for H was made in all possible channels, considering those in which the QCD backgrounds, much larger than in an e^+e^- machine, were manageable. The main production channels are shown in Fig. 9.43: (a) gluon fusion, (b) H bremsstrahlung and (c) vector boson fusion.

The search is very difficult for small M_H values just above the LEP limit. At Tevatron the b-quark jets can be distinguished from the quark and gluon jets produced by strong processes by detecting the b decay vertices, as we saw in Section 9.11, but even after that identification, called b tagging, the background from beauty particles produced by QCD processes is large.

The Tevatron search was effective for $M_H > 165$ GeV, where the channel $H \rightarrow WW$ opens up, soon reaching a branching ratio close to 100%. Both Ws can be detected and separated from backgrounds looking for the decays $W^+ \rightarrow l^+ \nu_l$ and $W^- \rightarrow l^- \bar{\nu}_l$.

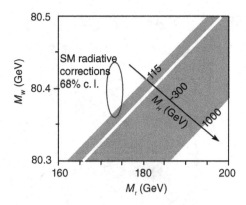

The grey bands show the non-excluded values of M_H at the start of LHC. Ellipsis is the Standard Model radiative corrections prediction.

At the start of LHC a Standard Model H boson was excluded by the Tevatron in the mass range 147 GeV $< M_H < 179$ GeV. Fig. 9.44 summarizes graphically the results of the H searches by LEP and Tevatron, together with the Standard Model expectations based on the radiative corrections at the time when the LHC started data-taking.

9.14 LHC, ATLAS and CMS

The large hadron collider LHC and its two general-purpose experiments ATLAS and CMS were designed in the 1990s with a vast experimental programme ranging from the search for the H boson, which we shall discuss in this book, to a comprehensive and profound exploration of physics beyond the Standard Model, which might appear at the TeV energy scale. It took more than 20 years to build them and to make them operational.

The energy of the collider should have been as high as possible. However, it had to be lodged in the 27 km long existing LEP tunnel, a vast civil work investment of the CERN member states. To increase the energy at the fixed curvature of the orbit one needs to increase as much as possible the dipoles' magnetic field. Substantial efforts and investments were made in collaboration amongst research agencies, CERN and specialized companies to push superconductive magnet technology beyond the state of the art.

The LHC dipoles have a maximum design field of $B = 8.33$ T, allowing, with some margin, the maximum beam energy $E_b = 7$ TeV to be reached, for a CM energy $\sqrt{s} = 14$ TeV. The luminosity must also have unprecedented levels. These levels cannot be obtained colliding protons against antiprotons because the feasible antiproton beam intensities are too low. In LHC there are two proton (or ion) beams circulating in opposite directions in the same dipole ring. The magnet structure, as shown in the cross-section in Fig. 9.45(a), is designed to obtain opposite directions of the field in the two beam pipes. Fig. 9.45(b) is a photo of one of the 1232, 14 m long, LHC dipoles.

LHC DIPOLE : STANDARD CROSS-SECTION

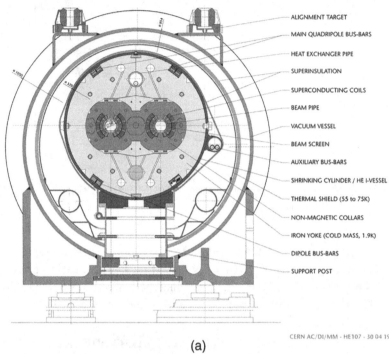

ALIGNMENT TARGET
MAIN QUADRIPOLE BUS-BARS
HEAT EXCHANGER PIPE
SUPERINSULATION
SUPERCONDUCTING COILS
BEAM PIPE
VACUUM VESSEL
BEAM SCREEN
AUXILIARY BUS-BARS
SHRINKING CYLINDER / HE I-VESSEL
THERMAL SHIELD (55 to 75K)
NON-MAGNETIC COLLARS
IRON YOKE (COLD MASS, 1.9K)
DIPOLE BUS-BARS
SUPPORT POST

CERN AC/DI/MM - HE107 - 30 04 1999

(a)

(b)

Fig. 9.45.   (a) LHC dipole cross-section (image © CERN). (b) LHC superconducting dipoles (photo © CERN).

All the magnets, the dipoles and the other magnets necessary to govern the beams, are electromagnets designed and built with frontier cryogenic techniques. Indeed, the needed current intensity, which in the dipoles is 11 850 A, imposes that resistance in the coils should be zero. They are cooled at 1.9 K by a 'river' of 120 t mass, 27 km long, liquid helium. This cryogenic system requires 40 000 leak-tight pipe seals.

The transverse beam size is of $200 - 300$ μm, which is reduced at the intersections to 16 μm. The energy stored in one beam is 362 MJ. To imagine what might happen if a beam were lost hitting the LHC structures, if all the necessary safety measures were not in place, consider that the energy needed to melt 1 kg of copper is 0.7 MJ.

The LHC was designed with energy and luminosity adequate to search for the H boson up to $M_H = 1000$ GeV. At these high mass values, the H boson is produced predominantly by vector boson fusion. As shown in Fig. 9.43(c), a quark from each of the protons radiates a W, the energy of which must be about $M_H / 2$, namely 500 GeV. To radiate such a W, the initial quark should have at least about twice that energy, namely 1 TeV. The valence quarks in a proton carry, on average, 1/6 of the proton momentum. Hence the proton energies should be about 6 TeV. This is well matched with the 7 TeV maximum energy design.

Consider now the luminosity. The production cross-section of a H with $M_H = 1$ TeV at $\sqrt{s} = 14$ TeV is predicted by the Standard Model to be $\sigma \approx 10$ fb $= 10^{-38}$ cm^2. The main search channel is $H \rightarrow ZZ \rightarrow 2l^+ 2l^-$, where l is an electron or a muon. Its branching ratio is about 10^{-3}. If we want to have, say, ten events in one running year, assumed in 10^7 s, the required luminosity is $\mathcal{L} = 10 / \left(10^{-38} \times 10^{-3} \times 10^{34} \right) = 10^{34}$ cm^{-2}s^{-1} This is two orders of magnitude larger than that reached by Tevatron, and corresponds to about 10^9 interactions per second.

To obtain this luminosity the machine was designed and built to lodge as many as 2808 proton bunches in each beam, with 10^{11} protons per bunch. They are separated by 7.5 m, hence the time between two collisions at the intersection points is only 25 ns. The detectors must cope with this tremendous rate, in which the tracks due to a given beam crossing appear in the central part of the detector, when those due to the previous crossing, 25 ns earlier, have not yet left its external components. Moreover, at maximum luminosity, there are as many as 35 proton–proton collisions per crossing. This is called pile-up. The detectors must be able to associate charged tracks, electromagnetic and hadronic showers to each of these collisions.

The first beams circulated in LHC in September 2008. However, soon after, a quenching in a dipole caused considerable damage to machine components. Investigation showed that the problem was in the normal conducting soldering between superconducting cables. There are tens of thousands of them, each of which should have nano-ohm resistance. In a few cases that resistance was found to be larger. All the connections had to be controlled again. A long shutdown of the machine followed, during which part of the ring was warmed to ambient temperature, while storing the precious He gas, and the most important repairs were done. After cooling the machine again, the first collisions at centre of mass energy prudently reduced to 7 TeV started in March 2010. In 2011, 4.7 fb^{-1} were delivered at $\sqrt{s} = 7$ TeV, and 23 fb^{-1} in 2012 when the energy was increased to $\sqrt{s} = 8$ TeV, and initial luminosities reached

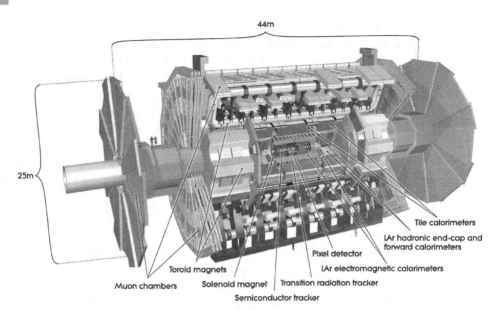

44m

25m

Tile calorimeters

LAr hadronic end-cap and forward calorimeters

Pixel detector

LAr electromagnetic calorimeters

Toroid magnets

Transition radiation tracker

Muon chambers

Solenoid magnet

Semiconductor tracker

Fig. 9.46. Artist's view of the ATLAS experiment (image © CERN and ATLAS).

values close to $10^{34}\,\mathrm{cm}^{-2}\mathrm{s}^{-1}$. In the same year, the discovery of the Higgs boson was announced (see next section). A long shutdown of LHC started in 2013 for the work necessary to reach the design energy $\sqrt{s} = 14$ TeV, and concluded in 2015. The machine restarted collisions at $\sqrt{s} = 13.6$ TeV in April 2015, delivering in the so-called Run-2 more than 150 fb^{-1} to both ATLAS and CMS. Run-2 was concluded at the end of 2018, followed by a long shutdown of the machine and experiments for further improvements. In April 2022 Run-3 started. Although the luminosity could now go well above $10^{34}\,\mathrm{cm}^{-2}\mathrm{s}^{-1}$, in practice it is limited to $2\times10^{34}\,\mathrm{cm}^{-2}\,\mathrm{s}^{-1}$ by the heat load induced on the high gradient quadrupoles used as focussing lenses in the crossing points. Under these conditions, the total cross-section of the SM Higgs production for $M_H = 125$ GeV being equal to 54 ± 2.6 pb, one Higgs boson is produced every second.

The two general-purpose LHC detectors, ATLAS and CMS, are shown respectively in Figs. 9.46 and 9.47. Even if we limit the discussion to the H boson, it should be kept in mind that the detectors were built for a much wider physics programme, to be able, in particular, to discover any sign of physics beyond the Standard Model that might appear in the new LHC energy regime.

The two experiments differ in many important details, but the general principles are similar. Both are roughly cylinders. The central cylindrical part is called 'the barrel', which is closed on the two sides by two 'end caps'. The structure is onion-like, in which the different layers have different functions; this is rather similar to our discussion of the UA1 experiment. The first layer is the precision tracking system, the second is the electromagnetic calorimeter, the third is the hadronic calorimeter, the fourth is the muon spectrometer. Fig. 9.48 is a sketch of a wedge of a cross-section of the barrel of CMS.

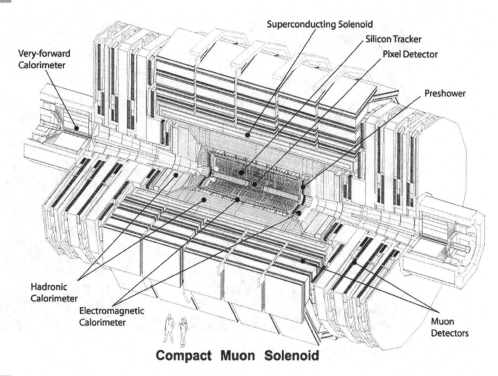

Very-forward
Calorimeter

Superconducting Solenoid

Silicon Tracker

Pixel Detector

Preshower

Hadronic
Calorimeter

Electromagnetic
Calorimeter

Muon
Detectors

Compact Muon Solenoid

Fig. 9.47. Artist's view of the CMS experiment (image © CERN and CMS). Scale different from that in Fig. 9.46.

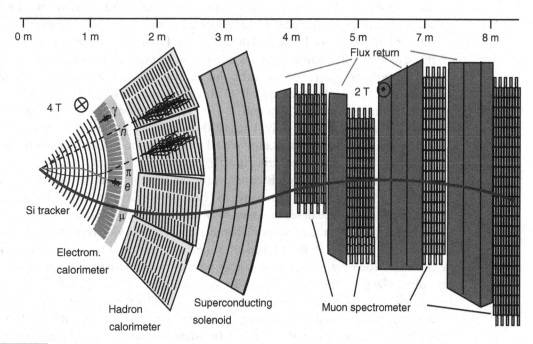

0 m   1 m   2 m   3 m   4 m   5 m   7 m   8 m

Flux return

4 T $\otimes$

γ

h

π

e

2 T $\odot$

Si tracker

μ

Electrom.
calorimeter

Hadron
calorimeter

Superconducting
solenoid

Muon spectrometer

Fig. 9.48. Cross-section of the CMS detectors in the barrel.

The ATLAS detector (photo © CERN and ATLAS).

The two experiments differ in the structure of their magnetic fields. In both cases the tracking is contained in a solenoid providing an uniform magnetic field in the direction of the beams. The solenoid of CMS has a 3 m radius and a high field $B = 3.8$ T. It also contains the electromagnetic and hadronic calorimeters. The solenoid of ATLAS is smaller both in radius, 1.5 m, and in field, $B = 2$ T. The muon spectrometer of CMS is made of four drift chamber systems interleaved by brass plates, which are also necessary to properly guide the magnetic flux return outside the solenoid. The muon spectrometer of ATLAS has a large level arm in a moderate magnetic field, which has a toroidal shape with $L = 4.5$ m and $B \approx 0.6$ T, being generated by eight superconducting coils, shown in Fig. 9.49. Each of them measures 26 m in length and 5 m in width. The corresponding sagitta for $p = 1$ TeV is $s \approx 0.5$ mm. The design momentum resolution is obtained by measuring it with a resolution of 50 μm.

One of the main challenges is coping with the very high luminosity. As we have already said, at the design luminosity, the time between two crossings is only 25 ns (50 ns in 2011–12). The detectors must respond within this time. In particular, traditional tracking with time projection chambers as in the previous collider experiments is no longer possible. Rather, layers of Si detectors are used (or other fast tracking elements). Moreover, about 1000 tracks emerge from the collisions at each beam crossing. The granularity of the tracking system must be high, especially in the innermost layers, to reduce the fraction of detector elements that contain information in a crossing, which is called 'occupancy'. The problem is solved by using, in the inner three layers, Si detectors with pixels (i.e. two-dimensional sensitive elements) in place of the traditional micro-strips. The latter are narrow but long, covering a relatively large area. The pixels are rectangles of $100-150\,\mu m$ long sides. The price is the large number, 70 million in CMS, of detector electronic channels. Both detectors and associated electronics must resist a high radioactivity environment for several years, they must be sufficiently 'radiation hard'.

The trackers of both experiments measure the momenta of the charged particles with a transverse momentum resolution $\delta p_T / p_T \approx 2\%-3\%$ for $p_T = 100$ GeV. For comparison, this is an order of magnitude better than in the LEP experiments. The inner part of the tracker, called the vertex detector, also gives a high spatial resolution picture of the event topology, showing, in particular, the presence of secondary vertices due, for example, to beauty particle decays.

Electromagnetic calorimeters must be radiation hard, have very good energy resolution and good granularity. To understand the importance of the latter, consider the decay of a particle into two gammas. That particle had been produced in, say, one of the 30 collisions of a given beam crossing. The calorimeters must provide enough information on the direction of the gammas to associate them with the correct interaction. The two experiments used different technologies. ATLAS used a sampling calorimeter with liquid argon as the detecting medium, containing lead plates as a radiator, with a novel mechanical structure (called an accordion), to provide high granularity, especially in the inner layers. CMS chose lead tungstenate crystals. This medium has a small Molière radius, $r_M = 21$ mm, due to its high density, hence producing narrow showers and allowing a high transverse granularity. The faces of the crystals measure 22×22 mm^2 in the barrel and 28.6×28.6 mm^2 in the end-caps, for a total of almost 76 000. The crystals are assembled in 'towers' pointing to the interaction region.

Narrow width particles decaying in two muons, two electrons or two photons can be more easily distinguished from the background if the di-muon, di-electron or di-photon mass resolution is high. The mass M of a particle decaying in two ones of energies E_1 and E_2 separated by the angle θ is

$$M = 2E_1 E_2 (1 - \cos\theta). \tag{9.123}$$

Consequently, the relative uncertainty for M is

$$\begin{aligned}\frac{\sigma_M}{M} &= \frac{1}{2}\sqrt{\left(\frac{\sigma(E_1)}{E_1}\right)^2 + \left(\frac{\sigma(E_2)}{E_2}\right)^2 + \left(\frac{\sigma(\theta)}{\tan\theta/2}\right)^2} \\ &\approx \frac{1}{\sqrt{2}}\left\langle\frac{\sigma(E)}{E}\right\rangle,\end{aligned} \tag{9.124}$$

where we have considered that, in practice, the last addendum is small and have expressed the result in terms of an average energy resolution. The experiments use the measured di-electron mass distribution at the Z to calibrate the energy scale and to check the resolution, which is found to be between 1.7% and 4%.

The hadron calorimeters must provide coverage of the largest fraction of the solid angle, down to about 1° with the beams' directions, to guarantee the hermeticity necessary to associate any missing transverse momentum with neutrinos. They also must have fine lateral segmentation to provide good di-jet mass resolution. The choice of

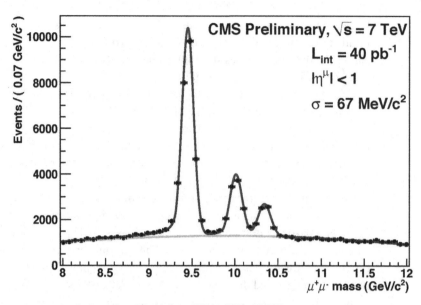

Fig. 9.50.   Di-muon mass spectrum in the region of the Υ's from CMS (© CERN and CMS).

CMS is for sampling calorimeters, both in the barrel and in the end caps, consisting of brass and scintillating plates. ATLAS chose an iron-scintillator calorimeter in the barrel and liquid argon technology in the end caps.

The muon system must be able to detect narrow states, such as the J/ψ or the Υ, decaying in a $\mu^+\mu^-$ pair. For that, the di-muon mass resolution and the ability to measure the charge sign of muons up to 1 TeV are essential. These requirements define a momentum resolution $\Delta p/p = 10\%$ for $p = 1$ TeV. Fig. 9.50 shows the di-muon mass spectrum as observed by CMS in the region around 10 GeV showing three narrow peaks at the Υ resonances. The di-muon mass resolution is 67 MeV (0.7%).

Question 9.9 How much is the average momentum resolution at the J/ψ and the Υ? □

High bending power is obtained by CMS with its compact solenoid and high field, $B = 3.8$ T, with the bending starting at the interaction point. The sagitta is measured with a precision of about 100 μm by four tracking stations at increasing distances.

As already mentioned, at the design luminosity there are approximately 10^9 proton–proton interactions per second, each producing about 1Mb of information. It is impossible to record all these data. On the other hand, the vast majority of the events are due to glancing, rather than central, collisions and are not very interesting. An online selection procedure, the 'trigger', is necessary. The trigger system, consisting of a combination of dedicated high-speed electronics and dedicated software, decreases the event rate to around 0.5 kHz before data storage, while preserving the acquisition of the physically interesting events.

9.15 The Discovery of the Higgs Boson

At LHC, a pp collider, the Higgs boson production mechanisms are different from those we have seen for the Tevatron, which is a $\bar{p}p$ collider. Since the proton beams do not contain many antiquarks, the dominant mechanism for producing the H is the collision of two gluons, called gluon–gluon fusion. It represents 87% of the production cross-section; Fig. 9.51 shows the process together with the other partons of the colliding protons. The process being searched for is accompanied by the fragmentation of these protons. The particles originating from the latter have, in general, small transverse momenta and can be eliminated with kinematical cuts similar to those we discussed in Section 9.7.

The H production cross-section has been calculated within the Standard Model. At $\sqrt{s} = 7$ TeV it ranges from about 20 pb for $M_H = 115$ GeV to about 0.4 pb for $M_H = 600$ GeV and down to 15 fb for $M_H = 1000$ GeV.

Question 9.10 What is the typical fraction of proton momentum x of the gluons to fuse in a H for $M_H = 200$ GeV and $\sqrt{s} = 8$ TeV? What is the typical fraction at $\sqrt{s} = 14$ TeV? Is the gluon distribution function large or small at these x values? □

Data for an integrated luminosity of about 5 fb^{-1} were delivered to the experiments by LHC in 2011. The search for the H as a function of its mass was performed by ATLAS and CMS with an analysis of all the different production and decay channels, exploring the range 110 GeV $< M_H < 600$ GeV. Powerful simulation programmes had been developed for an accurate calculation, based on perturbative QCD and electroweak theory, of the different reactions taking place in the collisions. The resulting distributions of several kinematic quantities were then compared with the experimentally measured ones. The latter were well reproduced by the simulation. The background in the search of the H had been understood.

The SM expectations for observing the H boson in each of the channels were similarly calculated. The best channel for the search in different M_H intervals depends both on the branching ratios shown in Fig. 9.42 and on the level of background.

$M_H < 130$ GeV is the most difficult region. The largest branching ratio is for $H \to b\bar{b}$ but the background due to the QCD production of beauty is overwhelming. A better channel is $H \to \gamma\gamma$. In this case the electromagnetic energy resolution of the

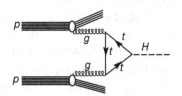

Fig. 9.51. The H production via gluon fusion at LHC.

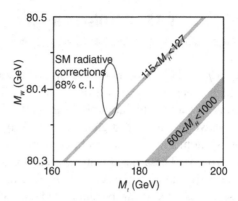

Fig. 9.52. The white region is the excluded one after 1 year of LHC data-taking. The ellipse is the Standard Model radiative corrections prediction.

calorimeters is vital for reconstructing the two-photon mass. The presence of the H would appear as a peak in the two-photon mass distribution. The problem is the very small branching ratio, about 2×10^{-3}.

For $125 < M_H < 180$ GeV, the channel $H \to W^{\pm} + W^{\mp(*)}$ is used, where (*) means that one of the W may be virtual. The final state in which the background is a minimum is when both W are decaying leptonically, $l^+ \nu_l l^- \bar{\nu}_l$, with l being an electron or a muon, which is identified in the detector.

For $125 < M_H < 300$ GeV, the channel $H \to Z^0 + Z^{0(*)}$ is available. It is very clean when both Z decay in a charged lepton pair, electrons or muons, because the four-lepton invariant mass can be reconstructed, to search for a H boson peak. Both lepton pairs, when above threshold, and one pair if below threshold, must have the mass of the Z.

For $300 < M_H < 600$ GeV the channel $H \to Z^0 + Z^0$, with both Z real, is fully available. However, the H production cross-section is now rather small and all the decay channels of the Z must be used to increase sensitivity.

The search by ATLAS and CMS required all the design characteristics of the detectors to perform perfectly. Without entering in any detail, we go directly to the result. In 2011, data were found to be consistent with the background-only hypothesis for 127 GeV $< M_H < 600$ GeV. Both experiments observed an excess of events above the expected Standard Model background at the low end of the explored mass range. The statistical significance of the effect was not enough to establish the existence of a new particle.

The situation is shown in Fig. 9.52.

In 2012 LHC worked at higher energy, $\sqrt{s} = 8$ TeV, and much higher luminosity, with typical initial values of $\mathcal{L} = 7 \times 10^{33}$ cm^{-2} s^{-1}. A total integrated luminosity of 23 fb^{-1} was collected by ATLAS and CMS. The search for the H boson was performed along the following lines: (a) establish whether a boson exists in the not yet excluded mass range and, if found, (b) measure its quantum numbers and couplings to establish if the boson is the Higgs or something else.

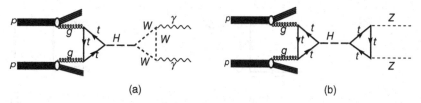

Fig. 9.53. Principal diagrams for the search of the H boson at low mass.

The first step was done in the channels in which the final state can be completely reconstructed, namely without neutrinos, in which the H should appear as a narrow peak in an invariant mass distribution, namely

$$H \to 2\gamma, \tag{9.125}$$

and

$$H \to Z + Z^* \to \left(l^+l^-\right) + \left(l^+l^-\right), \tag{9.126}$$

where the lepton pair can be e^+e^- or $\mu^+\mu^-$ and where one or both Z are off mass shell. Two important conditions for a successful search are a good energy resolution and a not too high background level. From the experimental point of view, the two channels are complementary, because in the di-gamma channel the energy resolution is very high but the background is large, while in the two-Z channel the background is low, but the energy resolution is a bit lower.

If found, the couplings to the γ and to the Z are also determined from the strength of the signal. Those to the W are determined in the channel $H \to WW^*$, in which a peak cannot be found due to the presence of undetected neutrinos. Measuring the Higgs coupling to fermions is more demanding. Amongst the quark–antiquark channels, the decay $H \to b\bar{b}$ is expected to have the largest branching ratio (see Fig. 9.41) due to the b having the largest mass – except top, a decay channel that opens only at twice its mass, about 345 MeV – but, as already said, the background of QCD processes is overwhelming. In the fermion channels the largest branching is for $H \to \tau\tau$, τ being the heaviest lepton.

Let us start from the first step. In both cases the dominant production mechanism is, as already mentioned, the gluon–gluon fusion. Fig. 9.53 shows the relevant first-order diagrams: (a) for $gg \to H \to \gamma\gamma$ and (b) for $gg \to H \to ZZ^* \to (l^+l^-)(l^+l^-)$. The total cross-sections for Standard Model Higgs boson production at the LHC with $M_H = 125$ GeV are predicted to be 17.5 pb at $\sqrt{s} = 7$ TeV and 22.3 pb at $\sqrt{s} = 8$ TeV.

In the $H \to \gamma\gamma$ analysis, a search is made for a narrow peak in the di-photon mass distribution. The cross-sections times branching ratio predicted by the Standard Model for $M_H = 125$ GeV are 39 fb at $\sqrt{s} = 7$ TeV and 50 fb at $\sqrt{s} = 8$ TeV. The peak should appear on a large irreducible background from QCD production of two photons, of which the principal diagram is shown in Fig. 9.54. Other backgrounds are due to when one or two reconstructed photons are fake, originating from misidentified jet fragments. This background is reduced as much as

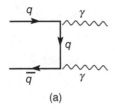
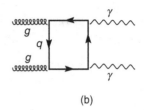

(a)   (b)

Fig. 9.54.   Processes producing gamma pairs.

possible, exploiting all the features of the detectors, applying suitable selection criteria to the events samples. Both background types are accurately calculated by using QCD and Monte Carlo simulations to go from the parton level to the hadron level.

In the presence of a large pile-up, as is the case when working at high luminosity, the selection of the right primary vertex, amongst all the reconstructed ones, is not trivial. It is performed by using the kinematic properties of the tracks associated with the vertex and their correlation with the di-photon kinematics. In addition, ATLAS employs the information provided by its longitudinally segmented calorimeter. In such a way, the direction of flight of the photon is measured by the calorimeter. This 'calorimeter pointing' feature proved to be a powerful tool resulting in a resolution of the vertex coordinate in the beams direction of about 15 mm, improving to about 6 mm when both photons are reconstructed. The di-photon mass resolution is dominated by the calorimetric measurements of the energies, as we saw in Eq. (9.119). It has similar values for both experiments with a sigma ranging between 1 and 2 GeV in the region of interest.

Fig. 9.55 shows a di-gamma event from ATLAS. The picture, an 'event display', is divided into five sectors (in addition to the sector with the logo of the experiment). Two of them are different views of the complete event, both tracking and calorimeters, one perpendicular and one parallel to the beams. The other three give the information from the EM calorimeter, showing two longitudinal views of the showers and one (the lego plot) transversal view.

Consider now the $H \to ZZ^* \to 4l$ channel, in which at least one of the two Z is off the mass shell. A search is made for a narrow peak in the four-lepton (invariant) mass distribution above a small continuous background. The analysis starts by selecting two pairs of isolated leptons at high transverse momentum. The leptons of a pair have the same flavour, electron or muon, and opposite charge. The expected cross-sections times branching ratio for the process with $M_H = 125$ GeV are 2.2 fb for $\sqrt{s} = 7$ TeV and 2.8 fb for $\sqrt{s} = 8$ TeV. The di-lepton with mass closer to the on-shell Z mass is denoted as 1 and its mass is denoted with m_1. The di-lepton is retained if $40 < m_1 < 120$ GeV. The second di-lepton, 2, is requested to have a mass $12 < m_2 < 120$ GeV.

Fig. 9.55. A di-gamma event from ATLAS (© CERN and ATLAS).

Fig. 9.56 shows an event from CMS with two $\mu^+\mu^-$ pair. One can see the large amount of hadronic tracks in the centre of the detector, having relatively low transverse momenta. The reconstructed muon tracks are the long penetrating ones.

A third channel was explored, namely $H \to WW^* \to l\nu l\nu$, in which, however, the reconstruction of the invariant mass does not show a narrow peak but a broad maximum due to the presence of the undetected neutrinos.

On 4 July 2012, the ATLAS and CMS experiments announced that they had independently observed a new particle in the mass range between 125 and 127 GeV. The statistical significance was above the 5 standard deviations threshold for both experiments (ATLAS > 5.9 σ, CMS > 5.0 σ). The observation of the new particle decaying

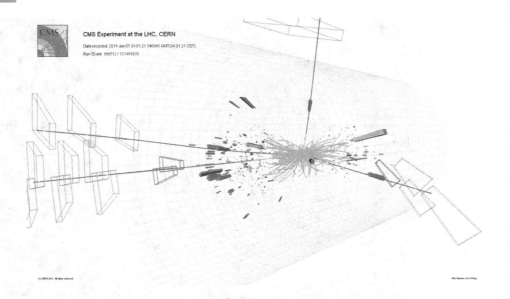

Fig. 9.56. An event from CMS with $2\mu^+$ and $2\mu^-$ (© CERN and CMS).

into two pairs of vector bosons whose net charge is zero identified it as a neutral boson. The observation of its decay into two photons excluded the spin one hypothesis $\left(\text{both } J^P = 1^- \text{ and } J^P = 1^+\right)$, as we discussed in the case of the π^0 (Section 3.5). The strengths of the signals in the different channels agree, within the uncertainties, with the predictions of the SM for the Higgs boson. Prudently, however, the experiments declared to have discovered a new boson compatible with the Higgs boson, more data being necessary to assess its nature in detail. The crucial issues remaining were whether the new particle had the unique properties of the Higgs boson to be a scalar particle and the dependence of its couplings on the mass of the particles. By the end of Run-1, in 2013, both experiments had found the $J^P = 0^+$ hypothesis consistent with the observations, while the other spin-parity assumptions up to spin two had been found to be unfavoured.

9.16 The Higgs Boson

In the subsequent data-taking period, Run-2, ATLAS and CMS collected (at $\sqrt{s} = 13$ TeV) an integrated luminosity of 150 fb^{-1}, more than five times the previous one. Approximately 8 million Higgs bosons were produced in this run. Improved methods of analysis were developed by both experiments. Many more final states could be studied, making it possible to separate the events by production mode and by decay channel, on the basis of their kinematic properties.

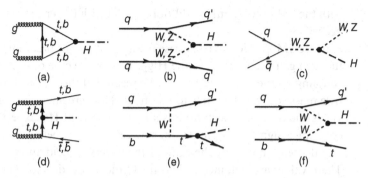

Fig. 9.57. Feynman diagrams for the leading Higgs boson production processes. (a) Gluon–gluon fusion (*ggH*), (b) vector boson fusion (*VBF*), (c) associated production with a *W* or *Z* boson (*VH*), (d) associated production with a top or bottom quark pair (*ttH* or *bbH*), (e) and (f) associated production with a single top quark (*tH*). The dots at the *H* boson vertices highlight the corresponding coupling.

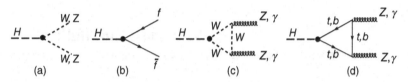

Fig. 9.58. Feynman diagrams for the leading Higgs boson decay processes. (a) Heavy vector boson pairs, (b) fermion–antifermion pairs, (c) and (d) photon or *Z* pairs. The dots at the *H* boson vertices highlight the corresponding coupling.

Fig. 9.57 shows the Feynman diagrams for the leading production processes. As already mentioned, the dominant one is the gluon–gluon proceeding via a virtual top loop (Fig. 9.57(a)). The next most important production mode is vector boson fusion (VBF), where a quark from each of the protons radiates a virtual vector boson (*W* or *Z*), which then fuse together to make an *H* boson (Fig. 9.57(b)). Next come, with decreasing importance, the *H* production in association with a vector boson pair, *W* or *Z* (*VH*) (Fig. 9.57(c)), or a heavy quark pair, bottom or top (*ttH* or *bbH*) (Fig. 9.57(d)) and, finally, in association with a single top (*tH*) (Fig. 9.57(e, f)). Notice that the measured quantity is the cross-section, namely the absolute square of the sum of the different diagrams. However, experiments have succeeded in separating the different production channels and in measuring their intensity, checking in this way, as we shall see, the different coupling strength of the *H* boson, graphically represented with dots in Fig. 9.57.

Fig. 9.58 shows the leading Feynman diagrams, which have been experimentally explored so far, for the leading *H* boson decay processes: the tree-level decays in two vector bosons, *W* or *Z* (Fig. 9.58(a)) or in a fermion–antifermion pair (Fig. 9.58(b)) and the one loop level diagrams for the decays in a pair of neutral vector bosons, γ or *Z* via virtual *W* (Fig. 9.58(c)) or heavy quark loop (Fig. 9.58(d)). The measurement of the strength of each decay, graphically highlighted with a dot, is compared with the SM prediction, in addition to, and with better precision than, the production processes.

In the SM, particle masses arise from spontaneous breaking of the gauge symmetry, through gauge couplings to the Higgs field in the case of vector bosons, and Yukawa couplings in the case of fermions. We recall that the former is proportional to the particle mass square, the latter to the mass. The agreement between the observed signal yields and the SM expectations is quantified in the LHC experiments defining for each coupling a signal-strength parameter, generically labelled μ, which is the normalized signal strength, namely the ratio of the observed yield and that predicted by the SM. The μ values are obtained as best fits to the model developed to implement the SM in the experimental conditions.

In the next sections we shall start with the measurements of the H boson mass and width. Next we shall consider the experimental verification of the two fundamental predictions of the BEH mechanism: the Higgs boson is a scalar particle, and its couplings to the mass.

9.17 Mass and Width

The H boson mass is measured in both $H \rightarrow \gamma\gamma$ and $H \rightarrow ZZ^* \rightarrow 2l^+ 2l^-$ channels, in which the final state invariant mass can be completely reconstructed.

The channel $H \rightarrow \gamma\gamma$ is very clean, but its search must pay the toll of a very small branching ratio, about 0.2% in the lower mass range of Fig. 9.52. Consequently, the boson is expected to appear in the $\gamma\gamma$ invariant mass distribution as a small peak over a smoothly decreasing background due to other sources of prompt photons and decays of strongly produced hadrons. The latter, which is dominant, can be reduced selecting high transverse energy photons and imposing on them isolation criteria (photons from QCD processes come as parts of hadronic jets). The procedure is called the selection of a 'fiducial region'. Much effort had been dedicated by both experiments to the development of electromagnetic calorimeters with both high energy resolution and granularity, in order to have an isolated $H \rightarrow \gamma\gamma$ peak as narrow as possible.

Fig. 9.59 shows the result of ATLAS at $\sqrt{s} = 13$ TeV with an integrated luminosity of 139 fb^{-1} (ATLAS Collaboration 2019). The one of CMS is completely similar.

The upper panel shows the data points of the di-photon mass distribution together with the fit (the curve) used to extract the size of the signal. One sees that the signal-to-background ratio is really small, but clear. The background shape at both sides of the peak is smooth; no other structure is visible. The data in these only-background regions are used to check if the model developed to simulate it accurately reproduces the data. The experiment then extracts values of the mass (see next section) and of the signal strength, shown in the lower panel after background subtraction. This results in a measurement of the

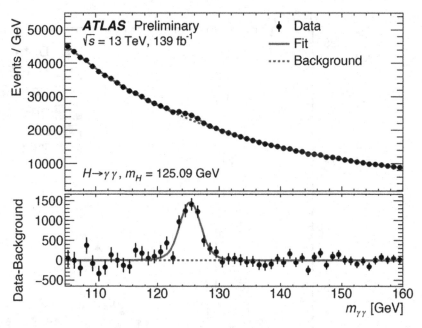

Fig. 9.59. ATLAS. Di-photon mass distribution, with the curve used in the fit to extract the signal (upper panel), and after background subtraction (lower panel). © CERN and ATLAS https://cds.cern.ch/images/ATLAS-PHOTO-2019-027-3.

production cross-section, in the fiducial region, times branching ratio in $\gamma\gamma$ of $\sigma_f = 65.2 \pm 4.5 (\text{stat.}) \pm 5.6 (\text{syst.}) \pm 0.3 (\text{theo.})$ fb, where the last uncertainty is due to the theoretical input. This is in agreement with the SM prediction $\sigma = 63.6 \pm 3.3$ fb. The same can be expressed in terms of the above defined normalized signal strength, in this case $\mu = (\sigma \cdot \text{BR})_{\text{obs}} / (\sigma \cdot \text{BR})_{\text{SM}}$. For the $H \to \gamma\gamma$ channel, the CMS Run-2 value is $\mu = 1.12 \pm 0.09$ and the ATLAS one $\mu = 1.02 \pm 0.14$.

The channel $H \to ZZ^* \to 2l^+2l^-$, where l is a charged lepton, e or μ has a small branching ratio too, about 2.7% $H \to ZZ^*$ times 6.7% for the decay of the Z into e^+e^- or $\mu^+\mu^-$ namely 0.18%, but a large signal-to-background ratio. The final states can be fully reconstructed with high resolution due, for electrons, to the already mentioned electromagnetic calorimeters, and, for muons, to the excellent momentum resolution of both experiments. This is the dominant process to measure the H boson mass.

Fig. 9.60 shows the CMS four leptons invariant mass distribution at $\sqrt{s} = 13$ TeV with an integrated luminosity of 137 fb^{-1} (CMS Collaboration 2021a). The one of ATLAS is completely similar. Points with error bars represent the data. Two peaks are visible, one for the Z, one for the H. The former is due to the rare decay of the Z in four leptons as in Fig. 9.61(a). Stacked histograms represent the calculated backgrounds. The main backgrounds to the H boson signal, which are irreducible with

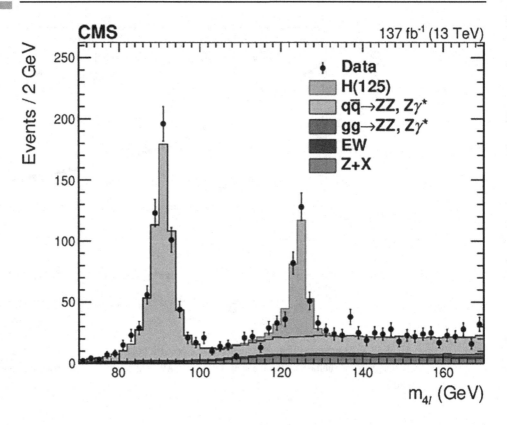

Fig. 9.60.   CMS. Four-lepton mass distribution, m_{4l}, up to 170 GeV. Points with error bars represent the data and stacked histograms represent the expected distributions for the signal and background processes. The SM Higgs boson signal with $m_H = 125$ GeV, denoted as H(125), the ZZ and rare electroweak backgrounds are normalized to the SM expectation, the $ZZ + X$ background to the estimation from data. © CMS Collaboration. A. M. Sirunyan *et al. Eur. Phys. J. C* 81 488 (2021), Fig. 4 third panel. Under CC BY 4.0 https://link.springer.com/content/pdf/10.1140/epjc/s10052-021-09200-x.

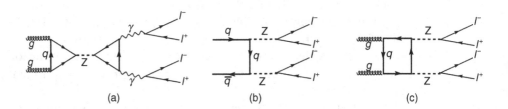

Fig. 9.61.   Examples of Feynman diagrams for (a) Z decay in four leptons, (b) $q\bar{q} \rightarrow ZZ$ and (c) $gg \rightarrow ZZ$.

kinematic cuts, come from the production of ZZ via $q\bar{q}$ annihilation, which happen already at tree level as in Fig. 9.61(b), and via gluon–gluon fusion at loop level, of which an example is shown in Fig. 9.61(c). Small electroweak contributions are also considered. Additional backgrounds are due to processes such as the production of a Z, which decays into two leptons, together with a hadronic jet experimentally misidentified as electrons. These backgrounds are reduced with suitable kinematic cuts. CMS then extracts a value of the normalized signal strength of $\mu = 0.94^{+0.12}_{-0.11}$; for ATLAS, $\mu = 1.01 \pm 0.11$.

The mass of the H boson is measured in these channels in which the invariant mass of the final states can be completely reconstructed. The results of the two experiments agree with one another: CMS finds $M_H = 125.38 \pm 0.14$ GeV (CMS collaboration 2020), ATLAS finds $M_H = 124.99 \pm 0.17 (\text{stat}) \pm 0.03 (\text{syst})$ GeV (ATLAS Collaboration 2022).

Coming now to the H boson width, we observe that, in the SM, its lifetime is $\tau_H \approx 1.6 \times 10^{-22}$ s, which is by far too short to be directly measurable. Measuring the corresponding width, $\Gamma_H = 4.14 \pm 0.02$ MeV, would require a relative energy resolution better than 3×10^{-5} and is also impossible. This value is much smaller than the experimental resolution and cannot be determined by the observed width of the peak, which is only instrumental. There is an indirect way, however, which has been used by CMS consisting of a comparison of the Higgs boson production both on-shell and off-shell. Let us see the principles.

Consider the Higgs boson production via the dominant processes, ggH and VBF shown in Fig. 9.57(a, b), followed by the decay in ZZ. In the analysis just considered, H is on mass shell and at least one of the Z is off-shell, because $M_H < 2M_Z$. We can also consider the case in which both the Z are on-shell and the H is off-shell, the more so the larger the $2Z$ invariant mass is.

We start from the production on-shell and use the 'narrow width approximation'. This is possible because Γ_H is so small that all the relevant quantities do not practically vary across the width. In practice we approximate the resonance curve with a Dirac delta function. The peak of the resonance curve is due to the denominator that is very small at resonance, approaching zero for decreasing width as we saw in Eq. (4.2). This factor, written in terms of the square centre of mass energy of the process as the parton level $\hat{s}$, is $1 / \left[\left(\hat{s} - M_H^2 \right)^2 + M_H^2 \Gamma_H^2 \right]$. We note that the total yield is proportional to

$$\int_{-\infty}^{+\infty} \frac{d\hat{s}}{\left(\hat{s} - M_H^2 \right)^2 + M_H^2 \Gamma_H^2} = \int_{-\infty}^{+\infty} \frac{d\hat{s}}{\hat{s}^2 + M_H^2 \Gamma_H^2}$$

$$= \frac{1}{M_H \Gamma_H} \int_{-\infty}^{+\infty} \frac{dx}{1 + x^2} = \frac{\pi}{M_H \Gamma_H}, \tag{9.127}$$

where we first neglected $M_H^2 \Gamma_H^2$ and then substituted $x = \hat{s} / (M_H \Gamma_H)$. Notice the inverse proportionality to the total width. This is easy to understand thinking that if we squeeze the resonance curve, keeping its area constant, its height increases in inverse proportionality to the width. The result shows that if we want to substitute the resonance curve with a delta function, the substitution is

$$\frac{1}{\left(\hat{s} - M_H^2 \right)^2 + M_H^2 \Gamma_H^2} \rightarrow \frac{\pi \delta \left(\hat{s} - M_H^2 \right)}{M_H \Gamma_H}. \tag{9.128}$$

We have now to multiply by the square couplings both in production and in the decay. For example if the production is via gluon–gluon fusion and the decay is into two Z, namely if the process is $gg \rightarrow H \rightarrow ZZ$, we have the couplings ggH and ZZH. Generically, we call the production and decay square coupling g_p^2 and g_d^2 respectively. The couplings of the principal channels have been extracted from the data, as we shall see soon. The cross-section on-shell is then

$$\sigma^{\text{on-shell}} \propto \frac{g_p^2 g_d^2}{\Gamma_H}. \tag{9.129}$$

On the other hand, the cross-section of the off-shell process does not resonate; rather it varies smoothly as a function of the ZZ invariant mass. But it is still proportional to the square couplings

$$\sigma^{\text{off-shell}} \propto g_p^2 g_d^2. \tag{9.130}$$

Comparing the two expressions, we see that their ratio, once properly normalized, is the searched-for total width.

In practice the main difficulty is that the Z pairs produced by other processes, like in Fig. 9.61, are much more frequent than those from off-shell Higgs decay, but CMS succeeded in extracting the signal, finding $\Gamma_H = 3.2^{+2.4}_{-1.7}$ MeV (CMS Collaboration 2022a).

9.18 Spin and Parity

We come now to the spin-parity. As already remarked, a crucial characteristic that the new boson must have to be identified with the H is to be a scalar particle. Both ATLAS and CMS in Run-2 were able to prove that the data are in agreement with $J^P = 0^+$, and exclude all the other assignments up to 2. We present here a simplified discussion. As already stated, the observation of its decays in $\gamma\gamma$ excludes the spin parities 1^- and 1^+. In addition, the decay in ZZ^* does not appear to be suppressed as it should be if the matrix element would contain a factor proportional to a power of the CM momentum q larger than 1, in view of this momentum being very small. These higher powers (orbital momentum $L > 1$) are present in several of the $J = 2$ matrix elements and we will ignore them all. We are left with two possibilities, $J^P = 0^+$ and $J^P = 0^-$. The situation is similar to what we did for the π^0 in Section 3.5.

Consider now the decay of a boson X, which we do not necessarily assume to be the Higgs, of spin parity J^P

$$X \rightarrow Z + Z^* \rightarrow \left(l^+ l^-\right) + \left(l^+ l^-\right). \tag{9.131}$$

The two Z have different masses, m_1 and m_2, because at least one of them is off mass shell. Take Z_1 as the one of larger mass, and Z_2 the other one. To make the issue as simple as possible, we work in the CM frame of the X boson and follow the same type of argument as in Sections 3.5. We call $\mathbf{q}$ the momentum of Z_1 (the momentum of Z_2 is obviously $-\mathbf{q}$), $\mathbf{e}_1$ and $\mathbf{e}_2$ their polarization vectors and E_1 and E_2 their energies, with $E_1 \leq E_2$.

We call L the orbital momentum and S the total spin of the two Z. Possible values of L and S are those that can sum to J, respecting the selection rules. Possible values of S are $S = 0, 1, 2$. The spin component on the line of flight of a massive vector boson can be $+1$, -1 like the photon, but also 0, unlike the photon. Consequently, a spin 1 particle may decay into two massive vectors. In other words, Eq. (3.23), namely $\mathbf{e}_1 \cdot \mathbf{q} = 0$ and $\mathbf{e}_2 \cdot \mathbf{q} = 0$, are not requested to hold.

In the case $J = 0$ that we are considering, we have $L = S$. If the parity of the final state is positive, L must be even, and we have two possibilities, $L = S = 0$ and $L = S = 2$. Bose statistics is satisfied because, in both cases, the spin and space wave functions are symmetrical under the exchange of the two Z, and hence the total wave functions are also symmetrical. We add together the two terms to give the most general expression:

$$M_S = a_1 \mathbf{e}_1 \cdot \mathbf{e}_2 + a_2 (\mathbf{e}_1 \cdot \mathbf{q})(\mathbf{e}_2 \cdot \mathbf{q}). \tag{9.132}$$

We see that in the first term the two polarization vectors combine to make a scalar, for $S = 0$, and $\mathbf{q}$ does not appear, corresponding to $L = 0$. Similarly, in the second term $\mathbf{q}$ appears twice, corresponding to $L = 2$ and the polarization vectors combine, corresponding to $S = 2$. As already observed, however, because the CM momentum is necessarily small in this case, the second term in (9.132) where it appears at the second power is kinematically strongly suppressed, and we can write in good approximation

$$M_S \approx a_1 \mathbf{e}_1 \cdot \mathbf{e}_2. \tag{9.133}$$

For the pseudoscalar final state, there is only one possibility, $L = S = 1$. We write

$$M_P = a_3 \mathbf{e}_1 \times \mathbf{e}_2 \cdot \mathbf{q}. \tag{9.134}$$

We see that the polarizations combine to make an axial vector, corresponding to $S = 1$, and $\mathbf{q}$ appears once, corresponding to $L = 1$. Both the spin and space wave functions are antisymmetric, making the total one symmetric, respecting Bose statistics.

If the decay conserves parity, only M_S is present in the decay of a scalar and only M_P is present in the decay of a pseudoscalar; both are present if parity is violated.

Even if the polarization vector of a Z is not directly measured, its direction is correlated with that of the normal to the Z decay plane, which is measured, as was the case for the virtual photons from the π^0 decay. The form factors a_i in the above expressions are functions of m_1 and m_2.

The Standard Model predicts completely the form factors for the Higgs. Parity is conserved, hence $a_3 = 0$. Also, in addition to the kinematical suppression, $a_2 = 0$ in the first order, with additional contributions from radiative corrections of the order of α.

The distribution of the angle ϕ between the planes of the two lepton pairs in the two Z decays when the four lepton invariant mass is in the peak of the Higgs, namely between 118 and 130 GeV, from the CMS experiment is shown in Fig. 9.62. The data are compared with the 0^+ and 0^- hypotheses above the calculated back-

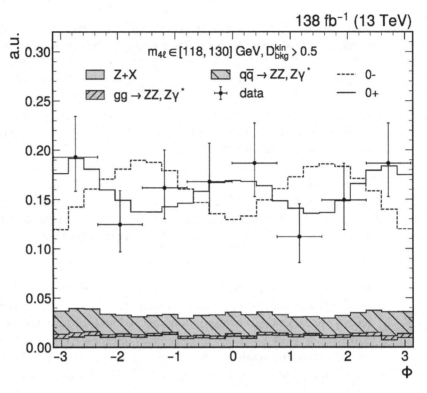

Fig. 9.62. Distribution of the angle (in rad) between the dilepton pairs of the two *Z* from the *H* boson from the CMS experiment. The data are compared with the scalar and pseudoscalar hypotheses. The calculated backgrounds, same as in Fig. 9.60, are shown. Kindly by A. Tarabini and CMS, https://cms-results.web.cern.ch/cms-results/public-results/publications/HIG-21-009/index.html.

grounds, which are those discussed in relation to Fig. 9.60. Clearly, 0^+ is in agreement with data, 0^- is not. Comparison with Fig. 3.1 shows how now the data are offset in phase of $\pi/2$.

We can also observe that, if the decay does not conserve parity, both M_S and M_P should be present. Data do not show any evidence for that. Within their accuracy, the particle decay conserves parity.

9.19 The Couplings

Once the mass of the Higgs boson is known, the Standard Model predicts precisely all of its coupling strengths. The Higgs couplings to bosons and to fermions correspond to different parts of the Lagrangian, as we discussed in Section 9.12, and both must be measured to check the theory.

Table 9.2. The largest branching ratios of the Higgs boson with $m_H = 125$ GeV as predicted by the SM from Workman *et al.* 2022. The relative uncertainties are 1–2%.

Decay mode	Branching ratio	Decay mode	Branching ratio
$H \to WW^*$	21.4%	$H \to b\bar{b}$	58.2%
$H \to gg$	8.6%	$H \to \tau^+\tau^-$	6.27%
$H \to ZZ^*$	2.62%	$H \to c\bar{c}$	2.89%
$H \to \gamma\gamma$	0.227%	$H \to \mu^+\mu^-$	0.0218%

Tests have been done both in production and in decay of the Higgs boson. For the former, CMS considered six production processes, gHH, VBF, VH, ZH, ttH and tH. All the coupling strengths were found to be compatible with 1 within uncertainties ranging from 8% to 25% (CMS Collaboration 2022b). For the decays, Table 9.2 gives the SM expectations for the branching ratios of a Higgs boson mass of 125 GeV. Let us see how they are measured.

We have already shown that ATLAS and CMS have measured the signal strength for $H \to \gamma\gamma$ and $H \to ZZ^* \to 2l^+2l^-$, finding them in agreement with the predictions of the SM within an accuracy of about 10%. We briefly consider the other channels measured up to now.

$H \to WW^* \to 2l2\nu$. The channel allows a further check of the Higgs coupling to the vector bosons. The branching ratio shown in Table 9.2 is sizeable, but the events cannot be completely reconstructed owing to the presence of two neutrinos. The boson can be observed as a wide bump in the mass distribution. ATLAS observed a signal with 6.1 σ significance, and CMS observed one with 4.3 σ significance. The corresponding signal strengths are $\mu = 0.76 \pm 0.21$ and $\mu = 1.5 \pm 0.6$, respectively.

$H \to \tau^+\tau^-$. The decay is searched in τ decays into electrons, muons and hadrons. The $\tau^+\tau^-$ invariant mass can be reconstructed with a kinetic fit from the measured momenta of the charged decay particles and the missing momentum due to the neutrinos, resulting in a poor (about 15%) resolution. A broad bump over the expected background is searched for. Both experiments have detected the signal. Its normalized strength (in Run-2) is measured by ATLAS in $\mu = 0.92 \pm 0.13$ and by CMS in $\mu = 0.85^{+0.12}_{-0.11}$.

$H \to b\bar{b}$. The expected branching ratio of this channel is large, (see Table 9.2), but the background is enormous. The high-resolution vertex detectors are able to identify and tag, in a large fraction of the cases, the beauty particles, but this is not enough to reduce sufficiently the background due to the production of beauty pairs via strong interactions. A more manageable signal-to-background ratio is obtained by searching for the H boson produced in association with a vector boson, W or Z, when the latter decays into a lepton pair. In Fig. 9.63 we show the relevant diagrams that we already

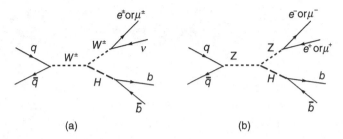

(a)   (b)

Fig. 9.63. The diagrams of H boson production by quark–antiquark annihilation in association with (a) a W boson and (b) a Z boson, as in Fig. 9.57(c), followed by the relevant decays.

considered in Fig. 9.43(b). The corresponding topologies are two b-tagged jets and one charged lepton and a neutrino or two charged opposite-sign leptons. The signal has been observed, the signal strength measured by ATLAS is $\mu = 1.02^{+0.12+0.14}_{-0.11-0.13}$ and by CMS is $\mu = 1.01 \pm 0.22$.

$H \to \mu^+\mu^-$. The channel is very clean because the $\mu^+\mu^-$ invariant mass can be reconstructed with very good resolution (ATLAS 1–3%, CMS 2–3%) due to the precise measurements of the muon momenta, but the branching ratio is extremely small (see Table 9.2), an order of magnitude smaller than for $H \to \gamma\gamma$. The dominant irreducible background arises from $Z \to \mu^+\mu^-$ and $\gamma^* \to \mu^+\mu^-$ that have rates several orders of magnitude larger than the expected signal. Both ATLAS and CMS, however, have already observed a small signal with significance of 2σ and 3σ, respectively. The corresponding signal strengths are $\mu = 1.2 \pm 0.6$ and are $\mu = 1.19 \pm 0.40\,(\text{stat}) \pm 0.15(\text{syst})$. In both cases the dominant uncertainty is statistics. Consequently, substantial improvement is expected from Run-3.

We can now turn to the test of the H to the vector boson couplings to vector bosons and to fermions; the couplings are expected to be proportional to the square of the mass m_V for vector bosons and to the mass m_F for the fermions. To take into account the difference, the CMS collaboration (CMS Collaboration 2021b), defines the 'reduced coupling strength modifiers' as $y_V = \sqrt{\kappa_V}\, m_V / \upsilon$ for the vector bosons and $y_F = \kappa_F m_F / \upsilon$ for the fermions, where υ is the VEV of the Higgs field ($\upsilon = 246.22$ GeV) and the constants κ are the Higgs coupling strengths to different particles. These constants are extracted with a fit to the measured quantities (with the Higgs boson mass fixed at $M_H = 125.38$ GeV).

Fig. 9.64 shows the reduced coupling strength modifiers as a function of the mass of the probed particles. The remarkable agreement with the predictions of the BEH mechanism over three orders of magnitude of mass is a powerful test of the validity of the theory. Statistical and systematic uncertainties contribute at the same level to all measurements, except for $\mu^+\mu^-$, which is dominated by the statistical uncertainty (CMS Collaboration 2022b).

The overall agreement of the measurements with the expectations of the SM can be estimated by the overall signal-strength parameter, which CMS has measured to

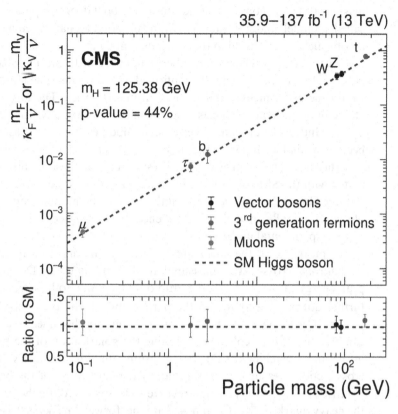

Fig. 9.64. The best-fit estimates for the reduced coupling modifiers extracted for fermions and weak bosons, compared with their corresponding prediction from the SM. The error bars represent 68% confidence level intervals for the measured parameters. In the lower panel, the ratios of the measured coupling modifiers' values to their SM predictions are shown (www.hepdata.net/record/99375).

be $\mu = 1.002 \pm 0.057$. In conclusion, we observe that, in several cases, the agreement between experiment and theory is within some 10% uncertainty.

9.20 The Global Fit

With the discovery of the Higgs boson, the existence of all the particle/field constituents of the Standard Model, spin-1/2 fermions in three families of two quarks and two leptons, spin-1 gauge fields (one for the electromagnetic, three for the weak, eight for the strong interactions) and the scalar boson that 'gives mass' to the other fundamental particles, have been experimentally established. The SM constraints the properties of its fields/particles and techniques have been developed to calculate the observables with precision comparable to the experimental one, in terms of the couplings and the masses of the fermions, quantities that the model does not predict (which is why it is called a 'model' and not 'theory'). Many observables have been measured, the majority of which we discussed in the previous chapters.

We can ask, in conclusion, whether or not the constraints imposed by the SM on the observables are experimentally confirmed (those we discussed and those we did not) within the experimental and theoretical uncertainties. The answer comes from the global fit of the available measurements of the relevant physical quantities performed by the Gfitter Group, which periodically updates the results as new measurements become available. Input to the fit are 21 measured quantities with their values and uncertainties. The fit provides the 'best values' of these quantities, those that best match with the SM constraints. Fig. 9.65 summarizes the results of the latest edition (Gfitter Group 2018). The horizontal bars and the numbers on the right side of the plot are the 'pulls', namely the differences between fitted and input quantity divided by the experimental uncertainty.

The meanings of the labels are as follows. The first five are self-evident. σ_{had}^0 is the hadronic cross-section measured at LEP and by the SLD detector at SLC at the peak energy of the Z. R_{lep}^0 is the ratio of the leptonic to hadronic cross-sections and, further down, R_c^0 and R_b^0 are the partial rates in the two heavy quarks, all at the peak energy of the Z. We have mentioned, but not discussed, the measurements that can be done at e^+e^- colliders exploiting the polarization of the beams, in particular in the linear machine (SLD). The measured quantity is the asymmetry between the cross-section measured with right and left polarization of the beam. These are the $A_l(\mathrm{LEP})$ and $A_l(\mathrm{SLD})$ entries, relative to leptons, and, further down, A_c and A_c for the heavy quarks. $A_{\mathrm{FB}}^{0,l}$, $A_{\mathrm{FB}}^{0,c}$ and $A_{\mathrm{FB}}^{0,b}$ are the 'forward-backward asymmetries' for leptons, charm and beauty, respectively, at the Z peak. The asymmetry is between the angular cross-section integrated on the forward hemisphere to that in the backwards one. Entries 11 and 12 are on the weak mixing angle. As we saw in Section 9.6, this quantity, being the ratio of coupling constants, depends on energy through radiative corrections, namely renormalization. The definition of $\sin^2\theta_W$ we gave in Section 9.3 and the corresponding relation with the Z and W masses 'promotes' this relation, obtained at tree level, to all orders. This is perfectly legitimate, but another definition exists more directly linked to some experimental data that we have not discussed. It is called 'effective' mixing angle and is defined for each fermion in terms of its charge, and the ratio of its vector and axial couplings $\sin^2_{eff,f} = \dfrac{1}{4|Q_f|}\left(1 - \dfrac{c_f^V}{c_f^A}\right)$. The two definitions are linked by the SM. This quantity was measured at LEP and at the Tevatron (in different channels). $\Delta\alpha_{\mathrm{had}}^{(5)}\left(M_Z^2\right)$ is the hadronic contribution to the evolution of the fine structure constant at the mass of the Z. Finally, the meanings of $\alpha_s\left(M_Z^2\right)$ and m_t are self-evident.

The plot summarizes the results of decades of experimental and theoretical efforts. The experimental results have been obtained in conditions as different as on neutrino beams, e^+e^-, ep, pp, $\bar{p}p$, heavy ion colliders, precision atomic physics experiments, etc. The theoretical efforts have used different techniques ranging from perturbative expansions to calculations on a lattice. The agreement of a single unified theory with

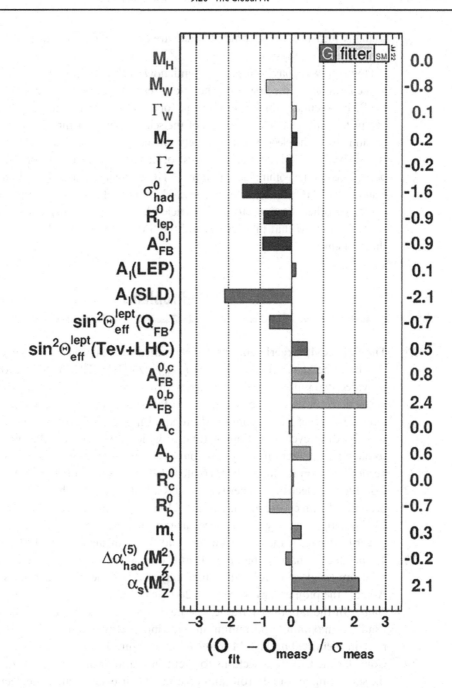

Fig. 9.65. Pull values of the fit, defined as deviations between measurements (O_{meas}) and predictions evaluated at the best-fit point (O_{fit}), divided by the experimental uncertainties. J. Haller *et al.* arXiv:2211.07665 From CERN documents service (https://cds.cern.ch/record/2847573/plots#2).

so many measurements, some of which are of very high precision, is certainly a major achievement of physics.

However, what has been explored until now is not at all complete. The Higgs couplings to the fermions of the first two generations remain almost untested. And the structure of the BEH potential, which is responsible for the electroweak symmetry breaking, remains completely untested. The strengths of the self-couplings of three and four Higgs bosons resulting from the electroweak symmetry breaking depends on a single constant, λ, which is given by the SM in terms of the Higgs boson mass and the Fermi constant. All of this will be explored in the coming decennium at LHC in the just started Run-3 and with the High Luminosity LHC scheduled to start in 2029, capable of collecting up to 4000 fb^{-1} per year.

On the other hand, sectors of particle physics, specifically with neutrinos, and in cosmology show that physics beyond the SM exists. We shall discuss some of this in the next chapters.

Problems

9.1   The CHARM2 experiment at CERN studied the reaction $\nu_\mu e^- \to \nu_\mu e^-$ using a 'narrow band beam' with mean energy $\langle E_{\nu\mu} \rangle = 24$ GeV. The Super Proton Synchrotron provided two pulses $\Delta t = 6$ ms apart, with a cycle of $T = 14.4$ s. The useful mass of the target detector was $M = 547$ t, the side of its square useful cross-section was $l = 3.2$ m long. The target nuclei contained an equal number of protons and neutrons. What was the duty cycle (fraction of time in which interactions take place)? If we want to have an interaction every four pulses on average, how large a neutrino flux Φ is needed? How much intensity I? (Note $\sigma / E_\nu = 1.7 \times 10^{-45} \text{m}^2 \text{ GeV}^{-1}$.)

9.2   Consider an electron of energy $E_e = 20$ GeV detected by the CHARM2 experiment, produced by an elastic $\nu_\mu e$ scattering. How large can the scattering angle be, at most? Evaluate the accuracy that is necessary in the measurement of the electron direction to verify this to be the case. Can we build the calorimeter using Fe?

9.3   To produce a narrow band ν_μ beam, one starts from an almost monochromatic π^+ beam (neglecting K^+ contamination) and lets the pions decay as $\pi^+ \to \mu^+ + \nu_\mu$. Assume the pion energy to be $E_\pi = 200$ GeV. Find the neutrino energy in the π^+ rest frame. In the laboratory frame, the neutrino energy depends on the decay angle θ. Find the maximum and minimum neutrino energy. Find the laboratory angle θ of neutrinos emitted in the CM frame at $\theta^* = 50$ mrad.

9.4   Consider the CC cross-sections for neutrinos and antineutrinos on nuclei containing the same number of neutrons and protons. Their masses can be neglected in comparison with their energies. Show that the neutrino total cross-section is three times larger than that of the antineutrino.

9.5   Give the values of the isospin, its third component and the hypercharge for $e_L, \nu_{\mu L}, u_L, d_R$, and for their antiparticles.

9.6 Give the values of the isospin, its third component and the hypercharge for $\mu_L^+, \tau_R^+, \bar{t}_L, \bar{b}_R$.

9.7 Establish which of the following processes (which might be virtual) are allowed and which are forbidden, and give the reasons: $W^- \to d_L + \bar{u}_L, W^- \to u_L + \bar{u}_L, Z \to W^- + W^+,$ $W^+ \to e_R^+ + \bar{v}_R$.

9.8 Establish which of the following processes (which might be virtual) are allowed and which are forbidden, and give the reasons: $d \to W^- + u_L, Z \to e_L^+ + e_L^-, W^+ \to Z + W^+,$ $W^+ \to e_L^- + \bar{v}_L$.

9.9 A 'grand unification' theory assumed the existence at very high energies of a symmetry larger than $SU(2) \otimes U(1)$, namely $SU(5)$. A prediction of the theory, which was experimentally proved to be false, was the value of the weak mixing angle, $\sin^2 \theta_W = 3/8$. Find the partial widths of the Z in this hypothesis. Find the value of Γ_μ / Γ_h.

9.10 Calculate the partial and total widths of the Z for $\sin^2 \theta_W = 1/4$. Find $\Gamma_{\mu\mu} / \Gamma_h$. (Note $M_Z = 91$ GeV.)

9.11 Evaluate the branching ratio for $W \to e^+ v_e$.

9.12 Evaluate the ratio g_{Zee}^2 / g_{Wev}^2 and the decay rates ratio $\Gamma(Z \to e^+ e^-)/\Gamma(W \to e^+ v_e)$.

9.13 Evaluate the ratio $g_{Zuu}^2 / g_{Wud}'^2$ and the decay rates ratio $\Gamma(Z \to u\bar{u})/\Gamma(W \to d'\bar{u})$.

9.14 Assume the cross-section value $\sigma(\bar{u}d \to e^+ v_e) = 10$ nb at the W resonance. Evaluate the total $\sigma(\bar{u}d \to \bar{q}q)$ cross-section at resonance.

9.15 Assume that the number of neutrinos with mass $<< M_Z$ is 3, 4 or 5 in turn, without changing anything else. Evaluate for each case the Z branching ratio into $\mu^+ \mu^-$ and the ratio $\Gamma_{\mu\mu} / \Gamma_Z$. Evaluate the ratio of the cross-sections at the peak for $e^+ e^-$ into hadrons for $N = 3, 4$ and 5.

9.16 Calculate the cross-section $\sigma(e^+ e^- \to \mu^+ \mu^-)$ at the Z peak and $\sigma(\bar{u}d \to e^+ v_e)$ at the W peak.

9.17 A Z is produced in a $\bar{p}p$ collider working at $\sqrt{s} = 540$ GeV. The Z moves in the direction of the beams with a momentum $p_Z = 140$ GeV. It decays as $Z \to e^+ e^-$ with electrons at 90° to the beams in the Z rest frame. Calculate the two electron energies in the laboratory frame.

9.18 Consider the Z production at a proton–antiproton collider and its decay channel $Z \to e^+ e^-$. The energies of the two electrons as measured by the electromagnetic calorimeters are $E_1 = 60$ GeV and $E_2 = 40$ GeV. The energy resolution is given by $\sigma(E)/E = 0.15/\sqrt{E}$. The measured angle between the tracks is $\theta = 140° \pm 1°$. Find the error on m_Z.

9.19 Consider the prediction for the W and Z before their discovery, in round numbers $M_Z = 90$ GeV and $M_W = 80$ GeV. If $\sin^2 \theta_W \approx 0.23$ with an uncertainty of 20%, what is the uncertainty on M_W? If we measure M_Z / M_W with 1% uncertainty, what is the uncertainty on $\sin^2 \theta_W$?

9.20 At 1 GeV of energy, the weak charge g is larger than the electric charge $\sqrt{a}$ by $\sqrt{4\pi} / \sin \theta_W \approx 7.4$. Why is the electrostatic force between two electrons at 1 fm distance so large compared with the weak force?

9.21 Consider the $\bar{p}p$ Tevatron collider working at $\sqrt{s} \approx 2$ TeV. For a Z produced at rest, what are the approximate momentum fractions of the annihilating quark and antiquark? Evaluate in which fraction they are sea quarks. If the Z is produced with a longitudinal momentum of 100 GeV, what are the approximate momentum fractions of the quark and of the antiquark?

9.22 LEP 2 was designed to study the process $e^+e^- \to W^+W^-$. If the cross-section at $\sqrt{s} = 200$ GeV is $\sigma = 17$ pb and the luminosity is $\mathcal{L} = 10^{32}$ cm^{-2} s^{-1}, find the number of events produced per day.

9.23 What is the percentage variation of the Z total width for an additional neutrino type? What is the variation of the peak hadronic cross-section?

9.24 Working at the Z with an electron–positron collider and assuming statistical uncertainty only, how many events are needed to exclude, at five standard deviations, the existence of a fourth neutrino?

9.25 If there were more than three families, the Z would decay into more neutrino–antineutrino channels. Given the existing limits on the masses, however, the charged leptons and quark channels of the new families would be closed. Consequently, the Z width would increase by Γ_{vv} for every extra family. The total width of the W would not increase, because the third family channel $W \to t + \bar{b}$ is already closed (as was established in 1990 when CDF gave the limit $m_t > 90$ GeV). UA1 and UA2 measured the ratio $R = \dfrac{\sigma_W BR(W \to ev_e)}{\sigma_Z BR(Z \to e^+e^-)}$ simply from the numbers of events observed in the two channels. Both theoretical and experimental systematic uncertainties cancel out in the ratio. Writing the ratio as $R = \dfrac{\Gamma(W \to lv_l)}{\Gamma(Z \to l\bar{l})} \dfrac{\Gamma_Z}{\Gamma_W} \dfrac{\sigma_W}{\sigma_Z}$, one sees that it increases with the number of neutrino types. The experimental upper limit established by joining the UA1 and UA2 data was $R < 10.1$ at 90% confidence level. Evaluate R for 3, 4 and 5 and establish an upper limit for the number of neutrinos. Take $\sigma_W / \sigma_Z = 3.1$.

9.26 Calculate the ratio between the CC cross-sections of neutrinos and antineutrinos on nuclei with the same numbers of neutrons and protons, considering only the valence quark contributions. Repeat the calculation for NC interactions.

9.27 The largest fraction of matter in our Galaxy is invisible. It might consist of particles similar to neutrinos but much more massive, the 'neutralinos'. Let us indicate them by χ and let m_χ be their mass. According to one theory, these particles coincide with their antiparticles. The annihilation processes $\chi + \chi \to \gamma + \gamma$ and $\chi + \chi \to Z^0 + \gamma$ would then take place. Assume initial kinetic energies to be negligible. If a gamma telescope observes a monochromatic signal $E_\gamma = 136$ GeV, find m_χ in both hypotheses.

9.28 (1) Consider a pair of quarks with colours R and B and a third quark G. Establish whether the force of the pair on G is attractive or repulsive for each of the combinations $RB + BR$ and $RB - BR$.

Considering quarks of the same colour, establish which of the following processes are allowed or forbidden, giving the reasons: (2) $W^+ \to \bar{b}_L + c_R$ and (3) $Z \to \tau_R^+ + \tau_R^-$.

9.29 (1) Establish for each of the following decays whether it is allowed or forbidden, giving the reasons. The left upper label is the colour, the right lower one the chirality.

(a) $W^- \to {}^B s_L + {}^B \bar{u}_R$

(b) $W^- \to {}^B d_R + {}^B \bar{u}_L$

(c) $W^- \to {}^R d_L + {}^B \bar{u}_R$

(d) $Z^0 \to {}^G u_L + {}^G \bar{u}_R$

(e) $Z^0 \to {}^G u_R + {}^G \bar{u}_R$

(f) $Z^0 \to {}^G u_R + {}^G \bar{c}_L$

(g) $Z^0 \to {}^G t_L + {}^G \bar{t}_R$.

(2) Is the three-quark status $\dfrac{1}{\sqrt{6}}\left[{}^R q\left({}^B q + {}^G q\right) + {}^B q\left({}^G q + {}^R q\right) + {}^G q\left({}^R q + {}^B q\right)\right]$ bound?

9.30   In 1960, S. Glashow pointed out the possibility of establishing charged weak bosons via the resonant reaction $v_e + e^+ \to W^+ \to \mu^+ + v_\mu$. Why did the idea not work in practice?

9.31   The largest fraction of the higher-energy electron neutrinos, the 'boron neutrinos' are produced in the Sun by the decay ${}^8 B \to 2\alpha + e^+ + v_e$. Their maximum energy is $E_{v,\,max} = 16$ MeV. Consider the elastic collision of a boron neutrino with a nucleus of a ^{76}Ge detector. Is the resolving power sufficient to resolve the nuclear structure, assuming a nuclear radius $R_A = 4$ fm? What is the maximum recoil energy of the nucleus? How does the interaction probability depend on the neutrino flavour?

9.32   Consider the NC scattering of v_μ on the valence quarks of a nucleus. Write down the Z-charge factors squared of the quarks and their values, assuming $\sin^2 \theta_W = 0.23$. The total v_μ cross-section is proportional to a combination of the Z-charge factors squared of the quarks. Write it and evaluate it for the valence quarks and for a nucleus containing the same number of u and d quarks.

9.33   Consider the NC scattering of $\bar{v}_\mu$ on the valence quarks of a nucleus. Write down the Z-charge factors squared of the quarks and their values, assuming $\sin^2 \theta_W = 0.23$. The total $\bar{v}_\mu$ cross-section is proportional to a combination of the Z-charge factors squared of the quarks. Write it and evaluate it for the valence quarks, and for a nucleus containing the same number of u and d quarks.

9.34   Compare the NC scattering of v_e and of v_μ of the same energy on the valence quarks of a nucleus. Both cross-sections are proportional to a combination of the Z-charge factors squared of the quarks. Consider the ratio of the contributions of the left and right quarks in the two cases. How large is the ratio between these ratios?

9.35   Find the relative values of the amplitude of couplings of the V and A currents to Z for v_e, e^-, u, d, assuming $\sin^2 \theta_W = 0.23$.

9.36   Atomic electrons interact with the nucleus not only with the electromagnetic interaction (photon exchange) but also with neutral current weak interactions (Z exchange). The latter contribution is extremely small but the interference of the two gives detectable parity-violation effects. The main contribution comes from the product of the axial electron current and the quark vector currents. Notice that the Z-charge of the nucleus is the sum of those of its nucleons and the charges of the nucleons are the sums of those of their quarks (consider only valence quarks in the following).

(a) Evaluate the axial Z-charges of the proton and of the neutron.

(b) Evaluate the axial Z-charge of a nucleus with Z protons and N neutrons. Is the larger contribution due to protons or neutrons?

(c) Explain why the product of the axial electron current and quark vector currents is less important.

Summary

In this chapter we have studied the electroweak unification in the Standard Model of particle physics. We have seen in particular

- the structure of the weak neutral currents
- the formalism of the electroweak unification
- the measurements of the weak mixing angle
- the discovery of the intermediate vector bosons
- the evolution of the weak mixing angle
- the precision tests of the electroweak theory at LEP and at the Tevatron
- in particular the precision measurements of the top and of the W masses
- the BEH mechanism for the spontaneous breaking of a gauge theory and the Higgs boson
- the LHC collider and its experiments
- the discovery of the H boson and the measurements of mass, width, spin, parity and couplings to vector bosons and fermions.

Further Reading

CMS Collaboration (2022) A portrait of the Higgs boson by the CMS experiment ten years after the discovery. *Nature* 607 60–68

Hollik, W. & Duckeck, G. (2000) *Electroweak Precision Tests at LEP*. Springer Tracts in Modern Physics

Llewellyn Smith C. (2015) Genesis of the Large Hadron Collider. *Phil. Trans. R. Soc. A* 373 20140037

Rubbia, C. *et al.* (1982) The search for the intermediate vector bosons. *Sci. Am.* March 38

Rubbia, C. (1984) Nobel Lecture; *Experimental Observation of the Intermediate Vector Bosons*

Salam, A. (1979) Nobel Lecture; *Gauge Unification of Fundamental Forces*

Steinberger, J. (1988) Nobel Lecture; *Experiments with High Energy Neutrino Beams*

Weinberg, S. (1974) Unified theories of elementary-particle interactions. *Sci. Am.* July 50

In this chapter we shall discuss neutrino physics. As anticipated in Sections 3.8 and 4.11, neutrinos produced with a certain flavour, v_e, v_μ or v_τ, may be detected later with a different one, if enough time is left between production and detection. Consequently v_e, v_μ and v_τ are not stationary states with definite mass, but quantum superpositions of them. We call v_1, v_2 and v_3 the states of definite mass. Neutrinos can change flavour by two mechanisms: oscillations and adiabatic flavour conversion in matter (AFC).

We start by presenting the formal description of neutrino mixing, showing, in particular, the analogies and the differences with the quarks. We then give the current values of the mixing angles and mass spectrum.

Neutrino oscillations are discussed in Section 10.2 together with the relevant experiments. These can be distinguished in *disappearance experiments* that measure the flux of neutrinos of a definite flavour at a distance from their source to check if it is less than that expected in the absence of oscillations, and *appearance experiments* in which neutrinos of a flavour not initially present are detected at a distance from the source. In any case the flight lengths must be long, of the order of kilometres or thousands of kilometres, depending on the energy.

We discuss the disappearance experiments on atmospheric neutrinos, on v_μ produced by proton accelerators, and on $\bar{v}_e$ produced by power reactors, as well as appearance experiments, searching for both v_e and v_τ, on accelerator-produced v_μ.

The formalism of flavour adiabatic conversion in matter is introduced in Section 10.3. We shall consider the v_e produced in the core of the Sun, where the effect has been discovered, and describe the low background experiments performed, with increasing accuracy, for almost half a century, in underground laboratories worldwide (Section 10.4).

We do not yet know the absolute scale of neutrino masses. Upper limits exist on different functions of the three neutrino masses, established by measuring the electron energy spectrum in the tritium beta decay and from cosmology. We shall present them in Section 10.5.

In Section 10.6 we discuss a fundamental question in neutrino physics. Are neutrinos completely neutral fermions? Like a completely neutral boson, as the γ and the Z^0, completely neutral fermions do not have any 'charge', including lepton number. They are described by the Majorana solution of the Dirac equation, which we studied in Section 2.9. Finally, in Section 10.7, we shall present how this hypothesis is tested, namely by searching for an extremely rare phenomenon, the neutrino-less double-beta decay.

10.1 Neutrino Mixing and Masses

Neutrinos have been found to change flavour by two mechanisms. Both phenomena are consequences of neutrino mixing.

- Oscillations, which are similar, but not identical, to the neutral meson oscillations. They occur both in vacuum and in matter. They have been discovered in the v_μ of the decays of mesons produced by cosmic rays in the atmosphere and in v_μ beams produced at proton accelerators. The energies of these neutrinos range from below 1 GeV to several dozen GeV. The distance from the production to the detection point varies from hundreds to several thousand kilometres.
- Adiabatic flavour conversion in matter (AFC) is a dynamical phenomenon due to the interaction of v_e beams with the matter electrons, similar to the refraction index of light. The phenomenon can most easily be observed if the flight length is large and if the density is high, as in a star. It has been observed in the v_e produced in the core of the Sun, which have energies of several MeV, and in atmospheric and artificial v_μ, with energies of several GeV, travelling in the Earth rock. Note that this phenomenon is often called oscillation, but it is not. Oscillation is an interference phenomenon, AFC is a dynamical one.

From the historical point of view, in 1957 Bruno Pontecorvo advanced the idea that the neutrino–antineutrino system might oscillate in analogy with the $K^0 \bar{K}^0$ system (Pontecorvo 1957), something that, it was later learnt, does not happen. At that time only one neutrino species was known; in fact, it had just been discovered. In 1962, just after the discovery of the second neutrino, a theoretical group of Kyoto (Katayama *et al.* 1962) and one of Nagoya (Maki *et al.* 1962) advanced the hypothesis of mixing between v_e and v_μ. Maki *et al.* also mentioned the possibility of 'transmutation' between neutrino flavours. Coming back to his idea in 1967, Pontecorvo (Pontecorvo 1967) considered oscillations not only between neutrinos and antineutrinos, but also between flavours. Analysing the experimental data, he reached the conclusion that ample room was left for the lepton number violation that the mixing implies. Discussing possible experiments, he concluded that the ideal neutrino source was the Sun. If the phenomenon existed, its effect on Earth would be to observe half of the expected electron neutrino flux (having assumed maximum mixing between two neutrino species; see later for the meaning of the term).

The propagation of neutrinos in matter of uniform density was studied by Wolfenstein in 1978 (Wolfenstein 1978) and their adiabatic flavour conversion in matter of density varying along the neutrino path by Mikheyev and Smirnov in 1985 (Mikheyev & Smirnov 1985). It is called the MSW effect (Mikheyev–Smirnov–Wolfenstein).

Neither oscillation nor flavour conversion in matter could happen if neutrinos were massless (unless they have presently unknown additional interactions). Consequently, both phenomena contradict the Standard Model in two ways: non-conservation of the lepton flavour and non-zero neutrino masses.

We now summarize the present status of our knowledge, leaving the experimental proofs of these conclusions for the following sections.

The definite flavour states ν_e, ν_μ and ν_τ are obtained from the stationary states ν_1, ν_2 and ν_3 with a transformation, which we assume to be unitary. We indicate the masses, which, as we know, are defined for the stationary states, by m_1, m_2 and m_3. The transformation, analogous to that of the quarks, is

$$\begin{pmatrix} \nu_e \\ \nu_\mu \\ \nu_\tau \end{pmatrix} = \begin{pmatrix} U_{e1} & U_{e2} & U_{e3} \\ U_{\mu1} & U_{\mu2} & U_{\mu3} \\ U_{\tau1} & U_{\tau2} & U_{\tau3} \end{pmatrix} \begin{pmatrix} \nu_1 \\ \nu_2 \\ \nu_3 \end{pmatrix}. \tag{10.1}$$

Having assumed unitarity, we can express the transformation in terms of three rotations, of angles that we shall again call θ_{12}, θ_{23} and θ_{13}, taken by convention in the first quadrant $0 \le \theta_{ij} \le \pi/2$ and of phases $\in (0, 2\pi)$. If neutrinos are Dirac particles, as assumed in the Standard Model, all but one of the phases can be absorbed, as in the case of quarks, in the wave functions, obtaining the same expression as (7.91). However, if neutrinos are completely neutral, two more phases, η_1 and η_2 cannot be absorbed and consequently are physical, but irrelevant for the oscillation and AFC. These phases are observable in the neutrino-less double-beta decay (Section 10.7). If different from 0 and π (2π being equivalent to 0) the phases induce CP violation, as we shall see in the following.

Experimentally, two masses are close to one another, while the third is more separate. The former ones are defined as ν_1 and ν_2, with ν_1 being the lighter. The sign of the other mass difference defines two possible mass orderings, either 'normal ordering' (NO) $m_3 > m_2 > m_1$ or 'inverted ordering' (IO) $m_2 > m_1 > m_3$. From the experiments, we have that $|U_{e1}| > |U_{e2}| > |U_{e3}|$, namely the electron neutrino component decreases from ν_1 to ν_2 to ν_3 The mixing matrix is

$$U = \begin{pmatrix} 1 & 0 & 0 \\ 0 & c_{23} & s_{23} \\ 0 & -s_{23} & c_{23} \end{pmatrix} \begin{pmatrix} c_{13} & 0 & s_{13}e^{-i\delta} \\ 0 & 1 & 0 \\ -s_{13}e^{i\delta} & 0 & c_{13} \end{pmatrix} \begin{pmatrix} c_{12} & -s_{12} & 0 \\ s_{12} & c_{12} & 0 \\ 0 & 0 & 1 \end{pmatrix} \begin{pmatrix} e^{i\eta_1} & 0 & 0 \\ 0 & e^{i\eta_2} & 0 \\ 0 & 0 & 1 \end{pmatrix}$$

$$= \begin{pmatrix} c_{12}c_{13} & s_{12}c_{13} & s_{13}e^{-i\delta} \\ -s_{12}c_{23}-c_{12}s_{23}s_{13}e^{i\delta} & c_{12}c_{23}-s_{12}s_{23}s_{13}e^{i\delta} & s_{23}c_{13} \\ s_{12}s_{23}-c_{12}c_{23}s_{13}e^{i\delta} & -c_{12}c_{23}-s_{12}c_{23}s_{13}e^{i\delta} & c_{23}c_{13} \end{pmatrix} \begin{pmatrix} e^{i\eta_1} & 0 & 0 \\ 0 & e^{i\eta_2} & 0 \\ 0 & 0 & 1 \end{pmatrix} \tag{10.2}$$

where $c_{ij} = \cos\theta_{ij}$, and $s_{ij} = \sin\theta_{ij}$.

As we shall discuss in the following sections, the experiments have measured different observables, such as cross-sections and energy spectra, relevant for the oscillation phenomena. Three different theoretical groups (NuFit, Capozzi et al. 2018; de Salas et al. 2018) periodically analyse all the available measurements and perform a 'global fit', considering the experimental uncertainties and their correlations, to extract the mass and mixing parameters, with almost identical results.

We anticipate the best-fit values of the angles (Capozzi et al. 2018) and discuss later how they have been measured.

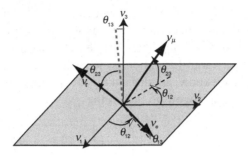

The rotations of the neutrino mixing.

$$\sin^2 \theta_{12} = 0.304^{+0.014}_{-0.013}; \quad \theta_{12} \approx 33.5°$$
$$\sin^2 \theta_{23} = 0.551^{+0.019}_{-0.080}; \quad \theta_{23} \approx 48° \tag{10.3}$$
$$\sin^2 \theta_{13} = 0.0214^{+0.0009}_{-0.0007}; \quad \theta_{13} \approx 8.4°$$

Comparing with (7.95), one sees that, unlike the case of quarks, the neutrino mixing angles are large. Fig. 10.1 schematically shows how different the stationary states are from the flavour states.

Equation (10.4) gives the three standard deviation ranges of the absolute values of the mixing matrix elements from NuFit (version 5.2):

$$|U| = \begin{pmatrix} 0.803-0.845 & 0.514-0.578 & 0.143-0.155 \\ 0.244-0.498 & 0.502-0.693 & 0.632-0.768 \\ 0.272-0.517 & 0.473-0.672 & 0.623-0.761 \end{pmatrix} \tag{10.4}$$

While in the quark sector the diagonal elements are large, those of the two near diagonals are much smaller and those of the next ones are very small, there is no similar hierarchy in the lepton sector.

The value of the CP-violating phase from the global fit of Capozzi *et al.* (2018) (those from the other ones are similar) is

$$\delta = \left(238^{+41}_{-33}\right)°, \tag{10.5}$$

which is quite a large value, close to the maximum in the third quadrant. However, caution is needed due to the still large uncertainty, the $\pm 3\sigma$ range $149° \leq \delta \leq 358°$, which is just at 3σ from the CP-conserving value of $360°$.

Coming to the mass spectrum, both oscillations and AFC are independent of the absolute values of the masses, giving information on the differences between their squares only. We call the smaller difference

$$\delta m^2 \equiv m_2^2 - m_1^2. \tag{10.6}$$

We define the larger square mass difference as

$$\Delta m^2 \equiv m_3^2 - \frac{m_1^2 + m_2^2}{2}. \tag{10.7}$$

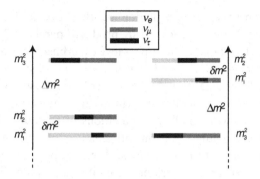

Fig. 10.2. Neutrino square-mass spectrum. Normal ordering (NO) on the left, inverted ordering (IO) on the right. The flavour contents of the eigenstates are also schematically shown.

It can be positive or negative. The longer periods are inversely proportional to the absolute value of these quantities. The above-mentioned global fit gives (Capozzi *et al.* 2018) the values

$$\delta m^2 = 73.4^{+1.7}_{-1.4} \text{ meV}^2; \qquad \left|\Delta m^2\right| = 2455^{+45}_{-32} \text{ meV}^2 \qquad (10.8)$$

The so-called hierarchy parameter α is small

$$\alpha \equiv \left|\delta m^2 / \Delta m^2\right| = 0.03. \qquad (10.9)$$

This circumstance decouples in first approximation the 'faster' and 'slower' oscillations.

Fig. 10.2 shows schematically the neutrino square-mass spectrum. As one can easily see from the values (10.3) of the mixing angles, ν_1 is about 70% ν_e and the rest is half ν_μ and half ν_τ. ν_2 contains about one third of each flavour. ν_3 is almost half ν_μ and half ν_τ; the fraction of ν_e is small.

We do not know the absolute scale of the mass. As the mass ordering is not known either, the global fits are performed separately for each of them. The resulting overall two χ^2 minima show that the NO is favoured over the IO by more than 3σ. However, IO cannot be rejected yet.

The experiments that have contributed, first, to the discovery, then to establish the values of the parameters, which have just been summarized, are of different types, each contributing to some of the parameters. Solar neutrinos originate from the core of the star, where they are produced as ν_e. They travel in the decreasing density solar matter and then, leaving the Sun, come to us travelling a distance of the order of 10^7 km; their energies are of the MeV scale. Disappearance experiments on solar neutrinos give dominant contribution on θ_{12} and important contribution to δm^2 and θ_{13}. The Sudbury Neutrino Observatory (SNO) appearance experiment established that the disappeared solar ν_e appear with the other flavours through AFC (not oscillation). Atmospheric neutrinos come from the decays of pions and K-mesons produced by cosmic ray collisions with atmospheric nuclei. At production, about two thirds of them are ν_μ and $\bar{\nu}_\mu$, about one third ν_e and $\bar{\nu}_e$. Their energy range is from 10^2 and 10^5 MeV and their flight lengths are between 10 and 10 000 km. Atmospheric neutrino disappearance experiments contribute predominantly to Δm^2 and θ_{23}, and to θ_{13} and the CP-violating phase δ. Reactor neutrinos are produced purely as $\bar{\nu}_e$ with energies in the several MeV range. Two types of disappearance experiment relevant for oscillations are performed: 'medium baseline' (MBL), over 1 km scale distance, and 'long baseline' (LBL) over 10–100 km. The former gives information mostly on Δm^2 and θ_{13}, the latter on Δm^2, θ_{13} and θ_{12}. Accelerator neutrino beams can be made both with neutrinos, which are mainly ν_μ with small contaminations of the other flavours, and with antineutrinos, which are mainly $\bar{\nu}_\mu$, again with small contaminations of the other flavours. The energies are in the $1-10$ GeV range. Both disappearance and appearance experiments are performed. The former ones give information mostly on Δm^2 and θ_{23}; the latter, in ν_e or $\bar{\nu}_e$ appearance, give the most important information on the CP-violating phase δ, and give important information on θ_{13} and θ_{23}. Accelerator neutrinos proved that the atmospheric ν_μ disappearance corresponds predominantly to ν_τ appearance.

10.2 The Longer Period Oscillation

The neutrino oscillation phenomenon is like that of the neutral mesons, but with three important differences: in the case of mesons, there are two eigenstates, there are three for neutrinos; for mesons the mixing is between particle and antiparticle, for neutrinos it is between flavours; and neutrinos do not decay, but mesons do.

Let us start by considering, for the sake of simplicity, an oscillation between two neutrino species. Neglecting effects proportional to $\sin^2\theta_{13}$, namely of a couple of per cent, this is the situation of the 'atmospheric oscillation', the longer period one, which involves in this approximation only the flavour states ν_μ and ν_τ and the stationary states ν_2 and ν_3. The mixing matrix of the problem is

$$\begin{pmatrix} \nu_\mu \\ \nu_\tau \end{pmatrix} = \begin{pmatrix} U_{\mu 2} & U_{\mu 3} \\ U_{\tau 2} & U_{\tau 2} \end{pmatrix} \begin{pmatrix} \nu_2 \\ \nu_3 \end{pmatrix},$$

which we can think of as a rotation in a plane, as in the two-quark case

$$\begin{pmatrix} \nu_\mu \\ \nu_\tau \end{pmatrix} = \begin{pmatrix} \cos\theta_{23} & \sin\theta_{23} \\ -\sin\theta_{23} & \cos\theta_{23} \end{pmatrix} \begin{pmatrix} \nu_2 \\ \nu_3 \end{pmatrix}. \tag{10.10}$$

Fig. 10.3(a), which is Fig. 7.14, recalls the Cabibbo rotation for comparison, which is about 13°. As stated above, a mixing angle can have any value between 0° and 90°. The rotated and non-rotated axes are close to each other both for small angles and for angles near 90°. The difference reaches a maximum at 45°, which is called maximal mixing. Actually, one of the mixing angles, the angle involved in the atmospheric oscillation θ_{23}, is close to 45°, as shown in Fig. 10.3(b).

By explicitly writing Eq. (10.10) we have

$$|\nu_\mu\rangle = \cos\theta_{23}|\nu_2\rangle + \sin\theta_{23}|\nu_3\rangle$$
$$|\nu_\tau\rangle = -\sin\theta_{23}|\nu_2\rangle + \cos\theta_{23}|\nu_3\rangle.$$

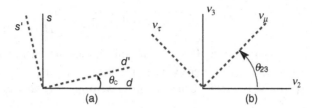

Fig. 10.3. Flavour rotation for (a) the (1,2) quark families and (b) for the (2,3) neutrino families.

In the case of maximal mixing, $\theta_{23} = 45°$, it would become

$$|\nu_\mu\rangle = \frac{1}{\sqrt{2}}(|\nu_2\rangle + |\nu_3\rangle); \qquad |\nu_\tau\rangle = \frac{1}{\sqrt{2}}(-|\nu_2\rangle + |\nu_3\rangle). \tag{10.11}$$

This transformation is equal to that of the K system, neglecting CP violation, Eq. (8.13).

Let us now consider a beam of neutrinos, all with the same momentum $\mathbf{p}$. The energies of the two stationary states are not equal owing to the difference of the masses. Considering that the masses are very small compared with these energies and letting E be the average of the two energies, we can write with very good approximation

$$E_i = \sqrt{p^2 + m_i^2} \simeq p + \frac{m_i^2}{2p} \simeq p + \frac{m_i^2}{2E}. \tag{10.12}$$

The evolution of the two states in a vacuum is given by the Schrödinger equation

$$i\frac{d}{dt}\begin{pmatrix} v_2(t) \\ v_3(t) \end{pmatrix} = H\begin{pmatrix} v_2(t) \\ v_3(t) \end{pmatrix}, \tag{10.13}$$

where the Hamiltonian is diagonal

$$H = \begin{pmatrix} E_2 & 0 \\ 0 & E_3 \end{pmatrix} \approx \begin{pmatrix} p + \dfrac{m_2^2}{2E} & 0 \\ 0 & p + \dfrac{m_3^2}{2E} \end{pmatrix}. \tag{10.14}$$

The evolution of the two flavour states is

$$i\frac{d}{dt}\begin{pmatrix} v_\mu(t) \\ v_\tau(t) \end{pmatrix} = H'\begin{pmatrix} v_\mu(t) \\ v_\tau(t) \end{pmatrix} = UHU^\dagger \begin{pmatrix} v_\mu(t) \\ v_\tau(t) \end{pmatrix}. \tag{10.15}$$

We easily find that

$$H' = p + \frac{m_2^2 + m_3^2}{4E} + \frac{\Delta m^2}{4E}\begin{pmatrix} -\cos 2\theta_{23} & \sin 2\theta_{23} \\ \sin 2\theta_{23} & \cos 2\theta_{23} \end{pmatrix}. \tag{10.16}$$

Here we make an observation, which will be useful in the following. The mixing angle, which we generically call θ, is given by the ratio between the non-diagonal element (the two are equal) and the difference between the two diagonal ones, namely by

$$\tan 2\theta = \frac{2H'_{12}}{H'_{22} - H'_{11}}. \tag{10.17}$$

Let $|v_\mu(0)\rangle$ and $|v_\tau(0)\rangle$ be the amplitudes at the initial time $t = 0$ of the definite flavour states. The time evolution of the stationary states is $|v_2(t)\rangle = |v_2(0)\rangle e^{-iE_2 t}$ and $|v_3(t)\rangle = |v_3(0)\rangle e^{-iE_3 t}$ Consequently, writing $c = \cos\theta$ and $s = \sin\theta$, we have

$$|v_\mu(t)\rangle = c|v_2(0)\rangle e^{-iE_2 t} + s|v_3(0)\rangle e^{-iE_3 t}$$

$$= \left[s^2 e^{-iE_2 t} + c^2 e^{-iE_3 t} \right]|v_\mu(0)\rangle + sc\left[e^{-iE_3 t} - e^{-iE_2 t} \right]|v_\tau(0)\rangle,$$

$$|v_\tau(t)\rangle = -s|v_2(t)\rangle e^{-iE_2 t} + c|v_3(0)\rangle e^{-iE_3 t}$$

$$= \left[c^2 e^{-iE_3 t} + s^2 e^{-iE_2 t} \right]|v_\tau(0)\rangle + s\left[ce^{-iE_3 t} - e^{-iE_2 t} \right]|v_\mu(0)\rangle.$$

Let us consider the case of an initially pure v_μ system, which occurs for cosmic ray and accelerator neutrinos. At time t we can observe the 'appearance' of v_τ with a probability given by

$$P(v_\mu \to v_\tau, t) = \left| \langle v_\tau(t)|v_\mu(0)\rangle \right|^2 = c^2 s^2 \left| e^{-iE_2 t} - e^{-iE_3 t} \right|^2$$

$$= c^2 s^2 \left| 2ie^{-i\frac{E_2 + E_3}{2} t} \sin\frac{E_3 - E_2}{2} t \right|^2,$$

namely

$$P(v_\mu \to v_\tau, t) = 4c^2 s^2 \sin^2 \frac{E_3 - E_2}{2} t = \sin^2 2\theta \sin^2 \frac{\Delta m^2}{4E} t. \tag{10.18}$$

Notice that the time t we are considering is the proper time, namely the time measured in the neutrino rest frame. In practice, we observe the phenomenon as a function of the distance L travelled by neutrinos in the reference frame of the laboratory. Writing Eq. (10.18) as a function of L, we have, with E in GeV, L in km and Δm in eV,

$$P(v_\mu \to v_\tau, t) = \sin^2 2\theta \sin^2 \left[1.27\Delta m^2 \left(\frac{L}{E} \right) \right]. \tag{10.19}$$

This is the probability measured by the 'appearance' experiments, which consist of a detector at a distance L from the source capable of detecting the possible presence of the initially non-existent flavour v_τ. In 'disappearance' experiments, the flux of a flavour initially present, say v_μ, is measured at a distance L and checked if it is what is expected in absence of oscillation. In both cases, the initial v_μ flux and energy spectrum must be known with an accuracy that depends on the precision aimed at by the experiment. It may be calculated, as in the case of atmospheric neutrinos, or measured, ensuring higher accuracy, for artificial neutrino beams from accelerators. In a disappearance experiment the ratio of the measured flux to the calculated one in the absence of oscillations gives the disappearance probability. This is the complement to the unit of the appearance probability

$$P(v_\mu \to v_\mu, t) = 1 - P(v_\mu \to v_\tau, t) = 1 - \sin^2 2\theta \sin^2 \left[1.27\Delta m^2 \left(\frac{L}{E} \right) \right]. \tag{10.20}$$

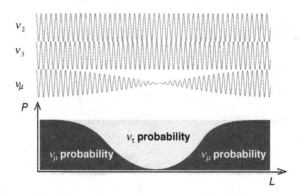

Fig. 10.4. Summing two monochromatic probability waves.

Notice that the evolution in time of both probabilities depends on the absolute value of the difference between the squares of the masses, not on its sign.

An analogy could be useful here. The initially pure v_μ state with definite momentum is the superposition of two monoenergetic stationary states of energies E_2 and E_3, analogous to a dichromatic signal, the sum of two monochromatic ones with angular frequencies, say ω_2 and ω_3. Their mixing is maximal if their amplitudes are equal, a situation shown in Fig. 10.4.

Initially the two monochromatic components are in phase, but their phase difference increases with time and the two components reach phase opposition at $t = 1/|\omega_3 - \omega_2|$, only to return to being in phase at $2/|\omega_3 - \omega_2|$, etc. The modulated amplitude varies in time as $\cos\left(\dfrac{\omega_3 - \omega_2}{2}t\right)$. The probability of observing a v_μ, which is proportional to the square of the amplitude, namely to $\cos^2\left(\dfrac{\omega_3 - \omega_2}{2}t\right)$, is 100% initially, decreases to zero (for the first time) at time $t = 1/|\omega_3 - \omega_2|$, then increases again, etc. The probability of observing the new flavour varies as $\sin^2\left(\dfrac{\omega_3 - \omega_2}{2}t\right)$, according to Eq. (10.19).

If the mixing is not maximal, or the amplitudes of the two monochromatic components are not equal, the original flavour never disappears completely and the appearance probability maximum $\sin^2 2\theta_{23}$ is smaller than one. This maximum is smaller and smaller the larger is the difference from 45°, either in the first or second octant. Indeed, as shown by Eqs. (10.19) and (10.20), the vacuum oscillations do not depend on the sign of $\pi/4 - \theta_{23}$. This fact is called 'octant ambiguity'.

It may also be useful to discuss a mechanical analogue of neutrino oscillations. Consider two identical pendulums of length L and mass m. Each of them, if initially excited, will vibrate in a harmonic motion. Remember that the square proper angular frequency of the harmonic oscillator is given by the restoring force per unit mass and per unit displacement, hence in this case $\omega^2 = g/L$. If we now connect them with a spring, constant k, the two pendulums are coupled as in Fig. 10.5(a). This system has two stationary states (or modes), which we call 2 and 3 in analogy to neutrinos. In the first mode the two vibrate in phase with the same amplitude (Fig. 10.5(b)); the square

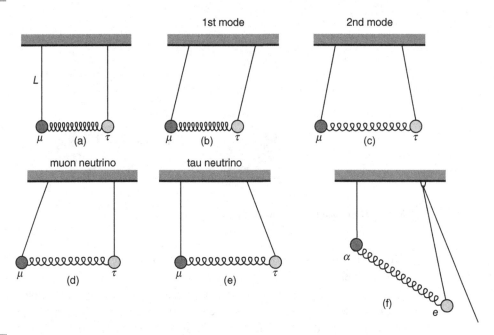

Coupled pendulums analogue to two neutrinos: (a) the double pendulum with maximum mixing, (b) the lower-frequency mode, (c) the higher-frequency mode, (d) the analogue of v_μ, (e) the analogue of v_τ, (f) non-maximal mixing, with the option to vary the mixing angle.

eigenfrequency of this mode is $\omega_2^2 = g / L$ (the spring remains in its natural length and does not exert any force). In the second mode the two pendulums vibrate, again in phase but with equal and opposite amplitudes (Fig. 10.5(c)); the square eigenfrequency is $\omega_3^2 = g / L + 2k / m$, where the factor 2 comes from being the contraction or expansion of the spring twice the displacement of each pendulum. These states are the analogous of v_2 and v_3 and the eigenfrequencies are analogous to the masses.

The states analogous to v_μ and v_τ are those in which one or the other pendulum vibrates. Let us bring the pendulum μ out of equilibrium, keeping the other one in its rest position, and let it go (Fig. 10.5(d)). We observe that the vibrations of pendulum μ gradually decrease in amplitude, to the point that it almost stops for a moment. The other pendulum, τ, starts to vibrate, with an initially small but gradually increasing amplitude to reach maximum when all the energy has been transferred from μ to τ (Fig. 10.5(e)). Then the process will invert itself and the energy will go back to μ. Energy, the analogue of probability, oscillates back and forth between the two pendulums with the angular frequency $\omega_3 - \omega_2$. Indeed, the motion of μ (for example) is

$$x_\mu(t) = \frac{a}{\sqrt{2}}\left(\cos \omega_2 t + \cos \omega_3 t\right)$$

$$= \frac{a}{\sqrt{2}} \cos\left(\frac{\omega_3 - \omega_2}{2} t\right) \cos\left(\frac{\omega_3 + \omega_2}{2} t\right),$$

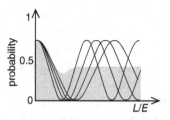

Fig. 10.6. Probability of observing the initial flavour for maximal mixing and with energy spread.

and its energy is proportional to the square of the amplitude, that is, to $\cos^2\left(\dfrac{\omega_3 - \omega_2}{2}t\right)$ whose frequency is $|\omega_3 - \omega_2|$. Notice that the measurement of the energy oscillation frequency determines the absolute value of the difference between the two eigenfrequencies and not its sign.

Question 10.1 Write down the equations of motion of the two coupled pendulums. For each take a coordinate with origin in its rest position, say x_μ and x_τ. (1) Find the modes and (2) solve the equations with the initial conditions: $x_\mu(0) = a, x_\tau(0) = 0$, $\dot{x}_\mu(0) = \dot{x}_\tau(0) = 0$. □

It is easy to see that if the two pendulums are not identical, if they have different lengths as in Fig. 10.5(f), then phenomenon will continue to happen, but with an important difference. Starting again with one pendulum excited, its energy will pass to the other one but only partially. The originally excited pendulum slows down but does not come to rest. The fraction of energy transfer between the pendulums decreases as the difference between their lengths increases. Maximum mixing is when the two pendulums are identical or, better, when, taken alone, have the same period. Only in this case is the energy transfer between them, when coupled, resonant and, consequently, fully efficient.

Let us come back to neutrinos. In practice no neutrino source is monochromatic as we have assumed. This can be partially compensated by measuring neutrino energy, with a certain energy resolution. In any case neutrino energy is known within a smaller or larger spread. Let us see how a disappearance experiment is affected by the neutrino energy spread.

We try to illustrate this in Fig. 10.6, taking maximal mixing as an example. The survival probabilities of the monochromatic components are in phase at $t = 0$ and remain such at short times, say in the first quarter period; then they gradually go out of phase and average out at constant value, sooner or later depending on their degree of monochromaticity. This average depends on the mixing angle and is equal to ½ in our example of maximal mixing.

Let us now go to the real situation of three neutrinos. We again assume the system at $t = 0$ to be composed purely of neutrinos of the same flavour and all with the same momentum. Let a detector capable of identifying the neutrino flavour be located at distance L. With three possible initial flavours and three possible detected flavours we have in total nine possibilities, which, however, are not all independent due to CPT

invariance. We shall not present here the calculation of the nine probabilities, which can be easily done starting from the mixing matrix. We only observe that, in general, there are three oscillation frequencies, corresponding to the three square-mass differences. The oscillation probability between a given pair of flavours is a sum of oscillating terms at these frequencies with maximum excursions that are functions, different in each case, of the three mixing angles.

In practice, two circumstances considerably simplify the situation.

(1) Two square-mass differences are equal to all practical effects: $m_3^2 - m_2^2 \approx m_3^2 - m_1^2 \approx \Delta m^2$. Consequently, there are only two oscillation periods.
(2) The two oscillation periods are very different, as we see from Eq. (10.7). Consequently, the experiments sensitive to the oscillation with shorter period (the atmospheric one) do not see the longer period (solar) oscillation because it has not yet started. On the other hand, the experiments sensitive to the longer period performed in practice average the signal on times that are much larger than the shorter period and are not sensitive to the first oscillation. However, detectors with large sensitive mass and very good energy resolution can be sensitive to the interference between the oscillations of the two periods. Such an experiment, JUNO, looking at the disappearance of reactor neutrinos at 50 km distance, is under construction in China, to measure the sign of Δm^2.

Let us now focus on the 'atmospheric' oscillation, for which we can use a two-flavour approximation. It was discovered by Super-Kamiokande in 1998 (Fukuda *et al.* 1998) located in the Kamioka underground observatory under the Japanese Alps. We saw in Section 1.12 that muon and electron neutrinos are present amongst the decay products of the hadrons produced by the cosmic ray collisions with atomic nuclei in the atmosphere. The oscillation probabilities between all flavour pairs have, in a first approximation, the same dependence on the flight-length-to-energy ratio L / E

$$P\left(v_x \to v_y, t\right) = A\left(v_x \to v_y\right) \sin^2\left[1.27 \Delta m^2 \left(L / E\right)\right]. \tag{10.21}$$

The constant $A\left(v_x \to v_y\right)$ is the maximum of the probability oscillation between flavours v_x and v_y. Let us see the values of the constants, using the values of the mixing angles (10.3).

The specific phenomenon discovered by Super-Kamiokande is the muon neutrino disappearance. For this phenomenon the maximum probability is

$$A\left(v_\mu \to v_x\right) = \sin^2\left(2\theta_{23}\right)\cos^2\left(\theta_{13}\right)\left(1 - \sin^2\theta_{23}\cos^2\theta_{13}\right) \approx \frac{1}{2}, \tag{10.22}$$

where, to the right in the middle expression, we have taken $\cos^2\theta_{13} \approx 1$ and $\sin^2\theta_{23} \approx 1/2$.

The disappeared muon neutrinos appear in part as electron neutrinos and in part as tau neutrinos. The corresponding probabilities at the maximum are

$$A\left(v_\mu \to v_e\right) = \sin^2\left(\theta_{23}\right)\sin^2\left(2\theta_{13}\right) \approx 2\theta_{13}^2, \tag{10.23}$$

which is very small, and

$$A\left(v_\mu \to v_\tau\right) = \sin^2\left(2\theta_{23}\right)\cos^4\left(\theta_{13}\right) \approx 1. \tag{10.24}$$

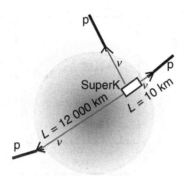

Fig. 10.7.   Flight lengths of atmospheric neutrinos.

Coming now to the experiment, we recall from Section 1.12 that Super-Kamiokande (Fig. 1.19) is a large water Cherenkov with 22 500 t fiducial mass. Atmospheric neutrinos are detected by their charged current scattering processes

$$\nu_\mu + N \to \mu + N' \qquad \nu_e + N \to e + N'. \tag{10.25}$$

In both cases a unique ring signalling the charged lepton is observed. The ring is sharp in the case of the μ corresponding to its straight track, like the ring in Fig. 1.20(a), while it is diffuse in the case of the electron, like the ring in Fig. 1.20(b), which has a track that is wiggly due to bremsstrahlung. This allows the single-ring events to be classified as 'e-type' or 'μ-type'. Clearly, the charge remains unknown. Neutrino energies range from a few hundred MeV to several GeV. At these values the differential cross-sections are strongly forward peaked and, consequently, the measured final lepton direction is almost the same as the neutrino one. Knowing the incident neutrino direction, we also know the distance it has travelled from its production point in the atmosphere, as illustrated in Fig. 10.7. Keep in mind that neutrinos pass through the Earth without absorption. Calling θ the angle of the neutrino direction with the zenith, the flight length varies from about 10 km for $\theta = 0$ to more than 12 000 km for $\theta = \pi$.

The detector gives a rough measurement of the charged lepton energy, which is statistically correlated to the incident neutrino energy. Both e-type and μ-type events are then divided into a low-energy sample, less than about 1 GeV, and a high-energy sample, up to several GeV.

An essential component of the experiment is the calculation, based on a number of measurements, of the muon and electron neutrino fluxes as expected in the absence of oscillations, as functions of the energy and of the zenith angle.

We consider four categories of events of Super-Kamiokande, 'e-like' and 'μ-like', low energy (visible energy $<$ 1.33 GeV) and high energy (visible energy $>$ 1.33 GeV). Fig. 10.8 shows the distributions of the cosine of the zenith angle for each of them for an exposure of 328 kt yr. The dashed histograms are the predictions in the absence of oscillations; the continuous histograms were obtained assuming $\nu_\mu \to \nu_\tau$ oscillation,

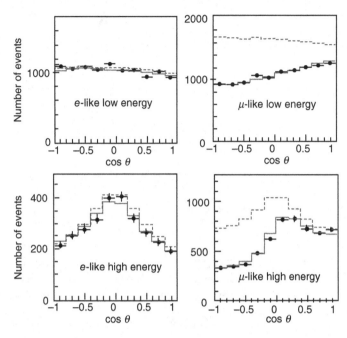

Fig. 10.8. Electron and muon neutrino fluxes versus zenith angle of Super-Kamiokande in a 328 kt yr exposure.

and fitting to the data with Δm^2 and θ_{23} as free parameters. The fitting procedure determines their best values.

We observe that the ν_e data do not show large signs of oscillations, implying that θ_{13} is small. Looking at the high-energy ν_μ, we see that, at small zenith angles, corresponding to short flight lengths, all of them reach the detector; however, above a certain distance the muon neutrino flux is one half of the expected flux. From Fig. 10.6 we understand that the value of that distance determines Δm^2, while the value $1/2$ of the reduction factor says that $\theta_{23} \approx \pi/4$. Finally, the low-energy ν_μ do oscillate even at small distances.

The ν_μ disappearance phenomenon was confirmed by two other experiments on atmospheric neutrinos, MACRO at the INFN Laboratori Nazionali del Gran Sasso (LNGS) in Italy and Soudan 2 in the USA.

The first long baseline experiment with an artificial neutrino beam was K2K, active from 1999 to 2004. The neutrino beam was produced at the proton synchrotron of the KEK laboratory at Tsukuba in Japan. The far detector was Super-Kamiokande at a distance of 250 km. The peak energy was 1.5 GeV chosen to have the first maximum of the oscillation amplitude at the far detector (Ahn *et al.* 2003). This was a disappearance experiment, in which, for the first time, the neutrino flux both at production, near the accelerator, and at the distant detector were under experimental control. K2K fully confirmed the Super-Kamiokande discovery.

In the second experiment, again a disappearance one, the neutrino source was the 'NuMi' beam at the main injector proton accelerator at FNAL, which provided a much more intense neutrino beam. The far detector was MINOS, which was a tracking calorimeter at a distance of 735 km. The beam energy spectrum was chosen to peak at $2-3$ GeV, chosen with the same criterion as above (Michael *et al.* 2006). MINOS took data from 2005 to 2012 providing an accurate measurement of Δm^2.

Now consider θ_{13}, the smallest of the mixing angles and, consequently, the last to have been measured, after some theorists had assumed it to be zero. Full evidence for $\theta_{13} \neq 0$ was obtained only in 2012. There are two ways to study θ_{13}: appearance experiments searching for ν_e in an initially ν_μ beam produced by an accelerator over a baseline a few hundred kilometres long, and disappearance experiments on $\bar{\nu}_e$ originated on a power nuclear reactor. The oscillation amplitude is given for the former, in a first approximation, by Eq. (10.23); for the latter, it is

$$A\left(\nu_e \to \nu_x\right) = \sin^2\left(2\theta_{13}\right) \approx 4\theta_{13}^2. \tag{10.26}$$

Notice that the same value of L/E, the relevant variable for the oscillations, is reached in the case of reactors at distances much shorter than those of accelerators, because energies are much smaller in the former case.

The discovery of $\theta_{13} \neq 0$ is due to the Daya Bay disappearance experiment (An *et al.* 2012) in China with near and far detectors, as we shall now see. The discovery was immediately confirmed by a similar disappearance reactor experiment, RENO (Ahn *et al.* 2012) in Korea, also with near and far detectors, and an indication was reported by the Double CHOOZ in France (Abe *et al.* 2012a) with, however, only the far detector. The T2K experiment in Japan had already obtained the first indications for the ν_e appearance in an initial ν_μ beam from the J-Park accelerator in 2011 (Abe *et al.* 2011), and firmly confirmed it in 2013 (Abe *et al.* 2013). We shall come back to it in Section 10.5.

The basic idea of a precision disappearance experiment is to measure the antineutrino flux and energy spectrum near an antineutrino source, where the oscillation still has a small effect, and at a far location, at which the oscillation has fully developed.

The energy spectrum of the antineutrinos from a reactor peaks at about 3 MeV. The optimum far distance is at the first oscillation maximum for that energy, which, for the oscillation corresponding to Δm^2, is about 1.5 km. In practice, in order to have better control on the systematics, redundancy is very important. It is obtained by employing a number of sources and of detectors.

The most precise experiment is Daya Bay, which we shall now briefly present. Fig. 10.9(a) shows schematically its basic elements: six commercial nuclear power cores (dots in the figure) grouped in pairs and eight antineutrino detectors (grey rectangles), that have been installed in three underground experimental halls (EH1, EH2, EH3), having a flux-averaged baseline of about 500, 500 and 1650 m from the reactors, respectively. The rock overburden of a few hundred metres provides the necessary shield from cosmic rays. The total maximum power of the cores is 17.4 GW. This large value is important for having a high antineutrino flux. Reactors produce about $2 \times 10^{20}\,\bar{\nu}_e\,\mathrm{s}^{-1}\,\mathrm{GW}^{-1}$.

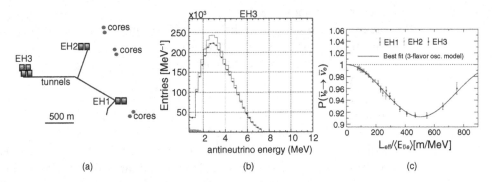

Fig. 10.9.   (a) The Daya Bay infrastructures. (b) The measured antineutrino energy spectrum measured at EH3 with the best-fit and no-oscillation curves superimposed. The error bars are statistical. The small grey area near the lower left corner is the background. (c) Measure of disappearance probability as a function of the ratio of the effective baseline length to the mean neutrino energy. Adapted from An *et al.* (2023) *Phys. Rev. Lett.* 130 161802 Fig. 3 (https://doi.org/10.1103/PhysRevLett.130.161802). Creative Commons Attribution 4.0 International.

$\bar{\nu}_e$ are detected via the inverse beta decay (IVB) in liquid scintillator

$$\bar{\nu}_e + p \rightarrow e^+ + n. \tag{10.27}$$

Each detectors contains 20 t of liquid scintillator, surrounded by photomultipliers that measure the light produced by the positron, which gives a measurement of its energy. The energy of the antineutrino is inferred from the measured positron energy using the relation $E_{\bar{\nu}_e} \approx E_{e^+} + 0.78$ MeV. Gadolinium (Gd) salt (0.1%) dissolved in the scintillator is a powerful tool for discriminating the signal from the background. Once the neutron produced in the inverse beta decay is moderated, it is captured by a Gd nucleus, a process with a very high cross-section, 49 kb. The capture leads to an excited state of the nucleus, which immediately decays, releasing a total of 8 MeV in gamma rays. This radiation is detected as a second light signal with a delay of the order of 100 μs after the electron. The six detectors provide an overall target mass of 120 t. They are immersed in water pools that shield them from the environmental radioactivity.

After the above-mentioned discovery in 2012, with only 55 days of data-taking, the collaboration (An *et al.* 2023) published its final precision measurement in 2023. With 3158 days of operation, a final sample of 5.55×10^6 IVB candidates was collected. Fig. 10.9(b) shows the antineutrino energy spectrum measured at the farthest location (EH3), compared with the no-oscillation prediction calculated with the information from the measurements in the nearer halls. A clear disappearance signal can be seen in the spectrum. The size of the effect gives θ_{13}^2 directly. Fig. 10.9(a) shows also the best-fit curve, corresponding to $\theta_{13} = 8.5° \pm 0.1°$, which has about 1% accuracy. The experiment is very clean, as shown by the extremely small background, shown as a grey area in Fig. 10.9(b).

The oscillation probability, we recall, is a function of neutrino energy to its flight length. None of them can be measured for a single event (the length not being known because it is unknown from which of the reactors the neutrino was coming), but one can define an 'effective baseline' for each detector, L_{eff}, and a mean antineutrino

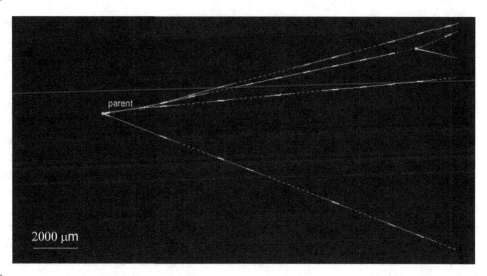

parent

2000 μm

A ν_τ event observed by OPERA. Courtesy INFN.

energy, $\langle E_{\bar{\nu}_e} \rangle$. Fig. 10.9(c) depicts the measured disappearance probability as a function of $L_{\mathrm{eff}}/\langle E_{\bar{\nu}_e} \rangle$ together with the best-fit curve. The oscillation pattern is clear.

The muon neutrino disappearance experiments, both with atmospheric and accelerator neutrinos, would lead to the conclusion that the vast majority of disappeared ν_μ should appear as ν_τ. This fact is implied in Eq. (10.26). However, this conclusion has to be experimentally tested. This was the principal goal of the CERN-INFN project, called CNGS (CERN Neutrinos to Gran Sasso). A new ν_μ beam was constructed at the CERN Super Proton Synchrotron aiming through the Earth's crust to the LNGS at 737 km distance (CNGS 1998). At LNGS the OPERA detector (OPERA 2000) searched for the ν_τ appearance, as identified by the reaction

$$\nu_\tau + N \to \tau + N'. \tag{10.28}$$

The project was optimized for tau neutrino appearance, implying the following main characteristics. The neutrino energy must be high enough for the reaction (10.28) to be substantially above threshold, in practice more than 10 GeV. This implies that the 737 km flight length is small compared with the distance of the oscillation maximum, corresponding to a small expected number of tau neutrinos. Consequently, the detector must have a large mass and at the same time an extremely fine granularity to be able to distinguish between the production and decay vertices of the τ. In practice, a micrometre-scale resolution over kiloton-scale mass is obtained by OPERA with a combination of emulsion and electronic techniques (Agafonova *et al.* 2010). The former is an evolution of the emulsion chamber mentioned in Section 4.9. Neutrinos interact in a target, made of 'bricks'. A brick is a sandwich of emulsion layers and Pb sheets 1 mm thick. In total there are about 150 000 bricks, with 110 000 m² of emulsion films and 105 000 m² of lead plates, for about 1250 t. The data collection lasted from 2008 to 2012 for a total exposure of 1.8×10^{20} p.o.t. The second observed ν_τ candidate event is shown in Fig. 10.10. The τ candidate is observed to decay into three charged particles. In total, OPERA observed $10\,\nu_t$ candidates, which include 2.0 ± 0.2 estimated background events, mainly charm particle decays.

The appearance of ν_τ has also been observed by the Super-Kamiokande collaboration (Abe *et al.* 2012b) in atmospheric neutrinos.

10.3 Adiabatic Flavour Conversion in Matter

The phenomenon of adiabatic flavour conversion in matter (AFC) has been observed for neutrinos produced in the centre of the Sun that have reached us after crossing the high-density solar matter to reach the surface.

The Sun is a main-sequence star in the stable hydrogen-burning stage. Its density is very high in the centre, $\rho_0 = 10^5 \text{kg m}^{-3}$, and gradually diminishes towards the surface. The overall reaction that produces 95% of the energy is the fusion of four protons into a helium nucleus

$$4p \rightarrow \text{He}^{++} + 2\nu_e + 2e^+.$$

The two positrons immediately annihilate with two electrons. Therefore the energy generation process is

$$4p + 2e^- \rightarrow \text{He}^{++} + 2\nu_e + 26.7\,\text{MeV}. \qquad (10.29)$$

The basic elementary reaction is the 'pp fusion'

$$p + p \rightarrow\,^2\text{H} + e^+ + \nu_e. \qquad (10.30)$$

The thermonuclear reactions take place in the central part, the core, of the star where the thermal energy is of the order of tens of keV. These energies are much smaller than the Coulomb barriers of the interacting nuclei and, consequently, the cross-sections are very small; however, they are large enough for the reactions mentioned above to proceed.

Only a small part of the energy released by reaction (10.29) is taken by neutrinos, while the largest fraction is transported by photons. The original MeV-energy photons interact with the solar medium producing other photons of decreasing energies and in increasing number. The energy leaves the surface of the Sun as light only after a long delay that can be up to some 100 000 years. While the light of the Sun and of the stars is a surface phenomenon, neutrinos reach us directly from the centre of the Sun, without any absorption. However, even if the solar medium is transparent to neutrinos, something happens to them.

Observation of solar neutrinos gives fundamental information about stellar structure and evolution and about the properties of neutrinos, due to the wide range of matter densities in the Sun. The neutrino flux and energy spectrum at the source cannot obviously be measured. But we have a reliable 'solar standard model' (SSM) due to 40 years of work by John Bahcall and collaborators (Bahcall *et al.* 1964, 2005). Fig. 10.11 shows the principal components of the pp cycle and the corresponding contributions to the neutrino energy spectrum, in a simplified form. We have neglected the contribution of the carbon–nitrogen–oxygen (CNO) cycle, which is small in the Sun.

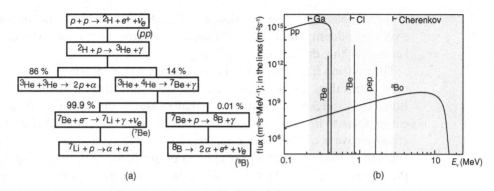

Fig. 10.11. (a) The *pp* cycle. (b) The principal components of the neutrino energy spectrum (from Bahcall *et al*. 2005). *pep* neutrinos come from the reaction $e^- + p + p \rightarrow d + v_e$. The sensitive regions of different experimental techniques are also shown.

We see in Fig. 10.11 that the largest fraction of the energy flux is due to the '*pp* neutrinos', namely the elementary reaction (10.30). This component of the flux is obtained from the measured luminous flux and is almost independent of the details of the solar model. However, it is the most difficult experimentally because of its very low neutrino energies, which are below 420 keV.

Two processes produce higher-energy neutrinos. The first process gives the 'beryllium neutrinos' through the reaction

$$^7\text{Be} + e^- \rightarrow {}^7\text{Li} + \gamma + v_e. \tag{10.31}$$

The neutrino flux is dichromatic, with the principal line at 0.86 MeV. The second process gives the 'boron neutrinos'

$$^8\text{B} \rightarrow 2\alpha + e^+ + v_e. \tag{10.32}$$

This is the higher-energy component with a spectrum reaching 14 MeV (neglecting much weaker components). As such it is the least difficult to detect. Notice that the boron is produced by the beryllium via the reaction

$$^7\text{Be} + p \rightarrow {}^8\text{B} + \gamma. \tag{10.33}$$

The last two processes make very small contributions to the electromagnetic energy flux we measure and, as a consequence, our knowledge of the corresponding fluxes is heavily based on the solar model and its parameters.

Example 10.1 Knowing that the solar constant, that is the flux of electromagnetic energy from the Sun on the surface of the Earth, is 1.3 kW m^{-2}, evaluate the total neutrino flux.

The energy produced by reaction (10.29) transported by photons is 26.1 MeV for every two neutrinos. The energy per neutrino is $26.1/2 = 13.05 \text{ MeV} = 2.1 \times 10^{-12} \text{ J}$.

The neutrino flux is then

$$\Phi_v = \left(1.3 \times 10^3 \text{ J m}^{-2} \text{ s}^{-1}\right) / \left(2.1 \times 10^{-12} \text{ J}\right) = 6.2 \times 10^{14} \text{ m}^{-2} \text{ s}^{-1}. \quad \square$$

We now observe that all reactions produce electron neutrinos and that they do so in a very high-density medium. Neutrinos will then cross a medium of decreasing density before reaching the surface of the Sun.

The neutrino stationary states in matter are not v_1, v_2 and v_3 and the corresponding mass eigenvalues are not m_1, m_2 and m_3. We shall call them $\tilde{v}_i$ and $\tilde{m}_i$, respectively. Both Earth and Sun, the mediums we shall consider, appear to neutrinos as transparent as water to light. As for light, the phase velocity of a monochromatic (monoenergetic) wave is different from that in vacuum, the ratio of the latter to the former being the refraction index n. The effect is proportional to the scattering amplitude in the forward direction. We parameterize the average interaction of neutrinos with the medium as an effective potential V. The refraction index for neutrinos of energy E is $n = 1 + V / E$. Its difference from the value in a vacuum is quite small; for example, for $E = 10$ MeV $n - 1 = 10^{-20}$ in the Earth and $n - 1 = 10^{-18}$ in the centre of the Sun. The potential differences are physically meaningful, not the potentials themselves. These differences may be different for neutrinos of different flavour due to their different interaction with matter. All neutrinos interact with electrons and quarks by NC weak interactions, independently of their flavour. Only electron neutrinos interact with electrons and quarks also by CC weak interactions. Consequently, their potential $V_e(r)$ is different from that of the other flavours $V_{\mu,\tau}(r)$. It can be shown that this difference is

$$\Delta V(r) \equiv V_e(r) - V_{\mu,\tau}(r) = \sqrt{2} G_F N_e(r), \tag{10.34}$$

where $N_e(r)$ is the electron number density at distance r from the centre of the Sun. Notice that, differently from the cross-section that is proportional to G_F^2, here the effect is proportional to G_F, making it more sizeable.

In the energy range of the solar neutrinos, v_μ and v_τ are indistinguishable because their interactions, which are of neutral current only, are identical. If we also assume $\theta_{13} = 0$ in a first approximation, we reduce the problem to that of two neutrino species, v_e and, say, v_α, where the latter is a superposition of v_μ and v_τ that we do not need to define.

Let us start by considering the evolution of the system in a uniform density medium. It is given by

$$i \frac{d}{dt} \begin{pmatrix} v_e(t) \\ v_\alpha(t) \end{pmatrix} = H_m \begin{pmatrix} v_e(t) \\ v_\alpha(t) \end{pmatrix}. \tag{10.35}$$

The important element in the Hamiltonian (see Mohapatra & Pal 2004) is

$$H_m = \begin{pmatrix} -\dfrac{\delta m^2}{4E} \cos 2\theta_{12} + \sqrt{2}\, G_F N_e & \dfrac{\delta m^2}{4E} \sin 2\theta_{12} \\[2ex] \dfrac{\delta m^2}{4E} \sin 2\theta_{12} & \dfrac{\delta m^2}{4E} \cos 2\theta_{12} - \sqrt{2}\, G_F N_e \end{pmatrix}. \tag{10.36}$$

Notice that the scattering produced by the potential does not change the flavour. As a consequence, the potential appears only in the diagonal terms of the Hamiltonian. The flavour is changed by the off-diagonal terms. These do not depend on density.

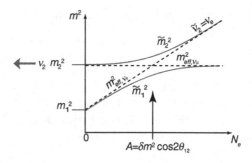

Fig. 10.12. Continuous curves are the eigenvalues of the squares of the masses as functions of A. Dashed lines are the effective squared masses of the flavour states.

Recalling Eq. (10.17), we see that the mixing is determined in matter by an 'effective mixing angle' $\theta_{12,m}$ given by

$$\tan 2\theta_{12,m} = \frac{2H_{m,12}}{H_{m,22} - H_{m,11}} = \frac{\delta m^2 \sin 2\theta_{12}}{\delta m^2 \cos 2\theta_{12} - A}, \tag{10.37}$$

where

$$A = 2\sqrt{2}G_F N_e E. \tag{10.38}$$

The effect of matter becomes dramatic for a particular value of the electron density:

$$N_e = \frac{1}{E} \frac{\delta m^2 \cos 2\theta_{12}}{2\sqrt{2}G_F.}. \tag{10.39}$$

At this density, Eq. (10.37) diverges, meaning that the effective mixing angle becomes $\theta_{12,m} = \pi / 4$. Consequently, the mixing becomes maximal even if the (vacuum) mixing angle is small. This is called 'resonance condition'.

To understand the phenomenon we must consider the stationary states of the system in the medium ($\tilde{\nu}_i$) and the corresponding eigenvalues ($\tilde{m}_i$), the effective masses, by diagonalizing the Hamiltonian (10.36). Skipping the calculation, we give the result graphically in Fig. 10.12.

We observe that, at high enough densities, electron neutrinos have an effective mass, due to their interaction with the electrons of the medium, larger than the other flavours. The opposite is true at low densities and, in particular, in a vacuum. We have a level-crossing phenomenon, a situation also found in other fields of physics.

As observed by Mikheyev and Smirnov (Mikheyev & Smirnov 1985), the crossing of the resonance can induce a change of the neutrino flavour. The eigenstates and their eigenvalues in a non-uniform medium differ from point to point; consequently, the description of the propagation of the system is not simple. Actually, the evolution of the system has different characteristics, depending on the various physical quantities of the problem. Nature has chosen the simplest situation in the Sun (but we did not know that at the beginning) in which the flavour conversion is adiabatic. Under these conditions, the electron neutrino state evolves following the upper curve in Fig. 10.12.

First consider the mixing in vacuum, where the mixing angle is θ_{12}. We have

$$\begin{aligned}\nu_1 &= \cos\theta_{12} \times \nu_e - \sin\theta_{12} \times \nu_\alpha \\ \nu_2 &= \sin\theta_{12} \times \nu_e + \cos\theta_{12} \times \nu_\alpha.\end{aligned} \tag{10.40}$$

In the very high-density core regions, where neutrinos are produced, $A \gg \delta m^2 \cos 2\theta_{12}$, the diagonal elements of the Hamiltonian (10.36) dominate and therefore mixing is suppressed. Initially, $\tan 2\theta_m$ is negative, then tends to zero with decreasing N_e, that is, $\theta_m \to \pi/2$. Consequently, we have

$$\begin{pmatrix} \nu_e \\ \nu_\alpha \end{pmatrix} = \begin{pmatrix} \cos\theta_{12,m} & \sin\theta_{12,m} \\ -\sin\theta_{12,m} & \cos\theta_{12,m} \end{pmatrix} \begin{pmatrix} \tilde{\nu}_1 \\ \tilde{\nu}_2 \end{pmatrix} \approx \begin{pmatrix} 0 & 1 \\ -1 & 0 \end{pmatrix} \begin{pmatrix} \tilde{\nu}_1 \\ \tilde{\nu}_2 \end{pmatrix} \to \begin{matrix} \tilde{\nu}_1 \approx \nu_\alpha \\ \tilde{\nu}_2 \approx \nu_e \end{matrix}. \tag{10.41}$$

The important conclusion is that electron neutrinos are produced in a mass eigenstate, to be precise the one with the larger mass, $\tilde{\nu}_2$. The other eigenstate $\tilde{\nu}_1$ is not produced. The $\tilde{\nu}_2$ eigenstate then propagates towards lower-density regions and may encounter a layer in which the resonance conditions are satisfied. The density varies slowly enough to satisfy the adiabaticity condition. Consequently, the state follows the upper curve in Fig. 10.12. Finally, when neutrinos reach the surface, they leave the Sun, still in the mass eigenstate, which is now ν_2. Notice that here a single state propagates; there is no interference, because there is nothing to interfere with. The phase of the state is irrelevant because only phase differences are physical. No oscillation can happen under these conditions; the AFC is an irreversible phenomenon.

We must now check whether neutrinos do meet the resonance or not. On their journey, neutrinos encounter all the electron densities smaller than the central one N_0. Its value is

$$N_0 \approx 6 \times 10^{33} \text{ m}^{-3}. \tag{10.42}$$

Neutrinos will meet the resonance if there is a density smaller than N_0 satisfying Eq. (10.39). Having fixed all the other quantities, this is a condition on the neutrino energy, which, with the known values of δm^2 and θ_{12} is

$$E > \frac{\delta m^2 \cos 2\theta_{12}}{2\sqrt{2} G_F N_0} \approx \delta m^2 \cos 2\theta_{12} \times 6.7 \times 10^{10} \text{ eV} \approx 2 \text{ MeV}. \tag{10.43}$$

In practice, the transition between vacuum oscillation and high-density conversion takes place smoothly in an energy interval of a few MeV, as we shall see in Section 10.4. At the highest energies, neutrinos emitted at the Sun surface are ν_2 and will remain in this state until they propagate in a vacuum. A detector on Earth sensitive to ν_e will observe only the component of amplitude $\sin\theta_{12}$, as from Eq. (10.40). The survival probability of electron neutrinos from the production to the detection points is

$$P_{ee} = \sin^2\theta_{12} \quad E \gg 2 \text{ MeV}. \tag{10.44}$$

Neutrinos of lower energy do not encounter the resonance, and propagate in the Sun as in a vacuum. They oscillate with a maximum excursion

$$A(v_e \to v_\alpha) = \sin^2(2\theta_{12}). \tag{10.45}$$

This factor multiplies the oscillating term. Our detectors take the average value of the oscillation term on times much longer than the oscillation periods, which is $1/2$. The survival probability is, in conclusion,

$$P_{ee} = 1 - \frac{1}{2}\sin^2 2\theta_{12} \quad E \ll 2\,\text{MeV}. \tag{10.46}$$

We now notice that the resonance corresponding to the larger neutrino square-mass difference Δm^2 does not exist in the Sun, because for that to occur neutrino energies should be about 33 times larger than the limit (10.43). Both resonances can exist in supernovae, where the densities are much larger.

The analogy with pendulums, even if not exact now, is useful to illustrate the adiabatic conversion too. With refence to Fig. 10.5(f), consider the pendulum on the right, called e, as a proxy of an electron neutrino in the centre of the Sun. It is vibrating with a certain amplitude. If the lengths of the two pendulums are very different (small mixing angle), the other pendulum, α, will not move much. If we now slowly pull the thread, gradually shortening the length of e, when the two lengths are almost equal, the resonance condition, pendulum α, will start oscillating with increasing amplitude. If the resonance condition is maintained for the right time, namely if the process is adiabatic, the entire oscillation energy is transferred from e to α. Then, continuing to shorten the length of e, the difference between the two pendulums will be again so large that α will keep oscillating at full amplitude. The process is one way; there is no oscillation back and forth of energy (no oscillation of the probability to observe α).

10.4 The Shorter Period Oscillation

The historical process leading to the discovery of the neutrino flavour conversion in the Sun is not due to a single experiment, but rather to a series of experimental and theoretical developments. In Section 2.4 we mentioned the radiochemical method to detect electron neutrinos put forward by B. Pontecorvo in 1946 and made quantitative by L. Alvarez in 1949, via the 'Cl–Ar reaction':

$$v_e + {}^{37}\text{Cl} \to e^- + {}^{37}\text{Ar}. \tag{10.47}$$

In 1962, J. Bahcall (Bahcall *et al.* 1964) started the construction of a solar model, which will be called the solar standard model (SSM), and the calculation of the expected reaction rate in the reaction (10.47). Unlike a reactor that produces antineutrinos, with solar neutrinos, which are v_e, the reaction can proceed. The initial result was discouraging: the rate was too small to be detectable. However, soon afterwards, Bahcall noticed the presence of a super-allowed transition to an analogue state of ${}^{37}\text{Ar}$ at about 5 MeV of excitation energy. This increased the estimated rate by almost a factor of 20. The experiment was feasible.

R. Davis used 615 t of perchloroethylene (C_2Cl_4) as the target detector medium, in which about one Ar nucleus per day produced by reaction (10.47) was expected. Every few weeks the metastable Ar nuclei within the atoms they had formed were extracted by using a helium gas stream. After suitable chemical processing, the extracted gas was introduced into a counter to detect the ^{37}Ar decays. Since the signal rate was only of a few counts per month, it was mandatory to work deep underground to be shielded from the cosmic rays and to use only materials extremely free of radioactive components. The experiment took place deep underground in the Homestake Mine in South Dakota at 1600 m depth, producing the first results in 1968 (Davis *et al.* 1968), and continued up to 1994.

The energy threshold of the experiment, 814 keV as shown in Fig. 10.11, allows the detection of the highest-energy neutrinos, the beryllium and boron neutrinos. In 1968 the measured flux already appeared to be substantially lower than the expected one. It was the beginning of the 'solar neutrino puzzle'. The value of the solar neutrino capture rate *per atom* resulting from running the experiment for a quarter of a century from 1970 to 1994 (Cleveland *et al.* 1998) is

$$R(Cl, exp.) = (2.56 \pm 0.16 \pm 0.16) \times 10^{-36} \text{ s}^{-1}. \tag{10.48}$$

This value is about 1/3 of the SSM prediction (Bahcall *et al.* 2005)

$$R(Cl, SSM) = (8.1 \pm 1.3) \times 10^{-36} \text{ s}^{-1}. \tag{10.49}$$

The first confirmation came from the Kamiokande experiment in 1987, which, like its larger successor Super-Kamiokande, was a smaller water Cherenkov detector. The experiment detected the electrons hit by neutrinos from the Sun in an elastic scattering

$$v_x + e^- \rightarrow v_x + e^-. \tag{10.50}$$

All neutrino flavours contribute; however, while electron neutrinos scatter via both NC and CC, the other two flavours scatter only via NC and, consequently, with a cross-section about 1/6 of the former. Given its energy threshold, the experiment was sensitive to the boron neutrinos only. The rate measured by Kamiokande (Hirata *et al.* 1989) was about one half of the expected rate, a value that was later confirmed by Super-Kamiokande (Hosaka *et al.* 2006). The measured flux is

$$\Phi_{exp} = (2.35 \pm 0.02 \pm 0.08) \times 10^{10} \text{ m}^{-2}\text{s}^{-1}, \tag{10.51}$$

while the theoretical one is

$$\Phi_{SSM} = (5.69 \pm 0.91) \times 10^{10} \text{ m}^{-2}\text{s}^{-1}. \tag{10.52}$$

By measuring the direction of the electron hit by the neutrino in the elastic scattering, Super-Kamiokande also established that neutrinos were coming from the Sun (the first neutrino telescope). At this point the existence of a problem was well established. However, was the problem due to some flaw in the solar model or to an anomalous behaviour of the neutrinos? Indeed, as anticipated, the high-energy neutrino flux is very sensitive to the values of the parameters of the model. For example, it depends on the temperature of the core as T^{18}.

The answer could come only from the measurement of the pp neutrino flux, which can be calculated from the solar luminosity with a 2% uncertainty. Two radiochemical experiments were built for this purpose: GALLEX, in the LNGS in Italy, and SAGE, in the Baksan Neutrino Observatory (BNO) in Russia, both deep underground. Both employ gallium as the target, 30 t and 60 t respectively. They are sensitive to electron neutrinos via the inverse beta decay reaction

$$v_e + {}^{17}\text{Ga} \rightarrow e^- + {}^{71}\text{Ge}, \tag{10.53}$$

with 233 keV energy threshold. GALLEX published the first results in 1992 (Anselmann *et al.* 1992): the ratio between the expected and measured rate was about 60%. SAGE soon confirmed this value. GALLEX ended in 1997, becoming, in an improved version, GNO, which ended in 2003. SAGE ended in 2014. The values, measured in solar neutrino units (SNU) (1 SNU = 10^{-36} captures per atom per second), of the electron neutrinos capture rates (Abdurashitov *et al.* 2002; Altman *et al.* 2005) are

$$\begin{aligned} R(\text{Ga}, \text{GALLEX} + \text{GNO}) &= 69.3 \pm 4.1 \pm 3.6 \, \text{SNU} \\ R(\text{Ga}, \text{SAGE}) &= 70.8^{+5.3+3.7}_{-5.2-3.2} \, \text{SNU}. \end{aligned} \tag{10.54}$$

These values, in mutual agreement, are again much smaller than the SSM prediction (Bahcall *et al.* 2005)

$$R(\text{Ga}, \text{SSM}) = 126 \pm 10 \, \text{SNU}. \tag{10.55}$$

Actually, by 1995, GALLEX (Anselmann *et al.* 1995) had reached such a precision that they could exclude the 'solar solution' of the puzzle by the following argument. The rate measured by GALLEX is the sum of three main contributions: from pp, from boron and from beryllium (see Fig. 10.11(a)). However, the sum of the first, as evaluated from the solar luminosity, and of the second, as measured by Super-Kamiokande, was already larger than the rate measured by GALLEX. Consequently, no space is left in the budget for beryllium neutrinos, which, on the other hand, must be present, because the boron, which exists as observed by Super-Kamiokande, is a daughter of beryllium (see Fig. 10.11(b)). This result was also later confirmed by SAGE. Further controls, calibrations and independent measurements of the relevant nuclear cross-sections led to the conclusion in 1997 that the solution of the puzzle was in the anomalous behaviour of neutrinos. In other words, neutrinos do not behave as assumed in the Standard Model. This was historically the first experimental evidence of physics beyond the SM.

The experiments were sensitive only (or almost so in the case of the Cherenkov experiments) to electron neutrinos. Apparently, electron neutrinos were disappearing in a large fraction on their way from the solar centre to the Earth. The most probable hypothesis was the AFC we have described in the previous sections.

The final proof came from the SNO experiment in 2002. The experiment was sensitive to electron neutrinos via CC interactions. In addition, it detected NC interactions, which are induced by all the neutrino flavours. In other words, the experiment detected both the neutrinos that did not oscillate and those produced by the oscillation, even if it did not distinguish their flavours.

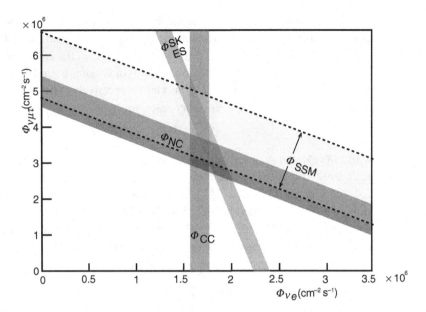

Fig. 10.13.   Fluxes of ^{8}B neutrinos $\Phi(\nu_e)$ and $\Phi(\nu_{\mu,\tau})$ deduced from SNO CC and NC results and from Super-Kamiokande elastic scattering (ES) results. Bands represent 1σ uncertainties. SSM predictions are also shown. By permission of Particle Data Group, Lawrence Berkeley National Lab.

The SNO experiment was a heavy-water Cherenkov detector with 1000 t of D_2O, located in a clean experimental hall built in a mine 2000 m deep in Canada. The observations started in 1999. More specifically, electron neutrinos are detected through the charged current reaction

$$\nu_e + d \to p + p + e^- \tag{10.56}$$

and all flavours through the neutral current reaction

$$\nu_x + d \to p + n + \nu_x. \tag{10.57}$$

The rate of (10.57) measures the total neutrino rate independently of the flavour, while the rate of (10.56) gives the contribution of electron neutrinos. SNO also measured the elastic cross-section, with results in agreement with Super-Kamiokande, with much poorer statistics, however. The experiment is sensitive in the higher energy part of the spectrum, namely to boron neutrinos.

We now extract from the measured rates the fluxes of electron neutrinos and of muon or tau neutrinos (which are indistinguishable) with the help of Fig. 10.13. The CC (10.56) event rate gives directly the electron neutrino flux, independently of the muon/tau neutrino flux. It is the vertical band in the figure. The NC rate (10.57) gives the sum of the three fluxes, the three cross-sections being equal. It is the band along the second diagonal of the figure. The elastic scattering (ES) rate was measured also by SNO, but much more accurately by Super-Kamiokande. The latter is shown in Fig. 10.13, appearing as an almost vertical band, taking into account that

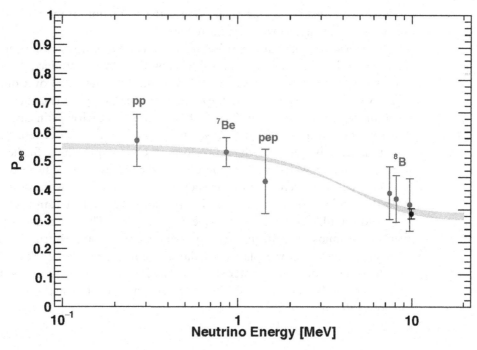

Fig. 10.14. Electron neutrino survival probability as a function of neutrino energy. The points represent, from left to right, the Borexino pp, ^{7}Be, pep and ^{8}B data and the SNO + SK^{8}B data. The error bars represent the $\pm 1\sigma$ experimental plus theoretical uncertainties. The curve corresponds to the $\pm 1\sigma$ MSW theoretical prediction. (Courtesy of Aldo Ianni).

$\sigma\left(\nu_{\mu,\tau} + e\right) \approx 0.16\sigma\left(\nu_e + e\right)$. Having three relationships between two unknown quantities, we can check for consistency. This is proven by the fact that the three bands cross in the same area. The values of the fluxes (Aharmin *et al.* 2005) are

$$\Phi_{CC}\left(\nu_e\right) = \left(1.68 \pm 0.06^{+0.08}_{-0.09}\right) \times 10^8 \text{ m}^{-2} \text{ s}^{-1} \tag{10.58}$$

and

$$\Phi_{NC}\left(\nu_x\right) = \left(4.94 \pm 0.21^{+0.38}_{-0.34}\right) \times 10^8 \text{ m}^{-2} \text{ s}^{-1}. \tag{10.59}$$

The total neutrino flux agrees with the predictions of the SSM. This proves that the missing electron neutrinos have indeed transformed into muon and/or tau neutrinos via adiabatic flavour conversion (not via oscillation).

We saw in Section 10.3 that electron neutrinos produced in the core of the Sun do not meet the MSW resonance if their energy is low, whereas they do meet it at higher energies. Their survival probability P_{ee} well below and above the resonance is given by (10.46) and (10.45), respectively. The experimental study of P_{ee} in the transition region between vacuum oscillation and MSW effect is an important test of the theory. This was done at LNGS by Borexino, an experiment using a hyper-pure liquid scintillator detector of 100 t fiducial mass. Borexino measured, in particular, the monoenergetic ν_e fluxes from ^{7}Be and pep, namely from the reaction $e^- + p + p \rightarrow d + \nu_e$ (see Fig. 10.11). Fig. 10.14 shows the corresponding survival probability below and above the transition region. The curve is the prediction of the SSM and the MSW process. It is in agreement with the data.

Solar neutrino data are sensitive mainly to two parameters of the neutrino system: δm^2 and θ_{12}. We now briefly consider how these are determined. The experiments on solar neutrinos have measured the neutrino flux integrated over different energy intervals. To each value, and to the associated uncertainty, corresponds a region in the parameter plane, δm^2 versus θ_{12}, which is different for each experiment. To be compatible with all the measurements, the solution must lie in the intersection of these regions. In practice, a best-fit procedure is performed, obtaining, with the further contribution of KamLAND, which we shall now discuss, the values reported in Section 10.1.

The AFC discovered with solar neutrinos was confirmed by the observation of the corresponding oscillation in the disappearance experiment KamLAND (Kamioka Liquid Scintillator Anti-Neutrino Detector) in 2002. The experiment was installed in the old Kamiokande site in Japan. It used the electron antineutrinos produced by 55 Japanese nuclear power plants. The latter are located near the ocean shores, owing to their need for nearby water sources. The fortunate consequence is that their distances from the detector, which is well inland, are, for their majority, roughly equal, about 180 km.

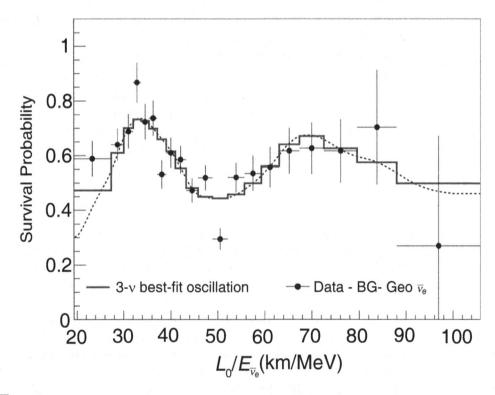

Fig. 10.15. Ratio of the background-and-geoneutrinos-subtracted $\bar{\nu}_e$ spectrum to the expectation for no-oscillation as a function of $L_0 / E_{\bar{\nu}_e}$. Particle Data Group, Lawrence Berkeley National Lab.

KamLAND is a 1 kt ultra-pure scintillator detector, utilizing the inverse beta decay reaction (10.27) to detect the $\bar{\nu}_e$. The positrons, with 2.8 MeV threshold, are detected and their energy measured with the scintillation light. The neutrino energy is given by $E_{\bar{\nu}_e} \approx E_{e^+} + 0.8$ MeV as we saw in Section 10.2. The neutron, once moderated, joins to a proton to produce a deuterium nucleus (99% of time), emitting a 2.2 MeV gamma ray. The detection of this gamma ray in a delayed coincidence is a powerful tool for reducing background.

The total counting rate and the energy spectrum expected in the absence of oscillations were calculated, taking into account the powers (which vary with time) and the distances of all the power stations. Comparison with the measured ones showed that neutrinos had disappeared (Eguchi *et al.* 2003). Fig. 10.15 (Abe *et al.* 2008) shows the ratio between the $\bar{\nu}_e$ measured spectrum, after subtraction of the instrumental background and of the contribution of the 'geoneutrinos', namely the antineutrinos produced by the radioactive decays in the Earth, to the expectation for no oscillation, as a function of $L_0/E_{\bar{\nu}_e}$ (Abe *et al.* 2008). Here L_0 is the effective baseline taken as a flux-weighted average ($L_0 = 180$ km). Data are then fitted by using the oscillation hypothesis with δm^2 and θ_{12} as free parameters. The histogram and curve show the expectations for the fitted values. The oscillation behaviour is clearly visible.

10.5 Search for CP Violation in Leptons

In this section we discuss two experiments on accelerator muon neutrino beams. Both run alternatively with ν_μ and $\bar{\nu}_\mu$ beams, for reasons that we shall explain. They look at the disappearance of the original flavour and at the appearance of ν_e or $\bar{\nu}_e$. Both aim at a precision measurement of the smallest mixing angle θ_{13}, and, even more importantly, to search for CP violation in the neutrino sector, namely if the phase δ is different from $0, \pi, 2\pi$. These experiments contribute to the value in Eq. (10.5). Notice that only appearance experiments are sensitive to CP violation. Indeed, any disappearance probability is the complement to 1 of the survival probability of the initial flavour, a process coincident with its opposite and hence not changing under time reversal. Consequently, CPT invariance implies independence of the CP-violating Dirac phase δ.

Consider now the electron neutrino appearance in an initially pure ν_μ beam produced at an accelerator. As we shall be looking at second-order effects in the hierarchy parameter α of Eq. (10.7), we must consider also terms that we neglected so far. Using a series expansion in α given by M. Freund (Freund 2001) we write

$$\Delta = \Delta m^2, \quad \Delta m_{21}^2 = \alpha\Delta \approx 0.03\Delta, \quad \Delta m_{32}^2 = (1 - \alpha)\Delta \tag{10.60}$$

For the oscillation probability in a vacuum at the first order in α, we have

$$P(\nu_\mu \to \nu_e) = P_0 + P_{\sin\delta} + P_{\cos\delta}, \tag{10.61}$$

where

$$P_0 = \sin^2 \theta_{23} \sin^2 2\theta_{13} \sin^2 \left(1.27 \Delta \frac{L}{E} \right) \tag{10.62}$$

is the zeroth-order term considered so far. There are two first-order terms in α. One is

$$P_{\sin \delta} = \pm \alpha \sin \delta \cos \theta_{13} \sin 2\theta_{12} \sin 2\theta_{23} \sin 2\theta_{13} \sin^2 \left(1.27 \Delta \frac{L}{E} \right) \tag{10.63}$$

where $+$ is for neutrinos and $-$ for antineutrinos. This is why it is useful for an experiment looking for CP violation to run both in neutrino and antineutrino modes. The second one is

$$P_{\cos \delta} = \alpha \cos \delta \cos \theta_{13} \sin 2\theta_{12} \sin 2\theta_{23} \sin 2\theta_{13} \cos \left(1.27 \Delta \frac{L}{E} \right) \sin^2 \left(1.27 \Delta \frac{L}{E} \right). \tag{10.64}$$

Further terms must be considered when the beam travels in matter on a long enough distance. Neutrinos travel in the Earth crust in a constant density medium, which simply changes the refractive index of the neutrino wave. The matter effects are coded in the following quantity (the two signs are for neutrinos and antineutrinos)

$$x = \pm \frac{2\sqrt{2} G_F N_e E}{\Delta m^2} \tag{10.65}$$

where G_F is the Fermi constant and N_e the neutrino number density. For both T2K and NOνA, this quantity is quite small. In both of them, the energy is chosen to have the oscillation maximum at its baseline: for T2K, $L = 295$ km, peak energy $E = 0.6$ GeV and $x = 0.05$; for NOνA, $L = 810$ km, $E = 2$ GeV and $x = 0.17$. Matter effects result in the following substitutions of the just mentioned terms of the electron neutrino appearance probability. In Eq. (10.62)

$$\sin^2 \left(1.27 \Delta \frac{L}{E} \right) \Rightarrow \frac{\sin^2 \left((1-x)1.27 \Delta \frac{L}{E} \right)}{(1-x)^2}, \tag{10.66}$$

and in both Eqs. (10.63) and (10.64)

$$\sin^2 \left(1.27 \Delta \frac{L}{E} \right) \Rightarrow \frac{\sin \left(x1.27 \Delta \frac{L}{E} \right) \sin \left((1-x)1.27 \Delta \frac{L}{E} \right)}{x(1-x)}. \tag{10.67}$$

The T2K muon neutrino beam is produced using the extracted high-intensity proton beam from the J-PARC 30 GeV rapid cycling proton synchrotron in Japan. The neutrino beam line is 2.5° off-axis to the proton beam. This 'trick' exploits the kinematics to produce a narrow energy band beam, centred, for T2K, at 600 MeV. The energy is tuned to have the oscillation maximum at the far detector, which is Super-Kamiokande (SK) at a distance of 295 km.

In SK, electron neutrinos and antineutrinos are detected by observing the electron or positron produced in the scatterings $\nu_e + n \rightarrow e^- + p$ or $\bar{\nu}_e + p \rightarrow e^+ + n$. The arrival time, direction and energy are measured. The process has no threshold; consequently, the neutrino beam energy can be relatively small. This has the advantage of reducing the background due to neutral current interactions of non-oscillated (the vast majority) v_μ, producing $\pi^0 \rightarrow 2\gamma$ with a photon simulating an electron.

The system near the detectors, located at 280 m from the target, provides the crucial information on the neutrino flux, angular distribution and energy spectrum before oscillation. All of that must be accurately measured in a precision experiment.

T2K runs alternatively with neutrinos and antineutrinos to optimize its sensitivity to CP violation. In particular, values of the phase in the second quadrant (maximally for $\delta = -\pi / 2$) enhance $v_\mu \rightarrow v_e$ and suppress $\bar{v}_\mu \rightarrow \bar{v}_e$, while those in the first quadrant (maximally for $\delta = +\pi / 2$) suppress $v_\mu \rightarrow v_e$ and enhance $\bar{v}_\mu \rightarrow \bar{v}_e$. Furthermore, NO enhances $v_\mu \rightarrow v_e$ and suppresses $\bar{v}_\mu \rightarrow \bar{v}_e$, and IO does the opposite.

For normal mass ordering, the best-fit value and 1σ uncertainties are $\delta = -1.89^{+0.70}_{-0.58}$, and for inverted ordering $\delta = -1.38^{+0.48}_{-0.54}$; in both cases the CP conserving points $\delta = 0$ and $\delta = \pi$ are excluded at the 95% confidence level (T2K Collaboration 2020). This is a strong hint, but not yet a discovery, which usually can be claimed at 5σ. Considering that the result is limited by statistical uncertainty, it will hopefully become more precise when more data are collected.

The second long baseline accelerator experiment is NOvA, on the NuMI muon neutrino beam from FNAL, which has reached the record beam intensity on target of 700 kW. The beam is narrow-band 14 mrad off-axis peaked at 2 GeV tuned to have the oscillation maximum at the far detector. The near detector at FNAL and the far detector at 810 km at Ash River (North Dakota) are functionally identical. The far detector is a tracking calorimeter made of liquid scintillator cells, 15.5 m long having 6.6×3.9 cm^2 cross-section located on the surface (rather than deep underground). The detector mass is 14 kt, of which 10.3 kt is fiducial. Up to now NOvA has excluded some value of δ for the inverted mass ordering, but no significant limit for the normal ordering.

Finally, the matter effects described by Eqs. (10.66) and (10.67) in both experiments, and Super-Kamiokande in addition, contribute, in the global fit, to making the NO favoured over the IO.

In conclusion, we mention that two large-scale long baseline experiments are in preparation, DUNE in the USA with a 1 300 km baseline and a 40 kt liquid Argon TPC far detector, and Hyper-Kamiokande in Japan, a water Cherenkov detector of 260 kt total mass.

10.6 Limits on Neutrino Mass

One physical quantity, or more for redundancy, independent of oscillations must be measured to know the neutrino mass spectrum. There are three experimental or observational possibilities.

- Beta decay experiments that probe the weighted average $m_{\nu_e}^2 \equiv \sum_{i=1}^{3} |U_{ei}|^2 \, m_i^2$.

- Cosmological observations that probe the sum of neutrino masses $\sum_{i=1}^{3} m_i$.

- Experiments on neutrino-less double-beta decay, a process that can exist only if neutrinos are Majorana particles. They probe the quantity $m_{ee} \equiv \left| \sum_{i=1}^{3} U_{ei}^2 m_i \right|$. Notice that the addenda are complex numbers.

We now discuss the first two measurements, which have given only upper limits up to now. We shall give a hint on double-beta decay in the next section.

We can give here only a few hints on cosmology, in relation to neutrino masses. Cosmology has made tremendous progress in the last few decades in the modelling, the quantity and, more importantly, the quality of the observational data. The different components of the mass–energy budget of the Universe have been determined within a few per cent accuracy. A standard cosmological model has been developed, called Lambda-Cold Dark Matter (ΛCDM), where lambda is the cosmological constant. The structures present in the Universe have been seeded by quantum fluctuations that took place soon after the Big Bang and grew during the following expansion. Since the gravitational interaction is only attractive, the regions of higher density attracted more and more mass into them.

The large-scale structures and their evolution are studied with a number of astrophysical observations, in different epochs, corresponding to different distances from us. Initially electrons and protons were mixed in a plasma, not being bound in atoms because the temperature was too high. Light was immediately absorbed by the plasma and could not reach us. When the Universe was 380 000 years old, the temperature had decreased enough and atoms were formed. The process is called *recombination*. The Universe became transparent and we have its earliest picture, the *cosmic microwave background* (CMB). Measuring accurately the temperature over all the sky, we see its fluctuations. The acoustic stationary waves that were present in the primordial plasma of photons and matter determined the distribution of the sizes of the fluctuations. The acoustic oscillation was due to the opposite actions of the gravitational attraction of the non-relativistic matter, both visible and invisible, and the repulsive pressure of the photons and other types of radiation. The high-accuracy measurement of the distribution in size of the fluctuations, called the CMB spectrum, has given enormous information to cosmology.

Masses between us and light sources farther away distort the light rays with their gravitational potential. The effect, called *gravitational lensing*, gives information on the total mass, including neutrinos, between us and the source.

More information is obtained by measuring the three coordinates of millions of galaxies. The *matter power spectrum* of the very large structures they form, clusters and super-clusters (collectively called large-scale structures (LSS)), is obtained. The spectrum is a function of the so-called wave number and is the Fourier transform of the correlation function of the matter density fluctuations in two points. The wave number is the Fourier conjugate of the distance between those points. An important feature is the baryon acoustic oscillation (BAO), a shell of baryon density larger than average of about 150 Mpc (1 Mpc=3.09×10^{22} m) radius. It evolved from the CMB fluctuations and is visible today amongst the clusters of galaxies.

From these and more data, the following conclusions are drawn.

The total mass–energy density is equal, within 1.5% uncertainty, to the critical one. This means that, on a large scale, the curvature of the Universe is zero. The density of the matter we know, called baryonic, is only $\Omega_b \approx 5\%$ of the total. Much more matter exists, which does not emit or absorb light, and this is called dark matter. It is non-relativistic and is called cold. Its density, as a ratio to the critical one, is $\Omega_{CDM} \approx 27\%$. We do not know what this 'cold dark matter' is. But there is more, something that accelerates the expansion rate of the Universe. Again, we do not know what it is. It is called dark energy and it might be an effect of the cosmological constant Λ, which appears in the Einstein equations, or something else. Its density is $\Omega_\Lambda \approx 68\%$. A contribution of neutrinos to the mass–energy budget has been searched for, but not (yet) found. Only an upper limit on Ω_ν has been established. Let us see how.

We start by noticing that radiation, like light, on one side, and matter, visible or dark, on the other, have different effects on the evolution of the Universe. Neutrinos are still relativistic at the epoch of recombination, if the sum of their masses is smaller than about 500 meV, and consequently behave like radiation. The shape of the CMB spectrum is mainly due to the physical evolution before recombination and, therefore, is only marginally affected by the neutrino mass. However, the lensing effect on the CMB, measured for the first time by the Planck satellite in 2013, provides important information on neutrino masses.

When the temperature had decreased so much that the neutrinos' kinetic energy became smaller than their mass, neutrinos became non-relativistic and had an influence on the growth of the large-scale structures. Owing to their small masses, they had speeds larger than the escape velocity from the smaller structures. Consequently, neutrinos could 'free stream' out of those structures, diminishing their total mass and changing the shape of the mass spectrum. To give an order of magnitude, this happens at scales of hundreds of Mpc for neutrino masses of the order of 100 meV. A galaxy, for comparison, is, in order of magnitude, 10 kpc across, while a supercluster may extend above 100 Mpc.

With these observations, and more that we do not discuss here, only upper limits on Ω_ν have been obtained, as already mentioned. Cosmology gives the relation between Ω_ν and the sum of neutrino masses as

$$\Omega_\nu = \frac{h^2}{93.14 \, \text{eV}} \sum_{i=1}^{3} m_i \approx \frac{0.5}{93.14 \, \text{eV}} \sum_{i=1}^{3} m_i \tag{10.68}$$

where $h \approx 0.7$ is the Hubble constant in units of 100 km $(\text{s kpc})^{-1}$ (not to be confused with the Planck constant). The total average neutrino number density at present is $n_\nu = 339.5 \, \text{cm}^{-3}$.

The limits on Ω_ν are model dependent and vary with the data combination adopted. A very robust upper limit comes from the CMB temperature and fluctuations alone

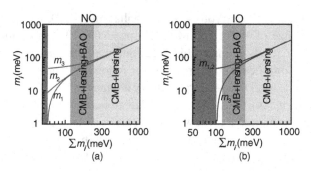

Fig. 10.16. Neutrino masses versus their sum and limits from cosmology: (a) normal mass ordering and (b) inverted mass ordering.

Palnck 2018 (Aghanim *et al.* 2020): $\sum_{i=1}^{3} m_i < 260$ meV. Adding the measurement of the BAO scale, namely the angular diameter distance at later epochs, allows us to break parameter degeneracies and, consequently, to reach a more stringent constraint (Abbott *et al.* 2022): $\sum_{i=1}^{3} m_i < 130$ meV. From neutrino oscillations, we know that the mass differences are quite smaller than this limit. Consequently, we can say that each mass must be

$$m_i < 130 \text{ meV}. \tag{10.69}$$

Fig. 10.16(a, b) shows the three neutrino masses as functions of their sum, for $\Delta m^2 > 0$ and $\Delta m^2 < 0$ respectively, and the limits from cosmology.

We now consider the beta decay of a nucleus. Non-zero neutrino mass can be detected by observing a distortion in the electron energy spectrum, just before its end point. Clearly, the sensitivity is higher if the end-point energy is lower. The most sensitive choice is the tritium decay

$$^3\text{H} \rightarrow \,^3\text{He} + e^- + \bar{\nu}_e, \tag{10.70}$$

owing to its very small Q value, $Q = m_{^3\text{H}} - m_{^3\text{He}} = 18.6$ keV. Let E_e, p_e and E_ν, p_ν be the energy and the momentum of the electron and of the neutrino, respectively. The electron energy spectrum was calculated by Fermi in his effective four-fermions interaction. We shall give only the result here, which, if neutrinos are massless, is

$$\frac{dN_e}{dE_e} \sim F\left(Z, E_e\right) p_e^2 E_\nu p_\nu = F\left(Z, E_e\right) p_e^2 \left(Q - E_e\right)^2. \tag{10.71}$$

In the last expression we have set $p_\nu = E_\nu$ since the neutrino is massless. F is a function of the electron energy characteristic of the nucleus (called the Fermi function). It may be considered a constant in the very small energy range near to the end point that we are considering. If we plot the quantity $K\left(E_e\right) \equiv \dfrac{\sqrt{dN_e / dE_e}}{p_e}$ versus E_e we obtain a straight line crossing the energy axis at Q. This diagram is called the Kurie plot (Kurie *et al.* 1936) and is shown in Fig. 10.17 as a dashed line.

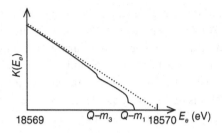

Fig. 10.17. Tritium Kurie plot with three neutrino types; mass values larger than in reality.

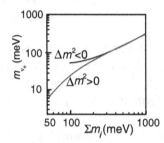

Fig. 10.18. Effective electron neutrino mass versus sum of neutrino masses.

Let us now suppose that neutrinos have a single mass m_ν. The factor $E_\nu p_\nu$ in (10.71) becomes $(Q - E_e)\sqrt{(Q - E_e)^2 - m_\nu^2}$. In the Kurie plot the end point moves to the left to $Q - m_\nu$ and the slope of the spectrum in the end point becomes perpendicular to the energy axis.

In the actual situation, with three neutrino types, the spectrum is given by

$$\frac{dN_e}{dE_e} \approx p_e^2 (Q - E_e) \sum_{i=1}^{3} |U_{ei}|^2 \sqrt{(Q - E_e)^2 - m_i^2}. \tag{10.72}$$

There are now three steps at $Q - m_i$, corresponding to the three eigenstates. Their 'heights' are proportional to $|U_{ei}|^2$. In Fig. 10.17 we have drawn the qualitative behaviour of (10.72) as a continuous curve for hypothetical values of the masses, assuming $m_3 > m_2$. Notice that the expected effects appear in the last eV of the spectrum.

In practice the energy differences between the steps are so small that they cannot be resolved, and the measured, or limited, observable is the weighted average, called the effective electron neutrino mass, even if, as we know, the electron neutrino is not a mass eigenstate:

$$m_{\nu_e} = \left(\sum_{i=1}^{3} |U_{ei}|^2 m_i^2 \right)^{1/2} \approx \left(0.68 m_1^2 + 0.30 m_2^2 + 0.02 m_3^2 \right)^{1/2}. \tag{10.73}$$

The effective electron neutrino mass is shown as a function of the sum of the neutrino masses, for both signs of Δm^2 in Fig. 10.18.

The experiment is extremely difficult. First, a very intense and pure tritium source is needed. Second, the spectrometer must be able to reject the largest part of the spectrum and to provide superior energy resolution. The best limit has been obtained by the KATRIN experiment at Karlsruhe in Germany. Its main component is the world's largest spectrometer, measuring 23 m length and 10 m diameter. The electrons from the tritium beta decay are guided into the spectrometer without changing their energy, which is then measured with sub-electronvolt accuracy. The design sensitivity of KATRIN is $m_{\nu_e} \leq 200$ meV. It started in 2018 and is still running. The present limit (Aker *et al.* 2022) is $m_{\nu_e} < 800$ meV.

10.7 Majorana Neutrinos

In the Standard Model, neutrinos are assumed to be solutions of the Dirac equation with mass rigorously equal to zero. An alternative was proposed, as we have seen in Section 2.9, in 1937 by E. Majorana (Majorana 1937), in which neutrinos are completely neutral fermions. After almost a century, no experimental proof exists of any of the two possibilities.

The Majorana field transforms into itself under charge conjugation and we write for the ν_i neutrino

$$\nu_i^C = \nu_i. \tag{10.74}$$

While completely neutral bosons may be massless, like the photon, or massive, like the Z^0, a completely neutral fermion cannot be massless, as we saw in Section 2.9. Considering the chiral projections, as we saw in Section 7.7, the charge conjugate of a left (negative chirality) field is right (positive chirality). Consequently, we can build non-zero mass terms of the form

$$m_i \overline{\nu_i^C} \nu_i.$$

Dirac neutrinos and antineutrinos are distinguished by the opposite value of the lepton number. On the contrary, Majorana neutrinos have no lepton number. As such they may induce processes violating lepton number conservation. Let us see how. Soon after the Majorana paper, G. Racah (1937) further developed his theory and proposed a way of distinguishing whether neutrinos are Dirac or Majorana particles. Consider a neutrino emitted in the beta decay of a nucleus that is later absorbed by a second nucleus via an inverse beta process. If the decay is β^-, as in the example of Fig. 10.19(a), then the absorption process is always inverse β^+. In other words, being produced with an electron, it is an antineutrino, and as such it can produce positrons, never electrons. This is the case, for example, of the Davis experiment at the Savannah River reactor we saw in Section 2.4. The opposite is true for β^+ decays. However, every neutrino, if a Majorana particle, whether produced in a β^- or β^+ decay, can produce both electrons and positrons.

The processes predicted by Racah for Majorana neutrinos have never been observed and one might conclude, as in the Standard Model, that neutrinos are Dirac particles and that their interactions conserve the lepton number. However this conclusion is not justified.

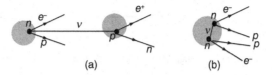

Fig. 10.19.   (a) The β^- decay followed by absorption for a Dirac neutrino. (b) Neutrino-less double-beta decay.

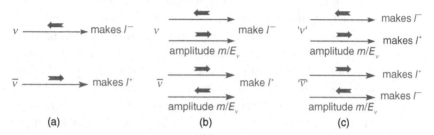

Fig. 10.20.   Helicity components of neutrinos if they are (a) massless, (b) massive Dirac and (c) massive Majorana. Long arrows are velocities; short, thick ones are the spins.

The V–A structure of the CC weak interaction and the smallness of neutrino masses, something that was not known to Majorana and Racah, are sufficient to explain the observations.

We define, in both Dirac and Majorana cases, 'electron neutrino' as the neutral particle produced with the positron in a β^+ decay, and 'electron antineutrino' as the neutral particle produced with an electron in a β^- decay. More generally, a neutrino of a certain flavour is, by definition, the neutral particle produced together with a positive lepton of that flavour in a W^- decay, and an antineutrino is the neutral particle produced with a negative lepton. The positive and negative helicity components of neutrino and antineutrino are given by Eqs. (2.63) and (2.65), that we write as

$$v \approx \frac{1}{2}\frac{m}{E}v^+ + v^- \qquad \bar{v} \approx \bar{v}^+ + \frac{1}{2}\frac{m}{E}\bar{v}^- \qquad (10.75)$$

Fig. 10.20 shows schematically three different cases, where $l = e,\ \mu,\tau$ is the lepton flavour. The charged current $\bar{v}_{iL}\gamma^\mu l_L$ creates and absorbs negative chirality neutrinos and positive chirality antineutrinos due to its V–A structure in all the cases we shall consider. Fig. 10.20(a) represents massless neutrinos. Left chirality neutrinos are eigenstates of both lepton number ($\mathcal{L} = +1$) and helicity ($h = -1$), similarly for right antineutrinos, with $\mathcal{L} = -1$ and $h = +1$. Fig. 10.20(b) represents Dirac massive neutrinos, with mass, say, m. Now neutrinos are in an eigenstate of the lepton number, with $\mathcal{L} = +1$, but not of the helicity; rather they are superpositions of a (predominant) negative helicity component and a (small) fraction, with amplitude m/E, of positive helicity (see Eq. (10.75)), and vice versa for antineutrinos. When such neutrinos interact, both helicity components produce electrons, or positrons if they are antineutrinos. Fig. 10.20(c) shows the case of Majorana neutrinos. Lepton numbers do not

(a)   (b)

Fig. 10.21.   Helicity can change by changing the reference frame.

exist for them, the two states being distinguished by helicity only. A Majorana particle produced with a positron has predominantly negative helicity with a small m/E
amplitude of a positive one, exactly as in the Dirac case, but now when the particle
interacts, its negative helicity component produces electrons, while the positive helicity one produces positrons, and vice versa for the particle produced with an electron.

Let us consider, as an example, the pion decay. The positive pion decay $\pi^+ \to \mu^+ + v_\mu$
produces an antimuon $\left(\mu^+\right)$ and a muon neutrino $\left(v_\mu\right)$. The latter particle will produce a μ^- when interacting. On the other hand, the neutral particle produced in the
decay of the π^- produces μ^+ when interacting. There is no need to invoke lepton number conservation because helicity is sufficient to distinguish between the two cases: the
neutral particle in the decay of π^+ has predominantly $h = -1$, the one in the decay of π^-
has predominantly $h = +1$. When they interact they produce predominantly μ^- and μ^+,
respectively, due to the V–A character of the interaction. So, the difference between
the Dirac and the Majorana cases, or between conservation or not of the lepton number, is all in the way the 'wrong' helicity component interacts. In the case of Dirac, in
which lepton number is conserved, the very small component of the neutral particle
in the decay of π^+ produces μ^-, in the case of Majorana, in which it is the helicity that
matters, it produces μ^+, violating the lepton number by $\Delta \mathcal{L} = \pm 2$.

However, in all the practical cases, the difference is too small to be detectable.
Consider for example the typical values of $E = 1$ GeV and $m = 100$ meV. The lepton
number violation would appear in $(m/E)^2 = 10^{-20}$ of the collisions, which is far too
small to be detectable.

Here, the reader might raise the following issue. Consider a particle of helicity $h = -1$
moving with velocity v in the positive x direction, as in Fig. 10.21(a). It is always possible to find another frame, as in Fig. 10.21(b), which moves at a relative speed larger
than v to the first one. The speed v' of the particle is now in the negative x direction, but
the direction of the spin is the same. So, for the second observer $h = +1$. Does a Majorana particle, when it interacts with a target, produce charged leptons of opposite signs,
depending on the frame? The answer is obviously no. The reason is that what matters
is the relative velocity between the neutrino and the target, which is frame independent.
The same conclusion is reached by considering that the interaction probability amplitude is given by the product of two currents, one for the leptons and one for the target.
Both currents depend on the frame, but their scalar product does not.

In 1939, W. H. Furry (Furry 1939) proposed a possible way of discovering whether
neutrinos are absolutely neutral. If they are, the process shown in Fig. 10.19(b), the
neutrino-less double-beta $(0v2\beta)$ decay, should be allowed. The possibility of observing a decay can be much larger than that of observing a decay followed by an absorption, because in the former case we can take under observation a very large number of
nuclides as those present in a macroscopic mass of matter.

The neutrino-less double-beta decay is connected to the two-neutrino double-beta decay ($2\nu 2\beta$). This is very rare, but allowed in the Standard Model. It was predicted theoretically by Maria Göppert Mayer (Göppert Mayer 1935). Both decays may happen when beta decay is energetically forbidden, as in the even–even nuclei. In the $2\nu 2\beta$ decay, two nucleons beta decay simultaneously in a second-order weak process:

$$(Z, A) \rightarrow 2e^- + 2\bar{\nu}_e + (Z + 2, A). \tag{10.76}$$

The underlying process at nucleon level is

$$2n \rightarrow 2e^- + 2\bar{\nu}_e + 2p \tag{10.77}$$

and, at the quark level, it is

$$2d \rightarrow 2e^- + 2\bar{\nu}_e + 2u. \tag{10.78}$$

The corresponding lowest-order diagram is shown in Fig. 10.21(a). The $2\nu 2\beta$ decay has been observed for ^{76}Ge, ^{100}Mo, ^{130}Te, ^{136}Xe and several other nuclides with very long lifetimes, in the range of $10^{19} - 10^{21}$ yr.

If neutrinos are completely neutral, the $0\nu 2\beta$ decay is possible too

$$(Z, A) \rightarrow 2e^- + (Z + 2, A) \tag{10.79}$$

or, at the quark level,

$$2d \rightarrow 2e^- + 2u. \tag{10.80}$$

The lowest-order diagram is shown in Fig. 10.21(b). The process violates the lepton number by two units. The Lagrangian describing Majorana neutrinos contains the same interaction term as for Dirac neutrinos, but in addition we have the mass term (10.74), which violates the lepton number.

We discuss now the meaning of the propagator. At the lower vertex, the neutrino is generated as a $\bar{\nu}_e$ as in the Standard Model. However, in the propagator the mass eigenstates appear and we must consider their coherent superposition as given by the mixing matrix. This is true both for Dirac and Majorana neutrinos. In addition, in both cases the particle is also a superposition of a component with $h = +1$ and one with $h = -1$ (of amplitude m_i/E). In the Majorana case, the latter produces electrons. It is this component that is to be selected by the interaction at the second vertex.

The factor just described reduces the probability of the $0\nu 2\beta$ decay, compared with that of the $2\nu 2\beta$ decay, by many orders of magnitude, making it proportional to the neutrino mass squared. Fortunately a different effect exists acting in the opposite direction. It is due to the fact that neutrinos in Fig. 10.22(a) are real, namely on the mass shell, while the one in Fig. 10.22(b) is virtual, off mass shell. Explicit calculations show, as found already by Furry (1939), that the effect is to enhance the decay rate by orders of magnitude. The $0\nu 2\beta$ and $2\nu 2\beta$ decay rates are expected to be similar in order of magnitude for neutrino masses of the order of 10 eV. However, neutrino masses are much smaller than that. This is a very rough evaluation, which does not take into account all the elements of the problem. Let us look into that more precisely.

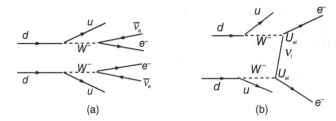

Fig. 10.22.   Lowest-order diagrams at quark level for (a) the $2\nu 2\beta$ decay and (b) the $0\nu 2\beta$ decay.

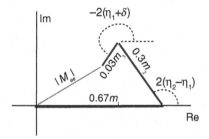

Fig. 10.23.   An illustration of the addenda in the effective electron neutrino Majorana mass; the (unknown) phases are arbitrary.

The half-life of the decay is given by

$$1/T_{1/2} = G M_{ee}^2 \, | M_{\mathrm{nucl}} |^2, \tag{10.81}$$

where M_{ee} is called 'effective Majorana mass' (or better, as this is a matrix in the three flavours space, its ee element). This is the quantity we are interested in. G is the phase-space factor, which can be calculated, and M_{nucl} is the nuclear matrix element present because the decay happens inside a nucleus. These matrix elements have not yet been precisely calculated, but are known within a factor of 2–3.

The Majorana effective mass is the coherent sum of the contributions of the three mass eigenstates

$$
\begin{aligned}
M_{ee} &= \left| \sum_{i=1}^{3} U_{ei}^2 m_i \right| \\
&= \left| \cos^2 \theta_{13} \cos^2 \theta_{12} m_1 + e^{2i(\eta_2 - \eta_1)} \cos^2 \theta_{13} \sin^2 \theta_{12} m_2 + e^{-2i(\eta_1 + \delta)} \sin^2 \theta_{13} m_3 \right| \\
&\approx \left| 0.68\, m_1 + 0.30\, e^{2i(\eta_2 - \eta_1)} m_2 + 0.02\, e^{-2i(\eta_1 + \delta)} m_3 \right|
\end{aligned} \tag{10.82}
$$

Notice that the presence of the phase factors may induce cancellations between the addenda, as shown in the illustration in Fig. 10.23.

The experiments measure the total energy E of the two electrons. Detectors are made of the double beta active nuclide to be studied; in other words, source and detector coincide, for example as germanium detectors made of enriched ^{76}Ge. The Q value of the decay, $Q_{\beta\beta}$, is known with high precision by measuring the ground-state energies of the initial and final nuclei. It depends on the isotope, ranging between about 1.5 and 3 MeV.

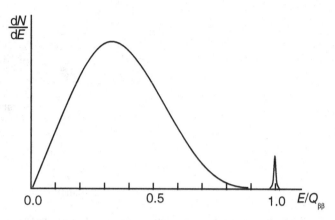

Fig. 10.24   Idealized sum electron spectrum for the double-beta decay.

In the case of $2\nu 2\beta$, which is always present, the distribution is continuous, with an end point at $Q_{\beta\beta}$, because it is a four-body decay, as shown in Fig. 10.24. On the other hand, the $0\nu 2\beta$, if present, would produce a peak at $Q_{\beta\beta}$ with a width determined by the energy resolution. The most sensitive experiments employ a mass of the active isotope of hundreds of kilos. The detector is built to measure the energy with the highest possible resolution, in practice a few to several keV (full width half maximum) in the best cases.

Fig. 10.24 is an idealization. In practice, traces of radioactive isotopes, which are present everywhere, including the detector itself, give background counts that add to the spectrum. These may contain lines, some of which might be in the signal region. Consequently, the search for the effect needs to eliminate as much as possible all the sources of natural radioactivity that can simulate the signal. Several experiments in underground laboratories are being performed and planned on a number of double beta active isotopes. The background levels of the best ones are of the order of below one count per keV per year per ton of isotope mass. No positive signal has been observed so far. The most sensitive lower limits on the half-lives, at 90% confidence level, are for ^{136}Xe, $T_{1/2}^{0\nu 2\beta}\left(^{136}\text{Xe}\right)\geq 2.3\times 10^{26}$ yr, for ^{76}Ge, $T_{1/2}^{0\nu 2\beta}\left(^{76}\text{Ge}\right)\geq 1.8\times 10^{26}$ yr and for ^{130}Te, $T_{1/2}^{0\nu 2\beta}\left(^{130}\text{Te}\right)\geq 2.2\times 10^{25}$ yr. Using the available calculations of the nuclear matrix elements, the corresponding bounds on the effective Majorana mass are $M_{ee}<36-156$ meV, $M_{ee}<79-180$ meV and $M_{ee}<90-305$ meV respectively.

The results are summarized in Fig. 10.25. The horizontal axis is the sum of neutrino masses Σm_i, on which the above-mentioned upper limits from cosmology exist, dependent on the choice of priors. The two bands of dots show the possible values of M_{ee} for IO (disfavoured by the oscillation data) and NO. Their widths are a consequence of the Majorana phases being unknown. The NO band extends, in principle, to even lower values, down almost to zero, but that can happen only for very few combinations of the phases. The dots have been obtained by Aldo Ianni uniformly populating the parameter space. The experiments give their limit on M_{ee} in bands, from the most optimistic to the most pessimistic nuclear matrix elements (NME).

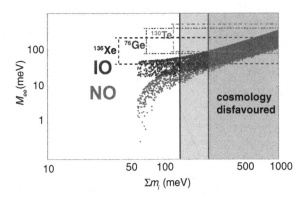

Fig. 10.25. Effective Majorana mass versus sum of neutrino masses, and the existing limits (based on a plot by A. Ianni).

Problems

10.1 An important reaction producing electron neutrinos in the Sun is $e^- + {}^7Be \rightarrow \nu_e + {}^7Li$. In the vast majority of the cases (90%), the Li nucleus is produced in its ground state. Consequently, the 'Be neutrinos' energy spectrum is monoenergetic with $E_\nu = 0.862$ MeV. The corresponding total neutrino flux at the Earth is $\Phi_\nu = 4.6 \times 10^{13}$ m^{-2}. The Borexino experiment at LNGS detects neutrinos via the reaction $\nu_e + e^- \rightarrow \nu_e + e^-$. Its fiducial volume contains 100 t of liquid scintillator. The liquid is pseudocumene C_9H_{12}. The light produced by the final electron is detected by an array of photomultipliers covering the surface surrounding the volume. Assume $\sigma(\nu_e e) = 0.6 \times 10^{-48}$ m^2, $\sigma(\nu_{\mu,\tau} e) ; \frac{1}{6} \sigma(\nu_e e), \theta_{12} = 34°, \delta m^2 = 80$ meV^2.

(1) If electron neutrinos did not change flavour, how many events would be expected per day in the 100 t target mass?

(2) Which is the principal mechanism of flavour conversion for Be neutrinos from the Sun, vacuum oscillation or MSW effect?

(3) Under these conditions, calculate the expected number of events per day in Borexino of Be neutrinos.

10.2 In the T2K experiment the proton beam accelerated by the J-PARC high-intensity proton synchrotron is driven on a target. The emerging pions are focussed in the forward direction and then drift in a vacuum pipe in which they decay. At the end of the decay volume, all the charged and neutral particles are absorbed, with the exclusion of neutrinos. The neutrino beam is aimed to the Super-Kamiokande detector located at Kamioka at a distance $L = 295$ km. The neutrino flux at Kamioka, averaged over the duty cycle, will be $\Phi = 2 \times 10^{11}$ m^{-2}yr^{-1}. Super-Kamiokande is a water Cherenkov detector with a fiducial mass of 22.5 kt. Consider the detection of the ν_μs by observing the μ produced via the CC processes on protons and neutrons and similarly of the of the ν_es by observing the e. Assume a cross-section per nucleon $\sigma = 3 \times 10^{-43}$ m^{-2} in both cases.

(1) Which is the dominant neutrino flavour in the resulting neutrino beam?

(2) Considering pions of energy $E_\pi = 5$ GeV; what is the energy E_ν of the neutrinos emitted at the angle $\theta = 2.5°$? What is the corresponding angle $\theta°$ in the pion CM frame?

(3) What is the energy E_ν of the neutrinos emitted at the angle $\theta = 0°$?

(4) Assuming (unrealistically) the neutrino beam to be monoenergetic with the energy of question (2) and that neutrinos do not oscillate, how many CC neutrino interactions per year would take place in the Super-Kamiokande fiducial volume?

(5) How much is the disappearance probability with
$\theta_{23} = 45°, \theta_{13} = 0°, \Delta m^2 = 2500$ meV^2 for the energy in question (2)?

(6) Calculate the ν_e appearance probability at the same energy, assuming $\theta_{23} = 5°$.

10.3   In the T2K experiment a ν_μ is produced at the J-PARC proton accelerator and detected in the water Cherenkov detector, Super-Kamiokande. The direction of the incoming neutrino is known, but not its energy E_ν. Consider the 'quasi-elastic' interaction $\nu_\mu + n \rightarrow \mu^- + p$. The energy E_μ of the muon and its direction θ_μ relative to that of the incoming neutrino are measured. How much is E_ν if $E_\mu = 0.5$ GeV and $\theta_\mu = 30°$?

10.4   Consider the muon neutrinos generated by the decays of the mesons produced by the collisions of cosmic rays in the atmosphere ('atmospheric neutrinos'). Their energy spectrum at the surface of the Earth extends over several orders of magnitude, decreasing with energy roughly as E_ν^{-3} and with important dependence on the angle to the zenith. At $E_\nu \approx 1$ GeV, their flux around the zenith is approximately $\Phi_{\nu_\mu} \approx 130 \, \text{m}^{-2}\text{s}^{-1}\text{sr}^{-1}\text{GeV}^{-1}$.

The Super-Kamiokande detector is a 22.5 kt fiducial mass water Cherenkov detector in the Kamioka underground observatory. Muon neutrinos (and antineutrinos, but we will not consider them) are mainly detected via their CC interactions on ^{16}O nuclei $\nu_\mu + ^{16}$O $\rightarrow \mu^- + X$. Assume $\sigma\left(\nu_\mu^{16}\text{O}\right) \approx 10^{-42}$ m$^2, \theta_{23} = 45°, \theta_{13} = 0°$, $\Delta m^2 = 2500$ meV^2.

(1) How many interactions per year will happen induced by muon neutrinos arriving with directions within $\Delta\Omega = 1$ sr around the zenith in a 1 GeV energy interval? (For the purpose of this problem, assume, unrealistically, all quantities to be constant in these intervals.)

(2) What is the fraction of surviving muon neutrinos coming vertically upwards?

(3) What is the fraction of surviving muon neutrinos incoming at 90° to the zenith?

10.5   One of the source of neutrinos in the Sun is the decay ^{8}B $\rightarrow 2\alpha + e^+ + \nu_e$. The boron neutrinos dominate the energy spectrum around 10 MeV. Consider the energy interval 9 MeV$< E_\nu <$11 MeV. In the absence of oscillations, their flux at the Earth should be $\Phi = 10^{10}$ m^{-2} s^{-1}. The Super-Kamiokande detector is a 22.5 kt fiducial mass water Cherenkov detector in the Kamioka underground observatory. It detects these neutrinos by observing the electron produced in the elastic scattering $\nu_e + e^- \rightarrow \nu_e + e^-$. Assume the cross-section $\sigma(\nu_e e) = 10^{-47}$ m^2 and a detection efficiency $\varepsilon = 50\%$. How many events per year are expected? How much is the observed rate in comparison? What is the reason for the difference, if there is one?

10.6   Consider a neutrino beam produced in a proton accelerator facility. The accelerated proton beam is driven on a target. The emerging pions are focussed in the forward direction and then drift in a vacuum pipe in which they decay. At the end of the decay volume, all the charged and neutral particles are absorbed, with the exclusion of neutrinos. In the 'long baseline' experiments studying neutrino oscillations, the detector is located at several hundred kilometres from the source, at $L = 730$ km in the OPERA experiment at Gran Sasso on the CNGS beam from the SPS accelerator at CERN and $L = 295$ km in the T2K experiment with the Super-Kamiokande detector on the beam from the J-PARC 50 GeV high-intensity accelerator in Japan. Assume a typical pion beam $E_\pi = 80$ GeV in the first case and $E_\pi = 7$ GeV in the second. In both cases, calculate the following.

(a) Calculate the pion decay length.
(b) Calculate the maximum and minimum neutrino energy.
(c) In the CM frame, neutrinos are emitted isotropically in θ^*. Hence half of them are emitted forward, namely with $\theta * \geq 0$. Find the corresponding angle in the L frame and the beam 'radius' at the far detectors.

10.7 In the OPERA experiment a v_μ beam is produced at CERN and aimed to the detector located at Gran Sasso at a distance $L = 730$ km. The experiment is designed to detect the appearance of tau neutrino by detecting the CC reaction $v_\tau + n \to \tau^- + p$. Neutrino energies are spread in a wide spectrum (see Problem 10.6), but we will assume all the neutrinos to have the same energy $E_v = 18$ GeV.

(a) Calculate the neutrino energy threshold.
(b) Assuming a neutrino flux at Gran Sasso integrated over one year of running $N_v = 4.3 \times 10^8 \, \text{m}^{-2}$, calculate the number of v_μ CC interactions per year in a Pb target of mass $M_t = 2000$ t, assuming the cross-section $\sigma_{CC} = 10^{-41} \text{m}^2$.
(c) How many v_τ CC interactions per year are expected in the presence of oscillations with $\theta_{23} = 45°, \theta_{13} = 0°, \Delta m^2 = 2500$ meV^2?
(d) If it were $\theta_{13} = 7°$, how many v_e CC interactions would be expected per year?

10.8 Electron antineutrinos from natural or artificial sources have typical energies E_v of a few MeV. They can be detected by observing the reaction $v_e + p \to e^+ + n$ in a detector medium containing free protons. The process is immediately followed by the positron annihilation $e^+ + e^- \to 2\gamma$ and by the deposit of the energy of the gammas. The medium is surrounded by photomultipliers and this energy, called visible energy E_{vis}, is measured and the neutrino energy is inferred.

(a) State the dominant process(es) of gamma energy deposit.
(b) Calculate the maximum kinetic energy of the recoiling neutron (neglect proton neutron mass difference) for the typical value $E_v = 3$ MeV.
(c) Taking into account the answer to (b), find the relation between E_v and E_{vis} (neglect the neutron recoil energy).
(d) What is the minimum detectable neutrino energy?

10.9 Consider a possible experiment looking for neutrino oscillations using as a source a nuclear reactor complex of 3 GW, producing $6 \times 10^{20} \text{s}^{-1}$ electron antineutrinos. You are planning a detector containing free protons (such as a liquid scintillator) to observe the reaction $\bar{v}_e + p \to e^+ + n$. In the MeV energy range, the cross-section of the process varies strongly with energy, but for a rough calculation assume the value $\sigma = 10^{-47}\,\mathrm{m}^2$. You want to measure the $\bar{\nu}_e$ flux at two distances, $L_1 = 100$ m and $L_2 = 2$ km. Calculate the necessary proton masses in the two detectors to have 100 counts per day.

10.10 The power radiated by the Earth through its surface is about 40 TW. A large fraction of this energy is due the decay of radioactive isotopes. The most important are ^{238}U, ^{232}Th and ^{40}K. In the 'standard' model built by geologists, the Bulk Silicate Earth Model, radioactivity is assumed to contribute 50% of the total power (i.e. 20 TW). The model contains important uncertainties and can be tested by detecting the electron antineutrinos produced in the decays. Their flux can be measured by observing the process $\bar{\nu}_e + p \rightarrow e^+ + n$ in a medium containing free protons such as a liquid scintillator. The energy threshold is $E_\nu \geq 1.8$ MeV (see Problem 10.8). The maximum antineutrino energy is $E_{\nu,\,\mathrm{max}} = 3.26$ MeV for ^{238}U, $E_{\nu,\,\mathrm{max}} = 2.25$ MeV for ^{232}Th and $E_{\nu,\,\mathrm{max}} = 1.31$ MeV for ^{40}K, so only U and Th produce antineutrinos above threshold. The total antineutrino flux at the surface due to ^{238}U and ^{232}Th is expected to be $\Phi_\nu = 3.5 \times 10^{10}\,\mathrm{m}^{-2}\mathrm{s}^{-1}$. It is evaluated that the fraction $P_{ee} \approx 0.6$ will reach our detector as $\bar{\nu}_e$ due to the oscillations. Moreover, only the fraction $f \approx 0.05$ of the flux is above detection threshold. Assume an average cross-section value $\sigma \approx 10^{-47}\,\mathrm{m}^2$.

We take as scintillator a blend of 20% $C_{16}H_{18}$ (PXE) and 80% dodecane $\left(C_{12}H_{26}\right)$ as proposed by the LENA proposal. Calculate the sensitive scintillator mass necessary to observe 1000 events.

10.11 The highest-energy cosmic ray protons produce charged pions colliding with the microwave background with the reaction $p + \gamma \rightarrow n + \pi^+$ (Greisen–Zatsepin–Kuzmin effect). The neutrinos from the decays of the pions have extremely high energies, say of the order of 10 EeV (Berezinsky and Zatsepin neutrinos). Their flux, however, is expected to be extremely small and they have not been detected yet. The neutrino–nucleon cross-sections are relatively large due to the huge number of open channels. Calculations give $\sigma_{NC} = 3 \times 10^{-36}\,\mathrm{m}^2$. Calculate the average distance to interact in the Earth $\left[\rho = 5 \times 10^3\,\mathrm{kg}\,\mathrm{m}^{-3}\right]$.

10.12 ^{76}Ge, ^{130}Te and ^{136}Xe are very stable nuclei. If a neutrino is a Majorana particle, it might decay via the double-beta mechanism without neutrinos. Assume a half-life $T_{1/2} = 10^{27}$ for each. What is the average number of decays expected in 1 year in 1 t of each isotope?

10.13 The cosmic radiation also contains neutrinos and antineutrinos in an extremely wide energy range. Neutrinos propagate on cosmological distances without attenuation due to their very small cross-sections with any target. However, at extremely high energies the resonant process '$\nu_x + \bar{\nu}_x \rightarrow Z \rightarrow$ anything' happens with a rather large cross-section on the cosmic antineutrino background (and vice versa) for any neutrino flavour x. The average kinetic energy of the 'relic' neutrino background is $E_{k,B} = 0.25$ meV and their density $n = 5.6 \times 10^7\,\mathrm{m}^{-3}$ each flavour. Assume all neutrinos to have the same mass $m_\nu = 100$ meV.

(a) Evaluate the mean square speed of the relic neutrinos.

(b) Evaluate the incoming neutrino energy E_ν for the resonant process.

(c) Evaluate the mean free path recalling that $\sigma\left(e^+e^- \to \mu^+\mu^-\right) = 2.1$ nb at the Z peak.

Summary

In this chapter we have studied neutrino physics. For the first time we have encountered phenomena beyond the Standard Model. Contrary to the model, neutrinos are massive and their flavours are not good quantum numbers. The 'real' neutrinos, namely the stationary states, are not the neutrinos produced in the CC weak interactions. We have seen in particular:

- the formalism of neutrino mixing
- what we know and what we do not
- neutrino oscillations and the experiments measuring their characteristics
- neutrino adiabatic flavour conversion in matter (AFC) and the experiments measuring their characteristics
- the limits on neutrino masses
- what a completely neutral fermion is and the possibility that neutrinos are such.

Further Reading

Bahcall, J. N. (1989) *Neutrino Astrophysics*. Cambridge University Press

Bahcall, J. N. (2002) Solar models: An historical overview. *AAPPS Bull.* 12N4 12–19; *Nucl. Phys. Proc. Suppl.* 118 (2003) 77–86

Bettini, A. (2018) A brief history of neutrino. In *The State of Art of Neutrino Physics*. A. Ereditato editor. World Scientific

Davis, R. Jr. (2002) Nobel Lecture; *A Half-Century with Solar Neutrinos*

Gómez Cadenas, J. J. *et al.* (2012) The search for neutrinoless double beta decay. *Riv. Nuovo Cim.* 35 29

Kajita, T. & Totsuka, Y. (2001) *Rev. Mod. Phys.* 73 85

Kirsten, T. A. (1999) *Rev. Mod. Phys.* 71 1213

Koshiba, M. (2002) Nobel Lecture; *Birth of Neutrino Astronomy*

Mohapatra, R. N. & Pal, P. B. (2004) *Massive Neutrinos in Physics and Astrophysics*, 3rd edn. World Scientific

11 Gravitational Waves

The Standard Model does not include one of the fundamental interactions, gravity. General relativity (GR) is a macroscopic approximation, just as Maxwell's equations are the macroscopic approximation of the Standard Model. The structures of the Universe at all scales, superclusters and clusters of galaxies, galaxies, stars and their planets, had their origin in and evolved from the primordial quantum fluctuations that took place when the Universe was very small. However, if GR were correct, nothing of this would exist, including us. The construction of the quantum theory of gravitation needs experimental and observational input. This is something extremely challenging, considering how many orders of magnitude separate the scale of our present knowledge from the Planck scale, 10^{19} GeV, at which, we presume, quantum effects should be dominant. The Universe became transparent to electromagnetic waves when electrons and protons combined into atoms, when it was 'only' a factor of about 1000 smaller than it is now. Gravitational waves (GWs) appear to be the only messengers reaching us from previous epochs, because they propagate freely everywhere. After several decennia of development, on 11 February 2016, the LIGO Scientific Collaboration and the Virgo Collaboration published the discovery of a GW. After briefly discussing, in Section 11.1, the elements of GR relevant for GW production, propagation and detection, the instruments and the discovery will be discussed in Section 11.2. After the first observation, dozens of gravitational signals have been detected, the vast majority from merging black holes and one, on 17 August 2017, from the merger of neutron stars. In this case, electromagnetic signals are expected, and have been detected, providing unique information to astrophysics and to fundamental physics as well. Here we limit the discussion to the latter, the measurement of the speed of the GW and the establishment of a bound on the mass of the graviton, the possibly existing quantum mediator of gravity.

11.1 Gravitational Waves

We start with a few historical facts, starting from 1905, the year of the special theory of relativity. On 5 June, H. Poincaré presented the relativity theory in a memoir to the Académie des Sciences in Paris, titled *Sur la dynamique de l'electron*, in which he already considered that not only the equations of the electromagnetism, but all the physics laws, including those of gravity, had to be Lorentz covariant. Gravity, Poincaré pointed out, cannot propagate instantaneously, as it would according to the

Newton law of universal gravitation, but with a finite speed, the same as light. Gravity waves, *ondes gravifiques* in his words, had to exist. Poincaré then tried, but did not succeed, to construct a relativistic theory of gravity. Gravitational waves, as solutions of the GR equations, were proposed by A. Einstein in 1916. Here we give only the few elements necessary for the following.

In GR, the field of the gravitational interaction is the metric tensor $g_{\mu\nu}$, which defines the space-time interval as

$$ds = g_{\mu\nu}dx^{\mu}dx^{\nu}.$$

$g_{\mu\nu}$ is four-dimensional and symmetric; hence it has 10 independent components, and the field equations are 10 in number. Here we notice an important difference with the other fundamental interactions. Their fields, for example the four-potential A_{μ} of electromagnetism, have the physical dimensions of energy–momentum per unit charge (different in the different interactions), while $g_{\mu\nu}$ is dimensionless.

The source and receptor of the gravitational potential in Newtonian physics are the (gravitational) masses. In GR the source of gravity is the energy–momentum tensor $T_{\mu\nu}$. Its physical effect is inducing a curvature in the space-time. A mass free from any other force, namely a freely falling object, moves following a geodesic of the curved space-time. The mirrors of the interferometers used to detect gravitational waves are, for the relevant purpose, free-falling bodies.

Gravitational waves are a solution of the Einstein equations for a weak gravitational field. Under these conditions the metric tensor differs only slightly from the pseudo-Euclidean one, $g_{ij}^{(0)} = \delta_{ik}$, $g_{i0}^{(0)} = 0$, $g_{00}^{(0)} = -1$, and we write

$$g_{\mu\nu} = g_{\mu\nu}^{(0)} + h_{\mu\nu}, \tag{11.1}$$

where $h_{\mu\nu}$ is a small correction due to radiation of the static, or slowly varying, background field.

Consider for comparison the electromagnetic wave emitted by a certain charge distribution in a volume V of limited dimensions R. At a distance $D \gg R$, the amplitude of the radiation field is proportional to the second time derivative of the dipole moment **P** of the charge distribution, and decreases as $1/D$. As expected, the energy flux, which is proportional to the square of the amplitude, decreases as $1/D^2$, inversely to the area on which it is distributed. Notice that the monopole moment, namely the integral of the charge density, which is the total charge, does not contribute because it is a constant.

In the gravitational case, the radiation field depends on the second time derivative of the quadrupole moment of the mass distribution $(i, j = 1, 2. 3)$

$$Q_{ij} = \int \mu(\mathbf{r})\left(3x_i x_j - \delta_{ij}r^2\right)dV \tag{11.2}$$

where μ is the mass density, as

$$h_{ij} = \frac{2G}{3c^4}\frac{1}{D}\frac{d^2 Q_{ij}}{dt^2} \tag{11.3}$$

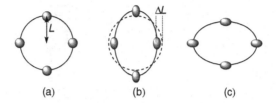

(a)   (b)   (c)

Fig. 11.1.   (a) Four free masses at equal distances. (b) Maximum displacement in one direction. (c) Maximum displacement in perpendicular direction after 1/4 period.

where G is the Newton constant. In this case, the monopole moment, which is the mass of the system, does not contribute because it does not vary with time, analogously to the electromagnetic case. But now, the dipole moment also does not contribute, being proportional to the total momentum, which is also constant. The first term that can contribute to a GW in the multipole expansion of the source is the quadrupole.

A GW is ripples of the space-time that stretch and compress the space-time itself, in planes perpendicular to the wave propagation direction. A typical amplitude is $h = 10^{-21}$. Consider four equal masses free of any force but gravity as in Fig. 11.1(a), at equal distance L from a point. An incoming GW will increase at a certain instant the distance between two opposite masses and decrease that between the other two, as in Fig. 11.1(b), and vice versa, after half a period, as in Fig. 11.1(c). The relative change in distance, called the strain, is $\Delta L / L = h$. If $L = 1$ m, then $\Delta L = 10^{-21}$ m. In the next section we shall show how it can be measured.

Coming to the detectors, let us consider again for comparison light, for which we have square law detectors. They are sensitive to the light intensity, namely the square of the amplitude. To detect light the detector must absorb a certain amount of energy, integrating the incident instantaneous flux over a certain area and a certain time, much longer than the period of the wave. Consequently, the signal is inversely proportional to the square distance from the source, as $1/D^2$. Otherwise, the GW detectors are sensitive to the amplitude of the wave. The periods are of the order of milliseconds or larger, sufficiently long to allow the test masses to follow the time evolution of the metric tensor. As a consequence, the signal decreases with increasing distance from the source as $1/D$ only. Consequently, improving the sensitivity by a factor of 10 results in an observable volume of the Universe 10^3 times larger.

11.2 The Discovery

All the data on GW have been published by a single worldwide effort, the LIGO–Virgo Collaboration, with three observatories, based on laser Michelson interferometers. LIGO has two interferometers each with 4 km long arms, located in the USA at a distance of 3002 km from one another, one at Hanford (WA) and one at Livingston (LA). Virgo has one interferometer with 3 km long arms located at Cascina near Pisa

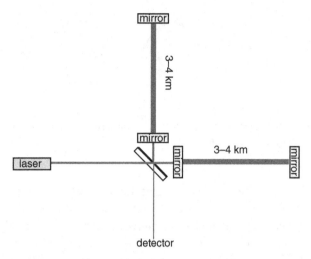

mirror
3–4 km
mirror
laser
mirror
3–4 km
mirror
detector

Fig. 11.2.   The basic structure of the LIGO (4 km) and Virgo (3 km) Michelson interferometers.

in Italy. The LIGO observatories, funded by the USA National Science Foundation (NSF), were built and are operated by Caltech and MIT. Data were collected from 2002 to 2010. No signal was observed, but this phase was of fundamental importance to learn from experience how to increase the sensitivity by a factor of 10. The Advanced LIGO detectors began operation in 2015. The detection of the first GW was reported by the LIGO–Virgo Collaboration in 2016. The Virgo observatory, funded and led by the Italian INFN and the French National Centre for Scientific Research (CNRS), started observations in 2003; as with LIGO, after a learning phase without detections, the interferometer was shut down for improvements, restarting as Advanced Virgo in 2017, when it joined back with LIGO.

Fig. 11.2 shows the basic structure of the interferometers. A beam splitter tilted at 45° divides the laser beam in two equal beams sent in the two arms of the interferometer. In each arm, two mirrors forming a Fabry–Pérot cavity reflect back and forth the light hundreds of times, extending in proportion the effective optical length, up to 2000 km in Advanced LIGO. This is approximately one half of the typical GW wavelength. The mirrors, which have the role of test-masses similar to those in Fig. 11.1, are suspended from multi-stage pendulum systems, which reduce their undesired motion induced by ground motion by about a factor of 10^{12}, from 1Hz to 10 Hz. Above the pendulums' resonant frequency, which is about 1 Hz, the mirrors can be effectively considered as in free fall in the direction of the laser beam. The oscillation amplitude due to a GW is typically $\frac{\Delta L}{L} \approx 10^{-18}$ m. A GW crossing the interferometer stretches one arm and shortens the other. The process reverses itself after half a period of the wave. The phase difference between the light beams in the two arms varies as a consequence when they join back at the beam splitter, and the intensity of the resulting beam at the photodetector varies from a maximum when the two beams are in phase to a minimum when in phase opposition.

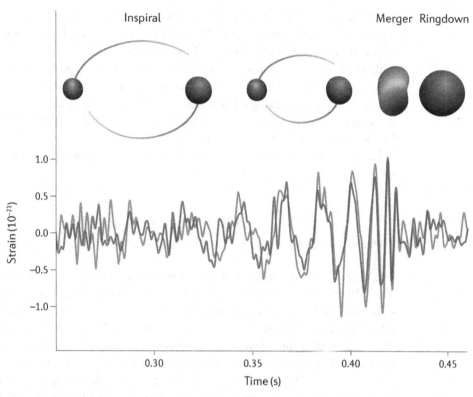

GW 150914 observed by LIGO Hanford and Livingstone. The waveforms are shifted and inverted to compensate for the slightly different arrival times and different orientations of the detectors. The upper inset is a simulation of the merger produced using numerical relativity to illustrate the evolution of the black hole event horizons as the system coalesces and merges. From Bailes, M. *et al. Nat Rev Phys* 3 344–366 (2021) (https://doi.org/10.1038/s42254-021-00303-8) under CC BY 3.0 license.

The sources observed so far are compact binary inspirals, two black holes (BHs) or two neutron stars or a BH and a neutron star. In any case the two bodies orbit one another at high orbital frequency Ω. This depends on the distance between the objects and their masses m_1 and m_2. The energy of the emitted GW is taken from the orbit causing the two bodies to spiral towards one another. With decreasing separation, Ω increases, which causes increased energy loss due to radiation, which causes further decrease of the orbit radius. The process, called 'inspiral', continues until the merging of the two partners. If they merge in a BH, as in Fig. 11.3, the final process is called 'ringdown'. The signal can be detected when Ω has decreased enough to enter in the sensitive frequency band of the detectors, which can then observe the final phases of the phenomenon. The temporal shape of the signal depends on the characteristics of the source, such as the parameters of the orbit, including its inclination on the line of sight, the spins of the two bodies and their masses. The experiments calculate with GR a large quantity of templates, for all possible combinations of the parameters, which are used to extract the signal from the background in which it is often buried.

'On September 14, 2015 at 09:50:45 UTC the two detectors of the Laser Interferometer Gravitational-Wave Observatory simultaneously observed a transient gravitational-wave signal. The signal sweeps upwards in frequency from 35 to 250 Hz with a peak gravitational-wave strain of 1.0×10^{-21}. It matches the waveform predicted by general relativity for the inspiral and merger of a pair of black holes and the ringdown of the resulting single black hole.' These are the first lines of the report of the discovery of the gravitational wave by the LIGO Scientific Collaboration and Virgo Collaboration (Abbott *et al.* 2016).

Fig. 11.3 shows (Bailes *et al.* 2021) the signals as detected by the two LIGO observatories.

11.3 The Speed of Gravity

The date 17 August 2017 is historical for both astronomy and fundamental physics. The GW antennas, two of Advanced LIGO and one of Advanced Virgo, observed a signal several minutes long, completely different from all the previous ones, labelled GW 170817. A short gamma ray burst (SGRB), GRB 170817A, was independently observed, with a time delay of $+1.74 \pm 0.05$ s, by two detectors in orbit, the Fermi Gamma-ray Space Telescope and the International Gamma-Ray Astrophysics Laboratory (INTEGRAL), as shown in Fig. 11.4 (Abbott *et al.* 2017). The probability of the near simultaneous temporal and spatial observations occurring by chance was estimated as 5.0×10^{-8}.

These observations made outstanding contributions to astrophysics and to fundamental physics. Indeed, the detection triggered a campaign of follow-up observations of electromagnetic signals, in all possible wavelengths, from visible, to infrared, to X and γ rays, to microwaves, etc. from observatories both on ground and in space. Immediately a 'new' star, of the kilonova type (so called for having a luminosity about 1000 times larger than a common nova), was discovered, whose progenitor had been a binary neutron star merger, hosted in NGC 4993 galaxy. The electromagnetic observations of the subsequent months provided fundamental information on the mechanisms of creation of the heavy elements and much more. Here we shall deal only with the GW velocity.

An important feature of GWs emitted by binary inspirals is that the time evolution of the detected signal allows us to determine the distance of the source, with the only assumption being the validity of GR. The frequency Ω and its variation in time $d\Omega / dt$ can be measured with good accuracy from the waveform. The quadrupole moment is simply calculated for a two-mass system. Then, GR gives the following relation

$$\frac{d\Omega}{dt} = \frac{96}{3}\left(\frac{GM_c}{c^3}\right)^{5/3} \Omega^{11/3} \tag{11.4}$$

(plus some corrections that are irrelevant here) where M_c is called the 'chirp mass' defined as $M_c = (m_1 m_2)^{3/5} / (m_1 + m_2)^{1/5}$. We see that the change of frequency

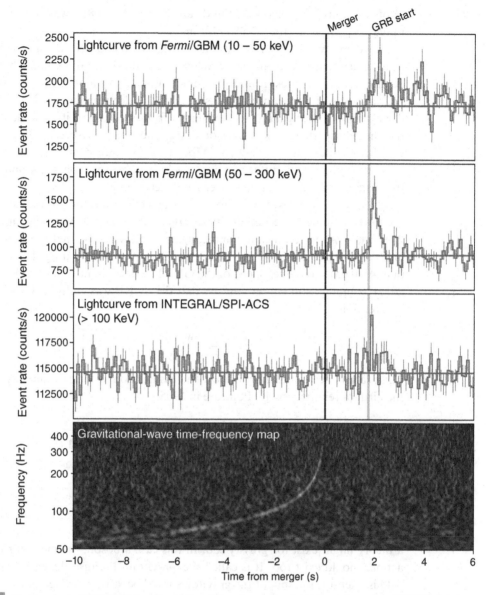

Fig. 11.4.   Top two panels: GRB 170817A from Fermi in two different energy bands; third panel: same from INTEGRAL. Bottom panel: the time-frequency map of GW170817 from LIGO-Hanford and LIGO-Livingston combined. From Abbott *et al.* (2017) *ApJL* 848 L13. DOI 10.3847/2041-8213/aa920c. Fig. 2 under CC BY 3.0 license.

depends only on the chirp mass. All binaries with the same chirp mass will have, to the leading order, the same evolution in time. This is a consequence of the fact that the intrinsic gravitational luminosity depends only on the number of orbits remaining before merging. Hence, measuring Ω and $d\Omega/dt$ we have the instantaneous luminosity at the source. Measuring the amplitude then gives the distance

D of the source. In this way, the distance of GW 170817 was found to be in the range 40^{+8}_{-14} Mpc $\left(1 \text{ Mpc} = 3.086 \times 10^{22} \text{ m}\right)$.

The relativity principle requires that gravitational and electromagnetic waves propagate in a vacuum with the same velocity. In addition, GR requires both waves to be affected by the background gravitational potentials along their common path in the same way. The distance of the source of the electromagnetic signals, namely the distance of the host galaxy, was evaluated from the ratio of its Hubble flow velocity (i.e. the velocity due to the expansion of the Universe), measured as 3017 ± 166 km s^{-1}, and the Hubble constant, 42.9 ± 3.2 Mpc, in perfect agreement with the above value.

If Δt is the small difference in travel time between photons and GWs over the distance D, the fractional speed difference during the trip can be written as $\left(v_{EM} - v_{GW}\right)/v_{EM} \approx v_{EM}\Delta t / D$. For the distance, the authors conservatively assume the lower bound of the luminosity distance range, $D = 26$ Mpc. The upper limit of $\Delta v = v_{GW} - v_{EM}$ is obtained assuming, again conservatively, that the peak of the GW and the first photons were emitted at the same time, attributing the entire delay of $\left(\pm 1.74 \pm 0.05\right)$ s to higher speed of the GW. The lower limit is obtained (Abbott *et al.* 2017) assuming, still conservatively, that the SGRB was emitted 10 s after the GW. The result is

$$-3 \times 10^{-15} \leq \frac{v_{GW} - v_{EM}}{v_{EM}} \leq +1.7 \times 10^{-16}. \tag{11.5}$$

This is a very accurate confirmation of an assumption historically already advanced at the birth of relativity by H. Poincaré in 1905.

11.4 The Mass of the Graviton

As already recalled, we do not have a quantum theory of gravitation, having as a classical limit GR. However, considering the success of the SM in describing all the other fundamental interactions with a quantum gauge theory, we can think that such a theory might exist for gravity too. In this case a boson, the mediator of the interaction, should exist too. It is called the 'graviton'. Because the classic gravitational field is a tensor, the spin of the graviton should be 2. Both in Newtonian and relativistic dynamics, the range of the gravitational interaction is infinite; the mass of the graviton, m_g, should consequently be zero. This is also proven by the speed of the gravitational waves being equal to that of the electromagnetic waves, as we saw in the previous section, provided that the photon mass m_γ is zero. Here we describe how an independent upper limit on m_g has been experimentally determined by the LIGO–Virgo Collaboration (Abbott *et al.* 2021). The limit is very stringent, even more than the existing one on m_γ.

The analysis is based on the measured waveforms of 23 signals from binary black holes collected from 2019. As already discussed, for each event the best-fitting template calculated with GR is obtained and subsequently carefully tested for consistency and for absence of significant experimental distortions. Templates are calculated assuming $m_g = 0$, namely the dispersion relation $E^2 = p^2 c^2$, where we do not use natural units. If $m_g \neq 0$, the dispersion relation becomes

$$E^2 = p^2 c^2 + m_g^2 c^4. \tag{11.6}$$

This corresponds to the GW phase velocity

$$\upsilon = pc^2 / E \tag{11.7}$$

where p is the momentum and E the energy. Because the phase velocity energy (frequency) is dependent, the relative phases of the Fourier components of the GW signal shift from one another more and more as they propagate towards Earth. The resulting difference from the GR prediction of the signal shape is calculated, for each event, as a function of m_g, and its maximum value of m_g compatible with the uncertainties is evaluated. The result of the statistical analysis is the upper bound of the graviton mass

$$m_g \leq 1.73 \times 10^{-23} \text{ eV}. \tag{11.8}$$

For comparison, the present limit on the photon mass (Workman *et al.* 2022) is

$$m_\gamma \leq 1 \times 10^{-18} \text{ eV}. \tag{11.9}$$

Further Reading

Bailes, M. *et al.* (2021) Gravitational-wave physics and astronomy in the 2020s and 2030s. *Nat Rev Phys* 3 344–366. https://doi.org/10.1038/s42254-021-00303-8

Barish, B. (2017) Nobel Lecture; *LIGO and the Discovery of Gravitational Waves, II*

12 Epilogue

The SM describes the world at elementary level with a few conceptual elements: three families of spin 1/2 fermions, and the gauge and Higgs bosons. In this book we have been faithful to the rule of sticking to the facts, namely to experimentally confirmed theoretical statements. We have seen how successful the SM is in describing nature. However, it leaves a number of open questions, as all the observations are not represented, and for theoretical reasons summarized below under Structural Problems. We shall briefly mention these problems.

12.1 Gravity

The SM is incomplete because it does not include gravity; indeed, we do not have a quantum theory of gravitation. Gravity is rather different from the other interactions. In addition to what we saw in Chapter 11, gravity needs two fundamental quantities rather than only one. In addition to the coupling, the Newton constant, it needs the 'cosmological constant' Λ, which has a completely different role. As we have seen in Chapter 11, the field, the metric tensor, is dimensionless, pure numbers. Its effect, described by the Einstein equations, is inducing curvature in space-time. Hence, the curvature tensor must be expressed in terms of the metric one, but it is not dimensionless, having the dimensions of the inverse of an area $\left[L^{-2} \right]$. As noticed by Dadhich (2011), the mismatch can be solved only by a proportionality constant having physical dimensions $\left[L^{-2} \right]$. This is the cosmological constant. Similarly to the speed of light c, Λ must be a universal, frame-independent, quantity.

12.2 Dark Matter

As we mentioned in Section 10.4, overwhelming evidence from astrophysical observations shows that the matter we know represents only a small fraction, about 5%, of the total mass–energy budget of the Universe. The largest fraction of matter, 27%, is 'dark' or, better said, invisible. It neither emits nor absorbs electromagnetic radiation and so it has been inferred indirectly through its gravitational effects. The contribution of neutrinos is very small, not larger than 1%. Consequently, dark matter should be composed of unknown particles. Many different types have been proposed, but none observed, leading to experimentally constrain at different levels. A possibility is weakly interacting massive particles (WIMPs). We think they have rather large masses, tens or hundreds of GeV. The direct search for WIMPs is extremely challenging. They should be around us in very large numbers, but their energy is small, of several keV and they interact only very rarely, somewhat like neutrinos. The search for them is performed in underground laboratories, well shielded from cosmic rays and other natural backgrounds. The detector acts also as the target, in which one tries to detect the small recoil energy of a nucleus hit by a WIMP. Detectors having sensitive masses of the order of several hundred kilos to tons are in operation and under development. Techniques have been developed to reduce by orders of magnitude the traces of natural radioactive isotopes that, present in the detector as in any material, produce background signals that hide the signal. The field is steadily progressing.

12.3 Dark Energy

The remaining 68% of the Universe is a density of negative pressure, or energy, which forces the expansion rate to accelerate, in our cosmological epoch, and is called 'dark energy'. The direct observational evidence is as follows. If the Universe contained only matter, its expansion should have slowed down in all the past epochs under the influence of gravity. Supernovae of type 1a, which are visible even at pretty large distances, can be used as standard candles because their absolute luminosity is known. If expansion is slowing down, distant supernovae should appear brighter and closer than their high redshifts might otherwise suggest. On the contrary, supernovae at distances between about 1 Gpc and 3.5 Gpc $\left(1 \text{ Gpc} = 3.09 \times 10^{25} \text{ m}\right)$, which exploded between 2.8 and 8 Gyr in the past, are dimmer and farther away than expected. This can only be explained if the expansion rate of the Universe has been accelerating since then. However, looking at larger distances, corresponding to epochs between 8 and 10 Gyr, the opposite effect appears, showing that the expansion rate then was decreasing. Dark energy density appears not to vary with the expansion, while matter density clearly decreases. This and all the other properties of dark energy observed so far are exactly those expected for the cosmological constant. However, this explanation, obviously of fundamental importance, must be checked more precisely. Astronomers worldwide are developing programmes to this purpose. Telescopes in the visible and in the infrared on the surface and in orbit will try to map the historical development of the Universe by measuring thousands of type 1a supernovae and the mass distribution at large scales with a number of techniques.

12.4 SUSY

Symmetries under groups of transformations relating fermions with bosons have a rather different mathematics from those we met, and are called supersymmetries

(SUSY). In particle physics, SUSY is not motivated by experimental observations or by contradictions in the SM, but as a means to cancel possible quadratic divergences in radiative corrections to the Higgs boson mass. The theoretical (aesthetical) question, called the hierarchy problem, is why the Higgs mass is so small. In supersymmetric extension of the SM, a partner is supposed to exist for each fermion and for each gauge boson with 1/2 spin difference. The partners of the fermions are called squarks, sleptons, etc.; the partners of the bosons are higgsino, gluinos, neutralinos, etc. The lightest of these hypothetical particles, the neutralino, is supposed to be stable and massive and might be the dark matter in the Universe. Since none of the supposed supersymmetric partners had been observed, it was concluded that they should have masses larger than their known partners and, consequently, SUSY must be broken. However, reasonably, at least some of the SUSY particles should have masses not larger than a few hundred GeV, because otherwise the hierarchy problem would reappear.

The search for SUSY particles produced in the highest-energy collisions available today has been done, and continues, at LHC. No evidence has been found, placing lower limits to the masses in the range 2–3 TeV for the gluinos, 1–2 TeV for the 1st and 2nd family squarks, 0.6–12 TeV for the 3rd family squarks and around 600 GeV for sleptons. Even if the supersymmetric models have so many free parameters that no prediction on the mass spectrum is possible, several versions of the theory appear to be under strong pressure.

12.5 Strong CP Violation

Logics seems to require that if a Lagrangian obeys a given symmetry, it should contain all the operators that respect that symmetry. However, in the QCD Lagrangian the term $G_{\mu\nu}\tilde{G}^{\mu\nu}$, where $\tilde{G}^{\mu\nu} = \epsilon^{\mu\nu\alpha\beta}G_{\alpha\beta}$, which is odd under CP, does not appear to be present, as strong interactions do not violate CP. In particular, if such an operator were present, the numerical constant in front of it should be $< 10^{-10}$ as a consequence of the experimental bound on the electric dipole moment of the neutron that is odd under CP. A possible explanation is the existence of the 'axion', a very light pseudoscalar meson that might also contribute to the dark matter in the Universe. The mass of the axion is very small, unknown, presumably in the μeV scale. However, we know, within an order of magnitude, the relation between its mass and its coupling to photons. Axions present around us are searched by converting them into photons in a strong magnetic field. Experiments are now reaching the sensitivity to explore the very small cross-section region for some values of the mass.

12.6 Matter–Antimatter Asymmetry in the Universe

The Universe is overwhelmingly made of matter; antimatter is almost absent, as far as we know. Strong limits on antimatter in space have been set by particle spectrometers in orbit, the Payload for Antimatter Matter Exploration and Light-nuclei Astrophysics (PAMELA) and the Alpha Magnetic Spectrometer (AMS). Sakharov showed that CP violation is a necessary ingredient for the matter–antimatter asymmetry, but the measured CP violation in the quark sector is too small to cause it. A remaining possibility is CP violation in the lepton sector. The necessary condition is that the product of the square sines of three neutrino mixing angles and of the Dirac phase δ is large enough. This is indeed the case for the former, and also for the latter if the presently favoured best-fit value of δ, or a value not far from it, is confirmed by future measurements.

12.7 Structural Problems

The Standard Model contains too many free parameters to be considered a complete theory. Of the particles of the SM, the photon and the gluons are massless due to gauge invariance. Neutrinos are also assumed to be massless. However, none of the masses of the remaining 12 particles (9 fermions and the W, Z and H bosons) is predicted by the SM. The values of the four elements of the CKM matrix are also not predicted. In addition, the SM should be corrected to include neutrino masses (three quantities) and mixing (four or six quantities depending on the neutrinos being Dirac or Majorana). In addition, the masses of the fermions, from neutrinos to the top quark, span 13 orders of magnitude. Why such big differences? And why is the mixing small in the quark sector, but large in the neutrino sector?

Even more ambitious questions are as follows: are the lepton and baryon numbers conserved? Is the proton really stable? Why do the proton and the electron have exactly equal (and opposite) charge (the limit on neutrality of matter is $< 10^{-21}$)? Why are there just three families? Are there any spatial dimensions beyond the three we know? These questions are not metaphysical, but physical, as they can be and are addressed experimentally.

alpha	α	A	iota	ι	I	rho	ρ	P
beta	β	B	kappa	κ	K	sigma	σ, ς	Σ
gamma	γ	Γ	lambda	λ	Λ	tau	τ	T
delta	δ	Δ	mu	μ	M	upsilon	υ	Υ, Υ
epsilon	ε	E	nu	ν	N	phi	ϕ, φ	Φ
zeta	ζ	Z	xi	ξ	Ξ	chi	χ	X
eta	η	H	omicron	o	O	psi	ψ	Ψ
theta	θ, ϑ	Θ	pi	π	Π	omega	ω	Ω

Appendix 2 Fundamental Constants

Quantity	Symbol	Value	Uncertainty (ppb)
Speed of light in vacuum	C	$299\ 792\ 458\ \text{ms}^{-1}$	exact
Planck constant	h	$6.626\ 070\ 15 \times 10^{-34}\,\text{J s}$	exact
Planck constant, reduced	$\hbar$	$1.054\ 571\ 817... \times 10^{-34}\,\text{J s}$	exact
		$6.582\ 119\ 569... \times 10^{-22}\,\text{MeV s}$	exact
Conversion constant	$\hbar c$	$197.326\ 980\ 4...\ \text{MeV fm}$	exact
Conversion constant	$(\hbar c)^2$	$389.379\ 327\ 1...\ \text{GeV}^2\mu\text{barn}$	exact
Elementary charge	q_e	$1.602\ 176\ 634 \times 10^{-19}\,\text{C}$	exact
Electron mass	m_e	$9.109\ 383\ 7015(28) \times 10^{-31}\,\text{kg}$	0.30
Proton mass	m_p	$1.672\ 621\ 923\ 69(51) \times 10^{-27}\,\text{kg}$	0.31
Bohr magneton	$\mu_B = \dfrac{q_e \hbar}{2m_e}$	$5.788\ 381\ 8060(17) \times 10^{-11}\,\text{MeV T}^{-1}$	0.30
Nuclear magneton	$\mu_N = \dfrac{q_e \hbar}{2m_p}$	$3.152\ 451\ 258\ 44(96) \times 10^{-14}\,\text{MeV T}^{-1}$	0.31
Bohr radius	$a = \dfrac{4\pi\varepsilon_0\hbar^2}{m_e q_e^2}$	$0.529\ 177\ 210\ 903(80) \times 10^{-10}\,\text{m}$	0.15
1/fine structure constant	$\alpha^{-1}(0)$	$137.035\ 999\ 166(15)$	0.11
Newton constant	G_N	$6.674\ 30(15) \times 10^{-11}\,\text{m}^3\text{kg}^{-1}\text{s}^{-2}$	2.2×10^4
Fermi constant	$G_F / (\hbar c)^3$	$1.166\ 378\ 8(6) \times 10^{-5}\,\text{GeV}^{-2}$	510
Weak mixing angle	$\sin^2\theta_W(M_Z)$	$0.231\ 21(4)$	1.7×10^5
Strong coupling constant	$\alpha_s(M_Z)$	$0.1179(9)$	7.6×10^6
Avogadro number	N_A	$6.022\ 140\ 76 \times 10^{23}\,\text{mol}^{-1}$	exact
Boltzmann constant	k	$1.380\ 649 \times 10^{-23}\,\text{J K}^{-1}$	exact

Values are mainly from CODATA (Committee on Data for Science and Technology) Recommended Values of the Fundamental Physical Constants: 2018, E. Tiesinga *et al. Rev. Mod. Phys.* **93** 025010 (2021) and https://physics.nist.gov/constants. Fine structure constant is from Fan *et al.* (2023). The Fermi constant, the strong coupling constant and the weak mixing angle are from the Particle Data Group (Workman *et al.* 2022) https://pdg.lbl.gov/2022/reviews/rpp2022-rev-phys-constants.pdf.
The figures in parentheses after the values give the one standard-deviation uncertainties in the last digits in ppb (parts per 10^9).

Appendix 3 Properties of Elementary Particles

Gauge Bosons

		Mass	Width	Main decays
Photon	γ	$< 1 \times 10^{-18}$ eV	stable
Gluon	g	0 (assumed)	stable
Weak	Z^0	91.1876 ± 0.0021 GeV	2.4952 ± 0.0023 GeV
Weak	$W^\pm$	80.377 ± 0.012 GeV	2.085 ± 0.042 GeV
Higgs	H	125.25 ± 0.17 GeV	$\Gamma_H = 3.2^{+2.4}_{-1.7}$

Data are from the Particle Data Group, Workman *et al. Progr. Theor. Exp. Phys.* 2022, 083C01; http://pdg.lbl.gov/.

Gauge Boson Couplings

Colour, electric charge (in elementary charge units), weak isospin and its third component and weak hypercharge of the gauge bosons.

	Colour	Q	I_W	I_{Wz}	Y_W
g	octet	0	0	0	0
γ	0	0	–	–	0
Z^0	0	0	–	–	0
W^+	0	+1	1	+1	0
W^-	0	−1	1	−1	0

Leptons

	Mass	Lifetime
e	$0.510\ 998\ 950\ 00 \pm 0.000\ 000\ 000\ 15$ MeV	$> 6.6 \times 10^{28}$ yr
μ	$105.658\ 3755 \pm 0.000\ 0023$ MeV	$2.196\ 9811 \pm 0.000\ 0022$ μs
τ	1776.86 ± 0.12 MeV	290.3 ± 0.5 fs
ν_1	$0 \le m_1 < 130$ meV for NO 50 meV $< m_1 < 130$ meV for IO	stable
ν_2	9 meV $< m_2 < 130$ meV for NO 50 meV $< m_2 < 130$ meV for IO	stable
ν_3	50 meV $< m_3 < 130$ meV for NO $0 \le m_3 < 130$ meV for IO	stable

Data for charged leptons are from the Particle Data Group, Workman *et al.* 2022; http://pdg.lbl.gov.
Upper limits on neutrino masses come from cosmology (Abbott *et al.* 2022), lower limits from oscillations and conversion in matter.

Quarks

	Q	I	I_z	S	C	B	T	B	Y	Mass
d	$-1/3$	$1/2$	$-1/2$	0	0	0	0	$1/3$	$1/3$	$4.67^{+0.48}_{-0.17}$ MeV
u	$+2/3$	$1/2$	$+1/2$	0	0	0	0	$1/3$	$1/3$	$2.16^{+0.49}_{-0.26}$ MeV
s	$-1/3$	0	0	-1	0	0	0	$1/3$	$-2/3$	$93.4^{+8.6}_{-3.4}$ MeV
c	$+2/3$	0	0	0	$+1$	0	0	$1/3$	$4/3$	1.27 ± 0.02 GeV
b	$-1/3$	0	0	0	0	-1	0	$1/3$	$-2/3$	$4.18^{+0.03}_{-0.02}$ GeV
t	$+2/3$	0	0	0	0	0	$+1$	$1/3$	$4/3$	172.69 ± 0.30 GeV

Electric charge Q (in unit of elementary charge), strong isospin I and its third component I_z, strangeness S, charm C, beauty B, top T, baryonic number and strong hypercharge Y of the quarks. Each quark can have red, blue or green colour.
Data are from the Particle Data Group, Workman *et al. Progr. Theor. Exp. Phys.* 2022, 083C01; http://pdg.lbl.gov.

Weak Couplings of the Fermions

	I_W	I_{Wz}	Q	Y_W	c_Z		I_W	I_{Wz}	Q	Y_W	c_Z
ν_{lL}	1/2	+1/2	0	−1	1/2	$\bar{\nu}_{lR}$	1/2	−1/2	0	1	−1/2
l_L^-	1/2	−1/2	−1	−1	$-1/2 + s^2$	l_R^+	1/2	+1/2	+1	1	$1/2 - s^2$
l_R^-	0	0	−1	−2	s^2	l_L^+	0	0	+1	2	$-s^2$
u_L	1/2	+1/2	2/3	1/3	$1/2 - (2/3)s^2$	$\bar{u}_R$	1/2	−1/2	−2/3	−1/3	$-1/2 + (2/3)s^2$
d'_L	1/2	−1/2	−1/3	1/3	$-1/2 + (1/3)s^2$	$\bar{d}'_R$	1/2	+1/2	1/3	−1/3	$1/2 - (1/3)s^2$
u_R	0	0	2/3	4/3	$-(2/3)s^2$	$\bar{u}_L$	0	0	−2/3	−4/3	$(2/3)s^2$
d'_R	0	0	−1/3	−2/3	$(1/3)s^2$	$\bar{d}'_L$	0	0	1/3	2/3	$-(1/3)s^2$

Weak isospin, hypercharge, electric charge and Z-charge factor $c_Z = I_{Wz} - s^2 Q$ of the fundamental fermions ($s^2 = \sin^2 \theta_W$). The values are identical for every colour.

Quark–Gluon Colour Factors

	R	G	B
$\bar{R}$	$\dfrac{1}{\sqrt{2}} \times g_7; \dfrac{1}{\sqrt{6}} \times g_8;$	$1 \times g_3$	$1 \times g_5$
$\bar{G}$	$1 \times g_1$	$-\dfrac{1}{\sqrt{2}} \times g_7; \dfrac{1}{\sqrt{6}} \times g_8;$	$1 \times g_6$
$\bar{B}$	$1 \times g_2$	$1 \times g_4$	$-\dfrac{2}{\sqrt{6}} \times g_8;$

Mesons (Lowest Levels)

Symbol	$q\bar{q}$	J_P	I_G	Mass (MeV)	Lifetime/width
$\pi^{\pm}$	$u\bar{d}, d\bar{u}$	0^-	1^-	139.57039(18)	26.033(5) ns
π^0	$u\bar{u}, d\bar{d}$	0^-	1^-	134.9768(5)	84.3 ± 1.3 as
η	$u\bar{u}, d\bar{d}, s\bar{s}$	0^-	0^+	547.862 ± 0.17	1.31 ± 0.05 keV
ρ	$u\bar{d}, u\bar{u}, d\bar{d}, d\bar{u}$	1^-	1^+	775.26 ± 0.23	149.14 ± 0.8 MeV
ω	$u\bar{u}, d\bar{d}$	1^-	0^-	782.66 ± 0.13	8.68 ± 0.13 MeV
η'	$u\bar{u}, d\bar{d}, s\bar{s}$	0^-	0^+	957.78 ± 0.06	188 ± 6 keV
φ	$s\bar{s}$	1^-	0^-	1019.461 ± 0.016	4.249 ± 0.013 MeV
K^+, K^-	$u\bar{s}, s\bar{u}$	0^-	1/2	493.677 ± 0.016	12.380 ± 0.020 ns
K_S^0		0^-		497.611 ± 0.013	89.54 ± 0.04 ps
K_L^0		0^-		497.611 ± 0.013	52.93 ± 0.09 ns
K^{*+}, K^{*-}	$u\bar{s}, s\bar{u}$	1^-	1/2	891.67 ± 0.26	51.4 ± 0.8 MeV
$K^{*0}, \bar{K}^{*0}$	$d\bar{s}, s\bar{d}$	1^-	1/2	895.55 ± 0.20	47.3 ± 0.5 MeV
D^+, D^-	$c\bar{d}, d\bar{c}$	0^-	1/2	1869.66 ± 0.05	1.033 ± 0.005 ps
$D^0, \bar{D}^0$	$c\bar{u}, u\bar{c}$	0^-	1/2	1864.84 ± 0.05	0.4103 ± 0.0010 ps
D_s^+, D_s^-	$c\bar{s}, s\bar{c}$	0^-	0	1968.35 ± 0.07	0.504 ± 0.004 ps
B^+, B^-	$u\bar{b}, b\bar{u}$	0^-	1/2	5279.34 ± 0.12	1.638 ± 0.004 ps
$B^0, \bar{B}^0$	$d\bar{b}, b\bar{d}$	0^-	1/2	5279.66 ± 0.12	1.519 ± 0.004 ps
$B_s^0, \bar{B}_s^0$	$s\bar{b}, b\bar{s}$	0^-	0	5366.92 ± 0.10	1.521 ± 0.005 ps
B_c^+, B_c^-	$c\bar{b}, b\bar{c}$	0^-	0	6274.47 ± 0.32	0.510 ± 0.009 ps
$\eta_c\,(1S)$	$c\bar{c}$	0^-	0	2983.9 ± 0.4	32.0 ± 0.7 MeV
$J/\psi\,(1S)$	$c\bar{c}$	1^-	0	3096.900 ± 0.006	92.6 ± 1.7 keV
$\chi_{c0}\,(1P)$	$c\bar{c}$	0^+	0	3414.71 ± 0.30	10.8 ± 0.6 MeV
$\chi_{c1}\,(1P)$	$c\bar{c}$	1^+	0	3510.67 ± 0.05	0.84 ± 0.04 MeV
$\chi_{c2}\,(1P)$	$c\bar{c}$	2^+	0	3556.17 ± 0.07	1.97 ± 0.09 MeV
$\psi\,(2S)$	$c\bar{c}$	1^-	0	3686.10 ± 0.06	294 ± 8 keV
$\psi\,(3S)$	$c\bar{c}$	1^-	0	3773.7 ± 0.4	27.2 ± 1.0 MeV
$\Upsilon\,(1S)$	$b\bar{b}$	1^-	0	9460.40 ± 0.10	54.02 ± 1.25 keV
$\Upsilon\,(2S)$	$b\bar{b}$	1^-	0	10023.4 ± 0.5	31.98 ± 2.63 keV
$\Upsilon\,(3S)$	$b\bar{b}$	1^-	0	10355.1 ± 0.5	20.32 ± 1.85 keV
$\Upsilon\,(4S)$	$b\bar{b}$	1^-	0	10579.4 ± 1.2	20.5 ± 2.5 MeV

Data from the Particle Data Group, Workman *et al. Progr. Theor. Exp. Phys.* 2022, 083C01; http://pdg.lbl.gov.

Baryons (Lowest Levels)

Symbol	qqq	J_P	I	Mass (MeV)	Lifetime/width
p	uud	$1/2^+$	$1/2$	$938.272\ 088\ 16 \pm 0.000\ 00029$	$> 2.4\ 10^{34}\,\text{yr}$
n	udd	$1/2^+$	$1/2$	$939.565\ 420\ 52 \pm 0.000\ 000\ 5$	$878.4 \pm 0.5\,\text{s}$
$\Delta^{++}(1232)$	uuu	$3/2^+$	$3/2$	1232 ± 2	$118 \pm 2\,\text{MeV}$
Λ	uds	$1/2^+$	0	1115.683 ± 0.006	$263 \pm 2\,\text{ps}$
Σ^+	uus	$1/2^+$	1	1189.37 ± 0.07	$80.18 \pm 0.26\,\text{ps}$
Σ^0	uds	$1/2^+$	1	1192.642 ± 0.024	$(7.4 \pm 0.7) \times 10^{-20}\,\text{s}$
Σ^-	dds	$1/2^+$	1	1197.449 ± 0.030	$147.9 \pm 1.1\,\text{ps}$
$\Sigma^+(1385)$	uus	$3/2^+$	1	1382.83 ± 0.34	$36.2 \pm 0.7\,\text{MeV}$
$\Sigma^0(1385)$	uds	$3/2^+$	1	1383.7 ± 1.0	$36 \pm 5\,\text{MeV}$
$\Sigma^-(1385)$	dds	$3/2^+$	1	1387.2 ± 0.5	$39.4 \pm 2.1\,\text{MeV}$
Ξ^0	uss	$1/2^+$	$1/2$	1314.86 ± 0.20	$290 \pm 9\,\text{ps}$
Ξ^-	dss	$1/2^+$	$1/2$	1321.71 ± 0.07	$163.9 \pm 1.5\,\text{ps}$
$\Xi^0(1530)$	uss	$3/2^+$	$1/2$	1531.80 ± 0.32	$9.1 \pm 0.5\,\text{MeV}$
$\Xi^-(1530)$	dss	$3/2^+$	$1/2$	1535.0 ± 0.6	$9.9 \pm 1.8\,\text{MeV}$
Ω^-	sss	$3/2^+$	0	1672.45 ± 0.29	$82.1 \pm 1.1\,\text{ps}$
Λ_c^+	udc	$1/2^+?$		2286.46 ± 0.14	$201.5 \pm 2.7\,\text{fs}$
Σ_c^{++}	uuc	$1/2^+?$	1	2453.97 ± 0.14	$1.89^{+0.09}_{-0.18}\,\text{MeV}$
Σ_c^+	udc	$1/2^+?$	1	$2452.65^{+0.22}_{-0.16}$	$2.3 \pm 0.4\,\text{MeV}$
Σ_c^0	ddc	$1/2^+?$	1	2453.75 ± 0.14	$1.83^{+0.11}_{-0.19}\,\text{MeV}$
Ξ_c^+	usc	$1/2^+?$	$1/2$	2467.71 ± 0.23	$453 \pm 5\,\text{fs}$
Ξ_c^0	dsc	$1/2^+?$	$1/2$	$2470.44 \pm 0.28\,\text{MeV}$	$151.9 \pm 2.4\,\text{fs}$
Ω_c^0	ssc	$1/2^+?$	0	2695.2 ± 1.7	$268 \pm 26\,\text{fs}$
Ξ_{cc}^+	dcc	$?$	$?$	3518.9 ± 0.9	$< 33\,\text{ps}$
Ξ_{cc}^{++}	ucc	$?$	$?$	3621.6 ± 0.4	$256^{+24}_{-22} \pm 14\,\text{ps}$
Λ_b^0	udb	$1/2^+$	0	5619.60 ± 0.17	$1.471 \pm 0.009\,\text{ps}$
Σ_b^-	ddb	$1/2^+?$	1	5815.64 ± 0.27	$5.3 \pm 0.5\,\text{MeV}$
Σ_b^+	uub	$1/2^+?$	1	5810.56 ± 0.25	$5.0 \pm 0.5\,\text{MeV}$
Ξ_b^-	dsb	$1/2^+?$	$1/2$	5797.0 ± 0.6	$1.572 \pm 0.040\,\text{ps}$
Ξ_b^0	usb	$1/2^+?$	$1/2$	5791.9 ± 0.5	$1.480 \pm 0.030\,\text{ps}$
Ω_b^-	ssb	$1/2^+?$	$1/2$	6045.2 ± 1.2	$1.65^{+0.18}_{-0.16}\,\text{ps}$

Workman *et al. Progr. Theor. Exp. Phys.* 2022, 083C01; http://pdg.lbl.gov.
Not yet observed are $\Sigma_b^0 = udb$, the strange-double charm $\Omega_{cc}^+ = scc$, the triple charm $\Omega_{ccc}^{++} = ccc$, the charm and beauties $\Xi_{bc}^+ = ubc, \Xi_{bc}^0 = dbc, \Omega_{bc}^0 = sbc, \Omega_{bcc}^+ = bcc$ and the double and triple beauties $\Xi_{bb}^0 = ubb, \Xi_{bb}^- = dbb, \Omega_{bb}^- = sbb, \Omega_{bbc}^0 = cbb$ and $\Omega_{bbb}^- = bbb$.

Appendix 4 Clebsch–Gordan Coefficients

$$\langle J_1, J_{z1}; J_2, J_{z2} | J, J_z; J_1, J_2 \rangle = (-1)^{J-J_1-J_2} \langle J_2, J_{z2}; J_1, J_{z1} | J, J_z; J_2, J_1 \rangle$$

$1/2 \otimes 1/2$		J, M			
m_1	m_2	$1, +1$	$1, 0$	$0, 0$	$1, -1$
$+1/2$	$+1/2$	1			
$+1/2$	$-1/2$		$\sqrt{1/2}$	$\sqrt{1/2}$	
$-1/2$	$+1/2$		$\sqrt{1/2}$	$-\sqrt{1/2}$	
$-1/2$	$-1/2$				1

$1 \otimes 1/2$		J, M					
m_1	m_2	$\frac{3}{2}, +\frac{3}{2}$	$\frac{3}{2}, +\frac{1}{2}$	$\frac{1}{2}, +\frac{1}{2}$	$\frac{3}{2}, -\frac{1}{2}$	$\frac{1}{2}, -\frac{1}{2}$	$\frac{3}{2}, -\frac{3}{2}$
$+1$	$+1/2$	1					
$+1$	$-1/2$		$\sqrt{1/3}$	$\sqrt{2/3}$			
0	$+1/2$		$\sqrt{2/3}$	$-\sqrt{1/3}$			
0	$-1/2$				$\sqrt{2/3}$	$\sqrt{1/3}$	
-1	$+1/2$				$\sqrt{1/3}$	$-\sqrt{2/3}$	
-1	$-1/2$						1

$1 \otimes 1$		J, M								
m_1	m_2	$2, +2$	$2, +1$	$1, +1$	$2, 0$	$1, 0$	$0, 0$	$2, -1$	$1, -1$	$2, -2$
$+1$	$+1$	1								
$+1$	0		$\sqrt{1/2}$	$\sqrt{1/2}$						
0	$+1$		$\sqrt{1/2}$	$-\sqrt{1/2}$						
$+1$	-1				$\sqrt{1/6}$	$\sqrt{1/2}$	$\sqrt{1/3}$			
0	0				$\sqrt{2/3}$	0	$-\sqrt{1/3}$			
-1	$+1$				$\sqrt{1/6}$	$-\sqrt{1/2}$	$\sqrt{1/3}$			
0	-1							$\sqrt{1/2}$	$\sqrt{1/2}$	
-1	0							$\sqrt{1/2}$	$-\sqrt{1/2}$	
-1	-1									1

Appendix 5 Spherical Harmonics and *d*-Functions

Spherical Harmonics

$$Y_l^m(\theta,\phi) = \sqrt{\frac{(2l+1)(l-m)!}{4\pi(l+m)!}} P_l^m(\cos\theta) e^{im\phi}.$$

$$P_l^{-m}(\cos\theta) = (-1)^m \frac{(l-m)!}{(l+m)!} P_l^m(\cos\theta).$$

$$Y_l^{-m}(\theta,\phi) = (-1)^m Y_l^{m*}(\theta,\phi).$$

$$Y_0^0 = \sqrt{\frac{1}{4\pi}}.$$

$$Y_1^0 = \sqrt{\frac{3}{4\pi}}\cos\theta \quad Y_1^1 = -\sqrt{\frac{3}{8\pi}}\sin\theta e^{i\phi}.$$

$$Y_2^0 = \sqrt{\frac{5}{4\pi}}\left(\frac{3}{2}\cos^2\theta - \frac{1}{2}\right) \quad Y_2^1 = -\sqrt{\frac{15}{8\pi}}\sin\theta\cos\theta e^{i\phi}$$

$$Y_2^2 = \frac{1}{4}\sqrt{\frac{15}{2\pi}}\sin^2\theta e^{i2\phi}.$$

d-Functions

$$d_{m,m'}^j(\theta,\phi) = (-1)^{m-m'} d_{m,m'}^j(\theta,\phi) = d_{-m,-m'}^j(\theta,\phi).$$

$$d_{0,0}^1 = \cos\theta.$$

$$d_{1,1}^1 = \frac{1+\cos\theta}{2} \quad d_{1,0}^1 = -\frac{\sin\theta}{\sqrt{2}} \quad d_{1,-1}^1 = \frac{1-\cos\theta}{2}.$$

$$d_{\frac{1}{2},\frac{1}{2}}^1 = \cos\frac{\theta}{2} \quad d_{\frac{1}{2},-\frac{1}{2}}^1 = -\sin\frac{\theta}{2}.$$

Appendix 6 Experimental and Theoretical Discoveries in Particle Physics

This table gives a brief timeline of the historical development of particle physics. However, the discoveries are rarely due to a single person and never happen instantaneously. The dates indicate the year of the most relevant publication(s); the names are those of the main contributors.

1896	H. Becquerel: discovery of radioactivity
1897	J. J. Thomson: discovery of the electron
1912	V. Hess: discovery of cosmic rays
	C. T. R. Wilson: cloud chamber
1924	S. N. Bose: quantum statistics – integer spins
1926	E. Fermi: quantum statistics – half-integer spins
1928	P. A. M. Dirac: relativistic wave equation for the electron
	H. Geiger: Geiger counter
1927	G. E. Lemaître: expansion of the Universe; also E. Hubble 1929
1930	W. Pauli: neutrino hypothesis
	E. O. Lawrence: cyclotron
1932	J. Chadwick: discovery of the neutron
	C. Anderson: discovery of the positron
1933	F. Zwicky: discovery of dark matter in the Universe
	E. Fermi: theory of weak interaction
1935	H. Yukawa: theory of strong nuclear forces
	P. Cherenkov and N. Vavilov: Cherenkov–Vavilov effect
1937	J. Street and E. Stevenson; C. Anderson and S. Neddermeyer: discovery of μ
	E. Majorana: theory of completely neutral fermions
1944/45	V. Veksler, E. McMillan: principle of phase stability in accelerators
1947	W. Lamb and R. Retherford: Lamb shift
	P. Kusch: electron magnetic moment anomaly
	M. Conversi, E. Pancini, O. Piccioni: leptonic character of the μ
	G. Occhialini, C. Powell *et al.*: discovery of the pion
	G. Rochester and C. Butler: discovery of V^0 particles
1948	S. Tomonaga, R. Feynman, J. Schwinger: quantum electrodynamics
1952	*Cosmotron operational at BNL at 3 GeV*
	D. Glaser: bubble chamber
	E. Fermi *et al.*: discovery of the baryon resonance $\Delta(1236)$
1953	Cosmic ray experiments: $\theta - \tau$ puzzle
	M. Gell-Mann, K. Nishijima: strangeness hypothesis
1954	*Bevatron operational at Berkeley at 7 GeV*

1955	O. Chamberlain *et al.*: discovery of the antiproton
	M. Conversi, A. Gozzini: flash chamber
	M. Gell-Mann, A. Pais: K^0 oscillation proposal
1956	C. L. Cowan, F. Reines: discovery of $\bar{v}_e$
	T. D. Lee and C. N. Yang: hypothesis of parity violation
1957	C. S. Wu *et al.*: discovery of parity violation
	Synchrophasotron operational at Dubna at 10 GeV
	G. Sudarshan and R. Marshak: *V–A* structure of CC weak interaction
	R. Feynman and M. Gell-Mann: *V–A* structure of CC weak interaction
1959	*Proton synchrotrons PS at CERN, AGS at BNL operational at 30 GeV*
	S. Fukui and S. Myamoto: spark chamber
1960	Y. Nambu: spontaneous symmetry breaking in particle physics
	B. Touschek: proposal of e^+e^- storage ring (ADA)
1961	L. Alvarez: discovery of meson resonances
1962	M. Schwartz, L. Lederman, J. Steinberger: discovery v_μ
	Z. Maki *et al.*: hypothesis of $v_\mu v_e$ neutrino mixing and oscillations
1963	N. Cabibbo: hadronic currents mixing
1964	V. Fitch and J. Cronin *et al.*: discovery of CP violation
	G. Zweig, M. Gell-Mann: quark model
	N. Samios *et al.*: discovery of the Ω^-
	Englert and Brout, Higgs: spontaneous breaking of gauge theories
1967	S. Glashow, A. Salam, S. Weinberg: electro-weak unification
	J. Friedman, H. Kendall, R. Taylor *et al.*: quark structure of the proton
	Proton synchrotron operational at Serpukhov at 76 GeV
	Electron linear accelerator operational at SLAC at 20 GeV
1968	C. Charpak *et al.*: multiwire proportional chamber
	R. Davis *et al.* and J. Bahcall: solar neutrino puzzle
	S. Glashow, I. Iliopoulos, L. Maiani: fourth quark hypothesis
1971	G. 't Hooft renormalizability of electroweak theory
	K. Niu *et al.*: discovery of charm
	Intersecting proton storage rings operational at CERN (30 + 30 GeV)
1972	*Fermilab proton synchrotron operational at 200 GeV, later at 500 GeV*
	SPEAR e^+e^- (4 + 4 GeV) storage ring operational at Stanford
	J. Heintze and A. H. Walenta: drift chamber
1973	Gargamelle bubble chamber: discovery of weak neutral currents
	D. Gross, D. Pulitzer, F. Wilczek, H. Fritzsch, M. Gell-Mann, G. 't Hooft: quantum chromodynamics
	M. Kobayashi and K. Maskawa: CP violation due to mixing of three hadron families
1974	B. Richter *et al.*, S. Ting *et al.*: discovery of J/ψ hidden charm particle
1975	M. Perl *et al.*: discovery of the τ lepton
1976	L. Lederman *et al.*: discovery of Υ hidden beauty particles
	Super proton synchrotron (SPS) operational at CERN at 400 GeV
1979	PETRA experiments at DESY: discovery of the gluon
1981	*First collisions in the SPS $p\bar{p}$ storage ring at CERN (270 + 270 GeV)*
1983	C. Rubbia *et al.*: discovery of the W and Z bosons
1985	S. Mikheyev and A. Smirnov: hypothesis of neutrino adiabatic flavour conversion in matter

1986	*TRISTAN e^+e^- (15 + 15 GeV) storage ring operational at KEK at Tsukuba*
1987	M. Koshiba: observation of neutrinos from a supernova
1989	*Stanford Linear Collider e^+e^- (50 + 50 GeV) operational*
	LEP e^+e^- storage ring operational at CERN (50 + 50 GeV, later 105 + 105)
1990	T. Berners-Lee, R. Cailliau (CERN): World Wide Web proposal
1991	*HERA ep collider operational at DESY (30 + 820 GeV). Later 30 + 920 GeV*
1992	GALLEX experiment: solar neutrino deficit at low energy
1995	CDF experiment: discovery of the top quark
1997	LEP experiments: W-bosons self-coupling
1998	Superkamiokande: discovery of neutrino oscillations
1999	*Beauty factories operational, KEKB at Tsukuba and PEP2 at Stanford*
2001	K. Niwa *et al.*: discovery of the tau neutrino
2002	A. McDonald *et al.*: discovery of neutrino adiabatic flavour conversion
2010	*LHC collider operational at CERN $\left(3.5 + 3.5\,TeV, 6.5 + 6.5\,TeV\ in\ 2014\right)$*
2012	ATLAS and CMS experiments: discovery of the Higgs boson
2015	LIGO Scientific Collaboration and Virgo Collaboration: discovery of Gravitational Waves

Solutions

1.2 $s = (3E)^2 - 0 = 9E^2 = 9(p^2 + m^2) = 88.9 \text{ GeV}^2$; $m = \sqrt{s} = 9.43 \text{ GeV}$.

1.3 $\Gamma_\pi^\pm = h/\tau_\pi^\pm = (6.6 \times 10^{-16} \text{eVs})/(2.6 \times 10^{-8}\text{s}) = 25 \text{ neV}$,

$\Gamma_K = 54 \text{ neV}$, $\Gamma_\Lambda = 2.5 \text{ μeV}$.

1.6 Our reaction is $p + p \rightarrow p + p + m$. In the CM frame the total momentum is zero. The lowest energy configuration of the system is when all particles in the final state are at rest.

(a) Let us write down the equality between the expressions of s in the CM and

L frames: $s = (E_p + m_p)^2 - p_p^2 = (2m_p + m)^2$. Recalling that $E_p^2 = m_p^2 + p_p^2$, we

have $E_p = \dfrac{(2m_p + m)^2 - 2m_p^2}{2m_p} = m_p + 2m + \dfrac{m^2}{2m_p}$.

(b) The two momenta are equal and opposite because the two particles have the same mass, hence we are in the CM frame. The threshold energy E_p^* is given by $s = (2E_p^*)^2 = (2m_p + m)^2$ which gives $E_p^* = m_p + m/2$.

(c) $E_p = 1.218 \text{ GeV}$; $p_p = 0.78 \text{ GeV}$; $T_p = 280 \text{ MeV}$; $E_p^* = 1.007 \text{ GeV}$;

$p_p^* = 0.36 \text{ GeV}$.

1.7 (a) $s = (E_\gamma + m_p)^2 - p_\gamma^2 = (E_\gamma + m_p)^2 - E_\gamma^2 = (m_p + m_\pi)^2 = 1.16 \text{ GeV}^2$, hence we have $E_\gamma = 149 \text{ MeV}$, hence we have

(b) $s = (E_\gamma + E_p)^2 - (\mathbf{p}_\gamma + \mathbf{p}_p)^2 = m_p^2 + 2E_\gamma E_p - 2\mathbf{p}_\gamma \cdot \mathbf{p}_p$. For a given proton energy, s reaches a maximum for a head-on collision. Consequently, $\mathbf{p}_\gamma \cdot \mathbf{p}_p = -E_\gamma p_p$ and, taking into account that the energies are very large, $s = m_p^2 + 2E_\gamma (E_p + p_p) \approx m_p^2 + 4E_\gamma E_p$. In conclusion,

$$E_p = \frac{s - m_p^2}{4E_\gamma} = \frac{(1.16 - 0.88) \times 10^{18} \text{eV}^2}{4 \times 10^{-3} \text{eV}} = 7 \times 10^{19} \text{eV} = 70 \text{ EeV}.$$

(c) The attenuation length is $\lambda = 1/(\sigma\rho) = 5.6 \times 10^{22} \text{ m} = 18 \text{ Mpc}$ (1 Mpc = 3.1×10^{21} m).

This is a short distance on the cosmological scale, only one order of magnitude larger than the distance to the closest galaxies, the Magellanic Clouds. The cosmic ray spectrum (Fig. 1.10) should not go beyond the above computed energy. This is called the Greisen–Zatsepin–Kuzmin (GZK) bound. The AUGER observatory is exploring this extreme energy region.

1.11 We must consider the reaction

$$M \rightarrow m_1 + m_2.$$

The figure defines the CM variables.

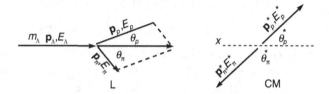

We can use equations (P1.5) and (P1.6) with $\sqrt{s} = M$, obtaining

$$E_{2f}^* = \frac{M^2 + m_2^2 - m_1^2}{2M}; \quad E_{1f}^* = \frac{M^2 + m_1^2 - m_2^2}{2M}.$$

The corresponding momenta are

$$p_f^* \equiv p_{1f}^* = -p_{2f}^* = \sqrt{E_{1f}^{*2} - m_1^2} = \sqrt{E_{2f}^{*2} - m_2^2}.$$

1.15 When dealing with a Lorentz transformation problem, the first step is the accurate drawing of the momenta in the two frames and the definition of the kinematic variables.

Using the expressions we found in the introduction, we have the following.

(a) $E_\pi^* = \dfrac{m_\Lambda^2 - m_p^2 + m_\pi^2}{2m_\Lambda} = 0.17$ GeV; $E_p^* = 0.95$ GeV;

$$p_\pi^* = p_p^* = \sqrt{E_\pi^{*2} - m_\pi^2} = 0.096 \text{ GeV}.$$

(b) We calculate the Lorentz factors for the transformation:

$$E_\Lambda = \sqrt{p_\Lambda^2 + m_\Lambda^2} = 2.29 \text{ GeV}; \quad \beta_\Lambda = \frac{p_\Lambda}{E_\Lambda} = 0.87; \quad \gamma_\Lambda = \frac{E_\Lambda}{m_\Lambda} = 2.05.$$

(c) We do the transformation and calculate the requested quantities

$$p_\pi \sin\theta_\pi = p_\pi^* \sin\theta_\pi^* = 0.096 \times \sin 210° = -0.048 \text{ GeV}$$

$$p_\pi \cos\theta_\pi = \gamma_\Lambda \left(p_\pi^* \cos\theta_\pi^* + \beta_\Lambda E_\pi^* \right) = 2.05 \left(0.096 \times \cos 210° + 0.87 \times 0.17 \right)$$
$$= 0.133 \text{ GeV}$$

$$\tan\theta_\pi = \frac{-0.048}{0.133} = -0.36 \quad \theta_\pi = -20°;$$

$$p_\pi = \sqrt{\left(p_\pi \sin\theta_\pi \right)^2 + \left(p_\pi \cos\theta_\pi \right)^2}$$
$$= 0.141 \text{ GeV}.$$

$$p_p \sin\theta_p = p_p^* \sin\theta_p^* = 0.048 \text{ GeV}$$

$$p_p\cos\theta_p = \gamma_\Lambda\left(p_p^*\cos\theta_p^* + \beta_\Lambda E_p^*\right) = 2.05(0.096\times\cos30° + 0.87\times0.95)$$
$$= 1.86 \text{ GeV}°$$

$$p_p\sin\theta_p = p_p^*\sin\theta_p^* = 0.048 \text{ GeV}$$

$$p_p\cos\theta_p = \gamma_\Lambda\left(p_p^*\cos\theta_p^* + \beta_\Lambda E_p^*\right) = 2.05(0.096\cos30° + 0.870.95)$$
$$= 1.86 \text{ GeV}$$

$$\tan\theta_p = \frac{0.048}{1.86} = 0.026 \quad \theta_p = 1.5°.$$

$$p_p = \sqrt{\left(p_p\sin\theta_p\right)^2 + \left(p_p\cos\theta_p\right)^2} = 1.9 \text{ GeV}; \theta = \theta_p - \theta_\pi = 21.5°.$$

1.17 We continue to refer to the figure of Problem 1.15. We shall solve our problem in two ways: by performing a Lorentz transformation and by using the Lorentz invariants.

We start with the first method. We calculate the Lorentz factors. The energy of the incident proton is $E_1 = \sqrt{p_1^2 + m_p^2} = 3.143$ GeV. Firstly, let us calculate the CM energy squared of the two-proton system (i. e. its mass squared).

$$p_{pp} = p_1 = 3 \text{ GeV}; \quad E_{pp} = E_1 + m_p = 4.081 \text{ GeV}.$$

Hence $s = 2m_p^2 + 2E_1 m_p = 7.656$ GeV^2.

The Lorentz factors are $\beta_{pp} = p_{pp}/E_{pp} = 0.735$ and $\gamma_{pp} = E_{pp}/\sqrt{s_{pp}} = 1.47$. Since all the particles are equal, we have

$$E_1^* = E_2^* = E_3^* = E_4^* = \frac{\sqrt{s}}{2} = 1.385 \text{ GeV}$$

$$p_1^* = p_2^* = p_3^* = p = \sqrt{E_1^{*2} - m_p^2} = 1.019 \text{ GeV}$$

We now perform the transformation. To calculate the angle we must calculate firstly the components of the momenta

$$p_3\sin\theta_{13} = p_3^*\sin\theta_{13}^* = 1.019\times\sin10° = 0.177 \text{ GeV}.$$
$$p_3\cos\theta_{13} = \gamma\left(p_3^*\cos\theta_{13}^* + \beta E_3^*\right) = 1.473\times(1.019\times\cos10° + 0.735\times1.385)$$
$$= 2.978 \text{ GeV}.$$
$$\tan\theta_{13} = \frac{0.177}{2.978} = 0.0594; \quad \theta_{13} = 3°.$$
$$-p_4\sin\theta_{14} = -p_4^*\sin\theta_{14}^* = -1.019\times\sin170° = -0.1769 \text{ GeV}.$$

$$p_4\cos\theta_{14} = \gamma\left(p_4^*\cos\theta_{14}^* + \beta E_4^*\right) = 1.473\times(1.019\times\cos170° + 0.735\times1.385)$$
$$= 0.0213 \text{ GeV}.$$

$$\tan\theta_{14} = -0.1769/0.0213 = -8.305 \quad \theta_{14} = -83° \quad \Rightarrow \quad \theta_{34} = \theta_{13} - \theta_{14} = 86°.$$

In relativistic conditions the angle between the final momenta in a collision between two equal particles is always, as in this example, smaller than 90°.

We now solve the problem using the invariants and the expressions in the introduction. We want the angle between the final particles in L. We then write down the expression of s in L in the initial state, which we have already calculated:

$$s = (E_3 + E_4)^2 - (\mathbf{p}_3 + \mathbf{p}_4)^2 = m_3^2 + m_4^2 + 2E_3E_4 - 2\mathbf{p}_3 \cdot \mathbf{p}_4,$$

that gives $\mathbf{p}_3 \cdot \mathbf{p}_4 = m_p^2 + E_3E_4 - s/2$ and hence $\cos\theta_{34} = \dfrac{m_p^2 + E_3E_4 - s/2}{p_3 p_4}$.

We need E_3 and E_4 (and their momenta); we can use (P.1.13) if we have t. With the data of the problem we can calculate t in the CM:

$$t = 2m_p^2 + 2p_i^{*2}\cos\theta_{13}^* - 2E_i^{*2} = 2p_i^{*2}\left(\cos\theta_{13}^* - 1\right) = 2 \times 1.019^2 \left(\cos 10° - 1\right)$$
$$= -0.0316 \text{ GeV}^2.$$

We then obtain

$$E_3 = \frac{s + t - 2m_p^2}{2m_p} = \frac{7.656 - 0.0316 - 2 \times 0.938^2}{2 \times 0.938} = 3.126 \text{ GeV}; \quad p_3 = 2.982 \text{ GeV}.$$

From energy conservation we have

$$E_4 = E_1 + m_p - E_3 = 3.143 + 0.938 - 3.126 = 0.955 \text{ GeV}; \quad p_4 = 0.179 \text{ GeV}.$$

Finally we obtain

$$\cos\theta_{34} = \frac{0.938^2 + 3.126 + 0.955 - 7.656/2}{2.982 \times 0.179} = 0.0696 \quad \Rightarrow \theta_{34} = 86°.$$

1.21 The maximum momentum transfer is at backwards scattering. In these conditions, $Q^2 = 4EE'$, where E' is the energy of the scattered electron. We have

$$Q_{max}^2 = \frac{4E^2 M}{M + 2E} = \frac{4 \times 4 \times 56}{56 + 4} = 15 \text{ GeV}^2.$$

1.25 $$\cos\theta = 1 - \frac{E/E' - 1}{E/M} = 1 - \frac{2.5 - 1}{20} = 0.925; \theta = 22°$$

1.27 The equation of motion is $q\mathbf{v} \times \mathbf{B} = \dfrac{d\mathbf{p}}{dt}$. Since in this case the Lorentz factor γ is constant, we can write $q\mathbf{v} \times \mathbf{B} = \gamma m \dfrac{d\mathbf{v}}{dt}$. The centripetal acceleration is then: $\left|\dfrac{d\mathbf{v}}{dt}\right| = \dfrac{qvB}{\gamma m} = \dfrac{v^2}{\rho}$. Simplifying, we obtain $p = qB\rho$. We now want pc in GeV, B in tesla and ρ in metres. Starting from $pc = qcB\rho$ we have

$$pc[\text{GeV}] \times 1.6 \times 10^{-10}[J/\text{GeV}] = 1.6 \times 10^{-19}[C] \times 3 \times 10^8[\text{m/s}] \times B[\text{T}] \times \rho[\text{m}].$$

Finally in NU: $p[\text{GeV}] = 0.3 \times B[\text{T}] \times \rho[\text{m}]$.

1.29 The Lorentz factor of the antiproton is $\gamma = \sqrt{p^2 + m^2}/m = 1.62$ and its velocity $\beta = \sqrt{1 - \gamma^{-2}} = 0.787$. The condition in order to have the antiproton above the Cherenkov threshold is that the index is $n \geq 1/\beta = 1.27$.

If the index is $n = 1.5$, the Cherenkov angle is given by $\cos\theta = 1/n\beta = 0.85$. Hence $\theta = 32°$.

1.30 The speed of a particle of momentum $p = m\gamma\beta$ is $\beta = \left(1 + \dfrac{m^2}{p^2}\right)^{-1/2} \approx 1 - \dfrac{m^2}{2p^2}$; that is a good approximation for speeds close to c. The difference between the flight times is $\Delta t = L \dfrac{m_2^2 - m_1^2}{2p^2}$ in NU. In order to have $\Delta t > 600$ ps, we need a base length $L > 26$ m.

1.32 Superman saw the light blue shifted due to Doppler effect. Taking for the wavelengths $\lambda_R = 650$ nm and $\lambda_G = 520$ nm, we have $v_G / v_R = 1.25$. Solving for β the Doppler shift expression $v_G = v_R \sqrt{\dfrac{1 + \beta}{1 - \beta}}$, we obtain $\beta = 0.22$.

1.36 (1) The total energy of the deuterons is $E_d = m_d + T_d = 1875.7$ MeV. The motion of the deuterons is not relativistic. Their momentum is $p_d = \sqrt{2 m_d T_d} = \sqrt{2 \times 1875.6 \times 0.13} = 61.25$ MeV. This is also the total momentum, which is so small that in this case the L frame is also, in practice, the CM frame.

The CM energy is $\sqrt{s} = \sqrt{(E_d + m_t)^2 - p_d^2}$; $E_d + m_t = 4684.6$. The result could be obtained by simply summing the two masses and the deuteron kinetic energy. This because the situation is non-relativistic. The total kinetic energy available after the reaction is $E_{kin,\,t} = E_d + m_t - m_\alpha - m_n = 17.6$ MeV, which is mainly taken by the lighter particle, the neutron. To be precise

$$T_n = \frac{s + m_n^2 - m_\alpha^2}{2\sqrt{s}} - m_n = \frac{4684.6^2 + 939.6^2 - 3727.4^2}{2 \times 4684.6} - 939.6$$
$$= 953.6 - 939.6 = 14.0 \text{ MeV}$$

and

$$T_\alpha = \frac{s + m_\alpha^2 - m_n^2}{2\sqrt{s}} - m_n = \frac{4684.6^2 + 3727.4^2 - 939.6^2}{2 \times 4684.6} - 3727.4$$
$$= 3.6 \text{ MeV}.$$

(2) The flux is $\Phi = \dfrac{I_n}{4\pi R^2} = \dfrac{3 \times 10^{10}}{4\pi \times 1^2} = 2.4 \times 10^9$ neutrons$/\left(\text{m}^2 \text{ s}\right)$.

(3) We can calculate the momentum of the neutron non-relativistically:
$p_n = \sqrt{2 m_n T_n} = \sqrt{2 \times 939.6 \times 14} = 162.2$ MeV, and its velocity
$\beta_n = \dfrac{p_n}{E_n} = \dfrac{162.2}{953.6} = 0.17$ $v_n = 5.1 \times 10^7$ m/s. We need 1 ns time resolution.

1.38 (a) $E_2 = \dfrac{h}{\lambda} = \dfrac{1240 \, \text{eV m}}{694 \, m} = 1.79 \, \text{eV}$

The CM energy for the head-on geometry is
$s = (E_1 + E_2)^2 - (\mathbf{p}_1 + \mathbf{p}_2)^2 = 2E_1 E_2 + 2E_1 E_2$.

At threshold $s = 4E_1 E_2 = (2m_e)^2$, that is $E_1 = \dfrac{m_e^2}{E_2} = \dfrac{(0.5)^2}{1.79 \times 10^{-6}} = 140$ GeV.

(b) $1-\beta = 1 - \dfrac{\mathbf{p}_1 + \mathbf{p}_2}{E_1 + E_2} = 1 - \dfrac{E_1 - E_2}{E_1 + E_2} = 1 - \dfrac{1 - \dfrac{E_2}{E_1}}{1 + \dfrac{E_2}{E_1}} \approx 2\dfrac{E_2}{E_1} = 2.6 \times 10^{-11}$

(c) $s = (E_1 + E_2)^2 - (\mathbf{p}_1 + \mathbf{p}_2)^2 = 2E_1E_2 - 2E_1E_2 = 0$. The mass is zero for any values of the two energies.

2.3 The second gamma moves backwards. The total energy is $E = E_1 + E_2$; the total momentum is $P = p_1 - p_2 = E_1 - E_2$. The square of the mass of the two-gamma system is equal to the square of the pion mass: $m_{\pi^0}^2 = (E_1 + E_2)^2 - (E_1 - E_2)^2 = 4E_1E_2$, from which we obtain $E_2 = \dfrac{m_{\pi^0}^2}{4E_1} = \dfrac{135^2}{4 \times 150} = 30.4$ MeV. The speed of the π^0 is $\beta = \dfrac{P}{E} = \dfrac{E_1 - E_2}{E_1 + E_2} = 0.662$.

2.8 Since the decay is isotropic, the probability of observing a photon is a constant $P(\cos \theta^*, \phi^*) = K$. We determine K by imposing that the probability of observing a photon at any angle is 2, that is, the number of photons.

We have $2 = \int K \sin \theta^* d\theta^* d\phi = \int_0^{2\pi} d\phi \int_0^\pi K d(\cos \theta^*) = K 4\pi$. Hence $K = 1/2\pi$ and $P(\cos \theta^*, \phi^*) = 1/2\pi$.

The distribution is isotropic in the azimuth in L too. To have the dependence of θ, that is given by $P(\cos \theta) \equiv \dfrac{dN}{d\cos \theta} = \dfrac{dN}{d\cos \theta^*} \dfrac{d\cos \theta^*}{d\cos \theta}$, we must calculate the 'Jacobian' $J = \dfrac{d\cos \theta^*}{d\cos \theta}$.

Calling β and γ the Lorentz factors of the transformation and taking into account that $p^* = E^{**}$, we have

$$p\cos\theta = \gamma(p^* \cos\theta^* + \beta E^*) = \gamma p^*(\cos\theta^* + \beta)$$
$$E = p = \gamma(E^* + \beta p^* \cos\theta^*) = \gamma p^*(1 + \beta \cos\theta^*).$$

We differentiate the first and third members of these relationships, taking into account that p^* is a constant. We obtain

$$dp \times \cos\theta + p \times d(\cos\theta) = \gamma p^* d(\cos\theta^*) \Rightarrow \dfrac{dp}{d\cos\theta^*}\cos\theta + p\dfrac{d\cos\theta}{d\cos\theta^*} = \gamma p^*,$$

$$dp = \gamma\beta p^* d(\cos\theta^*) \Rightarrow \dfrac{dp}{d\cos\theta^*} = \gamma\beta p^*$$

and $J^{-1} = \dfrac{d\cos\theta}{d\cos\theta^*} = \gamma\dfrac{p^*}{p}(1 - \beta\cos\theta).$

The inverse transformation is $E^* = \gamma(E - \beta p\cos\theta)$, that is, $p^* = \gamma p(1 - \beta\cos\theta)$, giving

$$J^{-1} = \dfrac{d\cos\theta}{d\cos\theta^*} = \gamma^2(1 - \beta\cos\theta)^2.$$

Finally, we obtain $P(\cos\theta) \equiv \dfrac{dN}{d\cos\theta} = \dfrac{1}{2\pi}\gamma^{-2}(1 - \beta\cos\theta)^{-2}.$

2.13 Considering the beam energy and the event topology, the event is probably an associate production of a K^0 and a Λ. Consequently, V^0 may be one of these two particles. The negative track is in both cases a π, while the positive track may be a π or a proton. We need to measure the mass of V. With the given data, we start by calculating the Cartesian components of the momenta

$$p_x^- = 121 \times \sin(-18.2°)\cos 15° = -36.5 \text{ MeV};$$
$$p_y^- = 121 \times \sin(-18.2°)\sin 15° = -9.8 \text{ MeV};$$
$$p_z^- = 121 \times \cos(-18.2°) = 115 \text{ MeV}.$$
$$p_x^+ = 1900 \times \sin(20.2°)\cos(-15°) = 633.7 \text{ MeV};$$
$$p_y^+ = 1900 \times \sin(20.2°)\sin(-15°) = -169.8 \text{ MeV};$$
$$p_z^+ = 1900 \times \cos(20.2°) = 1783.1 \text{ MeV}.$$

Summing the components, we obtain the momentum of V, that is, $p = 1998$ MeV.

The energy of the negative pion is $E^- = \sqrt{\left(p^-\right)^2 + m_\pi^2} = 185$ MeV. If the positive track is a π, its energy is $E_\pi^+ = \sqrt{\left(p^+\right)^2 + m_\pi^2} = 1905$ MeV, whereas if it is a proton, its energy is $E_p^+ = 2119$ MeV.

The energy of V is $E_\pi^V = 2090$ MeV in the first case; $E_p^V = 2304$ MeV in the second case. The mass of V is, consequently, $m_\pi^V = \sqrt{E^{V2} - p^2} = 620$ MeV in the first hypothesis, and $m_p^V = 1150$ MeV in the second. Within the ±4% uncertainty, the first hypothesis is incompatible with any known particle, while the second is compatible with the particle being a Λ.

2.14 (1) The CM energy squared is $s = \left(E_\nu + m_n\right)^2 - p_\nu^2 = m_n^2 + 2m_n E_\nu$. The threshold condition is

$$s = \left(m_e + m_p\right)^2 = m_p^2 + m_e^2 + 2m_e m_p.$$

Hence, the threshold condition is $E_\nu = \dfrac{\left(m_e + m_p\right)^2 - m_n^2}{2m_n} < 0$, meaning that there is no threshold; the reaction proceeds also at zero neutrino energy.

(2) The threshold condition is $s = \left(m_\mu + m_p\right)^2 = m_p^2 + m_\mu^2 + 2m_\mu m_p$. The threshold energy is

$$E_\nu = \frac{\left(m_\mu + m_p\right)^2 - m_n^2}{2m_n} = \frac{(105.7 + 938.3)^2 - 939.6^2}{2 \times 939.6} = 110 \text{ MeV}.$$

(3) The threshold energy is

$$E_\nu = \frac{\left(m_\tau + m_p\right)^2 - m_n^2}{2m_n} = \frac{(1777 + 938.3)^2 - 939.6^2}{2 \times 939.6} = 3.45 \text{ GeV}.$$

2.16 The laser photon energy is $E_{\gamma i} = \dfrac{h}{\lambda} = \dfrac{1240 \text{ eV nm}}{694 \text{ nm}} = 1.79 \text{ eV}$.

The electron initial momentum (we shall need its difference from energy) is

$$p_{ei} = \sqrt{E_{ei}^2 - m_e^2}; E_{ei} - \frac{m_e^2}{2E_{ei}}.$$

The total energy and momentum are

$$E_T = E_{ei} + E_{\gamma i} \qquad\qquad p_T = p_{ei} - E_{\gamma i}$$

Energy conservation gives

$$E_T = E_{ef} + E_{\gamma f} \qquad\qquad p_T = E_{\gamma f} - p_{ef}.$$

We can eliminate the final energy and momentum of the electron by imposing $E_{ef}^2 - p_{ef}^2 = m_e^2$.

$$E_{ef} = E_T - E_{\gamma f} \qquad p_{ef} = E_{\gamma f} - p_T.$$

Hence:

$$\left(E_T - E_{\gamma f}\right)^2 - \left(E_{\gamma f} - p_T\right)^2 = m_e^2.$$

Solving for $E_{\gamma f}$ we have $E_{\gamma f} = \dfrac{s - m_e^2}{2\left(E_T - p_T\right)}$.

$$E_T - p_T = \left(E_{ei} + E_{\gamma i}\right) - \left(p_{ei} - E_{\gamma i}\right); \frac{m_e^2}{2E_{ei}} + 2E_{\gamma i}$$

$$= \frac{0.5^2 \times^{-6}}{2 \times 20} + 2 \times 1.79 \times 10^{-9} = (6.25 + 3.58)10^{-9} \text{ GeV} = 9.83 \text{ eV}$$

$$s = \left(E_{\gamma i} + E_{ei}\right)^2 - \left(E_{\gamma i} - p_{ei}\right)^2 = m_e^2 + 4E_{\gamma i}E_{ei}.$$

Hence $s - m_e^2 = 4E_{\gamma i}E_{ei} = 4 \times 1.79 \times 20 \times 10^9 \times \text{eV}^2 = 14.3 \times 10^{10} \text{ eV}^2$, and

$$E_{\gamma f} = \frac{s - m_e^2}{2\left(E_T - p_T\right)} = \frac{14.3 \times 10^{10}}{2 \times 9.83} = 7.3 \text{ GeV}.$$

3.2 Strangeness conservation requires that a K^+ or a K^0 be produced together with the K^-. The third component of the isospin in the initial state is $-1/2$. Let us check if it is conserved in the two reactions. The answer is yes for $\pi^- + p \to K^- + K^+ + n$ because in the final state we have $I_z = -1/2 + 1/2 + 1/2 = +1/2$, and yes also for $\pi^- + p \to K^- + K^0 + p$ because in the final state we have $I_z = -1/2 - 1/2 + 1/2 = -1/2$. The threshold of the first reaction is just a little smaller than that of the second reaction because $m_n + m_{K^+} < m_p + m_{K^0}$ (1433 MeV < 1436 MeV). For the former we have

$$E_\pi = \frac{\left(2m_K + m_n\right)^2 - m_\pi^2 - m_p^2}{2m_p} = 1.5 \text{ GeV}.$$

3.4 (1) OK, S; (2) OK, W; (3) violates $\mathcal{L}_\mu$; (4) OK, EM; (5) violates C; (6) cannot conserve both energy and momentum; (7) violates $\mathcal{B}$ and S; (8) violates $\mathcal{B}$ and S; (9) violates J and $\mathcal{L}_e$; (10) violates energy conservation.

3.8 (a) NO for J and $\mathcal{L}$; (b) NO for J and $\mathcal{L}$; (c) YES; (d) NO for $\mathcal{L}$; (e) YES; (f) NO for $\mathcal{L}_e$ and $\mathcal{L}_\mu$; (g) NO for $\mathcal{L}$; (h) YES.

3.9

$$\left|\pi^- p\right\rangle = \left|1, -1\right\rangle \left|\frac{1}{2}, +\frac{1}{2}\right\rangle = \sqrt{\frac{1}{3}}\left|\frac{3}{2}, -\frac{1}{2}\right\rangle - \sqrt{\frac{2}{3}}\left|\frac{1}{2}, -\frac{1}{2}\right\rangle$$

$$\left|\pi^+ p\right\rangle = \left|1, -1\right\rangle \left|\frac{1}{2}, +\frac{1}{2}\right\rangle = \left|\frac{3}{2}, +\frac{3}{2}\right\rangle$$

$$\left|\Sigma^0 K^0\right\rangle = \left|1,0\right\rangle\left|\frac{1}{2},-\frac{1}{2}\right\rangle = \sqrt{\frac{2}{3}}\left|\frac{3}{2},-\frac{1}{2}\right\rangle + \sqrt{\frac{1}{3}}\left|\frac{1}{2},-\frac{1}{2}\right\rangle$$

$$\left|\Sigma^- K^+\right\rangle = \left|1,-1\right\rangle\left|\frac{1}{2},+\frac{1}{2}\right\rangle = \sqrt{\frac{1}{3}}\left|\frac{3}{2},-\frac{1}{2}\right\rangle + \sqrt{\frac{2}{3}}\left|\frac{1}{2},-\frac{1}{2}\right\rangle$$

$$\left|\Sigma^+ K^-\right\rangle = \left|1,+1\right\rangle\left|\frac{1}{2},+\frac{1}{2}\right\rangle = \left|\frac{3}{2},+\frac{3}{2}\right\rangle$$

$$\left\langle K^+\Sigma^+\middle|\pi^+ p\right\rangle = A_{3/2};\quad \left\langle K^-\Sigma^+\middle|\pi^- p\right\rangle = \sqrt{\frac{1}{3}}\sqrt{\frac{1}{3}}A_{3/2} = \frac{1}{3}A_{3/2}.$$

$$\left\langle K^0\Sigma^0\middle|\pi^- p\right\rangle = \sqrt{\frac{2}{3}}\sqrt{\frac{1}{3}}A_{3/2} = \frac{\sqrt{2}}{3}A_{3/2}.$$

Hence:

$$\sigma\left(\pi^+ p \to \Sigma^+ K^+\right):\sigma\left(\pi^- p \to \Sigma^- K^+\right):\sigma\left(\pi^- p \to \Sigma^0 K^0\right)=9:1:2.$$

3.15 $\sigma(1):\sigma(2):\sigma(3)=\left|-\dfrac{1}{\sqrt{6}}A_0 + \dfrac{1}{2}A_1\right|^2 : \left|\dfrac{1}{\sqrt{6}}A_0\right|^2 : \left|\dfrac{1}{\sqrt{6}}A_0 + \dfrac{1}{2}A_1\right|^2.$

3.19 (1) $C(\bar{p}p) = (-1)^{l+s} = C(n\pi^0) = +$. Then $l+s = $ even. The possible states are $^1S_0, ^3P_1, ^3P_2, ^3P_3$ and $^1D_2$.

(2) The orbital momentum is even, because the wave function of the $2\pi^0$ state must be symmetric. Since the total angular momentum is just orbital momentum, only the states $^1S_0, ^3P_2$ and $^1D_2$ are left. Parity conservation gives $P(2\pi^0) = + = P(\bar{p}p) = (-1)^{l+1}$. Hence, $l = $ odd, leaving only $^3P_2$.

3.21 It is convenient to prepare a table with the possible values of the initial J^{PC} and of the final l^{CP} with $l = J$ to satisfy angular momentum conservation. Only the cases with the same parity and charge conjugation are allowed. Recall that $P(\bar{p}p) = (-1)^{l+1}$ and $C(\bar{p}p) = (-1)^{l+s}$.

	$^1S_0$	$^3S_1$	$^1P_0$	$^3P_0$	$^3P_1$	$^3P_2$	$^1D_2$	$^3D_1$	$^3D_2$	$^3D_3$
J^{PC}	0^{-+}	1^{--}	1^{+-}	0^{++}	1^{++}	2^{++}	2^{-+}	1^{--}	2^{--}	3^{--}
J^{PC}	0^{++}	1^{--}	1^{--}	0^{++}	1^{--}	2^{++}	2^{++}	1^{--}	2^{++}	3^{--}
		Y		Y		Y		Y		Y

In conclusion: (1) $^1S_0$; (2) $^3S_1, ^3D_1$; (3) $^3P_2$.

3.23 Λ_b is neutral: (a) violates charm and beauty, (b) and (c) are allowed, (d) violates beauty, (e) violates baryon number.

3.26 (1) The minimum velocity is $\beta_{min} = \dfrac{1}{n} = 0.75$.

(2) The minimum kinetic energy for electrons is

$$E_{kin,min}(e) = m_e\left(\frac{1}{\sqrt{1-\beta_{min}^2}} - 1\right) = 0.511 \times 0.51 = 0.26 \text{ MeV}$$

and for K^+: $E_{kin,min}(K^+) = m_K\left(\dfrac{1}{\sqrt{1-\beta_{min}^2}} - 1\right) = 497.6 \times 0.51 = 254 \text{ MeV}.$

(3) In the decay $p \rightarrow e^+ + \pi^0$, the CM kinetic energy of the e^+ is

$$E_{kin}\left(e^+\right) = \frac{m_p^2 + m_e^2 - m_{\pi^0}^2}{2m_p} - m_e = \frac{938.3^2 + 0.51^2 - 0.135^2}{2 \times 938.3} - 0.51$$

$$= 469 \text{ MeV}, \quad \text{above threshold.}$$

In the decay $p \rightarrow K^+ + \nu$, the CM kinetic energy of the K is

$$E_{kin}\left(K^+\right) = \frac{m_p^2 + m_K^2 - m_\nu^2}{2m_p} - m_K = \frac{938.3^2 + 497.6^2}{2 \times 938.3} - 497.6$$

$$= 104 \text{ MeV}, \quad \text{below threshold.}$$

3.31 The beam energy is enough to produce strange particles, but not for heavier flavours. In order to conserve strangeness, the V^0s must be a K^0 and a Λ. The simplest reaction is $\pi^+ + p \rightarrow \pi^+ + \pi^+ + K^0 + \Lambda$.

We calculate the mass of each V^0, assuming it to be the K^0 or the a Λ in turn. If 1 is a Λ,

$$M^2 = m_p^2 + m_\pi^2 + 2\sqrt{p_1^2 + m_p^2}\sqrt{p_{1-}^2 + m_\pi^2} - 2p_{1+}p_{1-}\cos\theta_1$$

$$= 0.938^2 + 0.139^2 + 2 \times 1.02 \times 1.905 - 2 \times 0.4 \times 1.9 \times \cos 24.5°$$

$$= 3.38 \text{ GeV}^2$$

or $M = 1.83$ GeV, not compatible with being a Λ.

If 1 is a K^0,

$$M^2 = 2m_\pi^2 + 2\sqrt{p_1^2 + m_\pi^2}\sqrt{p_{1-}^2 + m_\pi^2} - 2p_{1+}p_{1-}\cos\theta_1$$

$$= 0.04 + 2 \times 0.424 \times 1.905 - 2 \times 0.4 \times 1.9 \times \cos 24.5°$$

$$= 0.246 \text{ GeV}^2$$

or $M = 0.495$ GeV, compatible, within the errors, with the mass of the K^0.

If 2 is a Λ,

$$M^2 = m_p^2 + m_\pi^2 + 2\sqrt{p_{2+}^2 + m_p^2}\sqrt{p_{2-}^2 + m_\pi^2} - 2p_{2+}p_{2-}\cos\theta_2$$

$$= 0.938^2 + 0.139^2 + 2 \times 1.20 \times 0.29 - 2 \times 0.75 \times 0.25 \times \cos 22°$$

$$= 1.59 - 0.35 = 1.24 \text{ GeV}^2$$

or $M = 1.11$ GeV, compatible, within the errors, with the mass of the Λ.

If 2 is a K^0,

$$M^2 = 2m_\pi^2 + 2\sqrt{p_{2+}^2 + m_\pi^2}\sqrt{p_{2-}^2 + m_\pi^2} - 2p_{2+}p_{2-}\cos\theta_2$$

$$= 0.04 + 2 \times 0.76 \times 0.29 - 0.35 = 0.138 \text{ GeV}^2$$

or $M = 0.371$ GeV, incompatible, within the errors, with the mass of the K^0.

4.3 The ρ decays strongly into 2π, hence $G = +$. The possible values of its isospin are 0, 1 and 2. In the three cases, the Clebsch–Gordan coefficients are $\langle 1,0|1,0;1,0\rangle = 0, \langle 0,0|1,0;1,0\rangle$. and $\langle 2,0|1,0;1,0\rangle \neq 0$. Hence $I = 1$.

Since $I = 1$, the isospin wave function is antisymmetric. The spatial wave function must consequently be antisymmetric, that is, the orbital momentum of the two π must be $l = $ odd. The ρ spin is equal to l. $C = (-1)^l = -1$. $P = (-1)^l = -1$.

4.7 (1) Two equal bosons cannot be in an antisymmetric state; (2) $C(2\pi^0) = +1$; (3) the Clebsch–Gordan coefficient $\langle 1,0|1,0;1,0\rangle = 0$.

4.11 It is useful to prepare a table with the quantum numbers of the relevant states.

	$\bar{p}p\,{}^3S_1$	$\bar{p}p\,{}^3S_1$	$\bar{p}p\,{}^1S_0$	$\bar{p}p\,{}^1S_0$	$\bar{p}n\,{}^3S_1$	$\bar{p}n\,{}^1S_0$
J^P	1^-	1^-	0^-	0^-	1^-	0^-
C	$-$	$-$	$+$	$+$	X	X
I	0	1	0	1	1	1
G	$-$	$+$	$+$	$-$	$+$	$-$

Here $\bar{p}n \to \pi^-\pi^-\pi^+$. Since $G = -1$ in the final state, there is only one possible initial state, that is, ${}^1S_0$

$$|\bar{p},n\rangle = |1,-1\rangle = \frac{1}{\sqrt{2}}|1,0;1,-1\rangle - \frac{1}{\sqrt{2}}|1,-1;1,0\rangle = \frac{1}{\sqrt{2}}|\rho^0;\pi^-\rangle - \frac{1}{\sqrt{2}}|\rho^-;\pi^0\rangle$$

hence $R(\bar{p}n \to \rho^0\pi^-)/R(\bar{p}n \to \rho^-\pi^0) = 1$.

$$|\bar{p},p\rangle = |1,0\rangle = \frac{1}{\sqrt{2}}|\rho^-,\pi^+\rangle + 0\frac{1}{\sqrt{2}}|\rho^0;\pi^0\rangle - \frac{1}{\sqrt{2}}|\rho^+;\pi^-\rangle$$

hence $R(\bar{p}p(I=1) \to \rho^+\pi^-) : R(\bar{p}p(I=1) \to \rho^0\pi^0) : R(\bar{p}p(I=1) \to \rho^-\pi^+) = 1:0:1$.

$$|\bar{p},p\rangle = |0,0\rangle = \frac{1}{\sqrt{3}}|\rho^-;\pi^+\rangle - \frac{1}{\sqrt{3}}|\rho^0,\pi^0\rangle + \frac{1}{\sqrt{3}}|\rho^+;\pi^-\rangle$$

hence $R(\bar{p}p(I=0) \to \rho^+\pi^-) : R(\bar{p}p(I=0) \to \rho^0\pi^0) : R(\bar{p}p(I=0) \to \rho^-\pi^+) = 1:1:1$.

4.13 The matrix element $\mathcal{M}$ must be symmetric under the exchange of each pair of pions. Consequently, we have the following:
(1) If $J^P = 0^-$, $\mathcal{M} = $ constant; there are no zeros.
(2) If $J^P = 1^-$, $\mathcal{M} \propto \mathbf{q}(E_1 - E_2)\,(E_2 - E_3)\,(E_3 - E_1)$; zeros on the diagonals and on the border.
(3) If $J^P = 1^+$, $\mathcal{M} \propto \mathbf{p}_1 E_1 + \mathbf{p}_2 E_2 + \mathbf{p}_3 E_3$; zero in the centre, where $E_1 = E_2 = E_3$; zero at $T_3 = 0$, where $\mathbf{p}_3 = 0$, $\mathbf{p}_2 = -\mathbf{p}_1$; $E_2 = E_1$.

4.15 A baryon can contain between 0 and 3 c valence quarks; therefore, the charm of a baryon can be $C = 0,1,2,3$. Since the charge of c is equal to 2/3, the baryons with $Q = +1$ can have charm $C = 2\,(ccd, ccs, ccb)$, $C = 1$ (e.g. cud) or $C = 0$ (e.g. uud). If $Q = 0$, one c can be present, as in cdd, or none as in udd. Hence $C = 1$ or $C = 0$.

4.24 We start from

$$\sigma(E) = \frac{3\pi}{E^2}\frac{\Gamma_e\Gamma_f}{(E-M_R)^2 + (\Gamma/2)^2} = \frac{12\pi\Gamma_e\Gamma_f}{\Gamma^2}\frac{1}{E^2}\frac{1}{[2(E-M_R)/\Gamma]^2 + 1}.$$

In the neighbourhood of the resonance peak, the factor $1/E^2$ varies only slowly, compared to the resonant factor, and we can approximate it with the constant $1/M_R^2$, that is:

$$\int_{-\infty}^{+\infty} \sigma(E)dE = \frac{12\pi\Gamma_e\Gamma_f}{\Gamma^2} \int_{-\infty}^{+\infty} \frac{1}{E^2} \frac{1}{\left[2(E - M_R)/\Gamma\right]^2 + 1} dE$$

$$\approx \frac{12\pi\Gamma_e\Gamma_f}{\Gamma^2 M_R^2} \int_{-\infty}^{+\infty} \frac{1}{\left[2(E - M_R)/\Gamma\right]^2 + 1} dE$$

Setting $\tan\theta = \dfrac{2(E - M_R)}{\Gamma}$, we have

$$\int_{-\infty}^{+\infty} \sigma(E)dE = \frac{12\pi\Gamma_e\Gamma_f}{\Gamma^2 M_R^2} \int_{-\infty}^{+\infty} \frac{1}{\tan^2\theta + 1} dE = \frac{12\pi\Gamma_e\Gamma_f}{\Gamma^2 M_R^2} \int_{-\infty}^{+\infty} \cos^2\theta\, dE$$

$$= \frac{12\pi\Gamma_e\Gamma_f}{\Gamma^2 M_R^2} \int_{-\pi/2}^{+\pi/2} \cos^2\theta \frac{dE}{d\theta} d\theta.$$

We find that $\dfrac{dE}{d\theta} = \dfrac{dE}{d\tan\theta} \dfrac{d\tan\theta}{d\theta} = \dfrac{\Gamma}{2} \dfrac{1}{\cos^2\theta}$, obtaining

$$\int_{-\infty}^{+\infty} \sigma(E)dE = \frac{6\pi\Gamma_e\Gamma_f}{\Gamma M_R^2} \int_{-\pi/2}^{+\pi/2} d\theta = \frac{6\pi^2\Gamma_e\Gamma_f}{\Gamma M_R^2}.$$

4.30 $G = +$, because there are overall 4π.

The isospin cannot be 1, because the Clebsch–Gordan coefficient $\langle 1,0;1,0|1,0\rangle = 0$.
It may then be $I = 0$ or $I = 2$.
$C = +$ because the two particles are identical. Check: $G = C(-1)^I = +$.
The two particles are identical bosons, hence L must be even, $L = 0, 2, 4, \ldots$.
The spin wave function must be symmetrical too, hence $S = 0, 2$.
It can be:
$J = 0$, with $L = 0$, $S = 0$ and with $L = 2$, $S = 2$
$J = 1$ with $L = 2$, $S = 2$
$J = 2$ with $L = 0$, $S = 2$, with $L = 2$, $S = 0$, with $L = 2$, $S = 2$ and with $L = 4$, $S = 2$.

4.34 With a π^- beam, to conserve strangeness and charm we need to produce together with Ω_c^0 (ssc) one particle containing $\bar{c}$, say D^- $(d\bar{c})$ and two containing $\bar{s}$, say, to conserve also the charge, K^+ $(u\bar{s})$ and K^0 $(d\bar{s})$. The reaction is $\pi^- p \rightarrow \Omega_c^0 D^- K^+ K^0$. Its threshold is

$$E_\pi = \frac{\left(m_{\Omega_c^0} + m_{D^-} + m_{K^+} + m_{K^0}\right)^2}{2m_p} = \frac{(2.698 + 1.869 + 0.494 + 0.498)^2}{2 \times 0.938}$$

$$= 16.5 \text{ GeV}.$$

With a K^- beam the initial strangeness is $S = +1$, hence only one K meson needs to be produced. The reaction is $K^- p \rightarrow \Omega_c^0 D^- K^+$. Its threshold is

$$E_K = \frac{\left(m_{\Omega_c^0} + m_{D^-} + m_{K^+}\right)^2}{2m_p} = \frac{(2.698 + 1.869 + 0.494)^2}{2 \times 0.938} = 13.7 \text{ GeV}.$$

With a K^+ beam the initial strangeness is $S = -1$, hence three K mesons must be produced. The reaction is $K^+ p \rightarrow \Omega_c^0 D^- K^+ K^+ K^+$. Its threshold is

$$E_K = \frac{\left(m_{\Omega_c^0} + m_{D^-} + m_{K^+}\right)^2}{2m_p} = \frac{(2.698 + 1.869 + 0.494)^2}{2 \times 0.938} = 19.5 \text{ GeV}.$$

5.2 Since the speeds are small enough, we can use non-relativistic concepts and expressions. The electron potential energy, which is negative, becomes smaller with its distance r from the proton as $-1/r$. The closer the electron is to the proton, the better its position is defined and consequently the larger is the uncertainty of its momentum p. Actually, the larger the uncertainty of p, the larger is its average value and, with it, the electron kinetic energy. The radius of the atom is the distance at which the sum of potential and kinetic energies is minimum.

Owing to its large mass, we consider the proton to be immobile. At distance r, the energy of the electron is

$$E = \frac{p^2}{2m_e} - \frac{1}{4\pi\varepsilon_0} \frac{q_e^2}{r}.$$

The uncertainty principle dictates $pr = \hbar$ and we have $E = \frac{\hbar}{2m_e r^2} - \frac{1}{4\pi\varepsilon_0} \frac{q_e^2}{r}$.

To find the minimal radius a, we set $\left(\frac{dE}{dr}\right)_a = 0 = -\frac{\hbar}{m_e a^2} + \frac{1}{4\pi\varepsilon_0} \frac{q_e^2}{a^2}$, obtaining

$a = \frac{4\pi\varepsilon_0 \hbar^2}{m_e q_e^2} = 52.8$ pm, which is the Bohr radius.

5.6 At the next-to-tree-level order in the t channel, there are the eight diagrams in the following figure.

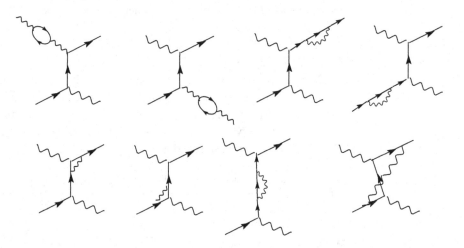

There are as many diagrams in the s channel. The last one is shown here.

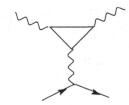

5.14 (a) In the case of the electron, the Bohr radius is $a_e = \dfrac{4\pi\epsilon_0\hbar^2}{q_e^2 m_e} = \dfrac{\hbar^2}{\alpha m_e} = 53\,\text{pm}$,

which is inversely proportional to the electron mass. To be precise, the reduced mass must be considered. The reduced mass of a system composed of a proton and a particle of mass m is $m_R \equiv \dfrac{mm_p}{m+m_p}$. We have: $m_{R\mu} = 95\,\text{MeV}$, $m_{R\pi} = 121\,\text{MeV}$, $m_{RK} = 325\,\text{MeV}$, $m_{R\bar{p}} = 469\,\text{MeV}$ and $a_\mu = a_e\dfrac{m_e}{m_{R\mu}} = 280\,\text{fm}$, $a_\pi = 220\,\text{fm}$, $a_K = 82\,\text{fm}$, $a_{\bar{p}} = 56\,\text{fm}$.

(b) It is $13.6\,\text{eV}$ for an electron and is proportional to the (reduced) mass. Hence

$$13.6\frac{121}{0.5} = 3.3\,\text{keV}.$$

5.17 (a) $C(\eta) = C(\pi\pi)C(\gamma)$ or $+ = -C(\pi\pi)$, hence $C(\pi\pi) = -$ and $l_{\pi\pi} = \text{odd}$, with minimum value $= 1$. The total wave function must be even. Because the spatial part is odd, the isospin part must be odd, hence $I_{\pi\pi} = 1$. The isospin violation is $\Delta I = 1$.

(b) $C(\omega) = C(\pi\pi)C(\gamma)$ or $- = -C(\pi\pi)$, hence $C(\pi\pi) = +$ and $l_{\pi\pi} = \text{even}$, with minimum value $= 0$. Because the spatial part is even, the isospin part must be even, hence $I_{\pi\pi} = 0$ or 2. Electromagnetic interaction has $\Delta I = 0$ or 1. Hence $I_{\pi\pi} = 0$ only.

(c) As (b) but with $I_{\pi\pi} = 0$ and 2 both allowed.

(d) A $\pi^0\pi^0$ cannot be in $|I,I_z\rangle = |1,0\rangle$: $\eta \to \pi^0\pi^0\gamma$ forbidden, $\omega \to \pi^0\pi^0\gamma$ allowed, $\rho^0 \to \pi^0\pi^0\gamma$ allowed.

6.2 The first case is below the charm threshold, hence $R(u,d,s) = 2$; the second case is above the charm threshold and below the beauty one, hence $R(u,d,s,c) = 10/3 = 3.3$.

6.5 From (6.12) with $W = m_p$, $2m_p\nu = Q^2$ follows and then, from (6.16), we have $x = 1$. Using (6.11), we then obtain $2m_p\nu = Q^2 = 2EE'(1-\cos\theta)$, and then we use (1.83) because $\nu = E - E'$.

6.8 For every x, the momentum transfer Q^2 varies from a minimum to a maximum value when the electron scattering angle varies from $0°$ to $180°$. From Eqs. (6.11) and (6.14), which are valid in the L frame, and (6.16), we obtain

$$Q^2 = \frac{2E^2(1-\cos\theta_f)}{1 + \dfrac{E}{xm_p}(1-\cos\theta_f)}.$$ Clearly, we have $Q^2 = 0$ in the forward direction

($\theta = 0$). The maximum momentum transfer is for backwards scattering ($\theta = 180°$), that is $Q^2_{\text{max}} = \dfrac{4E^2}{1 + \dfrac{2E}{xm_p}} \approx 2xm_p$

For $E = 100\,\text{GeV}$ and $x = 0.2$, we have $Q^2_{\text{max}} = 37.5\,\text{GeV}^2$, corresponding to a resolving power of $32\,\text{am}$.

6.11 (a) $\Lambda_c^+ = udc$ violates charm, (b) $D^- = d\bar{c}$ OK, (c) $\bar{D}^0 = u\bar{c}$, charm conserved but electric charge violated, (d) $D_s^- = s\bar{c}$, charm conserved but strangeness violated.

6.13 The colour wave function is $\dfrac{1}{\sqrt{6}}[RGB - RBG + GBR - GRB + BRG - BGR]$,

which is completely antisymmetric. Since the space wave function is symmetric, the product of the spin and isospin wave functions must be completely symmetric for any two-quark exchange. The system, uud, is obviously symmetric in the exchange within the u pair. Consider the ud exchange. The totally symmetric combination $uud + udu + duu$ has isospin 3/2 and is not the proton.

The isospin 1/2 wave function contains terms that are antisymmetric under the exchange of the second and third quark, like $uud - udu$. We obtain symmetry by multiplying by a term with the same antisymmetry in spin, namely $(\uparrow\uparrow\downarrow - \uparrow\downarrow\uparrow)$. We thus obtain a term symmetric under the exchange of the second and third quarks:

$$(u\uparrow)(u\uparrow)(d\downarrow) - (u\uparrow)(d\uparrow)(u\downarrow) - (u\uparrow)(u\downarrow)(d\uparrow) + (u\uparrow)(d\downarrow)(u\uparrow).$$

Similarly, for the first two quarks we have

$$(u\uparrow)(u\uparrow)(d\downarrow) - (d\uparrow)(u\uparrow)(u\downarrow) - (u\downarrow)(u\uparrow)(d\uparrow) + (d\downarrow)(u\uparrow)(u\uparrow).$$

And for the first and third

$$(d\downarrow)(u\uparrow)(u\uparrow) - (u\downarrow)(d\uparrow)(u\uparrow) - (d\uparrow)(u\downarrow)(u\uparrow) + (u\uparrow)(d\downarrow)(u\uparrow).$$

In total we have 12 terms. We take their sum and normalize, obtaining

$$\frac{1}{\sqrt{12}}\big[2(u\uparrow)(u\uparrow)(d\downarrow) + 2(d\downarrow)(u\uparrow)(u\uparrow) + 2(u\uparrow)(d\downarrow)(u\uparrow) - (u\uparrow)(d\uparrow)(u\downarrow)$$
$$- (u\uparrow)(u\downarrow)(d\uparrow) - (d\uparrow)(u\uparrow)(u\downarrow) - (u\downarrow)(u\uparrow)(d\uparrow) - (u\downarrow)(d\uparrow)(u\uparrow)$$
$$- (d\uparrow)(u\downarrow)(u\uparrow)\big]$$

that is, as required, completely antisymmetric for the exchange of any pair.

6.20 (a) $\sqrt{s} \approx 2\sqrt{E_p E_e} = 300$ GeV

(b) $Q^2 = 4E_e E_e' \sin^2(\theta/2) = 4 \times 28 \times 223 \times \sin^2 60° = 18\,732$ GeV^2. The four-momentum of the initial proton is $P_\mu = E_p, \mathbf{P}_p$ and the four-momentum transfer $q^\mu = (E_e' - E_e), (\mathbf{p}_e' - \mathbf{p}_e)$ and their (invariant) product

$$\begin{aligned}
P_\mu q^\mu &= E_p(E_e' - E_e) - (\mathbf{p}_e' - \mathbf{p}_e)\cdot \mathbf{P}_p \\
&= E_p(E_e' - E_e) - p_e' P_p \cos(180° - \theta) + p_e P_p \cos 180° \\
&\approx E_e'(E_p - 2E_e) + E_e' E_p \cos\theta \\
&= 223 \times (820 - 2 \times 28) + 223 \times 820 \times \cos 120° \\
&= 78942 \text{ GeV}^2
\end{aligned}$$

and

$$x = \frac{Q^2}{2P_\mu p^\mu} = \frac{18\,732}{2 \times 78\,942} = 0.11$$

$$v = \frac{P_\mu q^\mu}{m_p} = \frac{78\,942}{0.938} = 84\,160 \text{ GeV}^2$$

$$W = m_p^2 + 2m_p\nu - Q^2 = 0.938^2 + 2\times 0.938 \times 84\,160 - 18\,732$$
$$= 6.5\times 10^4 \text{ GeV} = (250 \text{ GeV})^2.$$

6.23 (a) The exchanged gluon is $g_2 = R\bar{B}$; the colour charges are $\left(\dfrac{\sqrt{\alpha_s}}{\sqrt{2}}\right)\left(\dfrac{\sqrt{\alpha_s}}{\sqrt{2}}\right) = \dfrac{\alpha_s}{2}.$

(b) The exchanged gluon is $g_2 = R\bar{B}$; the colour charges are
$$\left(\frac{\sqrt{\alpha_s}}{\sqrt{2}}\right)\left(-\frac{\sqrt{\alpha_s}}{\sqrt{2}}\right) = -\frac{\alpha_s}{2}.$$

(c) There are two possible gluons to be exchanged: $g_7 = \dfrac{1}{\sqrt{2}}\left(R\bar{R} - G\bar{G}\right)$

and $g_8 = \dfrac{1}{\sqrt{6}}\left(R\bar{R} + G\bar{G} - 2B\bar{B}\right)$; the colour charges are $\left(\dfrac{\sqrt{\alpha_s}}{\sqrt{2}}\dfrac{1}{\sqrt{2}}\right)$
$$\left(-\frac{\sqrt{\alpha_s}}{\sqrt{2}}\frac{1}{\sqrt{2}}\right) = -\frac{\alpha_s}{2}\frac{1}{2} \text{ and } \left(\frac{\sqrt{\alpha_s}}{\sqrt{2}}\frac{1}{\sqrt{6}}\right)\left(-\frac{\sqrt{\alpha_s}}{\sqrt{2}}\frac{1}{\sqrt{6}}\right) = -\frac{\alpha_s}{2}\frac{1}{6}. \text{ In total } -\frac{\alpha_s}{2}\frac{2}{3}.$$

6.26 Because the colour wave function is symmetric, the product of the spin and space wave functions must be symmetric.

The total spin and the corresponding symmetry are $S = 0$ symmetric, $S = 1$ anti-symmetric, $S = 2$ symmetric.

The total orbital momentum can be $L = 0$ symmetric, $L = 1$ antisymmetric.
Hence the following combinations are possible: $S, L = 0, 0$ or $1,1$ or $2,2$.
Recall that $P = (-1)^L, C = (-1)^{L+S}.$
For $S = 0, L = 0$, we have $J^{PC} = 0^{++}.$
For $S = 1, L = 1$, we have $J^{PC} = 0^{--}, J^{PC} = 1^{-+}$ and $J^{PC} = 2^{--}.$
For $S = 2, L = 0$, we have $J^{PC} = 2^{++}.$
For $S = 2, L = 2$, we have $J^{PC} = 0^{++}, J^{PC} = 1^{++}, J^{PC} = 2^{++}, J^{PC} = 3^{++}, J^{PC} = 4^{++}.$

7.1 $K^{*+} \to K^0 + \pi^+$. We start by writing the valence quark compositions of all the particles, that is $(u\bar{s}) \to (d\bar{s}) + (u\bar{d})$, and then draw the diagram (Fig. (a)). Since it is a strong process we do not draw any gauge boson.

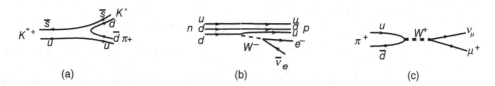

(a)   (b)   (c)

Here $n \to p + e^- + \bar{\nu}_e$. It is a weak process. In order to draw the diagram we consider two steps: the emission of a W, $(udd) \to (udu) + W^-$ and its decay $W^- \to e^- + \bar{\nu}_e$ (Fig. (b)). $\pi^+ \to \mu^+ + \nu_\mu$: We have $u\bar{d} \to W^+$ followed by $W^+ \to \mu^+ + \nu_\mu$ (Fig. (c)).

7.9 The quantity $\mathbf{p}_\Lambda \cdot \boldsymbol{\sigma}_\Lambda$ is a pseudoscalar. It must be zero if parity is conserved; therefore, the polarization must be perpendicular to $\mathbf{p}_\Lambda$.

7.18 The decay $c \to d + e^+ + \nu_e$ is disfavoured because its amplitude is proportional to $\sin\theta_C$. The decay $c \to s + e^+ + \nu_e$ is favoured because its amplitude is proportional to $\cos\theta_C$. We write down the valence quark compositions: $D^+ = c\bar{d}, K^+ = u\bar{s},$

$K^- = s\bar{u}$, $\bar{K}^0 = s\bar{d}$. Consequently, the decays of D^+ in final states containing a K^- or $\bar{K}^0$ are favoured. For example, $D^+ \to K^- + \pi^+ + e^+ + \nu_e$, $D^+ \to \bar{K}^0 + e^+ + \nu_e$, $D^+ \to \bar{K}^{*0} + e^+ + \nu_e$ are favoured. $D^+ \to \pi^- + \pi^+ + e^+ + \nu_e$, $D^+ \to \pi^0 + \pi^+$ and $D^+ \to \rho^0 + e^+ + \nu_e$ are disfavoured.

7.20 Because V_{tb} is very near to 1, the dominant decay is $t \to b\,W$. There are seven diagrams.

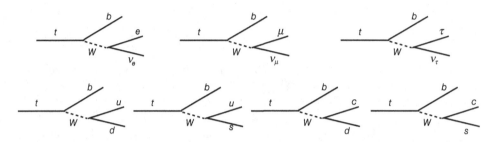

7.23 We start by writing the valence quark contents of the hadrons; we then identify the decay at the quark level and which quark acts as a spectator.
  (1) At the hadron level we have $c\bar{d} \to s\bar{d} + u\bar{d}$ and at the quark level $c \to su\bar{d}$ with a spectator $\bar{d}$. The decay rate is proportional to $|V_{cs}|^2 |V_{ud}|^2 \approx \cos^4\theta_C$.
  (2) At hadron level it is $c\bar{d} \to u\bar{s} + s\bar{d}$ and at the quark level it is $c \to su\bar{s}$ with a spectator $\bar{d}$. The decay probability is proportional to $|V_{cs}|^2 |V_{ud}|^2 \approx \cos^4\theta_C$.
  (3) The π^0 has a $u\bar{u}$ and a $d\bar{d}$ component. The decay picks up the latter. At hadron level it is $c\bar{d} \to u\bar{s} + d\bar{d}$ and at the quark level it is $c \to du\bar{s}$ with a spectator $\bar{d}$. The decay probability is proportional to $|V_{cd}|^2 |V_{us}|^2 \approx \sin^4\theta_C$.

7.28 There are $n_{Fe} = \rho \times N_A 10^3 / A = 10^{17} \times 6 \times 10^{23} \times 10^3 / 56 = 1.1 \times 10^{42}$ m^{-3} nucleons per unit volume. Consequently, the mean free path is

$$\lambda_\nu = \frac{1}{n_{Fe}\sigma} = \frac{1}{1.1 \times 10^{42} \times 3 \times 10^{-46}} = 3 \text{ km}.$$ This distance is smaller than the radius of the supernova core.

7.30 $\Sigma_c^{++}(uuc) \to \Sigma^+(uus)\pi^+; \Sigma^+(uus) \to p(uud)\pi^0$.

$$\Xi_c^+(usc) \to \Xi^0(uss)\pi^+; \Xi^0(uss) \to \Lambda(uds)\pi^0; \Lambda(uds) \to p(udu)\pi^-.$$

Notice that $\Xi^0(uss) \to \Sigma^+(uus)\pi^-$ is forbidden by energy conservation.

$$\Omega_c^0(ssc) \to \Omega^-(sss)\pi^+; \Omega^-(sss) \to \Xi^0(uss)\pi^-; \Xi^0(uss) \to \Lambda(uds)\pi^0;$$

$$\Lambda(uds) \to p(udu)\pi^-.$$

8.4 $\gamma = E / m_K = 4.05$ and $\beta = 0.97$; $t = \gamma\tau\ln10 = 1.1 \times 10^{-7}$ s and $d = \beta ct = 32$ m.

8.9 We invert the system of Eq. (8.32), that is $\sqrt{2}\left|K_L^0\right\rangle = (1+\varepsilon)\left|K^0\right\rangle + (1-\varepsilon)\left|\bar{K}^0\right\rangle$ and the corresponding one for K_S^0, that is $\sqrt{2}\left|K_S^0\right\rangle = (1+\varepsilon)\left|K^0\right\rangle - (1-\varepsilon)\left|\bar{K}^0\right\rangle$, taking into account that $|\varepsilon|$ is small. We obtain $\sqrt{2}\left|K^0\right\rangle = (1-\varepsilon)\left|K_S^0\right\rangle + (1-\varepsilon)\left|K_L^0\right\rangle$, $\sqrt{2}\left|K^0\right\rangle = (1+\varepsilon)\left|K_S^0\right\rangle - (1+\varepsilon)\left|K_L^0\right\rangle$. The decay amplitudes can be written as

$$\sqrt{2}A\left(\bar{K}^0 \to \pi^+\pi^-\right) = (1+\varepsilon)\left[A\left(K_S^0 \to \pi^+\pi^-\right) - A\left(K_L^0 \to \pi^+\pi^-\right)\right]$$
$$= A\left(K_S^0 \to \pi^+\pi^-\right)(1+\varepsilon)(1-\eta_{+-})$$

and similarly $\sqrt{2}A(K^0 \to \pi^+\pi^-) = A\left(K_S^0 \to \pi^+\pi^-\right)(1-\varepsilon)(1+\eta_{+-})$. We finally obtain

$$\left|\frac{A\left(\bar{K}^0 \to \pi^+\pi^-\right)}{A\left(K^0 \to \pi^+\pi^-\right)}\right| = \left|\frac{(1+\varepsilon)-(1+\varepsilon)\eta_{+-}}{(1-\varepsilon)+(1-\varepsilon)\eta_{+-}}\right| = \left|\frac{(1+\varepsilon)-(1+\varepsilon)(\varepsilon+\varepsilon')}{(1-\varepsilon)+(1-\varepsilon)(\varepsilon+\varepsilon')}\right| \approx 1-2\,\mathrm{Re}\,\varepsilon'.$$

8.12 If CP is conserved, in the decay $K_2^0 \to \pi^+\pi^-\pi^0\ L = l = 0$. The spatial wave function of the dipion is even. The total must be even. Consequently, the isospin wave function of the di-pion must be even, hence $I_{\pi\pi} = 0$ or $I_{\pi\pi} = 2$.

In the decay $K_1^0 \to \pi^+\pi^-\pi^0, L = l = $ odd. The isospin wave function of the dipion must be odd, hence $I_{\pi\pi} = 1$ or $I_{\pi\pi} = 3$.

9.3 In the rest frame of the pion, the neutrino energy is
$E_\nu^* = \left(m_\pi^2 - m_\mu^2\right)/(2m_\pi) = 30\ \mathrm{MeV}$.

The Lorentz factors are $\gamma = E_\pi/m_\pi = 1429$ and

$1-\beta = 1 - \sqrt{1-\gamma^{-2}} \approx \frac{1}{2}\gamma^{-2} = 2.4 \times 10^{-7}$.

The neutrino energy in the L frame is $E_\nu = \gamma(E_\nu^* + \beta p^* \cos\theta^*) = \gamma E_\nu^*\left(1 + \beta\cos\theta^*\right)$. Its maximum for $\theta^* = 0$ is $E_\nu^{\max} = \gamma E_\nu^*(1+1) = 1429 \times 30 \times 10^{-3}\,(\mathrm{GeV})2 = 85.7\ \mathrm{GeV}$.

Its minimum for $\theta^* = \pi$ is $E_\nu^{\min} = \gamma E_\nu^*(1-\beta) = 10\ \mathrm{keV}$.

We use the Lorentz transformations of the components of the neutrino momentum to find the relationship between the angle θ^* in CM and θ in L.

$p_\nu \sin\theta = p^* \sin\theta^*$; $p_\nu \cos\theta = \gamma\left(p^* \cos\theta^* + \beta E_\nu^*\right) \approx \gamma p^*\left(\cos\theta^* + 1\right)$, which gives

$$\tan\theta = \frac{\sin\theta^*}{\gamma\left(\cos\theta^* + 1\right)} \cong \frac{0.05}{1429 \times 2} = 22 \times 10^{-6} \Rightarrow \theta = 22\ \mu\mathrm{rad}.$$

9.7 For each reaction we check whether charge Q and hypercharge Y are conserved. We write explicitly the hypercharge values.

For $W^- \to d_L + \bar{u}_L$ we have $0 \to 1/3 - 4/3$. It violates Y.

For $W^- \to u_L + \bar{u}_L$ we have $0 \to 1/3 - 1/3$. OK.

For $Z \to W^- + W^+$ we have $0 \to 0 + 0$. OK.

For $W^+ \to e_R^+ + \bar{\nu}_R$ we have $0 \to 1 + 1$. It violates Y, and lepton number.

9.9

$$\Gamma_\nu = \frac{G_F M_Z^3}{3\sqrt{2}\pi}\left(\frac{1}{2}\right)^2 \approx 660 \times 1/4\ \mathrm{MeV} = 165\ \mathrm{MeV}.$$

$$\Gamma_l = \frac{G_F M_Z^3}{3\sqrt{2}\pi}\left[\left(-\frac{1}{2} + s^2\right)^2 + s^4\right] \approx 660 \times 0.148 \approx 98\ \mathrm{MeV}.$$

$$\Gamma_u = \Gamma_c = 3\frac{G_F M_Z^3}{3\sqrt{2}\pi}\left[\left(\frac{1}{2} - \frac{2}{3}s^2\right)^2 + \left(-\frac{2}{3}s^2\right)^2\right] \approx 3 \times 660 \times 0.173 \approx 342\ \mathrm{MeV}.$$

$$\Gamma_d = \Gamma_s = \Gamma_b = 3\frac{G_F M_Z^3}{3\sqrt{2}\pi}\left[\left(-\frac{1}{2} + \frac{1}{3}s^2\right)^2 + \left(\frac{1}{3}s^2\right)^2\right] \approx 3 \times 660 \times 0.207$$

$\approx 410\ \mathrm{MeV}.$

$$\Gamma_Z = 3 \times 165 + 3 \times 98 + 2 \times 342 + 3 \times 410 = 2.7\ \mathrm{GeV}.$$

$$\Gamma_h = 2 \times 342 + 3 \times 410 = 1910 \text{ MeV}; \Gamma_\mu / \Gamma_h = \frac{98}{1910} = 5.1\%.$$

9.18 $m^2 = 4E_1 E_2 \sin^2 \theta / 2$ fi $m = 92 \text{ GeV}$.

$$\frac{\sigma_{M_Z}}{M_Z} = \frac{1}{2} \sqrt{\left(\frac{\sigma(E_1)}{E_1}\right)^2 + \left(\frac{\sigma(E_2)}{E_2}\right)^2 + \left(\frac{\sigma(\theta)}{\tan \theta / 2}\right)^2} = \frac{1}{2} 10^{-2} \sqrt{2.4^2 + 2^2 + 0.6^2} = 1.6\%.$$

9.21 The energy squared in the quark–antiquark CM frame is $\hat{s} = x_q x_{\bar{q}} s$. Assuming, for the sake of our evaluation, $x_q = x_{\bar{q}}$, we have $x_q = x_{\bar{q}} = \sqrt{\hat{s}} / s = M_Z / \sqrt{s} = 0.045$.
The sea quarks' structure functions are about
$$x\bar{d}(0.045) \approx x\bar{u}(0.045) \approx xd(0.045) \approx 0.5xu(0.045).$$
The momentum fraction of the Z with longitudinal momentum
$P_Z = 100 \text{ GeV}$ is $x_Z = x_q - x_{\bar{q}} = p_z / p_{\text{beam}} = 0.1$. By substitution into
$m_Z^2 = x_q x_{\bar{q}} s$, we obtain $m_Z^2 = x_q (x_q - 0.1) s$ and $x_q^2 - 0.1x_q - \frac{m_Z^2}{s} = 0$ or, numerically, $x_q^2 - 0.1x_q - 0.02 = 0$. Its solution is $x_q = 0.1 + \sqrt{0.1^2 + 4 \times 0.002} = 0.234$.
The other solution is negative and therefore not physical.

9.33 The Z-charges squared of the u and d quarks have been calculated in Problem 9.3. The difference now is that antineutrinos have positive helicity; therefore, the factor $1/3$ is for the L quarks.

$$c_Z^2(u_L) = \left(\frac{1}{2} - \frac{2}{3}s^2\right)^2 = 0.12, \quad c_Z^2(d_L) = \left(-\frac{1}{2} + \frac{1}{3}s^2\right)^2 = 0.18,$$

$$c_Z^2(u_R) = \left(-\frac{2}{3}s^2\right)^2 = 0.024, \qquad c_Z^2(d_R) = \left(\frac{1}{3}s^2\right)^2 = 0.006.$$

The neutrino cross-section on an u quark is proportional to

$$\frac{1}{3}c_Z^2(u_L) + c_Z^2(u_R) = \frac{1}{3}0.12 + 0.024 = 0.064,$$

and that on a d quark to $\frac{1}{3}c_Z^2(d_L) + c_Z^2(d_R) = \frac{1}{3}0.18 + 0.006 = 0.066$.

The cross-section on a nucleus containing the same number of u and d quarks

(only) is proportional to $\frac{1}{3}\left[c_Z^2(u_L) + c_Z^2(d_L)\right] + c_Z^2(u_R) + c_Z^2(d_R) = 0.013$.

9.36 (a) We start from the results of Problem 9.6. The proton contains 2 u quarks and one d quark. Hence its axial Z-charge is

$$c_{ZV}(p) = 2c_{ZV}(u) + c_{ZV}(d) = 1 - \frac{8}{3}s^2 - \frac{1}{2} + \frac{2}{3}s^2 \approx 0.04$$

And that of the neutron is
$$c_{ZV}(n) = c_{ZV}(u) + 2c_{ZV}(d) = \frac{1}{2} - \frac{4}{3}s^2 - 1 + \frac{4}{3}s^2 = \frac{1}{2} = 0.5.$$

(b) The axial Z-charge of a nucleus with Z protons and N neutrons is
$\frac{1}{2}\left[N + (1 - 4s^2)Z\right]$ and is dominated by the neutron contribution.

(c) The vector Z-charge of the electron is very small (0.04) compared with its axial charge (-0.50).

10.2 (1) Muon neutrinos.

(2) In the π CM frame, energy and momentum of the neutrinos are

$$p_v^* = E_v^* = \frac{m_\pi^2 - m_\mu^2}{2m_\pi} = \frac{139.6^2 - 105.7^2}{2 \times 139.6} = 29.8 \text{ MeV.}$$

The Lorentz factor is $\gamma = \dfrac{E_\pi}{m_\pi} = \dfrac{5}{0.1396} = 35.8$.

The Lorentz transformations

$$p_v^* \sin\theta_v^* = p_v \sin\theta_v \approx p_v\theta_v$$
$$p_v^* \cos\theta_v^* = \gamma p_v \left(1 - \beta\cos\theta_v\right) \approx p_v\theta_v^2 / 2$$

Hence $\tan\theta_v^* = \dfrac{2}{\gamma\theta_v} = \dfrac{2}{35.8 \times 0.044} = 1.27$ and $\theta_v^* = 52°$.

$$p_v = p_v^* \frac{\sin\theta_v^*}{\sin\theta_v} = 540 \text{ MeV.}$$

(3) For $\theta_v = 0$ it is also $\theta_v^* = 0$, and $p_v^* = \gamma p_v \left(1 - \beta\right)$. Hence: $p_v = \dfrac{p_v^*}{\gamma\left(1-\beta\right)}$.

$$\gamma^2 = \frac{1}{1-\beta^2} = \frac{1}{\left(1-\beta\right)\left(1+\beta\right)}$$

$$\approx \frac{1}{2\left(1-\beta\right)} \Rightarrow \gamma\left(1-\beta\right) = \frac{1}{2\gamma}$$

And $p_v = 2\gamma p_v^* = 2130 \text{ MeV.}$

(4) The number of nucleons in $M = 22.5$ t of water is

$$N_N = M \times 10^3 \times N_A = 2.25 \times 10^{10} \times 6 \times 10^{23} = 1.35 \times 10^{34}.$$

(5) The number of CC v_μ interactions in absence of oscillations would be
$N_i = N_N \Phi\sigma = 1.35 \times 10^{34} \times 2 \times 10^{11} \times 3 \times 10^{-43} = 800$.

The disappearance probability is

$$P\left(v_\mu \to v_x\right) = \sin^2 2\theta_{23} \cos^2\theta_{13} \sin^2\left(1.27\Delta m^2 \frac{L}{E}\right)$$

$$\approx \sin^2\left(1.27 \times 2.5 \times 10^{-3} \times \frac{295}{0.54}\right)$$

$$= \sin^2\left(1.73\right) = 0.97.$$

(6) The v_e appearance probability is

$$P\left(v_\mu \to v_{xe}\right) = \sin^2\theta_{23} \sin^2 2\theta_{13} \sin^2\left(1.27\Delta m^2 \frac{L}{E}\right)$$

$$\approx 2\theta_{13}^2 \sin^2\left(1.27 \times 2.5 \times 10^{-3} \times \frac{295}{0.54}\right)$$

$$= 1.5 \times 10^{-2} \sin^2\left(1.73\right) = 1.5 \times 10^{-2}.$$

10.6 (a) CNGS. The Lorentz factor is $\gamma = \dfrac{E_\pi}{m_\pi} = \dfrac{80}{0.1396} = 573$ and the decay length

$$l_\pi = c\gamma\tau = 3 \times 10^8 \times 573 \times 1.6 \times 10^{-8} = 2.75 \text{ km.}$$

$$\text{T2K. } \gamma = \frac{E_\pi}{m_\pi} = \frac{7}{0.1396} = 50, \, l_\pi = 240 \, \text{m}.$$

(b) The CM momentum, which is also the neutrino energy in the pion decay, is

$$p^* = E_v^* = \frac{m_\pi^2 - m_\mu^2}{2m_\pi} = 29.8 \, \text{MeV}.$$

Consider a neutrino emitted at the angle θ^* to the beam in the CM frame and let us transform to the L frame

$$p_v \sin\theta = p_v^* \sin\theta^*$$
$$p_v = E_v = \gamma\left(E_v^* + \beta p_v^* \cos\theta^*\right) = \gamma p_v^*\left(1 + \beta\cos\theta^*\right)$$

and we have

$$p_{v,\text{max}} = \gamma\left(1+\beta\right)p_v^* \approx 2\gamma p_v^*, \text{for } \theta^* = 0$$
$$p_{v,\text{min}} = \gamma\left(1-\beta\right)p_v^* \approx \frac{1}{2\gamma}p_v^*$$

CNGS: $p_{v,\text{max}} = 33$ GeV; $p_{v,\text{min}} = 25$ keV
T2K: $p_{v,\text{max}} = 2.9$ GeV; $p_{v,\text{min}} = 300$ keV.

(c) The momentum components of a neutrino at $\theta^* = 90°$ are as follows:
        Transverse component $p_{vy} = p_v^*$; Longitudinal component $p_{vx} = \gamma p_v^*$;
$\theta \approx \tan\theta \approx 1/\gamma$.

The angle in the L frame is
CNGS $\theta = 0.9$ mrad mrad. Beam 'radius' @ OPERA
$R = 7.3\times10^5 \times 1.4\times10^{-3} \approx 0.6$ km
T2K $\theta = 20$ mrad mrad. Beam 'radius' @ SuperK
$R = 2.95\times10^5 \times 2\times10^{-2} \approx 5.9$ km

10.10 The number of interactions in one year on N_t target protons is $N_{\text{int}} = \Phi_v P_{ee}\sigma N_t$; hence the number of free protons needed is

$$N_t = \frac{N_{\text{int}}}{3.1\times10^7 \, (\text{s/yr})\times\Phi_v \times f \times P_{ee} \times \sigma}$$
$$= \frac{10^3}{3.1\times10^7 \times 3.5\times10^{10} \times 0.05\times0.6\times10^{-47}} = 3.1\times10^{33}.$$

An effective mole of the blend contains

$$N_{\text{free}} = \left(0.20\times18 + 0.80\times26\right)N_A = 24.4\times6\times10^{23}$$
$$= 1.46\times10^{25} \text{ protons/mol.}$$

Hence we need

$$\frac{N_t}{N_{\text{free}}} = 2.1\times10^8 \text{ effective mol.}$$

The weighted molar mass is $M_A = 0.20\times210 + 0.80\times218 = 266$. The blend mass needed is

$$M = \frac{N_t}{N_{\text{free}}} M_A \times 10^3 = 2.1 \times 10^8 \times 0.266$$

$$= 56 \times 10^6 \, \text{kg}.$$

10.13 (a) The kinetic energy $E_{k,B} = m_\nu$. Neutrinos are non-relativistic.

$\beta = \sqrt{\dfrac{2E_{k,B}}{m_\nu}} = 0.07$. Their energy is close to the mass–energy

$E_{\nu B} \approx m_\nu = 100 \, \text{meV}$.

(b) From $s \approx 2E_\nu E_{\nu B}$ we have $E_\nu = \dfrac{m_Z^2}{2E_{\nu B}} = \dfrac{91^2}{2 \times 10^{-10}} = 4.1 \times 10^{13} \, \text{GeV}$.

(c) We have
$$\sigma\left(\nu_x \bar{\nu}_x \to \nu_x \bar{\nu}_x\right) = \left(\frac{\Gamma_\nu}{\Gamma_l}\right)^2 \sigma\left(e^+ e^- \to \mu^+ \mu^-\right) = 1.99^2 \times 2.1 \, \text{nb} = 8.4 \, \text{nb}.$$

The mean free path is then

$$\lambda = \frac{1}{\sigma\left(\nu_x \bar{\nu}_x \to \nu_x \bar{\nu}_x\right) \rho} = \frac{1}{8.4 \times 10^{-37} \times 5.6 \times 10^7} = 2 \times 10^{28} \, \text{m}, \text{ which is larger}$$
than the radius of the Universe.

References

Aaij, R. *et al.* (2019a), *Phys. Rev. Lett.* 122, 211803

Aaij, R. *et al.* (2021a) *J. High Energ. Phys.* 2021 75

Aaij, R. *et al.* (2021b) *Phys. Rev. Lett.* 127 111801

Aaij, R. *et al.* (2021c) *J. High Energ. Phys.* 2021 169

Aaij, R. *et al.* (2022) *Nature Phys.* 18 (1–5) 54–58.

Aaltonen, T. *et al.* (2012) *Phys. Rev. Lett.* 108 151803

Abazov, V. M. *et al.* (2012) *Phys. Rev. Lett.* 108 151804

Abbiendi, G. *et al.* (2001) *Eur. Phys. J.* C21 411

Abbiendi, G. *et al.* (2004) *Eur. Phys. J.* C33 173

Abbiendi, G. *et al.* (2006) *Eur. Phys. J.* C45 1

Abbott, B. P. *et al.* (2016) *Phys. Rev. Lett.* 116 061192

Abbott, B. P. *et al.* (2017) *Astr. J. Lett.* 848 L13

Abbott, R. (2021) *Phys. Rev.* D103 122002

Abbott, T. M. C. *et al.* (2022) *Phys. Rev.* D105 023520

Abdallah, J. *et al.* (2006) *Eur. Phys. J.* C46 569

Abdurashitov, J. N. *et al.* (2002) *JETP* 95 181

Abe, F. *et al.* (1995) *Phys. Rev. Lett.* 74 2626

Abe, K. *et al.* (2005) *Phys. Rev.* D71 072003 (079903 erratum)

Abe, S. *et al.* (2008) *Phys. Rev. Lett.* 100 221803

Abe, K. *et al.* (2011) T2K collaboration *Phys. Rev. Lett.* 107 041801

Abe, K. *et al.* (2012a) DCHOOZ collaboration *Phys. Rev.* D86 052008

Abe, K. *et al.* (2012b) (Super-Kamiokande) *Phys. Rev. Lett.* 110 181802

Abe, K. *et al.* (2013) T2K collaboration *Phys. Rev.* D 88 032002

Abe, K. *et al.* (2017) (Super-Kamiokande) *Phys. Rev.* D95 (1) 012004

Abi, B. *et al.* (2021) *Phys. Rev. Lett.* 126 141601

Abouzaid, E. *et al.* (2008) *Phys. Rev. Lett.* 100 182001

Abrams, G. S. *et al.* (1974) *Phys. Rev. Lett.* 33 1453

Abreu, P. *et al.* (1998) *Phys. Lett.* B418 430

Achard, P. *et al.* (2005) (L3 Collaboration) *Phys. Lett.* B623 21

Adachi, I. *et al.* (2012) *Phys. Rev. Lett.* 108 171802

Agafonova, N. *et al.* (2010) (OPERA Collaboration) *Phys. Lett.* B691 138

Ageno, M. *et al.* (1950) *Phys. Rev.* 79 720

Aghanim, N. *et al.* (2020) *Astr & Astr.* 641 A6

Agostini, M. *et al.* (2015) *Phys. Rev. Lett.* 115 231802

Aharmin, B. *et al.* (2005) *Phys. Rev.* C72 055502

Ahn, M. H. *et al.* (2003) *Phys. Rev. Lett.* 94 081802; ibid. 94 (2005) 081802

Ahn, J. K. *et al.* (2012) *Phys. Rev. Lett.* 108, 191802

Aker, M. *et al.* (2022) *Nat. Phys.* 18 160–166

Alavi-Harati, A. *et al.* (2003) *Phys. Rev.* D67 012005; also D70 079904 (erratum)

Albajar, C. *et al.* (1987) *Z. Phys. C* 36 33

Albajar, C. *et al.* (1989) *Z. Phys. C* 44 15

Alff, C. *et al.* (1962) *Phys. Rev. Lett.* 9 325

ALICE Collaboration (2022) arXiv:2211.04384v1

Allison, W. W. M. (1972) Proc. Meet. on MWPC at Rutherford Lab RHEL/M/H21 81

Allison, W. W. M. *et al.* (1974a) *Nucl. Instr. Meth.* 119 499

Allison, W. W. M. *et al.* (1974b) Proposal CERN/SPSC 74–45; CERN/SPSC 75–15 and *Phys. Lett.* B93 (1980)

Alston, M. *et al.* (1961) *Phys. Rev. Lett.* 6 300

Altarelli, G. & Parisi, G. (1977) *Nucl. Phys.* B126 298

Altman, M. *et al.* (2005) *Phys. Lett.* B616 174

Alvarez, L. W. (1949) A proposed experimental test of the neutrino theory. Report UCRL 238. http://escholarship.org/uc/item/1sh4k6s2#page-4

Alvarez, L. W. *et al.* (1963) *Phys. Rev. Lett.* 10 184

Alvarez, L. W. (1972) In *Nobel Lectures, Physics 1963–1970*, Elsevier Publishing Company, Amsterdam

Amaldi, U. *et al.* (1987) *Phys. Rev.* D36 1385

Amato, G. & Petrucci, G. (1968) CERN Report Annuel, p. 32

Ambrosino, F. *et al.* (2006) *Phys. Lett.* B636 173

Amhis, Y. *et al.* AHFAG Collab. (2023) *Phys. Rev.* D107 052008

An, E. P. *et al.* (2012) *Phys. Rev. Lett.* 108 171803

An, E. P. *et al.* (2023) *Phys. Rev. Lett.* 130 161802

Anderson, C. D. (1933) *Phys. Rev.* 43 491

Anderson, C. D. & Neddermeyer, S. H. (1937) *Phys. Rev.* 51 884; *Phys. Rev.* 54 (1938) 88

Anderson, H. L. *et al.* (1952) *Phys. Rev.* 85 936 (also *ibid.* p. 934 and p. 935)

Anderson, P. W. (1963) *Phys. Rev.* 130 439–442

Andrews, A. *et al.* (1980) *Phys. Rev. Lett.* 44 1108

Andronic, A. *et al.* (2018) *Nature* 561 321–330

Anselmann, P. *et al.* (1992) *Phys. Lett.* B285 375 and 390

Anselmann, P. *et al.* (1995) *Phys. Lett.* B357 237

Anthony, P. L. *et al.* (2004) *Phys. Rev. Lett.* 92 181602

Aoyama, T. *et al.* (2018) *Phys. Rev.* D 97 036001

Arnison, G. *et al.* (1983a) *Phys. Lett.* B122 103

Arnison, G. *et al.* (1983b) *Phys. Lett.* B126 398

ATLAS Collaboration (2019) Proceedings of Science V. 364 -European Physical Society Conference on High Energy Physics (EPS-HEP 2019)

ATLAS Collaboration (2020) ATLAS-CONF-2020-027

ATLAS Collaboration (2022) arXiv:2207.00320

Aubert, B. *et al.* (2007) *Phys. Rev. Lett.* 98 211802

Aubert, B. *et al.* (2009) *Phys. Rev.* D79 072009

Aubert, J. J. *et al.* (1974) *Phys. Rev. Lett.* 33 1404

Augustin, J. E. *et al.* (1974) *Phys. Rev. Lett.* 33 1406

Bacci, C. *et al.* (1974) *Phys. Rev. Lett.* 33 1408

Bagnaia, P. *et al.* (1983) *Phys. Lett.* B129 130

Bahcall, J. N. *et al.* (1964) *Phys. Rev. Lett.* 12 300

Bahcall, J. N. *et al.* (2005) *Astroph. J.* 621 L85

Bailes, N. *et al.* (2021) *Nat. Phys.* 3 344–366

Banner, M. *et al.* (1982) *Phys. Lett.* B118 203

Banner, M. *et al.* (1983) *Phys. Lett.* B122 476

Barnes, V. E. *et al.* (1964) *Phys. Rev. Lett.* 12 204

Barr, G. D. *et al.* (1993) *Phys. Lett.* B317 233

Batlay, J. R. *et al.* (2002) *Phys. Lett.* B544 97

Beherend, H. J. *et al.* (1987) *Phys. Lett.* B183 400

Beringer, J. *et al.* (2012) *Phys. Rev.* D86 010001

Bernardini, M. *et al.* (1967) INFN/AE-67/3. V. Alles-Borelli *et al. Lett. Nuov. Cim.* 4 (1970) 1156.

Bethe, H. A. (1947) *Phys. Rev.* 72 339

Bethe, H. A. (1930) *Annalen d. Physik* 5 321

Bethe, H. & Peierls, R. (1934) *Nature* 133 532

Bettini, A. *et al.* (1966) *Nuov. Cim.* 42 695

Bjorken, J. D. (1969) *Phys. Rev.* 179 1547

Blackett, P. M. S. & Occhialini, G. P. S. (1933) *Proc. R. Soc. Lond.* A139 699

Borsanyi, Sz. *et al.* (2021) *Nature* 593 51

Boyarski, A. M. (1975) *Phys. Rev. Lett.* 34 1357

Brandelik, R. *et al.* (1980) *Phys. Lett.* B97 453

Broser, I. & Kallmann, H. (1947) *Z. f. Naturf.* 2a 439, 642

Burfening, J. *et al.* (1951) *Proc. Phys. Soc.* A64 175

Cabibbo, N. (1963) *Phys. Rev. Lett.* 10 531

Capozzi, F. *et al.* (2018) *Prog. Part. Nucl. Phys.* 102 48

Carrasco, R. *et al.* (2016) *Phys. Rev. D* 93 114512

Cartwright, W. F. *et al.* (1953) *Phys. Rev.* 91 677

Chamberlain, O. *et al.* (1950) *Phys. Rev.* 79 394

Chamberlain, O. *et al.* (1955) *Phys. Rev.* 100 947

Charpak, G. *et al.* (1968) *Nucl. Instr. Meth.* 62 262

Charpak, G. (1992) In *Nobel Lectures, Physics 1991–1995*, edited by G. Ekspong, World Scientific Publishing Co., Singapore (1997)

Chekanov, S. *et al.* (2001) *Eur. Phys. J.* C21 443

Cherenkov P. A. (1934) *C. R. Ac. Sci. USSR.* 8 451

Christenson, J. *et al.* (1964) *Phys. Rev. Lett.* 13 338

Clark, D. L. *et al.* (1951) *Phys. Rev.* 83 649

Clark, A. R. *et al.* (1976) Proposal for a PEP facility based on the Time Projection Chamber. PEP-PROPOSAL-004

Cleveland, B. T. *et al.* (1998) *Astrophys. J.* 496 505

CMS Collaboration (2019) *Eur. Phys. J.* C79 421

CMS Collaboration (2020) *Phys. Lett. B* 805, 135425

CMS Collaboration (2021a) *Europ. Phys. J. C* 81 488

CMS Collaboration (2021b) *J. High En. Phys.* 01 148

CMS Collaboration (2022a) *Nature Physics* 18 1329–1334

CMS Collaaboration (2022b) *Nature* 607 60–68

CNGS (1998) CERN 98–02, INFN/AE-98/05 and CERN-SL/99-034(DI), INFN/AE-99/05

Conde, C. A. N. & Policarpo, A. J. P. L. (1967) *Nucl. Instr. Meth.* 53 7

Connolly, P. L. *et al.* (1963) *Phys. Rev. Lett.* 10 114

Conversi, M. *et al.* (1947) *Phys. Rev.* 71 209 (L)

Conversi, M. & Gozzini, A. (1955) *Nuovo Cim.* 2 189

Costa, G. *et al.* (1988) *Nucl. Phys.* B297 244

Courant, E D. & Snyder, H. S. (1958) *Ann. Phys* 3 1

Cowan, C. L. *et al.* (1956) *Science* 124 103

Curran, S. C. *et al.* (1948) *Nature* 162 302

Curran, S. C. & Baker, W. (1944) Radiation Lab. Rep. 7.6.16, Nov.17. Contract W-7405-eng-48, Manhattan Project. Summarized in *Rev. Sci. Instrum.* 19 (1947) 116

Curran, S. C. & Craggs, J. D. (1949) *Counting Tubes, Theory and Applications.* Academic Press

Dadhich, N. (2011) *Int. J. Mod. Phys.* D20 2739–2747

Dalitz, R. H. (1956) Proceedings of the Rochester Conference; see also Dalitz, R. H. *Phil. Mag.* 44 (1953) 1068 and *ibid.* 94 (1954) 1046

Danby, G. *et al.* (1962) *Phys. Rev. Lett.* 9 36

Davies, J. H. *et al.* (1955) *Nuovo Cim. Ser. X* 2 1063

Davis, R. *et al.* (1968) *Phys. Rev. Lett.* 20 1205

Day T. B. *et al.* (1960) *Phys. Rev. Lett.* 3 61

de Salas, P. F. *et al.* (2018) *Phys. Lett.* B782 633

Dokshitzer, Y u. L. (1977) *Sov. Phys, JETP* 46 641

Durbin, R. *et al.* (1951) *Phys. Rev.* 83 646

Dürr, S. *et al.* (2008) *Science* 322 1224–1227

Eguchi, K. *et al.* (2003) *Phys. Rev. Lett.* 90 021802

Englert, F. & Brout, R. (1964) *Phys. Rev. Lett.* 13 321

Erwin, A. R. *et al.* (1961) *Phys. Rev. Lett.* 6 628

Fan, X. *et al.* (2023) *Phys. Rev. Lett.* 130 071801

Fermi, E. (1933) *La Ricerca Scientifica* 2 (12) and *Nuovo Cim.* 11 (1934) 1; *Z. Phys.* 88 161 [transl. into English by Wilson, F. L.; *Am. J. Phys.* 36 (1968) 1150]

Fermi, E. (1949) *Phys. Rev.* 75 1169

Feynman, R. P. (1948) *Phys. Rev.* 74 1430; 76 (1949) 749 and *ibid.* 769

Feynman, R. P. (1969) *Phys. Rev. Lett.* 23 1415

Feynman, R. P. & Gell-Mann, M. (1958) *Phys. Rev.* 109 193

Fock, V. (1926) *Z. Phys.* 39 226

Frank, I. M. & Tamm, I. E. (1937) *C. R. Ac. Sci. USSR* 14 105

Freund, M. (2001) *Phys. Rev.* D 64 053003

Fukuda, Y. *et al.* (1998) *Phys. Rev. Lett.* 81 1562

Fukui, S. & Myamoto, S. (1959) *Nuovo Cim.* 11 113

Furry, W. H. (1939) *Phys. Rev.* 56 1184

Gawrin, R. L. *et al.* (1960) *Phys. Rev. Lett.* 118 271–283

Geiger, H. & Mueller, W. (1928) *Naturwiss* 16 617

Geiregat, D. *et al.* (1991) *Phys. Lett.* B259 499

Gell-Mann, M. (1953) *Phys. Rev.* 92 833

Gell-Mann, M. & Pais, A. (1955) *Phys. Rev.* 97 1387

Gell-Mann, M. (1964) *Phys. Lett.* 8 214

Gfitter Group (2018) I. Haller *et al. Eur. Phys. J. C* 78 675

Gibbons, L. K. *et al.* (1993) *Phys. Rev. Lett.* 70 1203

Gjesdal, S. *et al.* (1974) *Phys. Lett.* 52B 113

Glaser, D. A. (1952) *Phys. Rev.* 87 665, *ibid.* 91 (1953) 496

Glashow, S. L. (1961) *Nucl. Phys.* 22 579

Glashow, S. L. *et al.* (1970) *Phys. Rev.* D2 1285

Goldberg, M. *et al.* (1964) *Phys. Rev. Lett.* 12 546

Goldhaber, M. *et al.* (1958) *Phys. Rev.* 109 1015

Goldhaber, G. *et al.* (1976) *Phys. Rev. Lett.* 37 255

Goldstone, J. (1961) *Nuov. Cim.* 19 154

Gómez-Cadenas, J. J. *et al.* (2012) *JINST* 7 C11007

Göppert Mayer, M. (1935) *Phys. Rev.* 48 512

Gribov V. N. & Lipatov, L. N. (1972) *Sov. J. Nucl. Phys.* 15 438, and Lipatov, L. N.
    ibid. 20 (1975) 95

Grodzins, L. (1958) *Phys. Rev.* 109 1014–15

Gross, D, J. & Wilczek, F. (1973) *Phys. Rev. Lett.* 30 1343

Grossman, Y. *et al.* (2005) *Phys. Rev.* D72 031501

Guralnik, G. S., Hagen, C. R. & Kibble, T. W. B. (1964) *Phys. Rev. Lett.* 13 585

Hasert, F. J. *et al.* (1973) *Phys. Lett.* B46 121 and ibid. 138; *Nucl. Phys* B73 (1974) 1

Heisenberg, W. (1932) *Z. Fur Phys.* 120 (513) 673

Herb, S. W. *et al.* (1977) *Phys. Rev. Lett.* 39 252

Hess, V. F. (1912) *Physik. Z.* 13 1084

Higgs, P. W. (1964a) *Phys. Lett.* 12 132

Higgs, P. W. (1964b) *Phys. Rev. Lett.* 13 508

Hirata, K. S. *et al.* (1989) *Phys. Rev. Lett.* 63 16

't Hooft, G. (1971) *Nucl. Phys.* B35 167 and 't Hooft, G. & Veltman, M. *ibid.* B44
    (1972) 189

't Hooft, G. (2000) The Creation of Quantum Chromodynamics and the Effective
    Energy, edited by L. N. Lipatov, p. 9, World Scientific

Hosaka, J. *et al.* (2006) *Phys. Rev.* D73 112001

Jegerlehner, F. (2017) arXiv:1711.06089

Kalbfleish, G. R. *et al.* (1964) *Phys. Rev. Lett.* 12 527

Kallmann, H. (1950) *Phys. Rev.* 78 621

Katayama, Y. *et al.* (1962) *Progr. Theor. Phys.* 28 675

Kaulard, J. *et al.* (1998) *Phys. Lett.* B422, 334

Kobayashi, M. & Maskawa, T. (1973) *Progr. Theor. Phys.* 49 652

Kodama, K. *et al.* (2001) *Phys. Lett.* B504 218

Koks, F. & van Klinken, J. (1976) *Nucl. Phys* A272 61

Kroll, N. M. & Wada, W. (1955) *Phys. Rev.* 98 1355

Kusch, P. & Foley, H. M. (1947) *Phys. Rev.* 72 1256; 73 (1948) 412; 74 (1948) 250

Kurie, F. N. D. *et al.* (1936) *Phys. Rev.* 49 368

Lamb, W. E. Jr. & Retherford, R. C. (1947) *Phys. Rev.* 72 241

Landau, L. D. (1948) On the angular momentum of a system of two photons *Dokl. Akad. Nauk SSSR* 60 (2) 207–209

Landau L. D. (1955) In *Niels Bohr and the Development of Physics*, ed. W. Pauli, Pergamon Press, Oxford

Landau, L. D. (1957) *Zh E T F* 32 405 [*J E T P* 5 1297]

Lattes, C. M. G. *et al.* (1947) *Nature* 159 694; *ibid.* 160 453 and 486

Lee, T. D. & Yang, C. N. (1956) *Phys. Rev. 104 254*; 105 (1957) 1671

LEP (2006) arXiv:hep-ex/0612034/

LEP & SLD (2006) *Phys. Rep.* 427 257

LEP EW Working Group (2012) March 2012 http://lepewwg.web.cern.ch/LEPEWWG/

Leprince-Ringuet, L. & L'Héritier, M. (1944) *Compt. Rend* 219 618

Lévy-Leblond, J. M. (1976) *Am. J. Phys.* 44, 271–277

Lorentz, H. A. (1904) *Versl. Kon. Akad. v. Wet., Amsterdam, Dl.* 12 (1904) 986 [English transl. *Proc. Acad. Sci. Amsterdam*, 6 (1904) 809]

Maglić, B. *et al.* (1961) *Phys. Rev. Lett.* 7 178

Majorana, E. (1937) *Nuovo Cim.* 14 171

Maki, Z. *et al.* (1962) *Progr. Theor. Phys.* 28 870

Markov, M. A. (1985) *Early Development of Weak Interactions in the USSR*, Nauka Publishers, Central Depart. Of Oriental Literature, Moscow

Marx, J. N. & Nygren, D. R. (1978) *Phys. Today* 31 (10) 46

McDonough, J. M. *et al.* (1988) *Phys. Rev.* D38 2121

McMillan, E. (1945) *Phys. Rev.* 68 143

Michael, D. G. *et al.* (2006) *Phys. Rev. Lett.* 97 191801

Mikheyev, S. P. & Smirnov A. Y. (1985) *Yad. Fiz.* 42 1441 [*Sov. J. Nucl. Phys.* 42 (1985) 913]

Mohapatra, R. & Pal, P. (2004) *Physics of Massive Neutrinos*, World Scientific

Muller, F. *et al.* (1960) *Phys. Rev. Lett.* 4 418

Nakato, T. & Nishijima, K. (1953) *Progr. Theoret. Phys.* 10 581

Nambu, Y. (1960) *Phys. Rev.* 117 648

Nambu, Y. & Jona-Lasinio, G. (1961) *Phys. Rev.* 122 345; *Phys. Rev.* 124 246

Naroska, B. (1987) *Phys. Rep.* 148 67

Nefkens, B. M. K. *et al.* (2005) *Phys. Rev.* C72 035212

Niu, K. *et al.* (1971) *Progr. Theor. Phys.* 46 1644. For a story of the charm discovery in Japan see K. Niu, *Proc. 1st Int. Workshop on Nucl. Em. Techn.*, Nagoya 1998, preprint DPNU-98-39

Nygren, D. R. (1981) *Phys. Scripta* 23 584

Nygren, D. R. (2009) *Nucl. Instr. Meth.* A603 337

Noecker, M. C. *et al.* (1988) *Phys. Rev. Lett.* 61 310

NuFit. Esteban, I. *et al.* www.nu-fit.org

OPERA (2000) Proposal. LNGS P25/2000; CERN/SPSC 2000-028; SPSC/P318

Orear, J. *et al.* (1956) *Phys. Rev.* 102 1676

Pacini, D. (1912) *Nuovo Cim.* VI-3 93 (English transl. arXiv:1002.1810)

Pais, A. & Piccioni, O. (1955) *Phys. Rev.* 100 1487

Pal, P. B. (2011) *Am. J. Phys.* 79 485–498

Panofsky, W. K. H. *et al.* (1951) *Phys. Rev.* 81 565

Peláez, J. R. *et al.* (2023) *Phys. Rev. Lett.* 130 051902

Pelissetto, A. & Testa, M. (2015) *Am. J. Phys.* 83 (4) 338–340

Perl, M. L. *et al.* (1975) *Phys. Rev. Lett.* 35 1489

Perkins, D. H. (2004) *Introduction to High Energy Physics*, 4th edn, Cambridge University Press

Peruzzi, I. *et al.* (1976) *Phys. Rev. Lett.* 37 569

Pevsner, A. *et al.* (1961) *Phys. Rev. Lett.* 7 421

Pjerrou, G. M. *et al.* (1962) *Phys. Rev Lett.* 9 180

Planck, M. (1906) *Verh. Deutsch. Phys. Ges.* 8 136

Pocanic, D. *et al.* (2004) *Phys. Rev. Lett.* 93 181803

Poincaré, H. (1905) *Comptes Rendues Ac. Sci. Paris* 140 1504

Poincaré, H. (1906) *Rend. Circolo Mat. Palermo* 21 129

Politzer, D. (1973) *Phys. Rev. Lett.* 30 1346

Pontecorvo, B. (1946) Chalk River Lab. PD-205 report

Pontecorvo, B. (1957) *Zh. Eksp. Teor. Fiz.* 33 549 [*Sov. Phys. JETP* 6 (1957) 429]

Pontecorvo, B. (1959) *Zh. Eksp. Teor. Fiz.* 37 1751 [*Sov. Phys. JETP* 10 (1960) 1236]

Pontecorvo, B. (1967) *Zh. Eksp. Teor. Fiz.* 53 1717 [*Sov. Phys. JETP* 6 (1968) 984]

Porsev, S. G. *et al.* (2009) *Phys. Rev. Lett.* 102 181601

Prescott, C. Y. *et al.* (1978) *Phys. Lett.* B77 347; *ibid.* B84 (1979) 524

Racah, G. (1937) *Nuov. Cim.* 14 322

Reines, F. *et al.* (1960) *Phys. Rev.* 117 159

Reynolds, G. T. *et al.* (1950) *Phys. Rev.* 78 448

Rochester, G. D. & Butler, C. C. (1947) *Nature* 160 855

Rohlf, J. W. (1994) *Modern Physics From α to Z°*. J. Wiley & Sons

Rossi, B. (1930) *Nature* 125 636

Rossi, B. (1933) *Zeits. f. Phys.* 82 151

Rossi, B. & Nereson, N. (1942) *Phys. Rev.* 62 418

Rossi, B. & Staub, H. (1949) *Ionization Chambers and Counters. Experimental Techniques*, McGraw-Hill, New York

Rossi, B. (1952) *High-Energy Particles*, Prentice-Hall

Rubbia, C. *et al.* (1976) *Proc. Int. Neutrino Conf.*, Aachen. (1977) p.683

Salam, A. & Ward, J. C. (1964) *Phys. Lett.* 13 168

Salam, A. (1968) Lecture at Eighth Nobel Symposium, ed. N. Svartholm, 367

Samios, N. P. *et al.* (1962) *Phys Rev.* 126 1844

Schlein, P. E. *et al.* (1963) *Phys. Rev. Lett.* 11 167

Schwartz, M. (1960) *Phys. Rev. Lett.* 4 306

Schwinger, J. (1948) *Phys. Rev. 73 416*, 74 (1948) 1439, 75 (1949) 651, 76 (1949) 790

Street, J. C. & Stevenson, E. C. (1937) *Phys. Rev.* 52 1003 (L)

Sudarshan, C. G. and Marshak, R. E. (1957) Proc. of the Padua-Venice Conference on 'Mesons and recently discovered particles'; *Phys. Rev.* 109 1860 (1958)

T2K Collaboration (2020) *Nature* 580 339–344

Takenaka, A. (2020) *Phys. Rev. D* 102, 112011

Taylor, R. E. (1991) *Rev. Mod. Phys.* 63 573

Tomonaga, S. (1946) *Progr. Theor. Phys. (Kyoto)* 27 1

Tonner, N. (1957) *Phys. Rev.* 107 1203

Thomson, J. J. (1897) Proc. Camb. Phil. Soc. 9; The Electrician, 21 May 1897; *Phil. Mag.* 48 (1899) 547.

Touschek, B. (1960) LNF, Int. Rep. 62

Ulmer S. *et al.* (2015) *Nature*, 524, 7564, 196

Vavilov, S. I. (1934) *C. R. Ac. Sci. USSR* 8 457

von Ignatowsky, W. (1911) *Arch. Math. Phys.* 17, 1–24

Veksler, V. I. (1944) *Compt. Rend. Acc. Sc. URSS (Doklady)* 43 329; 44 365

Vilain, P. *et al.* (1994) *Phys. Lett.* B335 246

Walenta, A. H. *et al.* (1971) *Nucl. Instr. Meth.* 92 373

Weinberg, S. (1967) *Phys. Rev. Lett.* 19 1264

Wilson, C. T. R. (1912) *Proc. R. Soc. Lond.* A87 277

Wilson, C. T. R. (1933) *Proc. R. Soc. Lond.* A142 88

Wolfenstein, L. (1978) *Phys. Rev.* D17, 2369; ibid D20 (1979) 2634

Workman, R. L. *et al.* (2022) (Particle Data Group) *Prog. Theor. Exp. Phys.* 083C01

Wu, C. S. & Shaknov, I. (1950) *Phys. Rev.* 77 136

Wu, C. S. *et al.* (1957) *Phys. Rev.* 105 1413 and *ibid.* 106 (1957) 1361

Wu, S. L. (1984) *Phys. Rep.* 197 324

Wu, T. T. and Yang, C. N. (1964) *Phys. Rev.* 85 947

Yao, W.-M. *et al.* (2006) *J. Phys. G.* 33 1

Yukawa, H. (1935) *Proc. Phys.-Math. Soc. Japan* 17 48

Zel'dovich, Ya. (1959) *Sov. Phys. JETP* 94 262

Zemach, C. (1964) *Phys. Rev.* B133 1202

Zweig, G. (1964) CERN report 8182/Th. 401

Index

Printed in the United States
by Baker & Taylor Publisher Services